AF358641

Power Electronics and Actuators

Power Electronics and Actuators

Guest Editor

Dong Jiang

Basel • Beijing • Wuhan • Barcelona • Belgrade • Novi Sad • Cluj • Manchester

Guest Editor
Dong Jiang
Huazhong University of
Science and Technology
Wuhan
China

Editorial Office
MDPI AG
Grosspeteranlage 5
4052 Basel, Switzerland

This is a reprint of the Special Issue, published open access by the journal *Actuators* (ISSN 2076-0825), freely accessible at: https://www.mdpi.com/journal/actuators/special_issues/Power_Electronics_Actuators.

For citation purposes, cite each article independently as indicated on the article page online and as indicated below:

Lastname, A.A.; Lastname, B.B. Article Title. *Journal Name* **Year**, *Volume Number*, Page Range.

ISBN 978-3-7258-2861-6 (Hbk)
ISBN 978-3-7258-2862-3 (PDF)
https://doi.org/10.3390/books978-3-7258-2862-3

Contents

Review

A Review of EMI Research of High Power Density Motor Drive Systems for Electric Actuator

Zhenyu Wang [1], Dong Jiang [1,*], Zicheng Liu [1], Xuan Zhao [1], Guang Yang [2] and Hongyang Liu [2]

[1] School of Electrical and Electronic Engineering, Huazhong University of Science and Technology, Wuhan 430074, China; wang_zhenyu@hust.edu.cn (Z.W.); liuzc@hust.edu.cn (Z.L.); zhaoxuan95@hust.edu.cn (X.Z.)

[2] Flight Automatic Control Research Institute, Xi'an 710076, China; yangg073@avic.com (G.Y.); liuhy076@avic.com (H.L.)

* Correspondence: jiangd@hust.edu.cn; Tel.: +86-186-2792-2372

Abstract: With the global attention given to energy issues, the electrification of aviation and the development of more electric aircraft (MEA) have become important trends in the modern aviation industry. The electric actuator plays multiple roles in aircraft such as flight control, making it a crucial technology for MEA. Given the limited space available inside an aircraft, the power density of electric actuators has become a critical design factor. However, the pursuit of high power density results in the need for larger rated power and higher switching frequency, which can lead to severe electromagnetic interference (EMI) issues. This, in turn, poses significant challenges to the overall reliability of the electric actuator. This paper provides a comprehensive review of EMI in high power density motor drive systems for electric actuator systems. Firstly, the state of the art of electric actuator systems are surveyed, pointing out the contradictory relationship between high power density and EMI. Subsequently, various EMI modeling approaches of motor control systems are reviewed. Additionally, the main EMI suppression methods are summarized. Active EMI mitigation methods are emphasized in this paper due to their advantages of higher power density compared with passive EMI filters. Finally, the paper concludes by summarizing the EMI research in motor drive systems and offering the prospects of electric actuators.

Keywords: more electric aircraft (MEA); electric actuator; high power density motor drive; electromagnetic interference (EMI); EMI suppression methods

Citation: Wang, Z.; Jiang, D.; Liu, Z.; Zhao, X.; Yang, G.; Liu, H. A Review of EMI Research of High Power Density Motor Drive Systems for Electric Actuator. *Actuators* **2023**, *12*, 411. https://doi.org/10.3390/act12110411

Academic Editor: Ioan Ursu

Received: 27 September 2023
Revised: 31 October 2023
Accepted: 2 November 2023
Published: 4 November 2023

1. Introduction

During the 21st century, with the increasing energy shortage crisis and the growing global climate issues, the United Nations has proposed taking urgent action to achieve sustainable development for human society [1,2]. The aviation industry, as an important mode of transportation and a crucial component of the global economy, also faces challenges in achieving sustainable development. In 2018, the aviation industry accounted for approximately 2.4% of CO_2 emissions. As air passenger traffic continues to grow at a rate of 4% to 5% annually, the issue of carbon emissions will become even more serious [3]. In response to the energy supply challenges and environmental issues, the European Union's Flightpath 2050 goals aim to reduce CO_2 emissions by 75%, NO_x emissions by 80%, and perceived noise emissions by 65% compared to the 2000 baseline [4,5]. Aviation electrification and the rise of more electric aircraft (MEA) have emerged as prominent solutions and are currently dominating the industry.

The concept of MEA has been around for a long time [6]. In recent decades, the rapid development of key technologies such as power electronics conversion, advanced motors, electrochemical energy storage, and high-temperature superconductivity has led to the rapid advancement of electrification [5,7]. The energy management architecture of traditional aircraft is complex, using jet fuel as the primary power for propulsion.

The remainder is then converted into four types of secondary power: pneumatic power, mechanical power, hydraulic power, and electrical power [8]. Meanwhile, a MEA mainly relies on electricity as the secondary power, from generators or batteries. Taking the Boeing 787 as an example [9], the electrical system has replaced most of the pneumatic system, eliminating the traditional bleed manifold. The electric compressors replace the traditional pneumatic systems of engines (expected to reduce energy loss by 35%), and the air conditioning packs and wing anti-icing systems are both driven by electric power. Overall, aerospace electrification can reduce the aircraft's reliance on fossil fuels, thereby reducing the size and weight of the aircraft [10]. However, complex electric power systems will place higher requirements on reliability and system control strategies [11].

With the advancement of MEA, the electric actuator, also called a power-by-wire (PBW) actuator, is gradually replacing the traditional hydraulic actuator [12,13]. As shown in Figure 1, the actuation systems perform multiple functions, including flight controls, landing gear controls, and engine actuation controls [14]. Compared to hydraulic actuators, electric actuators have advantages of a smaller size and weight, as well as higher efficiency. The electric actuators are primarily classified into two types: electro-hydraulic actuators (EHAs) and electro-mechanical actuators (EMAs), as shown in Figure 2. The EHA implements power control by adjusting the flow rate and power of a fix displacement pump using a variable speed motor. This eliminates the need for extensive piping systems and external hydraulic devices, resulting in a more compact design. The EMA implements power control by utilizing the motor and gearbox to directly move a ball screw. Compared to the EHA, the EMA offers evident advantages in terms of weight and efficiency as it does not require hydraulic devices. However, it is commonly believed that EMAs may not offer the same level of reliability as EHAs under the same power level [15]. Therefore, the EMA is more readily accepted for secondary flight controls that have lower power and safety requirements, such as flaps, slats, spoilers, and horizontal stabilizing surfaces. On the other hand, the primary flight controls mainly rely on EHAs.

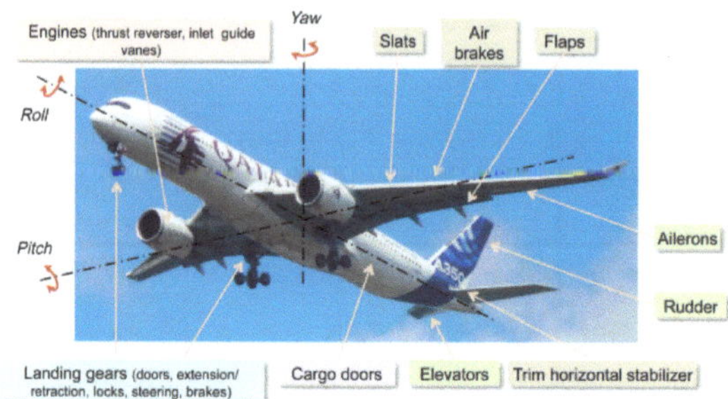

Figure 1. Different actuation needs on a commercial aircraft [14].

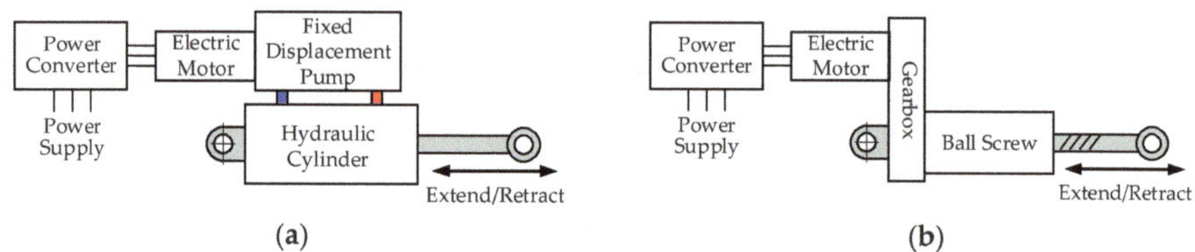

Figure 2. The schematic diagrams of EHA and EMA: (**a**) EHA; (**b**) EMA.

The reliability, weight, power density, and efficiency are crucial design indicators for electric actuators. On the one hand, electric actuators are evolving towards distributed design and redundant fault-tolerant design to enhance reliability [16]. As a result, open-winding, multi-phase, and multi-three-phase motors are gaining increasing attention [17].

On the other hand, electric actuators are evolving towards high-frequency and integrated design to increase power density. High-speed motors have distinct advantages in power density compared to traditional motors. As a result, they have been widely utilized. Additionally, wide-bandgap (WBG) power devices, represented by silicon carbide (SiC), present superior characteristics compared to conventional Si devices [18]. It has been proven that WBG devices have the potential to greatly enhance the power density and control performance of the system.

With the rapid progress of electrification and electric actuators, the MEA power system will inevitably encounter increasingly complex electromagnetic compatibility (EMC) challenges [4]. High-frequency switching actions of WBG power devices bring more serious electromagnetic interference (EMI). Firstly, high-frequency common-mode (CM) voltage can accelerate insulation aging of the motor, which can damage the motor insulation system and shorten the motor's service life [19]. Secondly, the presence of electromagnetic coupling can induce high-frequency voltage in the bearings and other mechanical and hydraulic devices of the actuator, increasing the risk of damage and maintenance costs [20]. Moreover, the controller faces crosstalk issues between the power circuit and signal circuit, which cause system malfunctions [21]. Lastly, the EMI of the electric actuation system can conduct along the power lines or radiate through confined spaces, interfering with the normal operation of other electronic devices.

In conclusion, the electric actuator is evolving towards higher frequencies and greater power density, which in turn brings about increasingly complex EMC issues. Therefore, predicting and suppressing EMI generated by motor drive systems is crucial for the EMC and electromagnetic safety [22]. There is a need to strike a balance between power density and EMI in order to meet EMC requirements [23]. This paper provides a comprehensive review of current research on EMC in motor drive systems. The objective is to offer guidance for the design of electric actuators. In Section 2, the challenge of the trade-off between high power density and EMC design is analyzed. Then, the EMI modeling and prediction methods are reviewed in Section 3. Section 4 summarizes the EMI suppression methods. Finally, Section 5 summarizes the state of art of EMC design and provides an outlook for electric actuators.

2. The Trade-Off between High Power Density and EMC Design

Whether it is an EHA or EMA, the motor drive system serves as the central power component. Optimizing the power density of the motor drive system is vital for the overall improvement in the power density of the electric actuator. To further enhance power density, two main areas of development are being pursued: the integrated motor drive (IMD) and the application of WBG power devices. However, striking a balance between high power density and EMC has become a challenging issue. Consequently, this section aims to discuss the current trends in the IMD and the impact of WBG power devices, while also highlighting the challenges in EMC design for electric actuators.

2.1. Integrated Electric Motor Drives

A traditional motor drive system connects the motor, power converter, and sensors through cables and signal lines. However, there are several drawbacks. Firstly, the use of cables increases the overall size and weight of the system, which consequently reduces the power density and efficiency. Additionally, cables can worsen EMI, which is undesirable. On the one hand, cables are the main propagation path for conducted EMI. Conducted EMI directly penetrates sensitive equipment through cables, or is coupled to adjacent cables or signal lines through crosstalk. On the other hand, cables are high-efficiency radiation antennas. Long cables will not only cause excessive radiated emissions, but also make the system susceptible to external radiated EMI, worsening electromagnetic sensitivity.

To suppress EMI problems caused by cables, EMI filters are usually used in commercial products and engineering applications, which unfortunately, reduce the power density. To suppress radiated EMI and improve the electromagnetic immunity of cables, electromagnetic shielding solutions such as shielding cables and enclosures are usually used, which can also alleviate crosstalk problems. However, the shielding properties are not only affected by the electrical and geometric configuration of the shielding cable and enclosures [24,25], but also depend on the grounding schemes [26]. The design of effective electromagnetic shielding requires not only a theoretical foundation but also extensive engineering experience. Electromagnetic shielding also reduces power density. Considering the limited space of MEA and strict EMC standards, cables should be shortened as much as possible, and the IMD will become the dominant trend.

Through the integration, the converter and the controller housing take up less space, and no longer need long cables, which can suppress the EMI caused by cables and improve power density with 10%~20% less volume [27]. Reference [28] provides an overview of the current status of IMD systems and elucidates the opportunities and challenges they face. In the early stages of the IMD, the drive electronics are built into a separate enclosure that is mounted on the side of the motor, thereby reducing the length of the cable [29]. This partial integration has been commercialized in industrial applications, as shown in Figure 3a. However, the power density and volume optimization are still limited [30]. To further improve power density, the IMD is moving towards overall design and modular design, as shown in Figure 3b,c. Among them, the integrated modulated motor drive (IMMD) has received widespread attention [31]. The IMMD consists of multiple modular components, with each modular unit comprising a single motor stator pole and matching drive electronics, as shown in Figure 3c. The IMMD shows higher reliability, which is crucial for MEA [17].

(a)

(b)

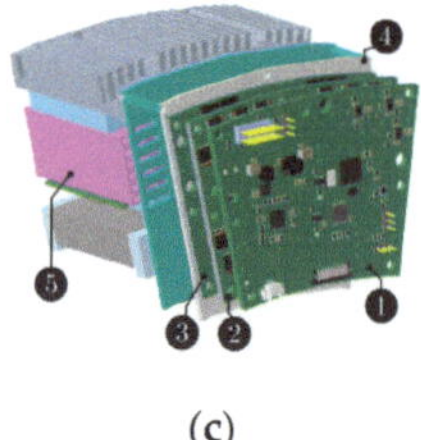

(c)

Figure 3. Examples of IMD: (**a**) Danfoss VLT FCM 300; (**b**) Siemens IMD technology for EV traction drives; (**c**) individual module of IMMD: (1) Communication board; (2) DC capacitor board; (3) Printed Circuit Board, PCB, of drive; (4) heatsink; (5) motor coils [32].

While the IMMD can indeed enhance power density and achieve fault-tolerant control of the actuator, it also faces several challenges, especially in heat dissipation and the EMI problem.

The heat in the IMD is generated not only from the copper loss and iron loss of the motor, but also from the switching loss and conduction loss of the power switching devices. The integrated design of the system makes it challenging to dissipate heat effectively, leading to overheating issues [33]. High temperatures can severely affect the performance of the system and may also cause damage to motor windings, switching devices, and drive circuits. The high-temperature working environment in the IMD has led to the application of WBG devices and posed higher requirements for system thermal management [34].

In addition, the IMD imposes stricter requirements for EMC design. EMC design aims to suppress EMI, with passive EMI filters being the most commonly used and most effective solution for reducing EMI. However, passive filters tend to be bulky and occupy up to 30% of the system volume, decreasing power density [35]. Moreover, the IMD puts forward new requirements for the trade-off between EMC and power density. Although, EMI can be reduced by filters, shortened cables and so on, the highly complex integrated design will bring new EMC challenges. Since the IMD controller is integrated inside the motor, the interference signal reflected by the motor housing can easily cause damage to sensitive devices [28]. This presents a significant challenge to the reliability of electric actuators [36]. Unfortunately, there is still insufficient research on the reliability of sensitive devices.

2.2. Applications of Wide-Bandgap Semiconductors

In recent years, the application of WBG devices in MEA has emerged as a prominent research area due to the rapid advancement of WBG semiconductor packaging and integration technology [37]. As shown in Table 1 [38], compared to conventional Si devices, WBG devices such as SiC and GaN offer significant advantages in terms of blocking voltage capability, on-state loss, switching frequency, and high-temperature characteristics. The application of WBG devices effectively enhances the power density of electric actuators. The high switching frequency capability of WBG devices is suitable for high-speed motors with low inductance and high fundamental frequency [39], and also reduces the volume of filters and DC-link capacitors [40]. High frequency is the most effective method of increasing the power density [41]. Furthermore, compared with conventional Si devices, WBG devices exhibit less on-state loss and switching loss, resulting in reduced heat generation and improved efficiency. Consequently, WBG devices are well-suited for actuators in MEA, simplifying the heat dissipation design for IMD [28].

Table 1. Properties of WBG devices.

Property	Si	GaN	SiC
Bandgap (eV)	1.1	3.4	3.2
Critical electric field (MV/cm)	0.3	3.5	3
Electron saturation velocity (10^7 cm/s)	1	2.5	2.2
Thermal conductivity (W/cm·°C)	1.5	1.3	5
Maximum operation temperature (°C)	200	300	600

However, with WBG switching devices, the characteristics of noise sources are worse than Si devices, which leads to more severe conducted and radiated EMI. WBG devices exhibit higher switching speeds, higher switching frequencies, and more severe ringing than Si devices. Reference [42] investigates and quantifies the increase in the conducted CM EMI of motor drives with SiC and GaN devices. It is indicated that the influence of dv/dt on the conducted CM emission is generally limited and the influence of switching frequency is more significant. Figure 4 illustrates the comparison of CM EMI between Si and WBG drives [42]. Due to the low parasitic capacitance and fast switch speed, the ringing or oscillation of WBG devices is much more severe. This can generate high-frequency EMI and seriously affect the reliability of electric actuators. References [43,44] investigate and model the mechanism of ringing and its impact on EMI.

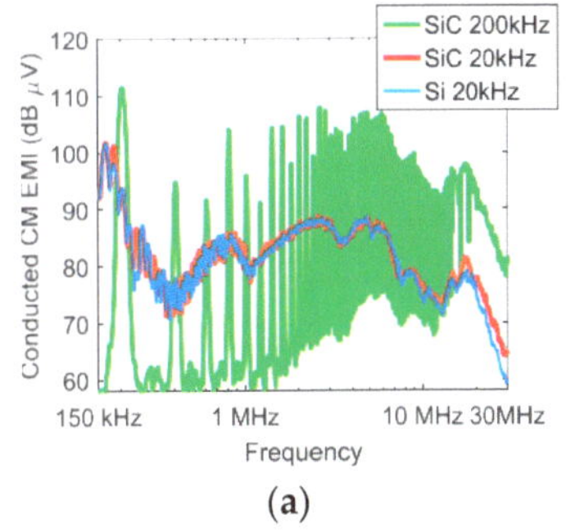

(a)

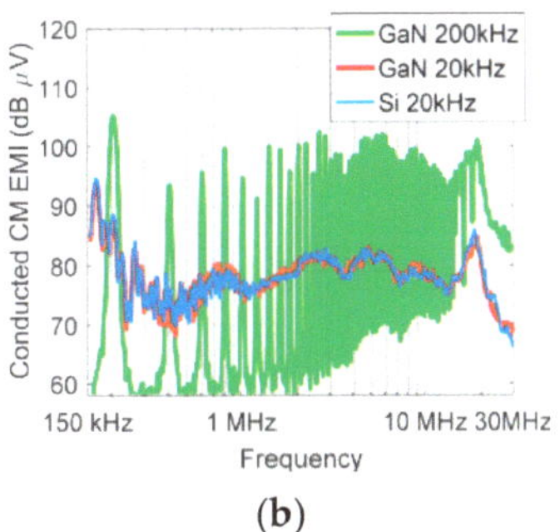

(b)

Figure 4. Comparison of CM EMI between the Si IGBT and WBG drives: (**a**) Si IGBT and SiC MOSFET; (**b**) Si MOSFET and GaN HEMT [42].

Although IMD technology and WBG technology can effectively improve the performance of electric actuators, it is essential to carefully consider the trade-off between power density and EMI and implement appropriate suppression strategies to meet EMC standards, such as DO160 or MIL-STD-461G. This paper aims to review the current state of EMI research on motor drive systems, including EMI modeling and suppression technology, to provide guidance for the design of high power density electric actuators.

3. EMI Modeling Methods

The purpose of EMI modeling is to reveal the mechanism of EMI generation and propagation, and to achieve accurate prediction, providing guidance for EMC design, thereby shortening the test time and cost. According to different propagation paths, EMI can be divided into conducted EMI and radiated EMI. The energy from conducted EMI is coupled with the power supply or other electrical equipment via the power line, while the energy from radiated EMI propagates through the electromagnetic field in space. The conducted EMI is defined with the frequency range from 150 kHz to 152 MHz, and the radiated EMI is from 100 MHz to 6 GHz in standard DO160. Although there is a lack of research on EMI modeling of electric actuators, the main objective of this paper is to provide a thorough overview of the research on EMI modeling of motor drives, thereby serving as a valuable reference.

3.1. Conducted EMI Modeling

At present, the conducted EMI modeling technology of a motor drive can mainly be divided into three types: the time-domain modeling method, the frequency-domain modeling method, and behavioral modeling.

3.1.1. Time-Domain Modeling

Time-domain modeling builds an equivalent circuit based on the physics model of the converter. The EMI spectrum is then obtained through time-domain simulation and frequency-domain calculation. Time-domain prediction relies on the accurate modeling of noise sources and propagation paths [45]. The core of noise source modeling lies in the precise description of the switching behavior of power devices. The equivalent circuit model is commonly used [46]. According to the datasheet provided by the manufacturer, the equivalent circuit can be obtained through software such as PSpice, Saber, and Simplore. The actual equivalent circuit needs to be validated and adjusted in combination with a double pulse test (DPT) [47]. To accurately model the propagation path, it is essential to consider the parasitic impedance of all components within the system [48]. As a result, broadband models of the line impedance stabilization network (LISN), passive components [49], power cables [50], and motor [51], and electromagnetic analysis of stray components generated in circuit layouts [52] should be investigated. Among them, the motor is the most critical and complicated. The motor is also typically modeled as an equivalent circuit model, which is more suitable for system-level simulation. Reference [51] proposed a behavior method based on series and parallel resonances in CM and DM impedance. The model considers the multiple resonances and low-impedance antiresonance. Therefore, the system-level EMI model has a high degree of accuracy, with a maximum prediction error of less than 5 dB across the frequency range of 100 MHz.

Although the equivalent circuit modeling can predict EMI over a wide frequency range, the electric actuator involves the coupling of electric, magnetic, and thermal multiphysics fields. The parasitic parameters are nonlinear, which affects the prediction accuracy. To address this issue, multi-physics co-simulation based on ANSYS is adopted in [53,54], which consider the time-varying, frequency-dependent, and thermal characteristics of key components.

Overall, time-domain modeling has the advantages of high accuracy, a clear physical meaning, strong portability, and ease of use. However, it should be noted that achieving precise time-domain modeling can be time-consuming and requires high CPU performance, which can limit its practical application.

3.1.2. Frequency-Domain Modeling

In frequency-domain modeling, the EMI sources are simplified using approximations of equivalent voltage/current sources that mimic switching characteristics, while the propagation paths are still modeled as equivalent circuits, as with time-domain modeling. Ultimately, the EMI spectrum is obtained by calculating the spectra of the EMI source and the conduction path. By separately modeling and independently calculating the differential-mode (DM) noise and common-mode (CM) noise, noise separation can be achieved [55], as shown in Figure 5a. However, if the converter is asymmetric with respect to the ground, there is coupling between DM and CM noise, and the mixed-mode (MM) noise will limit the effectiveness of the noise separation [56]. Therefore, a differential–common-mode mixed model was proposed in [57], as shown in Figure 5b.

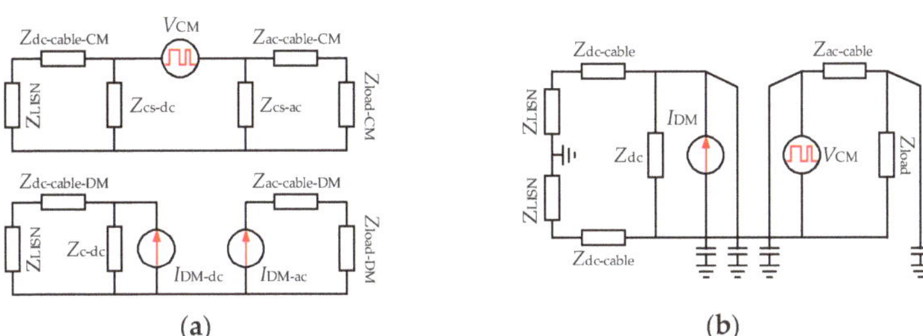

(a)   (b)

Figure 5. Frequency-domain EMI models. (**a**) Noise separation model; (**b**) differential–common-mode mixed model.

The accuracy of the frequency-domain model depends on how well the noise sources are approximated. Typically, the noise sources are modeled as ideal trapezoidal waves with fixed slopes [55,58]. However, this ideal model fails to consider parameters such as switch junction capacitance, busbar lead inductance, and power module lead inductance. As a result, it cannot accurately predict nonideal switching characteristics, such as ringing. In [57], a refined noise source model based on simulation and experimental measurements is proposed, considering voltage overshoot, ringing, and reverse recovery effects. It can accurately predict EMI within 10 MHz. However, this method relies on experimental measurements. Reference [59] proposes a fast and precise synthesis strategy for noise sources, according to the switching characteristics of Si IGBT. It can quickly and precisely predict the ringing frequency, which can be used to achieve accurate EMI prediction. Compared to Si devices, the switching characteristics of WBG devices are more complex. Due to smaller parasitic capacitance, the ringing issues are more severe. Additionally, the Miller plateau time is shorter, resulting in a higher switching rate, and the reverse conduction characteristics are also different. As a result, the EMI peak frequency is higher and the amplitude is larger. Reference [60] introduces a method that considers both ringing, the Miller plateau, and reverse conduction by analyzing the transient behavior of WBG switches. It proposes a precise calculation method for the spectrum envelope of the noise source. However, this method has not been experimentally verified in complex motor drive systems.

In conclusion, compared to time-domain methods, frequency-domain methods have a faster calculation speed but lower accuracy.

3.1.3. Behavioral Modeling

Both time-domain modeling and frequency-domain modeling require detailed modeling of circuit components and consideration of all parasitic parameters. However, when it comes to complex motor drive systems, it is very difficult to accurately predict EMI using detailed modeling [61]. To achieve a convenient and accurate prediction of EMI, behavioral modeling techniques have been developed [62]. Behavioral modeling treats the converter as a "black box" and uses a multi-port network (usually Thevenin or Norton circuit) to represent it. Through standardized measurements and numerical calculations, detailed parameters of the behavioral model can be obtained, enabling precise EMI prediction.

Behavioral models can be classified into terminated models and unterminated models. In terminal models, the load and inverter are represented as a single-port network. In [63], a three-terminal network is adopted to predict the EMI of the motor drive system, as shown in Figure 6a. The impact of the load on the noise source spectrum is also analyzed. This model can achieve precise EMI prediction within the frequency range of 30 MHz. The terminal model is relatively simple, but it is difficult to reflect changes in EMI caused by load variations. In unterminated modeling, the inverter is represented as a two-port network. Thus, the dc and ac side are independent, and the flexibility is strong. Reference [64] proposes a two-port network for the motor drive system to predict the CM EMI on the dc and ac sides, as shown in Figure 6b. For DM EMI, simplified single-port networks are used for predictions. This method accurately predicts EMI within the frequency range of 40 MHz. Reference [65] improved the extraction process of unterminated behavioral models, avoiding the influence of background noise during measurement on high-frequency EMI prediction. This method is suitable for motor drive systems that use WBG devices.

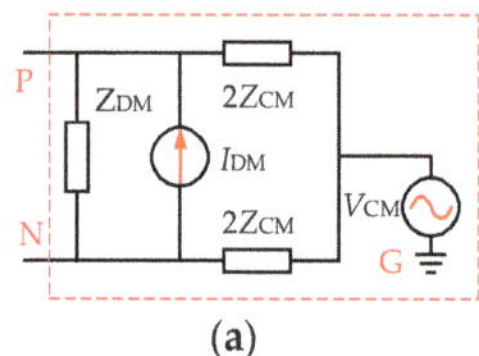

(a)

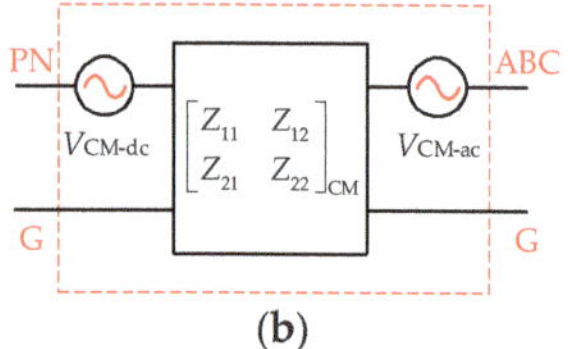

(b)

Figure 6. Behavioral models for motor drive systems. (**a**) Three-terminated network [63]; (**b**) two-port unterminated network [64].

From a fundamental perspective, behavioral models approximate the nonlinear time-varying power electronic converters as linear systems [66]. In order to satisfy the assumption of linearity and time invariance, and to improve the prediction accuracy of behavioral models, the mask impedance must meet certain conditions, which poses challenges for the application of behavioral models [67,68]. In practical applications, MM noise also poses a challenge for filter design in noise separation [69]. In conclusion, the "black box" in the behavioral model does not have practical physical significance and cannot reveal the generation and propagation mechanism of EMI. However, due to its simplicity and high accuracy, it has been widely applied in the design of EMC filters.

3.2. Radiated EMI Modeling

Compared to conducted EMI, there has been less research on radiated EMI in motor drive systems. Motor drive systems primarily consist of motors, cables, and inverters, and the coupling mechanism of radiated EMI is complex. Numerical simulation methods [70,71] are usually used to calculate the radiated electromagnetic field. Figure 7 shows a typical implementation procedure for the radiated EMI prediction [71]. Further research is still required for the radiated EMI modeling of electric actuators.

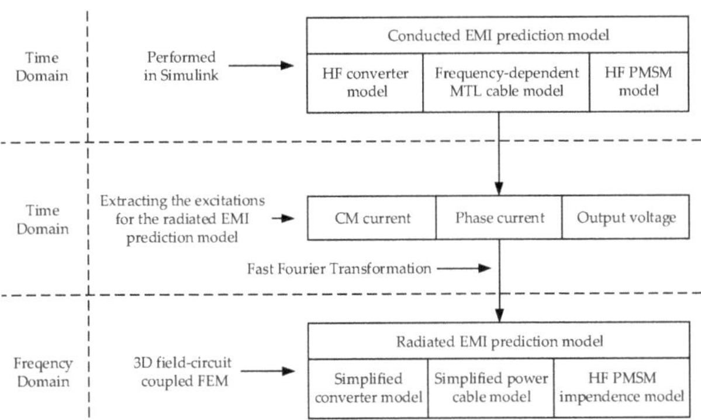

Figure 7. Implementation procedure for the radiated EMI prediction [71].

3.2.1. Radiated EMI Modeling of Cable

In a motor drive system, cables have a similar impact on radiation as high-efficiency antennas. They act as the primary source of radiated EMI. To reduce EMI, there is a current trend towards the IMD and thereby minimizing cable length. However, due to practical limitations in manufacturing and implementation, cables remain an integral part of motor drive systems in the current scenario.

The cable radiation modeling methods can be divided Into analytical methods based on multi-transmission line (MTL) theory [72] and numerical methods based on full-wave simulations. The numerical methods include the finite difference time domain (FDTD) [73], partial element equivalent circuit (PEEC) [74], and finite element method (FEM) [75]. Although numerical methods can evaluate and analyze the radiation effects of cables, the simulations are very time-consuming, especially for bent cables. Therefore, it is necessary to develop fast analytical modeling methods for cables.

In [76], the three-phase cable bundle is simplified to an equivalent single straight cable over the ground plane, as shown in Figure 8a. However, it is only applicable to ideal straight cables, without considering the tightly arranged and bent wiring of cables in engineering applications. Reference [77] improves upon the MTL theory by considering the impact of the proximity effect on charge distribution. The MTL matrixes are modified accordingly, and the Hertzian dipole method is then used to rapidly calculate the radiated electric field of arbitrarily bent cables, as shown in Figure 8b. This fast method offers high accuracy within the 1 GHz frequency range, comparable to numerical methods.

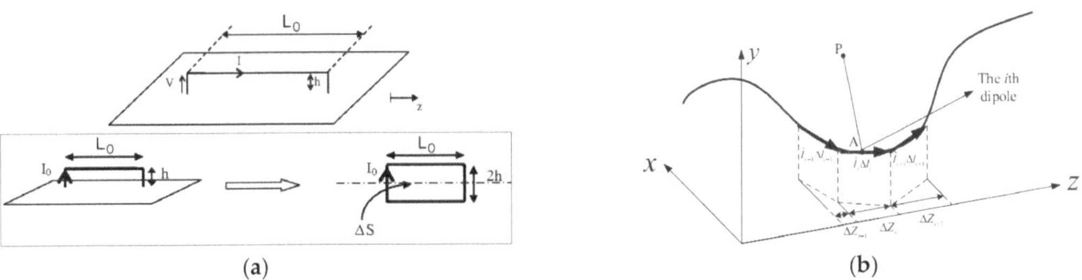

(a)   (b)

Figure 8. Fast cable model for calculating the radiation. (**a**) Single straight cable model [76]; (**b**) arbitrarily bent cable model [77].

3.2.2. Radiated EMI Modeling of Motor

As an important component in motor drive systems, most studies explore the impact of impedance characteristics of the motor on conducted CM currents, as well as the indirect effects on radiated EMI [71,77]. In fact, the contributions of radiated emissions from the motor itself are often neglected [78]. Engineering experiments have demonstrated that the use of high-power motors can lead to excessive radiated emissions. It is important to consider the impact of the structure of the motor, particularly the motor windings, on radiated EMI.

Currently, the motor radiation modeling method also involves analytical models or numerical methods. In [78], an analytical model based on a simplified winding structure is proposed, as shown in Figure 9a. The three-phase windings of the induction motor are equivalent to three Hertzian magnetic dipoles, and analytical expressions for the envelopes of the radiated emissions are deduced. Reference [70] develops a simplified numerical simulation method, where the three-phase windings are modeled as three 1D rectangular edge loops, as shown in the Figure 9b. Unfortunately, these simplified winding structures are only suitable for frequencies much lower than the resonant frequency of the motor. Based on a detailed winding structure as shown in Figure 9c, the radiation pattern of the stator winding is predicted using antenna array theory [79]. This can enable the prediction of the motor's radiation pattern before full-wave simulation. However, modeling a motor with many tightly packed winding wires in a complete full-wave model is challenging. To solve this problem, a reduction technique for modeling closely spaced wires was proposed in [80], which takes proximity effects into account. This approach considerably simplifies the full-wave model and accurately predicts the electric field radiation from 10 kHz to 30 MHz.

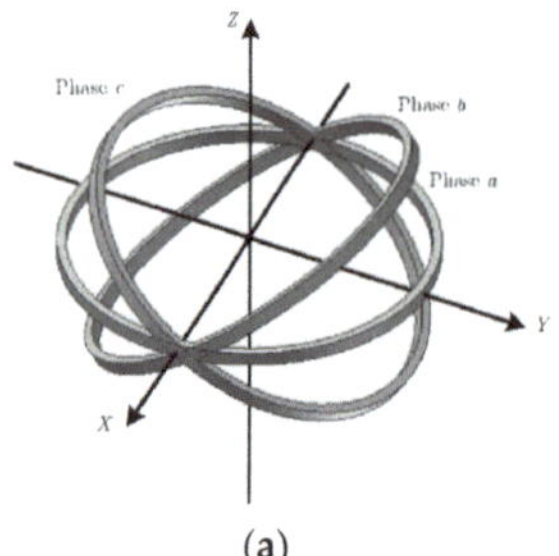

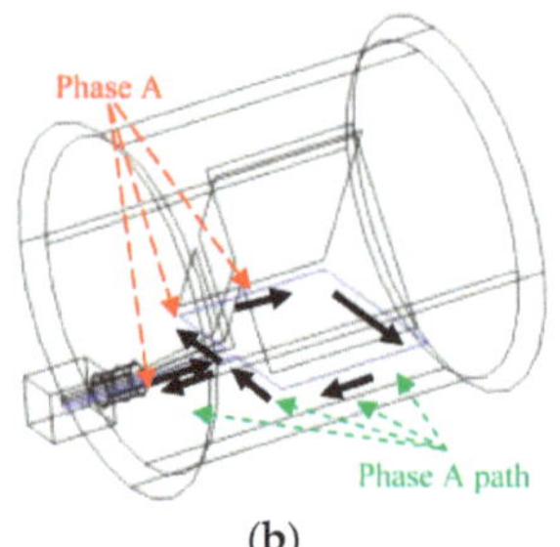

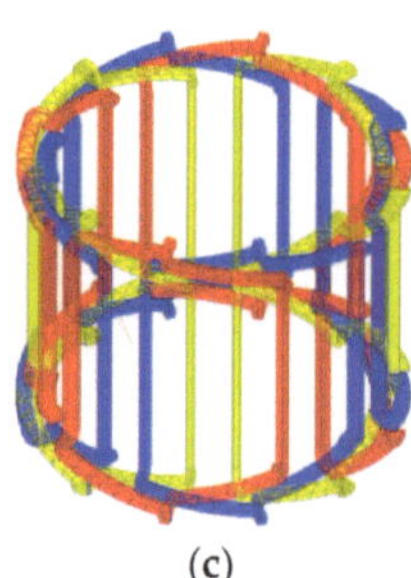

Figure 9. Winding structure of different radiation models. (**a**) Simplified winding structure based on Hertzian magnetic dipoles [78]; (**b**) simplified winding structure based on Hertzian magnetic dipoles [70]; (**c**) detailed winding structure [79].

3.2.3. Radiated EMI Modeling of Inverter

Radiation modeling of the inverter plays a crucial role in motor drive systems. The inverter, serving as the noise source, has a direct impact on the system's radiation, especially with the increasing applications of WBG devices. By acquiring the output signals of the inverter with different control parameters, utilizing the inverter port or circuit equivalent model in the radiated EMI model can facilitate a clearer and more intuitive investigation of the radiation mechanisms. It can also provide more targeted guidance for the suppression of radiated EMI.

A general radiated EMI model for a power converter is proposed in [81], as shown in the Figure 10. The input and output cables are treated as unintentional dipole antennas. The inverter is represented by a noise source and a source impedance in series. The current flowing through the antenna is the CM current of the system. The radiated electric field can be calculated using the developed model.

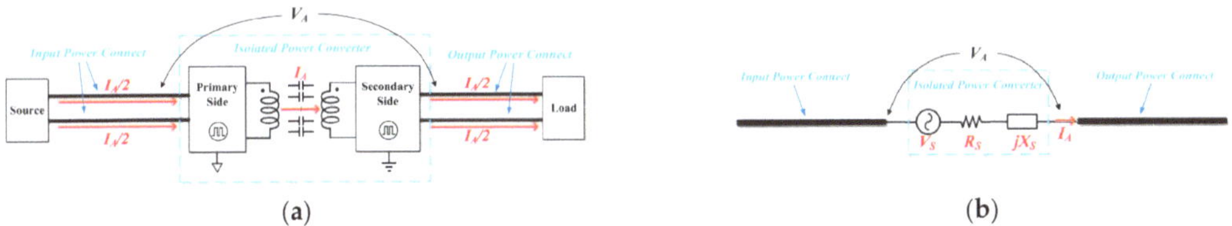

(a)   (b)

Figure 10. General radiated EMI model for power electronic converter [81]. (**a**) CM current flowing through the antenna. (**b**) Equivalent model.

In [82], the unintentional antenna is adopted to model the radiated EMI of the motor drive system with SiC devices. Time-domain analysis is adopted to obtain the accurate EMI spectrum of SiC devices. In addition, nonlinear parasitic capacitance, parasitic inductance, and load effects on the radiated EMI are taken into consideration. The predicted EMI results show good agreement with experimental results in the frequency range of 100 MHz to 1 GHz. With the applications of WBG devices, there has been a significant deterioration in radiated EMI within the frequency range of 150 kHz–30 MHz. At this moment, the radiation is exhibited as near-field coupling. According to Ampere's law, the total current in a cable is the sum of the conductive current and displacement current. However, in most studies, only the conductive current is considered, based on the assumption of far-field radiation. Reference [83] considers the coupling relationship between the displacement current and the antenna, effectively improving the prediction accuracy of radiated EMI in the low-frequency range.

4. EMI Suppression Method

The trend towards higher power density in electric actuators is expected to exacerbate EMI problems and worsen the overall electromagnetic environment of the system, presenting significant challenges for EMC design in MEA. EMI issues associated with electric actuators have become a major obstacle affecting the reliability of MEA. In this section, the state of art of EMI suppression methods for motor drive system is reviewed. As shown in Figure 11, EMI suppression methods are mainly divided into passive suppression and active suppression. Passive suppression utilizes additional passive devices or the optimization of parasitic parameters in circuits to achieve EMI suppression. Active suppression, on the other hand, suppress EMI by using active switch devices.

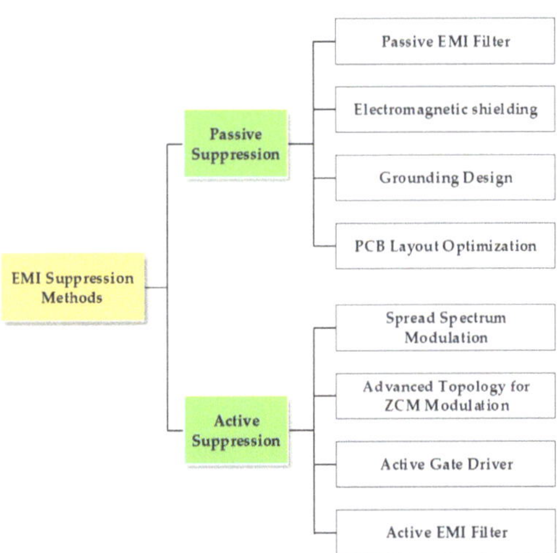

Figure 11. EMI suppression methods for motor drive systems.

4.1. Passive Suppression

The common passive suppression methods include four types: passive filter, electromagnetic shielding, grounding design, and PCB layout optimization. Grounding design serves as the basis of EMC, but relying solely on grounding makes it difficult to meet the EMI standards. Electromagnetic shielding [84] and PCB layout optimization [85] are the two main methods to suppress conducted and radiated EMI. However, these methods are primarily applied in DC/DC converters and are more intricate to implement in high power density motor drive systems. The passive filter is one of the most important and effective solutions for suppressing conducted and radiated EMI by improving propagation paths. The typical arrangement of a first-order CM/DM filter is shown in Figure 12.

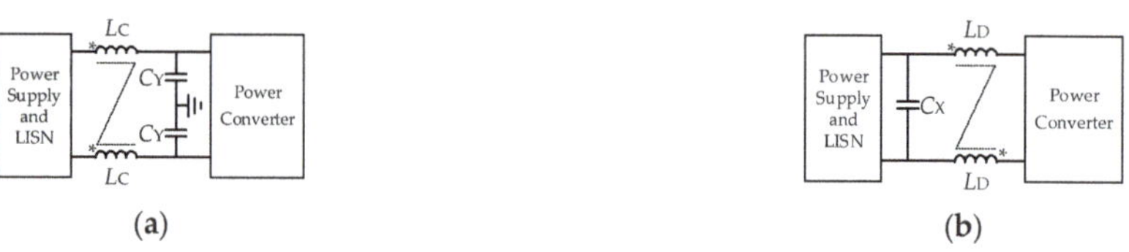

Figure 12. Basic structures of passive EMI filters. (**a**) CM filter; (**b**) DM filter.

In the common filter design, the EMI that exceeds the EMC standards is considered as the attenuation requirement or insertion loss requirement, without considering the influence of parasitic parameters. However, the insertion loss of passive filters is greatly affected by parasitic parameters. Figure 13 illustrates the equivalent circuit and insertion loss of a DM EMI filter considering these parasitic parameters [86]. It is evident from the figure that parasitic parameters significantly reduce the suppression performance of passive EMI filters in the high-frequency range above megahertz. In some cases, the presence of these parasitic parameters can even exacerbate the EMI in specific frequency ranges.

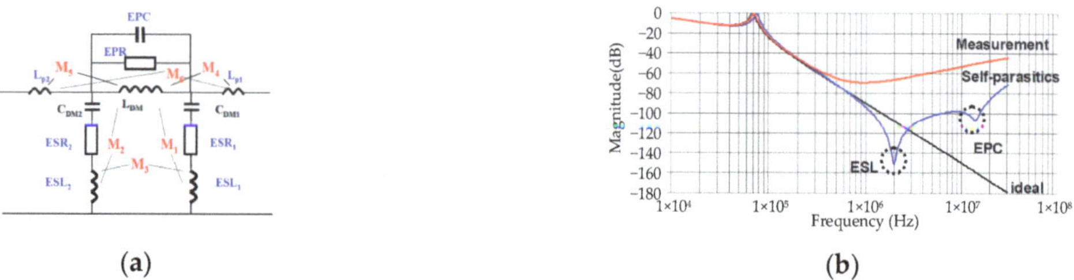

Figure 13. The influence of parasitic parameters on EMI filter suppression performance [86]. (**a**) Parasitic model of the passive filter; (**b**) effects of parasitic parameters on the insertion loss.

With the increasing application of WBG devices in power electronic systems, the switching frequency has greatly increased, resulting in a more serious EMI problem at high frequencies. The high-frequency performance of traditional passive EMI filters cannot meet the practical requirements. In addition, passive EMI filters are prone to being overdesigned to meet stringent EMC standards, and they tend to occupy a large volume in the system, which is not conducive to increasing power density.

4.1.1. Optimization of High-Frequency Performance

To enhance the high-frequency performance of an EMI filter, it is crucial to account for the influence of parasitic parameters on insertion loss during the design procedure. Reference [87] analyzes the relationship between the filter structure's size and parasitic parameters, and proposes a design procedure with consideration of both the low-frequency and high-frequency attenuation requirements. Moreover, efforts should be made to minimize parasitic parameters. Reference [86] optimizes the structural layout, effectively improving the self-parasitic and mutual parasitic problems. The insertion loss of the filters is significantly improved in the range of 1 MHz to 30 MHz.

In passive filters, near magnetic field emission from magnetic components is a crucial issue that can easily affect the high-frequency performance of the filter. In [88], a comprehensive analysis of magnetic coupling is proposed, considering the influence of the displacement current of parasitic capacitors on the stray flux. It reveals the frequency variation effect of a high-frequency magnetic field, and its influence on the performance of the EMI filter. The shielding plate is an effective solution to reduce magnetic coupling, but it is bulky and costly. As a result, a new topology of a CM inductor based on symmetric windings is proposed in [89]. It consists of two toroidal magnetic cores as shown in Figure 14. This topology reduces near-field emission due to DM currents and increases the DM inductance, at the expense of a reduced flux density. In the range above 2 MHz, the improved CM inductor reduces CM EMI by 10 dB compared to the conventional CM inductor, while DM EMI is reduced by up to 22 dB.

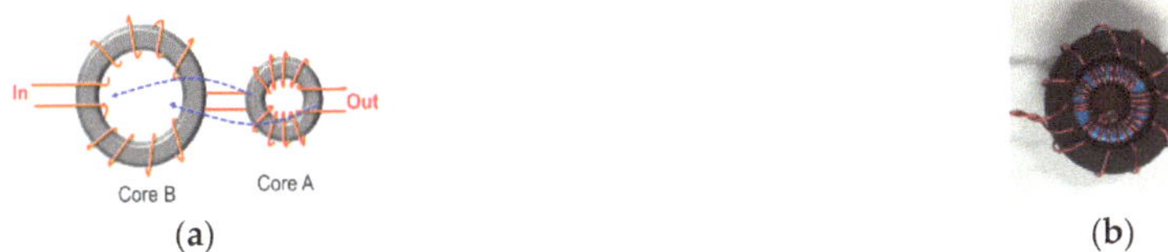

(a)   (b)

Figure 14. Novel CM inductor with reduced near-field coupling [89]. (**a**) Winding connection; (**b**) prototype.

For compact power electronic converters, the near-field coupling between the DC-link capacitors and the filter is also an important issue. Reference [90] studied the near-field coupling between the DC-link capacitor and the EMI filter in an active-clamp Flyback converter. It improves the single shielding by double shielding, as shown in Figure 15. Experiments have confirmed that double shielding effectively reduces CM conducted noise above 2 MHz. The double shielding technique can also be used to suppress radiated EMI [91].

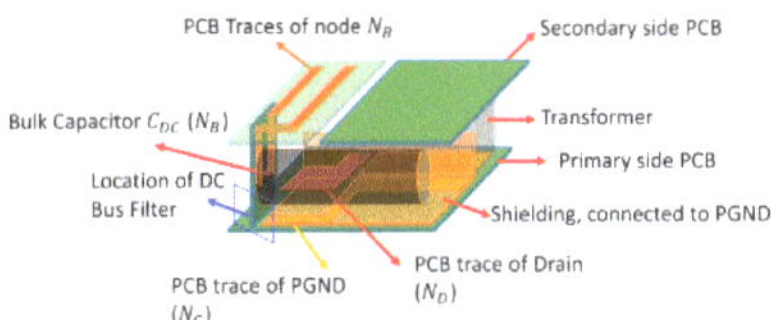

Figure 15. Double shielding technique to reduce near-field coupling [90].

4.1.2. Optimization of Power Density

Passive filters are one of the main limiting factors that hinder the improvement of a system's power density. Passive filter integration is a method for optimizing power density. It can be categorized into three levels: the functional level, material and technology level breakthroughs, and the system-level integration level.

Functional integration relies on magnetic integration technology. Figure 16b illustrates an integrated EMI inductor proposed in [92]. This design incorporates a low-permeability DM choke within the CM choke and improves the winding structure to achieve higher DM inductance. In [93], a DM solenoid is placed in the CM choke for magnetic integration, as shown in Figure 16c. This design allows for independent control of DM inductance, providing increased flexibility compared to the structure in [92]. In addition, reference [94] proposes a new method to realize the CM inductor on the AC and DC sides. Reference [95] realizes the integration of DM and CM mode inductors in a multi-stage filter.

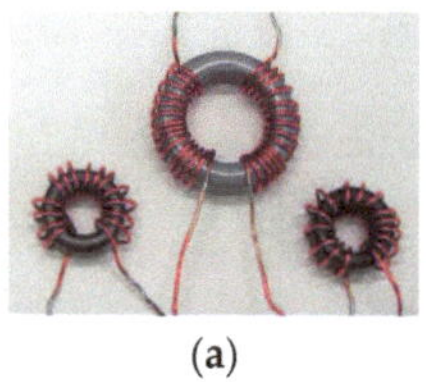

(a)

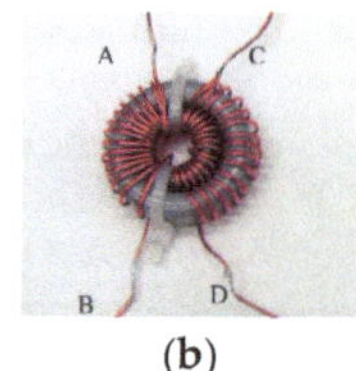

(b)

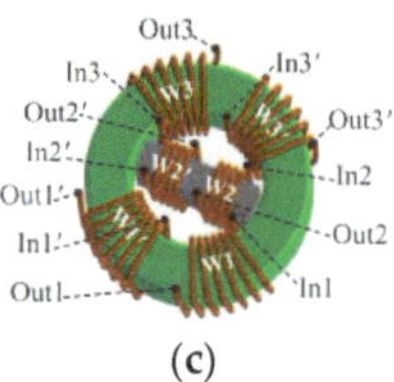

(c)

Figure 16. Structure of different EMI inductors. (**a**) Regular discrete EMI inductors; (**b**) integrated EMI inductor in [92]; (**c**) integrated EMI inductor in [93].

However, magnetic integration cannot realize the complete integration of capacitors and inductors, and the high-frequency performance of the filter is affected by the layout. To achieve a breakthrough in the power density of high-performance filters, advancements in materials and technology are required. Planar electromagnetic integration technology (EMIT) based on a PCB and flexible multi-layer foil (FMLF) has been developed. For instance, in [96], DM and CM filters are integrated using planar EMIT, and a ground layer is inserted to eliminate winding parasitic capacitance, enhancing power density and high-frequency performance. FMLF materials further reduce the loss and size of filters. Reference [97] proposes a fully integrated symmetrical filter based on FMIT, achieving a more compact design.

System-level integration enables the direct integration of the filter and power electronic converter. This is made possible through advanced packaging technology, which enables the integration of passive filters and power modules. Reference [98] integrates CM capacitors into SiC modules to reduce conducted EMI. Similarly, to minimize the filter parasitic effects, Reference [99] integrates the CM filter with the CaN half-bridge power module, as shown in Figure 17. Through this integration technology, up to 50 dB attenuation is achieved in the frequency range of 10 to 100 MHz. This modular integrated filter holds promising potential for application in electric actuators.

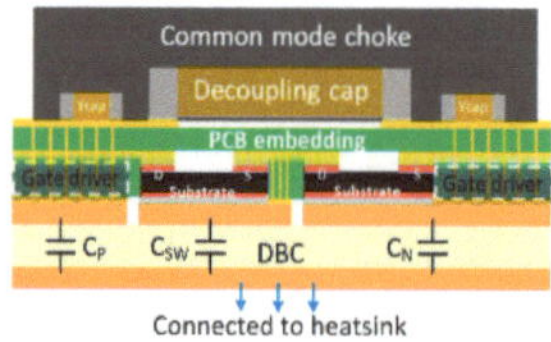

Figure 17. GaN half-bridge power module package with the integrated CM filter [99].

While passive EMI filters have seen advancements in high-frequency performance and power density, achieving a balance between power density, high-frequency performance, and manufacturing costs remains a challenge. The volume and weight of passive filters are still the main factors restricting the power density of an electric actuator.

4.2. Active Suppression

Active suppression technologies optimize EMI noise sources through strategies such as spread spectrum modulation, advanced topology, and an active gate driver, or actively detect and compensate EMI through active EMI filters. Compared with passive suppression, active suppression reduces the use of passive components and increases the power density of the system, so it has good application prospects in electric actuators.

4.2.1. Spread Spectrum Modulation

The high-frequency switching of power devices controlled by PWM is the primary source of EMI in power converters. Traditional constant switching frequency PWM (CSF-PWM) concentrates energy mainly at the switching frequency and its harmonics, resulting in EMI peaks at corresponding frequencies [100]. To reduce conducted EMI, spread spectrum modulation (SSM) is proposed by improving the noise source. Figure 18 illustrates that by adjusting the modulation, the narrow-band harmonic energy concentrated at the switching frequency and its multiples is dispersed over a wider spectrum range, leading to the attenuation of EMI peaks. Depending on the implementation, SSM is mainly divided into three categories: random PWM; periodic PWM; and programmed PWM.

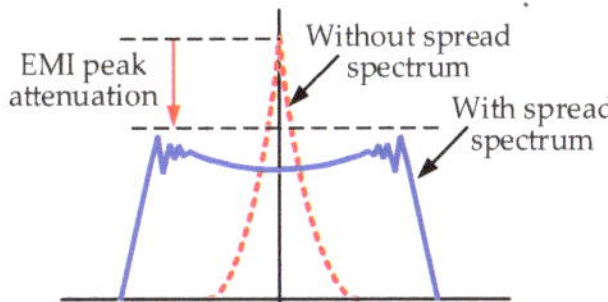

Figure 18. The principle of spread spectrum modulation [100].

Random PWM (RPWM) is a technique that utilizes the statistical properties of random numbers to modify the carrier characteristics in each switching cycle, thereby achieving the spread spectrum effect. RPWM is mainly divided into random carrier frequency PWM (RCFPWM) [101] and random pulse position PWM (RPPPWM) [102]. Reference [103] analyzes the impact of RCFPWM on the performance of power converters and verifies the effectiveness of RCFPWM in reducing conducted EMI. Reference [102] proposes an improved RPPPWM, which is easy to implement and integrate into motor drive systems. Then an improved RPWM that combines RCFPWM and RPPPWM is proposed, making the sampling frequency constant and harmonic cluster distribution more uniform [104]. RPWM has good performance in EMI suppression. However, the difficulty in generating random numbers hinders the practical application of RPWM.

In contrast to RCFPWM, periodic PWM changes the switching frequency periodically according to a certain law, such as sinusoidal wave, exponential wave, triangular wave [105,106], sawtooth pattern [107], or uniform distribution [108]. In [100], the influence of the peak deviation of the switching frequency on EMI suppression is analyzed. It is found that as the peak deviation increases, the EMI peak decreases. However, when the peak deviation reaches a certain level, harmonic overlap can occur, which counterproductively affects EMI suppression. Reference [108] analyzes the impact of the statistical distribution of switching frequency on EMI suppression, and proposes that the uniform distribution PWM improves the suppression effect. Since the switching frequency of periodic PWM changes periodically, inverter loss prediction can be performed to improve efficiency and optimize the thermal design [109].

RPWM and periodic PWM change the switching frequency with the goal of reducing the peak value of the EMI spectrum, which may have a negative impact on other performances of the motor and inverter. Programmed PWM, also known as model predictive PWM, uses the degree of freedom of the switching frequency to optimize specific indicators to improve the performance of the converter [110]. Variable switching frequency PWM (VSFPWM) uses the current ripple or torque ripple as the control object to change the switching frequency [111,112], achieving the multi-objective optimization of motor performances and EMI. Since EMI is not the primary target for programming PWM, the EMI suppression is unsatisfactory, compared with RPWM and periodic PWM.

4.2.2. Advanced Topology for ZCM Modulation

The CM voltage of the motor drive systems is determined by the switching actions of the inverter. However, the traditional three-phase two-level inverter topology lacks enough switching freedom, which prevents the improved modulation strategies [113] from maintaining a constant CM voltage. As a result, there still exist CM leakage of current issues. In order to eliminate CM current, several advanced topology solutions with multiple switching freedoms have been proposed, including three-phase four-leg inverters [114], three-level inverters [115], paralleled inverters [116], and dual three-phase motors [117]. Schematics of these advanced topologies are shown in Figure 19.

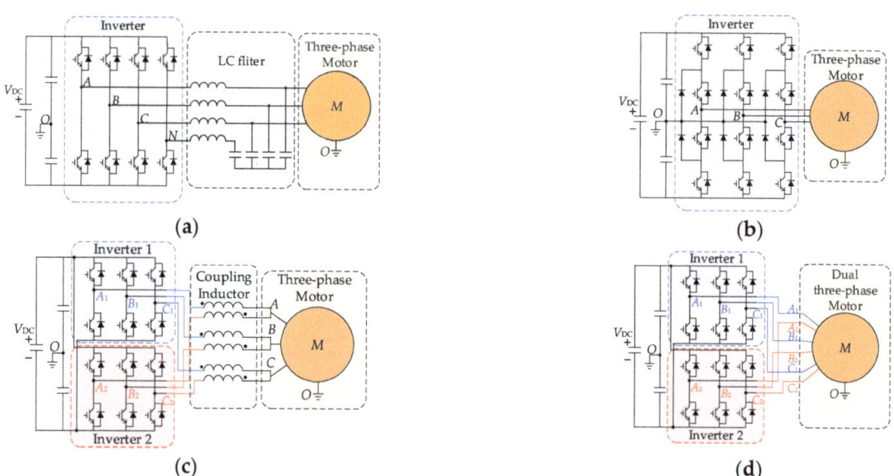

Figure 19. Schematics of advanced topologies for CM elimination. (**a**) Three-phase four-leg inverter; (**b**) three-level inverter; (**c**) paralleled inverter; (**d**) dual three-phase motor.

A zero common-mode (ZCM) PWM scheme, based on a four-leg inverter, is introduced for conventional three-phase motor drive systems in [114]. This modulation is based on the CM reduction modulation for a three-phase inverter [118]. By controlling the fourth bridge arm, the inverter output CM voltage is eliminated. However, the nonutilization of the zero vectors in the space vectors may lead to issues such as a reduced modulation index or increased total harmonic distortion (THD) of the output voltage.

Multi-level inverters have good application prospects in medium-voltage high-power motor drive systems [119], and they also have the switching freedom of the ZCM. A ZCM PWM scheme for a neutral-point-clamped (NPC) three-level inverter was proposed in [115]. ZCM modulation is achieved by selecting the switching vectors with a ZCM voltage state to synthesize the reference voltage. However, as with the four-leg topology, sacrificing part of the switching freedom to achieve ZCM leads to problems such as a deterioration in the output voltage, a reduction in the modulation index, and the unbalanced midpoint voltage of the DC-link. How to balance ZCM and other system performances is an important research direction.

Parallel inverters' topology is another topology used to achieve ZCM, which is suitable for high-power cases. By interleaving [120], the switching ripple between the two inverters, the THD, and output EMI can be reduced. However, interleaving cannot eliminate the CM voltage. A ZCM strategy for parallel inverters is proposed in [116]. The reference voltage is synthesized by paralleled ZCM vectors, and PWM timing optimization is performed to ensure the voltage balance and switching balance of the two inverters. This method not only achieves ZCM, but also reduces current ripple. In [121], a new dual-segment three-phase permanent magnet motor is introduced to eliminate the need for the coupling inductors. Without sacrificing switching freedom, the modulation index remains unchanged.

Due to the winding redundancy and reliability, modular winding motors such as dual three-phase motors have promising potential in electric actuators. Moreover, it is also beneficial to achieving ZCM modulation. A ZCM scheme is proposed for the dual three-phase motors with symmetrical windings in [117]. The output voltages of the two inverters are of the same magnitude but are opposite in phase. Moreover, a unified ZCM scheme is proposed for dual three-phase motors with asymmetrical windings in [122]. The universal method can achieve decoupling of the offset angle and reference voltage between the two sets of windings. In [123], an interleaving together with ZCM modulation is proposed for a four-module three-phase motor to reduce the vibration and CM current at the same time.

Although advanced topologies with ZCM can eliminate CM voltage, it may also reduce the modulation index, increase switching loss, or increase ripple and harmonics. The trade-off between EMI suppression and system performances is important. In addition, the corresponding topology should be selected according to different application cases. Table 2 summarizes the reviewed publications on advanced topologies for ZCM modulation.

Table 2. Summary of advanced topologies for ZCM modulation.

Topology	Ref.	Cases	Number of Power Devices	ZCM Performances	System Performances	
Three-phase four-leg inverter	[114]	Conventional three-phase motor drive	8 full control switches	The CM voltage is reduced by 20 dB up to 100 kHz.	1.	Reduced modulation index
					2.	Increased THD
three-level inverter	[115]	Medium-voltage high-power motor drive	8 full control switches and 6 clamping diodes	The CM voltage is almost up to 20 kHz.	1.	Reduced modulation index
					2.	Increased THD
					3.	Unbalanced midpoint voltage
paralleled inverter	[116]	High-power motor drive	12 full control switches	The CM voltage is reduced by 30 dB up to 200 kHz, and 10 dB up to 2 MHz.	1.	Unchanged modulation index
					2.	Improved current ripple
	[121]			The performance is the same as that in [121].	1.	Unchanged modulation index
					2.	Improved current ripple
					3.	Increased power density
dual three-phase motor	[117]	High-reliability motor drive		The CM leakage current is reduced by almost 20 dB up to 40 kHz.	1.	Unchanged modulation index
					2.	Only suitable for symmetrical windings
	[122]			The CM voltage is reduced by more than 20 dB between 150 kHz and 900 kHz, and 10 dB up to 2 MHz.	1.	Unchanged modulation index
					2.	Suitable for asymmetrical windings
four-module three-phase motor	[123]		24 full control switches	The CM voltage is reduced by more than 30 dB between 150 kHz and 1 MHz, and 10 dB up to 3 MHz.	1.	Unchanged modulation index
					2.	Improved vibration
					3.	Improved current ripple
					4.	Only suitable for reversed windings

4.2.3. Active Gate Driver

PWM determines the low-frequency spectrum of the noise source, and it has been proven that SSM and ZCM modulation can effectively suppress conducted EMI below several MHz. However, the high-frequency spectrum of the noise source is primarily determined by the switching transient of the power device, such as switching speed and ringing. The high-frequency conducted and radiated EMI can be reduced by increasing the gate resister (Rg) to slow down the switching speed [124]. Unfortunately, this method increases the switching loss [125]. To manage the trade-off between switching loss, device stress, and EMI, the active gate drive (AGD) technique has been developed. The AGD adds active devices to the conventional drive circuit (CGD), and flexibly optimizes the switching trajectory.

To suppress high-frequency EMI problems caused by overcurrent and ringing, the AGD circuit in [126] controls the gate voltage according to different stages of the IGBT switching transient, reducing the switching speed. An integrated AGD is designed for GaN transistors in [127]. Based on closed-loop control, the drive current is reduced during the switching transient, reducing dv/dt and switching losses. With this drive, the spectrum energy is effectively attenuated in the range of 30 MHz~200 MHz. In addition, a programmable AGD with sub-nanosecond resolution is designed for GaN transistors [128], which can accurately control the Rg and reduce the switching speed and oscillation. This method can reduce EMI noise in the range of 200 MHz~1 GHz. In fact, reducing the switching speed will still inevitably increase switching loss.

Usually the switching trajectory is a trapezoidal wave represented by the duty cycle and switching speed. It is proven that the higher the derivative order of the pulse wave, the faster its spectral attenuation speed [129]. The comparison of pulse waves with different derivative orders is indicated in Figure 20. Since the trapezoidal wave only has the one-order derivative (that is, the switching speed), its spectrum envelope decreases slowly. As a result, AGD methods based on Gaussian switching are proposed in [130,131]. The Gaussian S-shape has an infinite-order derivative, and since it does not change the switching speed, this method does not significantly increase switching loss. However, this method is difficult to implement and is not suitable for high switching frequency situations.

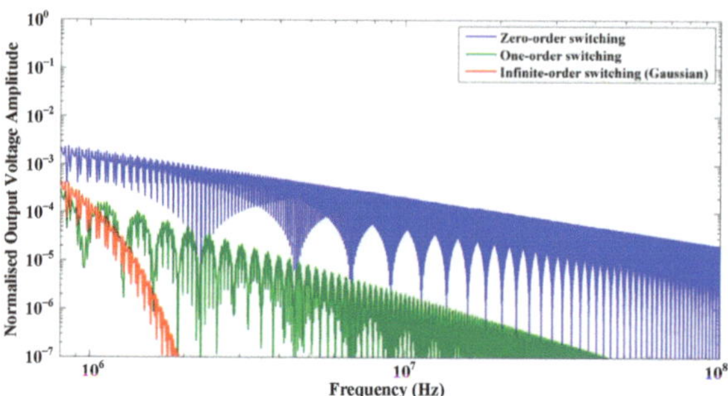

Figure 20. Comparison of pulse waves with different derivative orders: square wave with zero-order derivative (blue), trapezoidal wave with one-order derivative (green), Gaussian switching with infinite-order derivative (red) [130].

In summary, the AGD technique can effectively suppress both high-frequency conducted EMI and radiated EMI. The trade-off between switching loss and EMI is its main design basis. Currently, in the motor drive system, the application of AGD is insufficient and still needs to be explored and researched.

4.2.4. Active EMI Filter

Both the active EMI filter (AEF) and the passive EMI filter suppress EMI by reducing the effectiveness of the interference propagation path. However, AEFs actively sense the EMI and cancel it through integrated amplifiers, with a higher power density. The designs and implementations of different AEFs are summarized in [35]. According to the control method, AEFs can be divided into feedforward and feedback types. According to the signal types of sampling and compensation, they can be divided into voltage sampling voltage compensation (VSVC), voltage sampling current compensation (VSCC), current sampling current compensation (CSCC), and current sampling voltage compensation (CSVC).

The advantages of an AEF are reflected in two aspects. On the one hand, it is an effective way to suppress low-frequency EMI, which can increase the corner frequency of a passive filter and reduce the volume and weight of passive components. In [132], a feedforward VSVC AEF, also known as an active CM canceller (ACC), is applied in motor drive systems to eliminate the CM current on the output side. In [133], an AEF based on CSCC topology is proposed for CM suppression of motor drive systems. On the other hand, an AEF can increase the equivalent impedance of passive components, thereby reducing their volume and weight [134]. In [135], a feedback VSVC AEF is proposed to increase the equivalent capacitance of the compensation capacitor, which improves the attenuation characteristics of the passive filter in the high-frequency range.

In an AEF, the transformer is required to implement current sampling or voltage compensation, which results in increased costs and reduced power density. Therefore, the application of transformerless AEFs has been widely studied. A novel transformerless AEF based on VSCC is proposed in [136]. A CM impedance network with the same CM impedance as the motor is built on the AC side, and the compensation voltage is injected into the network to generate compensation current. In addition, Reference [137] designs the AEF between the motor and the ground, which eliminates the transformer and reduces the current stress of the AEF, further optimizing the power density.

The application of WBG devices in electric actuators places higher requirements on the attenuation capability of the filter in a wide frequency range. However, due to the amplifier bandwidth, transformer bandwidth, and parasitic parameters, an AEF is more suitable for EMI suppression in the range of tens of kHz to several MHz. Therefore, the concept of a hybrid EMI filter (HEF) is proposed to improve the power density of systems [133,138]. Low-frequency EMI is suppressed by the AEF, and high-frequency EMI is suppressed by the passive filter, as shown in Figure 21.

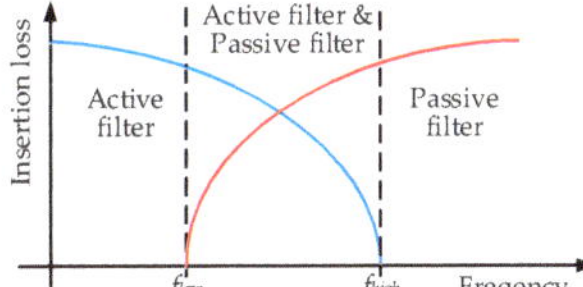

Figure 21. Insertion loss of the active and passive filters in a hybrid EMI filter [133].

5. Conclusions

This review introduces the current trends in electric actuator technology and the EMI problems it brings. The modeling and suppression methods for EMI in motor drive systems are summarized, providing guidance for the EMC design of electric actuators.

Due to the limited onboard space of MEA, high power density electric actuators have become the focus of research. As a result, electric actuators are developing towards integration and high frequency, with WBG devices being widely applied. However, the application of WBG semiconductors deteriorates the EMI of the system, posing higher requirements for EMC design, which in turn leads to an increase in the volume and weight of passive EMI filters, thereby reducing the power density of the system. Additionally, the complex electromagnetic environment of the IMD presents challenges to the reliability of sensitive devices such as micro-electronics components. Therefore, in future electric actuator designs, not only power density, efficiency, and reliability metrics must be considered but also the EMC.

EMI modeling and suppression are two important aspects of EMC design. This article provides a detailed review of the research on EMI in motor drive systems. EMI can be classified into conducted EMI and radiated EMI based on the different propagation paths. The research on modeling methods of conducted EMI is as follows:

(1)   The modeling methods for conducted EMI mainly include time-domain modeling, frequency-domain modeling, and behavioral modeling.

(2)   Time-domain modeling and frequency-domain modeling provide detailed modeling of converters, revealing the mechanism of EMI generation and propagation. They

provide a basis for the converter design. However, it is difficult to balance prediction accuracy, computational speed, and convergence.

(3) Behavioral modeling treats the converter as a "black box", which can achieve accurate and fast predictions. Behavioral modeling must rely on existing prototypes for modeling, meaning that it cannot provide guidance for EMC predesign. However, behavior modeling still plays a crucial role in guiding the design of filters.

The research on modeling methods of radiated EMI is as follows:

(1) Compared to conducted EMI, there is less research on radiated EMI modeling. Radiated EMI modeling can be mainly divided into three aspects: cable, motor, and inverter.

(2) With the development of IMDs, the impact of cables on radiation will gradually decrease, and the motor's radiation emissions will become the dominant factor in the future. Numerical methods are common modeling techniques, but they are time-consuming and computationally inefficient. Therefore, there is a growing research trend towards developing methods that balance accuracy and computational efficiency.

(3) Compared to the motor and cable, the inverter itself has lower levels of radiated EMI. However, as the noise source in motor drive systems, the research on the influence of the inverter on radiation is important.

The development of high power density electric actuators puts forward higher requirements for the trade-off between EMC and power density. At present, EMI suppression methods can be divided into two types: passive suppression and active suppression. The research on EMI suppression is as follows:

(1) Passive EMI filters are widely used and highly effective methods for suppressing electromagnetic interference (EMI). However, their bulky size poses a challenge to improving power density. Currently, research efforts are focused on optimizing the high-frequency performance and power density of these filters.

(2) Active suppression has gained increasing attention and research due to its advantages in high power density.

(3) SSM and ZCM modulation based on advanced topology are effective in suppressing electromagnetic interference for low-frequency conducted interference. However, using or sacrificing switch freedom to optimize EMI may lead to a deterioration in control performance, which is a crucial aspect to consider.

(4) AGD, which achieves noise attenuation by optimizing the switching trajectory, is effective in suppressing high-frequency EMI. However, compared to other suppression methods, AGD places higher demands on packaging technology and has limited applications in motor drive systems.

(5) Compared to passive EMI filters, AEFs can effectively increase power density. However, due to bandwidth limitations and parasitic parameters, AEFs are difficult to implement for effective EMI suppression across a wide frequency range. The combination of AEFs and passive EMI filters, known as hybrid filters, is currently a hot research direction.

Author Contributions: Conceptualization, Z.W. and D.J.; methodology, Z.W.; software, Z.W.; validation, Z.W., X.Z., Z.L. and G.Y.; formal analysis, Z.W.; investigation, Z.W.; resources, Z.W.; data curation, Z.W.; writing—original draft preparation, Z.W.and X.Z.; writing—review and editing, Z.W., G.Y. and H.L.; visualization, Z.W.; supervision, D.J. and Z.L.; project administration, D.J.; funding acquisition, D.J. All authors have read and agreed to the published version of the manuscript.

Funding: This research received no external funding.

Data Availability Statement: Not applicable.

Conflicts of Interest: The authors declare no conflict of interest.

References

1. Barzkar, A.; Ghassemi, M. Components of Electrical Power Systems in More and All-Electric Aircraft: A Review. *IEEE Trans. Transp. Electrif.* **2022**, *8*, 4037–4053. [CrossRef]
2. Sustainable Development Goals. Available online: https://www.un.org/sustainabledevelopment/sustainabledevelopment-goals/ (accessed on 4 June 2020).
3. Ansell, P.J.; Haran, K.S. Electrified Airplanes: A Path to Zero-Emission Air Travel. *IEEE Electrif. Mag.* **2020**, *8*, 18–26. [CrossRef]
4. Barzkar, A.; Ghassemi, M. Electric Power Systems in More and All Electric Aircraft: A Review. *IEEE Access* **2020**, *8*, 169314–169332. [CrossRef]
5. Berg, F.; Palmer, J.; Miller, P.; Dodds, G. HTS System and Component Targets for a Distributed Aircraft Propulsion System. *IEEE Trans. Appl. Supercond.* **2017**, *27*, 1–7. [CrossRef]
6. Rosero, J.A.; Ortega, J.A.; Aldabas, E.; Romeral, L. Moving towards a more electric aircraft. *IEEE Aerosp. Electron. Syst. Mag.* **2007**, *22*, 3–9. [CrossRef]
7. Sayed, E.; Abdalmagid, M.; Pietrini, G.; Sa'Adeh, N.; Callegaro, A.D.; Goldstein, C.; Emadi, A. Review of Electric Machines in More-/Hybrid-/Turbo-Electric Aircraft. *IEEE Trans. Transp. Electrif.* **2021**, *7*, 2976–3005. [CrossRef]
8. Ni, K.; Liu, Y.; Mei, Z.; Wu, T.; Hu, Y.; Wen, H.; Wang, Y. Electrical and Electronic Technologies in More-Electric Aircraft: A Review. *IEEE Access* **2019**, *7*, 76145–76166. [CrossRef]
9. 787 No-Bleed Systems: Saving Fuel and Enhancing Operational Efficiencies. Available online: https://www.boeing.com/commercial/aeromagazine/articles/qtr_4_07/AERO_Q407_article2.pdf (accessed on 1 September 2023).
10. Wheeler, P.; Bozhko, S. The More Electric Aircraft: Technology and challenges. *IEEE Electrif. Mag.* **2014**, *2*, 6–12. [CrossRef]
11. Cavallo, A.; Canciello, G.; Russo, A. Integrated supervised adaptive control for the more Electric Aircraft. *Automatica* **2020**, *117*, 108956. [CrossRef]
12. Qiao, G.; Liu, G.; Shi, Z.; Wang, Y.; Ma, S.; Lim, T.C. A review of electromechanical actuators for More/All Electric aircraft systems. *Proc. Inst. Mech. Eng. Part C J. Mech. Eng. Sci.* **2018**, *232*, 4128–4151. [CrossRef]
13. Russo, A.; Cavallo, A. Stability and Control for Buck–Boost Converter for Aeronautic Power Management. *Energies* **2023**, *16*, 988. [CrossRef]
14. Maré, J. *Aerospace Actuators 1: Needs, Reliability and Hydraulic Power Solutions*; ISTE Ltd.: London, UK, 2016; pp. 1–32.
15. Yin, Z.; Hu, N.; Chen, J.; Yang, Y.; Shen, G. A review of fault diagnosis, prognosis and health management for aircraft electromechanical actuators. *IET Electr. Power Appl.* **2022**, *11*, 1249–1272. [CrossRef]
16. Sarigiannidis, A.G.; Beniakar, M.E.; Kakosimos, P.E.; Kladas, A.G.; Papini, L.; Gerada, C. Fault Tolerant Design of Fractional Slot Winding Permanent Magnet Aerospace Actuator. *IEEE Trans. Transp. Electrif.* **2016**, *2*, 380–390. [CrossRef]
17. Cao, W.; Mecrow, B.C.; Atkinson, G.J.; Bennett, J.W.; Atkinson, D.J. Overview of Electric Motor Technologies Used for More Electric Aircraft (MEA). *IEEE Trans. Ind. Electron.* **2012**, *59*, 3523–3531.
18. Millan, J.; Godignon, P.; Perpina, X.; Perez-Tomas, A.; Rebollo, J. A Survey of Wide Bandgap Power Semiconductor Devices. *IEEE Trans. Power Electron.* **2014**, *29*, 2155–2163. [CrossRef]
19. Zhang, G.; Wang, S.; Li, C.; Li, X.; Gu, X. A Space Vector Based Zero Common-Mode Voltage Modulation Method for a Modular Multilevel Converter. *World Electr. Veh. J.* **2023**, *14*, 53. [CrossRef]
20. Tawfiq, K.B.; Güleç, M.; Sergeant, P. Bearing Current and Shaft Voltage in Electrical Machines: A Comprehensive Research Review. *Machines* **2023**, *11*, 550.
21. Umetani, K.; Matsumoto, R.; Hiraki, E. Prevention of Oscillatory False Triggering of GaN-FETs by Balancing Gate-Drain Capacitance and Common-Source Inductance. *IEEE Trans. Ind. Appl.* **2019**, *55*, 610–619. [CrossRef]
22. He, H.; Sun, F.; Wang, Z.; Lin, C.; Zhang, C.; Xiong, R.; Deng, J.; Zhu, X.; Xie, P.; Zhang, S.; et al. China's battery electric vehicles lead the world: Achievements in technology system architecture and technological breakthroughs. *Green Energy Intell. Transp.* **2022**, *1*, 100020. [CrossRef]
23. Perdikakis, W.; Scott, M.J.; Yost, K.J.; Kitzmiller, C.; Hall, B.; Sheets, K.A. Comparison of Si and SiC EMI and Efficiency in a Two-Level Aerospace Motor Drive Application. *IEEE Trans. Transp. Electrif.* **2020**, *6*, 1401–1411. [CrossRef]
24. Ren, D.; Du, P.; He, Y.; Chen, K.; Luo, J.; Michelson, D.G. A Fast Calculation Approach for the Shielding Effectiveness of an Enclosure With Numerous Small Apertures. *IEEE Trans. Electromagn. Compat.* **2016**, *58*, 1033–1041. [CrossRef]
25. Xiao, P.; Du, P.; Zhang, B. An Analytical Method for Radiated Electromagnetic and Shielding Effectiveness of Braided Coaxial Cable. *IEEE Trans. Electromagn. Compat.* **2019**, *61*, 121–127. [CrossRef]
26. Jia, X.; Hu, C.; Dong, B.; He, F.; Wang, H.; Xu, D. System-Level Conducted EMI Model for SiC Powertrain of Electric Vehicles. In Proceedings of the 2020 IEEE Applied Power Electronics Conference and Exposition (APEC), New Orleans, LA, USA, 15–19 March 2020.
27. Lee, W.; Li, S.; Minav, T.A.; Pietola, M.; Han, D.; Sarlioglu, B. Achieving high-performance electrified actuation system with integrated motor drive and wide bandgap power electronics. In Proceedings of the 2017 19th European Conference on Power Electronics and Applications (EPE'17 ECCE Europe), Warsaw, Poland, 11–14 September 2017.
28. Zhang, B.; Song, Z.; Liu, S.; Huang, R.; Liu, C. Overview of Integrated Electric Motor Drives: Opportunities and Challenges. *Energies* **2022**, *15*, 8299. [CrossRef]
29. Jahns, T.M.; Sarlioglu, B. The Incredible Shrinking Motor Drive: Accelerating the Transition to Integrated Motor Drives. *IEEE Power Electron. Mag.* **2020**, *7*, 18–27. [CrossRef]

30. Swanke, J.A.; Jahns, T.M. Reliability Analysis of a Fault-Tolerant Integrated Modular Motor Drive for an Urban Air Mobility Aircraft Using Markov Chains. *IEEE Trans. Transp. Electrif.* **2022**, *8*, 4523–4533. [CrossRef]
31. Lee, W.; Li, S.; Han, D.; Sarlioglu, B.; Minav, T.A.; Pietola, M. A Review of Integrated Motor Drive and Wide-Bandgap Power Electronics for High-Performance Electro-Hydrostatic Actuators. *IEEE Trans. Transp. Electrif.* **2018**, *4*, 684–693. [CrossRef]
32. Khan, W.A.; Ebrahimian, A.; Weise, N. Design of High Current, High Power Density GaN Based Motor Drive for All Electric Aircraft Application. In Proceedings of the 2022 IEEE 9th Workshop on Wide Bandgap Power Devices & Applications (WiPDA), Redondo Beach, CA, USA, 7 November 2022.
33. Mohamed, A.H.; Vansompel, H.; Sergeant, P. Electrothermal Design of a Discrete GaN-Based Converter for Integrated Modular Motor Drives. *IEEE J. Emerg. Sel. Top. Power Electron.* **2021**, *9*, 5390–5406. [CrossRef]
34. Gao, Z.; Jiang, D.; Kong, W.; Chen, C.; Fang, H.; Wang, C.; Li, D.; Zhang, Y.; Qu, R. A GaN-Based Integrated Modular Motor Drive for Open-Winding Permanent Magnet Synchronous Motor Application. In Proceedings of the 2018 1st Workshop on Wide Bandgap Power Devices and Applications in Asia (WiPDA Asia), Xi'an, China, 16–18 May 2018.
35. Narayanasamy, B.; Luo, F. A Survey of Active EMI Filters for Conducted EMI Noise Reduction in Power Electronic Converters. *IEEE Trans. Electromagn. Compat.* **2019**, *61*, 2040–2049. [CrossRef]
36. Chen, T.; Caicedo-Narvaez, C.; Wang, H.; Moallem, M.; Fahimi, B.; Kiani, M. Electromagnetic Compatibility Analysis of an Induction Motor Drive With Integrated Power Converter. *IEEE Trans. Magn.* **2020**, *56*, 1–4. [CrossRef]
37. Gui, H.; Zhang, Z.; Chen, R.; Ren, R.; Niu, J.; Li, H.; Dong, Z.; Timms, C.; Wang, F.; Tolbert, L.M.; et al. Development of High-Power High Switching Frequency Cryogenically Cooled Inverter for Aircraft Applications. *IEEE Trans. Power Electron.* **2020**, *35*, 5670–5682. [CrossRef]
38. Morya, A.K.; Gardner, M.C.; Anvari, B.; Liu, L.; Yepes, A.G.; Doval-Gandoy, J.; Toliyat, H.A. Wide Bandgap Devices in AC Electric Drives: Opportunities and Challenges. *IEEE Trans. Transp. Electrif.* **2019**, *5*, 3–20. [CrossRef]
39. Tuysuz, A.; Bosshard, R.; Kolar, J.W. Performance comparison of a GaN GIT and a Si IGBT for high-speed drive applications. In Proceedings of the 2014 International Power Electronics Conference (IPEC-Hiroshima 2014—ECCE ASIA), Hiroshima, Japan, 18–21 May 2014.
40. Hazra, S.; Madhusoodhanan, S.; Moghaddam, G.K.; Hatua, K.; Bhattacharya, S. Design Considerations and Performance Evaluation of 1200-V 100-A SiC MOSFET-Based Two-Level Voltage Source Converter. *IEEE Trans. Ind. Appl.* **2016**, *52*, 4257–4268. [CrossRef]
41. Wang, J.; Wang, B.; Zhang, L.; Wang, J.; Shchurov, N.I.; Malozyomov, B.V. Review of bidirectional DC–DC converter topologies for hybrid energy storage system of new energy vehicles. *Green Energy Intell. Transp.* **2022**, *1*, 100010. [CrossRef]
42. Han, D.; Li, S.; Wu, Y.; Choi, W.; Sarlioglu, B. Comparative Analysis on Conducted CM EMI Emission of Motor Drives: WBG Versus Si Devices. *IEEE Trans. Ind. Electron.* **2017**, *64*, 8353–8363. [CrossRef]
43. Chen, J.; Du, X.; Luo, Q.; Zhang, X.; Sun, P.; Zhou, L. A Review of Switching Oscillations of Wide Bandgap Semiconductor Devices. *IEEE Trans. Power Electron.* **2020**, *35*, 13182–13199. [CrossRef]
44. Liu, T.; Wong, T.T.Y.; Shen, Z.J. A Survey on Switching Oscillations in Power Converters. *IEEE J. Emerg. Sel. Top. Power Electron.* **2020**, *8*, 893–908. [CrossRef]
45. Wang, K.; Lu, H.; Chen, C.; Xiong, Y. Modeling of System-Level Conducted EMI of the High-Voltage Electric Drive System in Electric Vehicles. *IEEE Trans. Electromagn. Compat.* **2022**, *64*, 741–749. [CrossRef]
46. Zhuolin, D.; Dong, Z.; Tao, F.; Xuhui, W. Prediction of conducted EMI in three phase inverters by simulation method. In Proceedings of the 2017 IEEE Transportation Electrification Conference and Expo, Asia-Pacific (ITEC Asia-Pacific), Harbin, China, 7–10 August 2017.
47. Chen, L.; Xu, Z.; Zhou, T.; Chen, H. Modeling of Conducted EMI in Motor Drive System Based on Double Pulse Test and Impedance Measurement. In Proceedings of the 2022 Asia-Pacific International Symposium on Electromagnetic Compatibility (APEMC), Beijing, China, 1–4 September 2022.
48. Takahashi, S.; Wada, K.; Ayano, H.; Ogasawara, S.; Shimizu, T. Review of Modeling and Suppression Techniques for Electromagnetic Interference in Power Conversion Systems. *IEEJ J. Ind. Appl.* **2022**, *11*, 7–19. [CrossRef]
49. Hu, S.; Wang, M.; Liang, Z.; He, X. A Frequency-Based Stray Parameter Extraction Method Based on Oscillation in SiC MOSFET Dynamics. *IEEE Trans. Power Electron.* **2021**, *36*, 6153–6157. [CrossRef]
50. Wunsch, B.; Stevanovic, I.; Skibin, S. Length-Scalable Multiconductor Cable Modeling for EMI Simulations in Power Electronics. *IEEE Trans. Power Electron.* **2017**, *32*, 1908–1916. [CrossRef]
51. Wang, K.; Lu, H.; Li, X. High-Frequency Modeling of the High-Voltage Electric Drive System for Conducted EMI Simulation in Electric Vehicles. *IEEE Trans. Transp. Electrif.* **2023**, *9*, 2808–2819. [CrossRef]
52. Maekawa, S.; Tsuda, J.; Kuzumaki, A.; Matsumoto, S.; Mochikawa, H.; Kubota, H. EMI prediction method for SiC inverter by the modeling of structure and the accurate model of power device. In Proceedings of the 2014 International Power Electronics Conference (IPEC-Hiroshima 2014—ECCE ASIA), Hiroshima, Japan, 18–21 May 2014.
53. Tan, R.; Ye, S.; Deng, C.; Yu, C.; Zhou, A.; Du, C. Research on Prediction Modeling of EMI in Multi-in-one Electric Drive System. In Proceedings of the 2021 4th International Conference on Energy, Electrical and Power Engineering (CEEPE), Chongqing, China, 23–25 April 2021.
54. Safayet, A.; Islam, M. Modeling of Conducted Emission for a Three-Phase Motor Control Inverter. *IEEE Trans. Ind. Appl.* **2021**, *57*, 1202–1211. [CrossRef]

55. Ran, L.; Gokani, S.; Clare, J.; Bradley, K.J.; Christopoulos, C. Conducted Electromagnetic Emissions in Induction Motor Drive Systems Part II: Frequency Domain Models. *IEEE Trans. Power Electron.* **1998**, *4*, 768–776. [CrossRef]
56. Xue, J.; Wang, F. Mixed-mode EMI noise in three-phase DC-fed PWM motor drive system. In Proceedings of the 2013 IEEE Energy Conversion Congress and Exposition, Denver, CO, USA, 15–19 September 2013.
57. Touré, B.; Schanen, J.; Gerbaud, L.; Meynard, T.; Roudet, J.; Ruelland, R. EMC Modeling of Drives for Aircraft Applications: Modeling Process, EMI Filter Optimization, and Technological Choice. *IEEE Trans. Power Electron.* **2013**, *28*, 1145–1156. [CrossRef]
58. Jia, X. Influence of System Layout on CM EMI Noise of SiC Electric Vehicle Powertrains. *CPSS Trans. Power Electron. Appl.* **2021**, *6*, 298–309. [CrossRef]
59. Xiang, Y.; Pei, X.; Zhou, W.; Kang, Y.; Wang, H. A Fast and Precise Method for Modeling EMI Source in Two-Level Three-Phase Converter. *IEEE Trans. Power Electron.* **2019**, *34*, 10650–10664. [CrossRef]
60. Zhang, R.; Chen, W.; Zhou, Y.; Shi, Z.; Yan, R.; Yang, X. Mathematical Modeling of EMI Spectrum Envelope Based on Switching Transient Behavior. *IEEE J. Emerg. Sel. Top. Power Electron.* **2022**, *10*, 2497–2515. [CrossRef]
61. Kharanaq, F.A.; Emadi, A.; Bilgin, B. Modeling of Conducted Emissions for EMI Analysis of Power Converters: State-of-the-Art Review. *IEEE Access* **2020**, *8*, 189313–189325. [CrossRef]
62. Wan, L.; Beshir, A.H.; Wu, X.; Liu, X.; Grassi, F.; Spadacini, G.; Pignari, S.A. Black-box Modeling of Converters in Renewable Energy Systems for EMC Assessment: Overview and Discussion of Available Models. *Chin. J. Electr. Eng.* **2022**, *8*, 13–28. [CrossRef]
63. Amara, M.; Vollaire, C.; Ali, M.; Costa, F. Black box EMC modeling of a three phase inverter. In Proceedings of the 2018 International Symposium on Electromagnetic Compatibility (EMC EUROPE), Amsterdam, Netherlands, 27–30 August 2018.
64. Bishnoi, H.; Mattavelli, P.; Burgos, R.; Boroyevich, D. EMI Behavioral Models of DC-Fed Three-Phase Motor Drive Systems. *IEEE Trans. Power Electron.* **2014**, *29*, 4633–4645. [CrossRef]
65. Sun, B.; Burgos, R.; Boroyevich, D. Common-Mode EMI Unterminated Behavioral Model of Wide-Bandgap-Based Power Converters Operating at High Switching Frequency. *IEEE J. Emerg. Sel. Top. Power Electron.* **2019**, *7*, 2561–2570. [CrossRef]
66. Wan, L.; Beshir, A.H.; Wu, X.; Liu, X.; Grassi, F.; Spadacini, G.; Pignari, S.A.; Zanoni, M.; Tenti, L.; Chiumeo, R. Black-Box Modelling of Low-Switching-Frequency Power Inverters for EMC Analyses in Renewable Power Systems. *Energies* **2021**, *14*, 3413. [CrossRef]
67. Trinchero, R.; Stievano, I.S.; Canavero, F.G. Enhanced Time-Invariant Linear Model for the EMI Prediction of Switching Circuits. *IEEE Trans. Electromagn. Compat.* **2020**, *62*, 2294–2302. [CrossRef]
68. Zhou, P.; Pei, X.; Zhang, K.; Shan, Y. Improved EMI Behavioral Modeling Method of Three-Phase Inverter Based on the Noise-Source Phase Alignment. *IEEE Trans. Power Electron.* **2022**, *37*, 9333–9344. [CrossRef]
69. Zhou, W.; Pei, X.; Xiang, Y.; Kang, Y. A New EMI Modeling Method for Mixed-Mode Noise Analysis in Three-Phase Inverter System. *IEEE Access* **2020**, *8*, 71535–71547. [CrossRef]
70. Rosales, A.; Sarikhani, A.; Mohammed, O.A. Evaluation of Radiated Electromagnetic Field Interference Due to Frequency Swithcing in PWM Motor Drives by 3D Finite Elements. *IEEE Trans. Magn.* **2011**, *47*, 1474–1477. [CrossRef]
71. Huangfu, Y.; Wang, S.; Di Rienzo, L.; Zhu, J. Radiated EMI Modeling and Performance Analysis of a PWM PMSM Drive System Based on Field-Circuit Coupled FEM. *IEEE Trans. Magn.* **2017**, *53*, 1–4. [CrossRef]
72. Andrieu, G.; Reineix, A. On the Application of the "Equivalent Cable Bundle Method" to Cable Bundles in Presence of Complex Ground Structures. *IEEE Trans. Electromagn. Compat.* **2013**, *55*, 798–801. [CrossRef]
73. Ala, G.; Di Piazza, M.C.; Tine, G.; Viola, F.; Vitale, G. Evaluation of Radiated EMI in 42-V Vehicle Electrical Systems by FDTD Simulation. *IEEE Trans. Veh. Technol.* **2007**, *56*, 1477–1484. [CrossRef]
74. Yahyaoui, W.; Pichon, L.; Duval, F. A 3D PEEC Method for the Prediction of Radiated Fields From Automotive Cables. *IEEE Trans. Magn.* **2010**, *46*, 3053–3056. [CrossRef]
75. Barzegaran, M.R.; Mohammed, O.A. 3-D FE Wire Modeling and Analysis of Electromagnetic Signatures From Electric Power Drive Components and Systems. *IEEE Trans. Magn.* **2013**, *49*, 1937–1940. [CrossRef]
76. Costa, F.; Gautier, C.; Revol, B.; Genoulaz, J.; Demoulin, B. Modeling of the near-field electromagnetic radiation of power cables in automotives or aeronautics. *IEEE Trans. Power Electron.* **2013**, *28*, 4580–4593. [CrossRef]
77. Wei, S.; Pan, Z.; Yang, J.; Du, P. A Fast Prediction Approach of Radiated Emissions From Closely-Spaced Bent Cables in Motor Driving System. *IEEE Trans. Veh. Technol.* **2022**, *71*, 6100–6109. [CrossRef]
78. Della Torre, F.; Morando, A.P. Study on Far-Field Radiation From Three-Phase Induction Machines. *IEEE Trans. Electromagn. Compat.* **2009**, *51*, 928–936. [CrossRef]
79. Jo, H.; Han, K.J. Estimation of radiation patterns from the stator winding of AC motors using array model. In Proceedings of the 2016 IEEE International Symposium on Electromagnetic Compatibility (EMC), Ottawa, ON, Canada, 25–29 July 2016.
80. Wei, S.; Pan, Z.; Yang, J.; Du, P. A Reduction Technique for Modeling Closely Spaced Wires Considering Proximity Effects Applied to Predict Radiated Emission From Electric Motor. *IEEE Trans. Ind. Electron.* **2021**, *68*, 12535–12542. [CrossRef]
81. Zhang, Y.; Wang, S.; Chu, Y. Investigation of Radiated Electromagnetic Interference for an Isolated High-Frequency DC–DC Power Converter With Power Cables. *IEEE Trans. Power Electron.* **2019**, *34*, 9632–9643. [CrossRef]
82. Zhang, B.; Wang, S. Radiated Electromagnetic Interference Source Modeling for a Three Phase Motor Drive System with a SiC Power Module. In Proceedings of the 2021 IEEE International Joint EMC/SI/PI and EMC Europe Symposium, Raleigh, NC, USA, 26 July–13 August 2021.

83. Zhang, B.; Wang, S. Radiated Electromagnetic Interference Modeling for Three Phase Motor Drive Systems with SiC Power Modules. In Proceedings of the 2021 IEEE Applied Power Electronics Conference and Exposition (APEC), Phoenix, AZ, USA, 14–17 June 2021.

84. Fei, C.; Yang, Y.; Li, Q.; Lee, F.C. Shielding Technique for Planar Matrix Transformers to Suppress Common-Mode EMI Noise and Improve Efficiency. *IEEE Trans. Ind. Electron.* **2018**, *65*, 1263–1272. [CrossRef]

85. Yao, J.; Wang, S.; Luo, Z. Modeling, Analysis, and Reduction of Radiated EMI Due to the Voltage Across Input and Output Cables in an Automotive Non-Isolated Power Converter. *IEEE Trans. Power Electron.* **2022**, *37*, 5455–5465. [CrossRef]

86. Wang, S.; Lee, F.C.; van Wyk, J.D.; van Wyk, J.D. A Study of Integration of Parasitic Cancellation Techniques for EMI Filter Design With Discrete Components. *IEEE Trans. Power Electron.* **2008**, *23*, 3094–3102. [CrossRef]

87. Wang, R.; Boroyevich, D.; Blanchette, H.F.; Mattavelli, P. High power density EMI filter design with consideration of self-parasitic. In Proceedings of the 2012 Twenty-Seventh Annual IEEE Applied Power Electronics Conference and Exposition (APEC), Orlando, FL, USA, 5–9 February 2012.

88. Wang, R.; Blanchette, H.F.; Mu, M.; Boroyevich, D.; Mattavelli, P. Influence of High-Frequency Near-Field Coupling Between Magnetic Components on EMI Filter Design. *IEEE Trans. Power Electron.* **2013**, *28*, 4568–4579. [CrossRef]

89. Zhang, B.; Wang, S. Analysis and Reduction of the Near Magnetic Field Emission From Toroidal Inductors. *IEEE Trans. Power Electron.* **2020**, *35*, 6251–6268. [CrossRef]

90. Li, Y.; Wang, S.; Sheng, H.; Lakshmikanthan, S. Investigate and Reduce Capacitive Couplings in a Flyback Adapter With a DC-Bus Filter to Reduce EMI. *IEEE Trans. Power Electron.* **2020**, *35*, 6963–6973. [CrossRef]

91. Ma, Z.; Yao, J.; Wang, S.; Sheng, H.; Lakshmikanthan, S.; Osterhout, D. Radiated EMI Reduction with Double Shielding Techniques in Active-clamp Flyback Converters. In Proceedings of the 2021 IEEE International Joint EMC/SI/PI and EMC Europe Symposium, Raleigh, NC, USA, 26 July–13 August 2021.

92. Lai, R.; Maillet, Y.; Wang, F.; Wang, S.; Burgos, R.; Boroyevich, D. An Integrated EMI Choke for Differential-Mode and Common-Mode Noise Suppression. *IEEE Trans. Power Electron.* **2010**, *25*, 539–544.

93. Borsalani, J.; Dastfan, A.; Ghalibafan, J. An Integrated EMI Choke With Improved DM Inductance. *IEEE Trans. Power Electron.* **2021**, *36*, 1646–1658. [CrossRef]

94. Wang, A.; Zheng, F.; Gao, T.; Wu, Z.; Li, X. Integrated CM Inductor for Both DC and AC Noise Attenuation in DC-Fed Motor Drive Systems. *IEEE Trans. Power Electron.* **2023**, *38*, 510–522. [CrossRef]

95. Liu, Y.; Jiang, S.; Liang, W.; Wang, H.; Peng, J. Modeling and Design of the Magnetic Integration of Single- and Multi-Stage EMI Filters. *IEEE Trans. Power Electron.* **2020**, *35*, 276–288. [CrossRef]

96. Chen, R.; VanWyk, J.D.; Wang, S.; Odendaal, W.G. Improving the Characteristics of Integrated EMI Filters by Embedded Conductive Layers. *IEEE Trans. Power Electron.* **2005**, *20*, 611–619. [CrossRef]

97. Jiang, S.; Wang, P.; Wang, W.; Xu, D. Modeling and Design of Full Electromagnetic Integration of a Symmetrical EMI Filtering Circuit With Flexible Multi-Layer Foil Technique. *IEEE Trans. Electromagn. Compat.* **2023**, *65*, 414–424. [CrossRef]

98. Cougo, B.; Sathler, H.H.; Riva, R.; Santos, V.D.; Roux, N.; Sareni, B. Characterization of Low-Inductance SiC Module With Integrated Capacitors for Aircraft Applications Requiring Low Losses and Low EMI Issues. *IEEE Trans. Power Electron.* **2021**, *36*, 8230–8242. [CrossRef]

99. Jia, N.; Tian, X.; Xue, L.; Bai, H.; Tolbert, L.M.; Cui, H. Integrated Common-Mode Filter for GaN Power Module With Improved High-Frequency EMI Performance. *IEEE Trans. Power Electron.* **2023**, *38*, 6897–6901. [CrossRef]

100. Gonzalez, D.; Balcells, J.; Santolaria, A.; Le Bunetel, J.; Gago, J.; Magnon, D.; Brehaut, S. Conducted EMI Reduction in Power Converters by Means of Periodic Switching Frequency Modulation. *IEEE Trans. Power Electron.* **2007**, *22*, 2271–2281. [CrossRef]

101. Lezynski, P. Random Modulation in Inverters With Respect to Electromagnetic Compatibility and Power Quality. *IEEE J. Emerg. Sel. Top. Power Electron.* **2018**, *6*, 782–790. [CrossRef]

102. Mathe, L.; Lungeanu, F.; Sera, D.; Rasmussen, P.O.; Pedersen, J.K. Spread Spectrum Modulation by Using Asymmetric-Carrier Random PWM. *IEEE Trans. Ind. Electron.* **2012**, *59*, 3710–3718. [CrossRef]

103. Trzynadlowski, A.M.; Wang, Z.; Nagashima, J.; Stancu, C. Comparative investigation of PWM techniques for general motors' new drive for electric vehicles. In Proceedings of the Conference Record of the 2002 IEEE Industry Applications Conference. 37th IAS Annual Meeting (Cat. No.02CH37344), Pittsburgh, PA, USA, 13–18 October 2002.

104. Kiani Savadkoohi, H.; Arab Khaburi, D.; Sadr, S. A new switching method for PWM inverter with uniform distribution of output current's spectrum. In Proceedings of the 6th Power Electronics, Drive Systems & Technologies Conference (PEDSTC2015), Tehran, Iran, 3–4 February 2015.

105. Santolaria, A.; Balcells, J.; Gonzalez, D. Theoretical and experimental results of power converter frequency modulation. In Proceedings of the IEEE 2002 28th Annual Conference of the Industrial Electronics Society. IECON 02, Seville, Spain, 5–8 November 2002.

106. Gonzalez, D.; Bialasiewicz, J.T.; Balcells, J.; Gago, J. Wavelet-Based Performance Evaluation of Power Converters Operating With Modulated Switching Frequency. *IEEE Trans. Ind. Electron.* **2008**, *55*, 3167–3176. [CrossRef]

107. Yongxiang, X.; Qingbing, Y.; Jibin, Z.; Hao, W. Influence of periodic carrier frequency modulation on inverter loss of permanent magnet synchronous motor drive system. In Proceedings of the 2014 17th International Conference on Electrical Machines and Systems (ICEMS), Hangzhou, China, 22–25 October 2014.

108. Chen, J.; Jiang, D.; Sun, W.; Shen, Z.; Zhang, Y. A Family of Spread-Spectrum Modulation Schemes Based on Distribution Characteristics to Reduce Conducted EMI for Power Electronics Converters. *IEEE Trans. Ind. Appl.* **2020**, *56*, 5142–5157. [CrossRef]
109. Chen, J.; Jiang, D.; Shen, Z.; Sun, W.; Fang, Z. Uniform Distribution Pulsewidth Modulation Strategy for Three-Phase Converters to Reduce Conducted EMI and Switching Loss. *IEEE Trans. Ind. Electron.* **2020**, *67*, 6215–6226. [CrossRef]
110. Jiang, D.; Li, Q.; Shen, Z. Model predictive PWM for AC motor drives. *IET Electr. Power Appl.* **2017**, *11*, 815–822. [CrossRef]
111. Jiang, D.; Li, Q.; Han, X.; Qu, R. Variable switching frequency PWM for torque ripple control of AC motors. In Proceedings of the 2016 19th International Conference on Electrical Machines and Systems (ICEMS), Chiba, Japan, 2 February 2017.
112. Jiang, D.; Wang, F. Variable Switching Frequency PWM for Three-Phase Converters Based on Current Ripple Prediction. *IEEE Trans. Power Electron.* **2013**, *28*, 4951–4961. [CrossRef]
113. Hava, A.M.; Cetin, N.O.; Un, E. On the Contribution of PWM Methods to the Common Mode (Leakage) Current in Conventional Three-Phase Two-Level Inverters as Applied to AC Motor Drives. In Proceedings of the 2008 IEEE Industry Applications Society Annual Meeting, Edmonton, AB, Canada, 5–9 October 2008.
114. Julian, A.L.; Oriti, G.; Lipo, T.A. Elimination of Common-Mode Voltage in Three-Phase Sinusoidal Power Converters. *IEEE Trans. Power Electron.* **1999**, *14*, 982–989. [CrossRef]
115. Zhang, H.; Von Jouanne, A.; Dai, S.; Wallace, A.K.; Wang, F. Multilevel Inverter Modulation Schemes to Eliminate Common-Mode Voltages. *IEEE Trans. Ind. Appl.* **2000**, *36*, 1645–1653.
116. Jiang, D.; Shen, Z.; Wang, F. Common-Mode Voltage Reduction for Paralleled Inverters. *IEEE Trans. Power Electron.* **2018**, *33*, 3961–3974. [CrossRef]
117. von Jauanne, A.; Zhang, H. A Dual-Bridge Inverter Approach to Eliminating Common-Mode Voltages and Bearing and Leakage Currents. *IEEE Trans. Power Electron.* **1999**, *14*, 43–48. [CrossRef]
118. Chen, H.; Zhao, H. Review on pulse-width modulation strategies for common-mode voltage reduction in three-phase voltage-source inverters. *IET Power Electron.* **2016**, *9*, 2611–2620. [CrossRef]
119. Chudzik, P.; Steczek, M.; Tatar, K. Reduction in Selected Torque Harmonics in a Three-Level NPC Inverter-Fed Induction Motor Drive. *Energies* **2022**, *15*, 4078. [CrossRef]
120. Zhang, D.; Wang, F.; Burgos, R.; Boroyevich, D. Total Flux Minimization Control for Integrated Inter-Phase Inductors in Paralleled, Interleaved Three-Phase Two-Level Voltage-Source Converters With Discontinuous Space-Vector Modulation. *IEEE Trans. Power Electron.* **2012**, *27*, 1679–1688. [CrossRef]
121. Shen, Z.; Jiang, D.; Zou, T.; Qu, R. Dual-Segment Three-Phase PMSM With Dual Inverters for Leakage Current and Common-Mode EMI Reduction. *IEEE Trans. Power Electron.* **2019**, *34*, 5606–5619. [CrossRef]
122. Shen, Z.; Jiang, D.; Liu, Z.; Ye, D.; Li, J. Common-Mode Voltage Elimination for Dual Two-Level Inverter-Fed Asymmetrical Six-Phase PMSM. *IEEE Trans. Power Electron.* **2020**, *35*, 3828–3840. [CrossRef]
123. Jiang, D.; Liu, K.; Liu, Z.; Wang, Q.; He, Z.; Qu, R. Four-Module Three-Phase PMSM Drive for Suppressing Vibration and Common-Mode Current. *IEEE Trans. Ind. Appl.* **2021**, *57*, 4874–4883. [CrossRef]
124. Zhang, B.; Wang, S. A Survey of EMI Research in Power Electronics Systems With Wide-Bandgap Semiconductor Devices. *IEEE J. Emerg. Sel. Top. Power Electron.* **2020**, *8*, 626–643. [CrossRef]
125. Wang, R.; Liang, L.; Chen, Y.; Pan, Y.; Li, J.; Han, L.; Tan, G. Self-Adaptive Active Gate Driver for IGBT Switching Performance Optimization Based on Status Monitoring. *IEEE Trans. Power Electron.* **2020**, *35*, 6362–6372. [CrossRef]
126. Idir, N.; Bausiere, R.; Franchaud, J.J. Active gate voltage control of turn-on di/dt and turn-off dv/dt in insulated gate transistors. *IEEE Trans. Power Electron.* **2006**, *21*, 849–855. [CrossRef]
127. Bau, P.; Cousineau, M.; Cougo, B.; Richardeau, F.; Rouger, N. CMOS Active Gate Driver for Closed-Loop dv /dt Control of GaN Transistors. *IEEE Trans. Power Electron.* **2020**, *35*, 13322–13332. [CrossRef]
128. Dymond, H.C.P.; Wang, J.; Liu, D.; Dalton, J.J.O.; McNeill, N.; Pamunuwa, D.; Hollis, S.J.; Stark, B.H. A 6.7-GHz Active Gate Driver for GaN FETs to Combat Overshoot, Ringing, and EMI. *IEEE Trans. Power Electron.* **2018**, *33*, 581–594. [CrossRef]
129. Costa, F.; Magnon, D. Graphical Analysis of the Spectra of EMI Sources in Power Electronics. *IEEE Trans. Power Electron.* **2005**, *20*, 1491–1498. [CrossRef]
130. Yang, X.; Yuan, Y.; Zhang, X.; Palmer, P.R. Shaping High-Power IGBT Switching Transitions by Active Voltage Control for Reduced EMI Generation. *IEEE Trans. Ind. Appl.* **2015**, *51*, 1669–1677. [CrossRef]
131. Xu, C.; Ma, Q.; Xu, P.; Cui, T. Shaping SiC MOSFET Voltage and Current Transitions by Intelligent Control for Reduced EMI Generation. *Electronics* **2019**, *8*, 508. [CrossRef]
132. Ogasawara, S.; Ayano, H.; Akagi, H. An Active Circuit for Cancellation of Common-Mode Voltage Generated by a PWM Inverter. *IEEE Trans. Power Electron.* **1999**, *13*, 835–841. [CrossRef]
133. Wang, S.; Maillet, Y.Y.; Wang, F.; Boroyevich, D.; Burgos, R. Investigation of Hybrid EMI Filters for Common-Mode EMI Suppression in a Motor Drive System. *IEEE Trans. Power Electron.* **2010**, *25*, 1034–1045. [CrossRef]
134. Chen, W.; Yang, X.; Wang, Z. A Novel Hybrid Common-Mode EMI Filter With Active Impedance Multiplication. *IEEE Trans. Ind. Electron.* **2011**, *58*, 1826–1834. [CrossRef]
135. Takahashi, S.; Ogasawara, S.; Takemoto, M.; Orikawa, K.; Tamate, M. Common-Mode Voltage Attenuation of an Active Common-Mode Filter in a Motor Drive System Fed by a PWM Inverter. *IEEE Trans. Ind. Appl.* **2019**, *55*, 2721–2730. [CrossRef]
136. Zhang, Y.; Li, Q.; Jiang, D. A Motor CM Impedance Based Transformerless Active EMI Filter for DC-Side Common-Mode EMI Suppression in Motor Drive System. *IEEE Trans. Power Electron.* **2020**, *35*, 10238–10248. [CrossRef]

137. Zhang, Y.; Jiang, D. An Active EMI Filter in Grounding Circuit for DC Side CM EMI Suppression in Motor Drive System. *IEEE Trans. Power Electron.* **2022**, *37*, 2983–2992. [CrossRef]
138. Dai, L.; Chen, W.; Yang, X.; Zheng, M.; Yang, Y.; Wang, R. A Multi-Function Common Mode Choke Based on Active CM EMI Filters for AC/DC Power Converters. *IEEE Access* **2019**, *7*, 43534–43546. [CrossRef]

Article

Cogging Torque Minimization of Surface-Mounted Permanent Magnet Synchronous Motor Based on RSM and NSGA-II

Yinquan Yu [1,2,3,*], Yue Pan [1,2,3], Qiping Chen [1,2], Dequan Zeng [1,2,3], Yiming Hu [1,2,3], Hui-Hwang Goh [4], Shuangxia Niu [5] and Zhao Zhao [6]

1. School of Mechatronics and Vehicle Engineering, East China Jiaotong University, Nanchang 330013, China
2. Key Laboratory of Conveyance and Equipment of Ministry of Education, East China Jiaotong University, Nanchang 330013, China
3. Institute of Precision Machining and Intelligent Equipment Manufacturing, East China Jiaotong University, Nanchang 330013, China
4. School of Electrical Engineering, Guangxi University, Nanning 530004, China
5. Department of Electrical Engineering, The Hong Kong Polytechnic University, Hong Kong 999077, China
6. Faculty of Electrical Engineering and Information Technology, Otto-von-Guericke University Magdeburg, 39106 Magdeburg, Germany
* Correspondence: yinquan.yu@gmail.com

Abstract: A high-end permanent magnet (PM) synchronous motor's cogging torque is a significant performance measure (PMSM). During the running of the motor, excessive cogging torque will amplify noise and vibration. Therefore, the cogging torque must be taken into account while optimizing the design of motors with precise motion control. In this research, we proposed a local optimization-seeking approach (RSM+NSGA-II-LR) based on Response Surface Methodology (RSM) and Non-Dominated Sorting Genetic Algorithm-II (NSGA-II), which reduced the cogging torque of a permanent magnet synchronous motor (SPMSM). To reduce the complexity of optimization and increase its efficiency, the sensitivity analysis method was utilized to identify the structural parameters that had a significant impact on the torque performance. Second, RSM was utilized to fit the functional relationship between the structural parameters and each optimization objective, and NSGA-II was integrated to provide the Pareto solution for each optimization objective. The solution with a greater average torque than the initial motor and the lowest cogging torque was chosen, and a new finite element model (FEM) was created. On the basis of the sensitivity analysis, the structural factors that had the highest influence on the cogging torque were selected, and the RSM is utilized for local optimization to lower the cogging torque as much as feasible. The numerical results demonstrated that the optimization strategy presented in this study effectively reduced the cogging torque of the motor without diminishing the motor's average torque or increasing its torque ripple.

Keywords: SPMSM; cogging torque; torque ripple; RSM; multi-objective optimization

Citation: Yu, Y.; Pan, Y.; Chen, Q.; Zeng, D.; Hu, Y.; Goh, H.-H.; Niu, S.; Zhao, Z. Cogging Torque Minimization of Surface-Mounted Permanent Magnet Synchronous Motor Based on RSM and NSGA-II. *Actuators* **2023**, *11*, 379. https://doi.org/10.3390/act11120379

Academic Editor: Dong Jiang

Received: 29 October 2022
Accepted: 12 December 2022
Published: 16 December 2022

Publisher's Note: MDPI stays neutral with regard to jurisdictional claims in published maps and institutional affiliations.

1. Introduction

Due to their high power density, low loss, and small size, PMSMs are widely employed in transportation, aircraft, and industrial robots [1–4]. In order for the windings to fit into the stator slots, PMSMs are frequently designed with slotted magnetic circuit structures. However, the slotting can lead to interaction between the stator core and the magnets, which can have a significant impact on the stable operation of a PMSM. It is crucial to lower the cogging torque of the PMSM [5,6] because this influence is more pronounced in low-speed settings.

The cogging torque is an inherent property of slotted motors and cannot be abolished entirely. In prior research, the cogging torque was weakened primarily by motor control and construction. Regarding control, [7] proposed a solution based on harmonic torque to counteract the cogging torque. A speed- and position-adaptive controller was created to reduce the cogging torque in [8]. Although the aforementioned optimized controller methods could greatly reduce cogging torque, they failed to account for the effect on other motor performance characteristics.

A significant amount of research has centered on the structural design and optimization of electrical machines in an effort to enhance their output performance. In the field of structural design, researchers have altered the topology of electrical machines to enhance their electromagnetic performance [9,10]. In [11], a scheme of skew-toothed stator teeth was proposed to reduce the cogging torque, however the impact of the proposed improvement on other motor performance characteristics was not explored. In [12], a stator-tooth-notching method was used to reduce the cogging torque, but it also reduced the average torque. In [13], the method of offset poles was utilized to lower the cogging torque of the motor; however, the average torque also decreased, and the installation perfection for this approach required high precision. In [14], a PM radially unequal width layering technique was presented to reduce the cogging torque and torque pulsation as well as the average torque. However, the technique was too structurally changed and difficult to install. The preceding research has demonstrated that the weakening of cogging torque is frequently followed by a reduction in electromagnetic torque, and that altering a motor's construction may make the machining and installation of components more challenging.

In the field of optimization technology for electrical machines, the objective function, constraints, and boundaries are defined based on the electrical machine's optimization problem, and the design space is searched for the optimal combination of parameters to achieve a significant improvement in the electrical machine's performance [9,10,15]. With the advancement of optimization techniques, the study of the robustness of electrical machines design is gradually increasing [10,15]. Traditional approaches always optimize each structural parameter sequentially, neglecting structural parameter interactions. The Taguchi approach was proposed in [16–18] to optimize the major structural parameters of the motor. This method takes into account the interaction of the structural parameters based on an orthogonal test designed to successfully reduce the cogging torque of the motor. However, for a broad range of structural parameter values, the Taguchi method's optimization is limited by insufficient precision. In [19,20], a genetic algorithm was employed to optimize the cogging torque of the motor. However, while the cogging torque was effectively lessened, the average torque was also decreased.

In this paper, RSM and NSGA-II were applied to the structural parameter optimization design of a motor in order to optimize the torque performance of SPMSM to ensure that the average torque of the motor was not less than the average torque of the initial motor and minimize the cogging torque. This paper's outline is as follows: The second section describes the generation causes and analytical formulae for the SPMSM cogging torque. In the third section, a parameterization model for the SPMSM was created in order to examine the structural parameters' sensitivity. Using RSM and NSGA-II, the torque performance of the motor was optimized in the fourth section. The torque performance of the initial motor was compared to the torque performance of the optimized motor in the fifth section. Conclusions were drawn in the sixth section.

2. Mechanism and Analysis of Cogging Torque Generation in SPMSMs

Cogging torque is produced by the contact between the PM and the stator core, which is damaging to a motor's performance. During the rotation of the rotor, the magnetic field between the PM and the stator slot is nearly constant, and the PM's forces are balanced. However, the magnitude of magnetic conductivity along the margins of both sides of the PM is constantly changing, resulting in differential magnitudes of magnetic field strengths on the left and right sides in this region and imbalanced forces on the PM, which generate the cogging torque. The formula for calculating the cogging torque is [21]:

$$T_c = -\frac{\partial W}{\partial \alpha} \tag{1}$$

where W is the energy of the magnetic field in the motor when no current is applied to the winding and α is the relative position of the stator and rotor. Assuming that the magnetic conductivity of the armature core is infinite, the energy of the magnetic field inside the motor can be expressed as:

$$W \approx W_{PM} + W_{airgap} = \frac{1}{2\mu_0} \int_V B^2 dV \tag{2}$$

where μ_0 is the air magnetic conductivity, B is the air gap magnetic density, W_{PM} is the magnetic field energy of the PM and W_{airgap} is the magnetic field energy of the air gap.

The distribution, $B(\theta,\alpha)$, of the airgap magnetic density along the armature surface of the motor in (2) can be expressed as:

$$B(\theta,\alpha) = B_r(\theta)\frac{h_m(\theta)}{h_m(\theta) + h_g(\theta,\alpha)} \tag{3}$$

From Equations (2) and (3), it can be seen that:

$$W = \frac{1}{2\mu_0} \int_V B_r^2(\theta)\left[\frac{h_m(\theta)}{h_m(\theta) + h_g(\theta,\alpha)}\right]^2 dV \tag{4}$$

where $h_m(\theta)$ is the length of the PM in the magnetization direction, also known as the thickness of the PM, $h_g(\theta,\alpha)$ is the effective airgap length and $B_r(\theta)$ is the residual magnetic induction of the PM. To further calculate the magnetic field energy in the motor, a Fourier expansion of $\left[\frac{h_m(\theta)}{h_m(\theta)+h_g(\theta,\alpha)}\right]^2$ and $B_r^2(\theta)$ is required and can be expressed as:

$$B_r^2(\theta) = B_{r0} + \sum_{n=1}^{\infty} B_{rn} \cos 2np\theta \tag{5}$$

$$\left[\frac{h_m(\theta)}{h_m(\theta) + h_g(\theta,\alpha)}\right]^2 = G_0 + \sum_{n=1}^{\infty} G_n \cos ns(\theta + \alpha) \tag{6}$$

where, $B_{r0} = \alpha_p B_r^2$, $B_{rn} = \frac{2}{n\pi} B_r^2 \sin n\alpha_p \pi$, α_p is the pole arc coefficient, p is the number of pole pairs, B_{rn} is the Fourier decomposition coefficient of the square of the air-gap magnetic density generated by PM, G_n is the Fourier decomposition coefficient of the square of the relative air gap permeability and B_{r0} and G_0 are the constant terms of $B_r^2(\theta)$ and $\left[\frac{h_m(\theta)}{h_m(\theta)+h_g(\theta,\alpha)}\right]^2$ after the Fourier decomposition.

G_0 and G_n in Equation (6) can be expressed as:

$$G_0 = \left[\frac{h_m}{h_m + h_g}\right]^2 \tag{7}$$

$$G_n = \frac{2s}{\pi}\int_0^{\frac{\pi}{s}-\frac{\alpha}{2}}\left[\frac{h_m}{h_m + h_g}\right]^2 \cos ns\theta d\theta = \frac{2}{n\pi}\left[\frac{h_m}{h_m + h_g}\right]^2 \sin\left[n\pi - \frac{ns\theta_{s0}}{2}\right] \tag{8}$$

where, θ_{so} is the width of the slot opening of the stator expressed in radians and s is the number of slots.

Substituting Equations (5) and (6) into (4), which is then juggled with (1), the cogging torque, T_c, can be expressed as:

$$T_c(\alpha) = \frac{\pi s L_a}{4\mu_0}\left(D_2^2 - D_{i2}^2\right)\sum_{n=1}^{\infty} nG_n B_{r\frac{ns}{2p}} \sin ns\alpha \tag{9}$$

where L_a is the axial length of the stator core, D_2 and D_{i2} are the inner radius of the armature and the outer radius of the rotor, respectively, and n is the integer that makes $ns/2p$ an integer.

The following equation can be used to calculate the electromagnetic torque of the SPMSM:

$$T_{em} = \frac{\sqrt{2}P_{out}}{w_r}\cos(pw_r t) + T_L \tag{10}$$

where T_{em} is the electromagnetic torque, P_{out} is the output power of the motor, w_r is the angular speed of the rotor and T_L is the load torque, where the output torque power can be expressed as:

$$P_{out} = mE_1 I_{ph} \tag{11}$$

where m is the number of phases of the motor, I_{ph} is the effective value of the phase current and E_1 is the effective value of a counter electromotive force. E_1 can be expressed as [22]:

$$E_1 = 4.44 f N_{ph} B_1 K_{w1} \frac{2}{\pi} \frac{\pi D_2}{p} L_a \tag{12}$$

$$N_{ph} = N_s \frac{n_s}{m} \tag{13}$$

where K_{w1} is the fundamental winding factor, f is the current frequency of the stator, N_{ph} is the number of turns in series per winding, B_1 is the magnetic induction, N_s is the number of conductors per slot and n_s is the total number of slots. As N_s is related to the area of the stator slot and the wire gauge of the winding, it can be expressed as [23]:

$$N_s = \frac{A_{ef}}{A_{Cu}} = \frac{A_s - A_i}{A_{Cu}} \tag{14}$$

where A_s and A_i can be expressed as:

$$A_s = \frac{B_{s1} + B_{s2}}{2} H_{s2} + \frac{\pi B_{s2}^2}{8} \tag{15}$$

$$A_i = C_i(2H_{s2} + \frac{\pi B_{s2}}{2} + B_{s1} + B_{s2}) \tag{16}$$

where C_i is the insulation thickness of the slot, A_{ef} is the effective area of the slot, A_{Cu} is the wire gauge of the estimated winding, A_s and A_i are the area of the slot and the insulation occupied area of the slot, respectively, B_{s1} is the width of the slot, B_{s2} is the radius of the slot and H_{s2} is the depth of the slot.

Substituting (12) and (13) into (11), the expression for P_{out} is obtained as:

$$P_{out} = \frac{8.88 f n_s N_s B_1 K_{w1} D_2 L_a I_{ph}}{p} \tag{17}$$

Letting $\zeta = \frac{8.88 f n_s B_1 K_{w1} D_2 L_a I_{ph}}{p}$, (15) and (16) are thus substituted into (14), which is then juggled with (10), (11) and (12) to derive the expression for the electromagnetic torque T_{em}:

$$T_{em} = \frac{\sqrt{2}\zeta}{w_r A_{Cu}}\left(\frac{B_{s1} + B_{s2}}{2} H_{s2} + \frac{\pi B_{s2}^2}{8} - C_i(2H_{s2} + \frac{\pi B_{s2}}{2} + B_{s2} + B_{s1})\right)\cos(pw_r t) + T_L \tag{18}$$

The average torque magnitude (T_a) of the SPMSM can be defined as the peak-to-peak average of the electromagnetic torque (T_{em}) during steady operation of the motor as follows:

$$T_a = avg(T_{em}) \tag{19}$$

The torque ripple, T_r, can be defined as the ratio of the peak-to-peak difference in the electromagnetic torque to the average torque, as shown below:

$$T_r = \frac{T_{\max} - T_{\min}}{T_a} \tag{20}$$

where avg means the calculated average, T_{max} is the maximum peak of the electromagnetic torque and T_{min} is the minimum peak of the electromagnetic torque.

According to Equation (9), the cogging torque is dominated by the magnitudes of G_n and B_{rn}. It can be shown from Equation (5) that the pole arc coefficient has a direct effect on B_{rn}. The width of the slot opening of the stator and the thickness of the PM have a direct effect on G_n, as shown by Equation (8). The structural parameters to be optimized are, therefore, the pole arc coefficient, the width of the slot aperture and the thickness of the PM. Changes in the other structural parameters of the stator slot in the SPMSM will likewise impact the magnetic field distribution in the stator's teeth and yoke as well as the motor's electromagnetic torque. Therefore, the stator-slot-related structural characteristics were chosen as the structural parameters to be optimized. It can be shown from Equation (18) that variations in the depth, width and radius of the slot will affect the electromagnetic torque capabilities. Changes in the thickness of the PM will also impact the distribution of the flux density in the airgap, thus affecting the electromagnetic torque capability. Consequently, the electromagnetic torque performance must be taken into account when reducing the cogging torque so that the electromagnetic torque performance is not severely diminished.

3. Numerical Simulation Analysis and Optimal Design of SPMSM

3.1. Parameterization Model of a SPMSM

In this study, an SPMSM with eight poles and thirty-six slots served as the object of analysis. Figure 1a illustrates the motor topology. Figure 1b depicts the mesh split for the numerical simulation study. The fundamental motor characteristics are provided in Table 1.

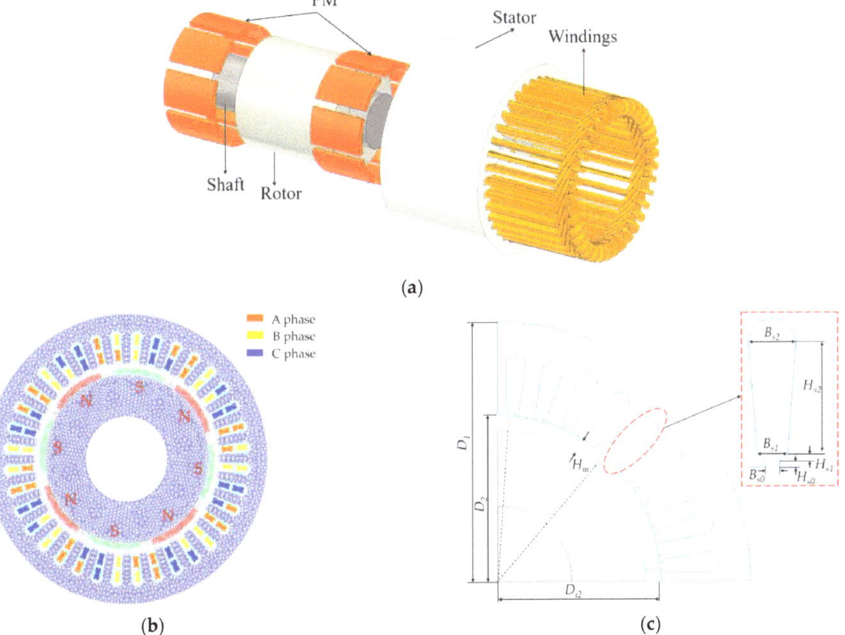

Figure 1. Model of SPMSM: (**a**) geometric model; (**b**) FEM; (**c**) parameterization model.

Table 1. Basic parameters of SPMSM.

Parameter	Value	Parameter	Value
Stator outer diameter/mm	165	Rated voltage/V	269
Stator inner diameter/mm	106	Number of rotor poles	8
Rotor outer diameter/mm	105	Number of stator slots	36
Rated power/kW	4.4	Rated speed/rpm	1500
Magnet type	NdFe35	Length/mm	100

Figure 1c shows the parameterization model of the motor. Table 2 displays the initial values and variation range for each structural parameter. This research focused on optimizing the stator slot and PM structure of the motor in order to reduce the cogging torque of the motor. Therefore, the pole arc coefficient (α_p) of the PM, the thickness (H_m) of the PM, the height (H_{s0}) of the slot opening, the height (H_{s1}) of the slot shoulder, the depth (H_{s2}) of the slot, the width (B_{s0}) of the slot opening, the width (B_{s1}) of the slot and the radius (B_{s2}) of the slot were selected for optimization. α_p is the ratio of the arc length spanned by each pole of PM to the pole pitch. The above structural parameters were taken within a reasonable range of variation.

Table 2. Initial values and range of values of structure parameters.

Parameter (Unit)	Range	Initial
α_p	0.7–0.9	0.82
H_m (mm)	3.5–5	4.5
H_{s0} (mm)	0.5–1	0.8
H_{s1} (mm)	0–1	0.9
H_{s2} (mm)	13.5–15.5	14.5
B_{s0} (mm)	1.5–3	2
B_{s1} (mm)	4–5	4.2
B_{s2} (mm)	6–7.5	6.6

3.2. Optimized Design of SPMSM

Figure 2 depicts the optimization process of this paper. It focused mostly on the sensitivity analysis of the structural parameters. RSM- and NSGA-II-based structural parameter optimizations were conducted.

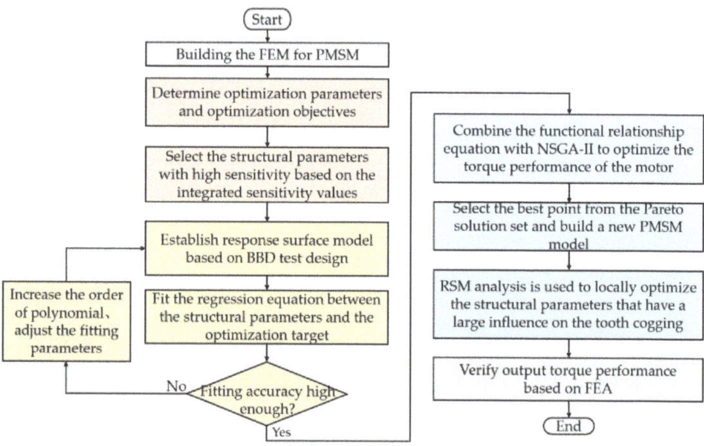

Figure 2. Optimized flow chart of SPMSM.

3.3. Sensitivity Analysis of Parameters

The sensitivity analysis method refers to an analysis technique in which the structural parameters are altered continually within a predetermined range during numerical simulation in order to observe the change rule and degree of the optimal objectives. A sensitivity analysis identifies the most significant factors among numerous others, thus minimizing the complexity and computational effort required for further optimization [24]. In this research, the Pearson correlation coefficient was utilized to calculate the influence index of each structural parameter on the optimal objective using the following equation:

$$\rho(X,Y) = \frac{COV(X,Y)}{\sigma_X \sigma_Y} \tag{21}$$

where X and Y are the structural parameter and the optimized objective, respectively. $COV(X,Y)$ is the covariance of X and Y, and σ_X and σ_Y are the standard deviations of X and Y, respectively.

After establishing the range of structural parameters and the optimum objectives, the Latin hypercube sampling approach in [25] was utilized to scan the design space for the sensitivity analysis. Then, a numerical simulation was used to determine the cogging torque, torque ripple and average torque values corresponding to each model. The minimum cogging torque, minimum torque ripple and highest average torque were the objectives of the parameter sensitivity analysis. Accordingly, the sensitivity of the selected structural parameters to the optimized objectives was then computed, as per Equation (21). The results are shown in Figure 3, where T_C is the cogging torque, T_r is the torque ripple and T_a is the average torque.

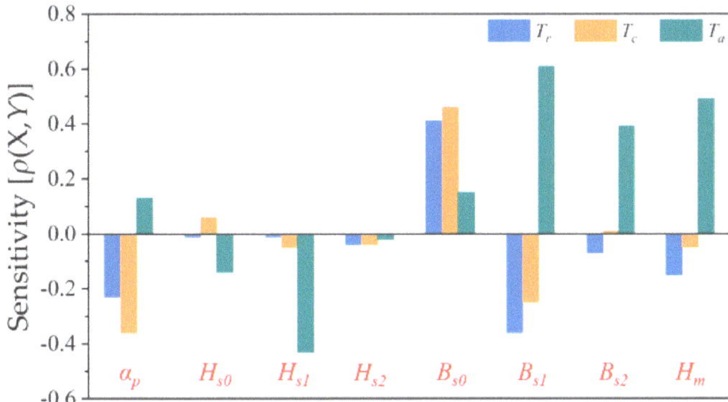

Figure 3. Analysis of sensitivity of structural parameters to three optimized objectives.

Figure 3 depicts the direction and magnitude of the association between each structural parameter and the optimization aim. B_{s0} correlated positively with the cogging torque, torque ripple and average torque. If B_{s0} changed in a way that promoted an increase in the average torque, the cogging torque and torque ripple likewise rose. This suggested that three optimized objectives did not always alter in a positive direction when B_{s0} was modified. Consequently, it was essential to implement the approach of trade-off optimization. As the number of optimization variables increases, so does the difficulty of optimization. Several structural characteristics that had a higher impact on the torque performance of the motor were selected for the optimization, according to Figure 3, in order to reduce the complexity of the optimization. Overall, the four structural parameters α_p, B_{s0}, B_{s1} and H_m had a greater impact on the torque performance of the motor. Therefore, the structural parameters for this optimization were α_p, B_{s0}, B_{s1} and H_m.

3.4. Torque Performance Optimization of SPMSM

3.4.1. Constructing RSM Models of Structural Parameters and Optimization Objectives

RSM is a technique for examining the relationship between numerous input factors and output targets. Suitable for studying multivariate problems, RSM can be used to identify optimal combinations by generating the response surfaces of input and output variables. Additionally, RSM can be utilized to develop polynomial equations for input and output variables. The Central Composite Design (CCD) and Box–Behnken Design (BBD) methods are two extensively utilized methods of test designing for response surfaces [26]. With the same number of selected parameters, the BBD design requires fewer trials. This research evaluated the nonlinear relationship between the three optimal objectives and the structural parameters using the BBD design approach. First, based on the number of structural parameters and the range of values, an appropriate test design method was established. Then, a numerical simulation was used to calculate the cogging torque, torque ripple and average torque for each combination of structural parameters. For the ensuing NSGA-II optimization, the polynomial regression equation between the structural parameters and the optimization objective was fitted using RSM. Due to the intricate interaction between the structural factors and the motor's optimal goals, In order to match their functional relationship, a second-order polynomial was used as the fitting function. The model of its response surface can be described by the following [27]:

$$y = \beta_0 + \sum_{i=1}^{k} \beta_{ii} x_i + \sum_{i=1}^{k} \beta_{ii} x_i^2 + \sum_{i=1}^{k} \beta_{ij} x_i x_j + \varepsilon \tag{22}$$

where ε is the fitting error, β is the coefficient to be determined, y is the predicted value of the optimized objective and x_i denotes the *i-th* structural parameter.

The number of test points was determined in accordance with the BBD design. There were twenty-five test locations and four specified structural parameters. By building a numerical simulation model for each test point in order to establish a polynomial regression model, the values of the relevant objective were calculated. Table 3 displays the BBD test design table and the numerical analysis findings. The table's results are kept to two decimal places.

Table 3. BBD test design table and numerical analysis results.

Number	α_p	B_{s0}	B_{s1}	H_m	T_c(N·m)	T_a(N·m)	T_r (%)
1	0.7	1.5	4.5	4.25	4.29	27.33	18.38
2	0.9	1.5	4.5	4.25	2.61	28.19	15.83
3	0.7	3	4.5	4.25	4.86	30.34	22.38
4	0.9	3	4.5	4.25	1.20	30.81	12.27
5	0.8	2.25	4	3.5	1.78	25.26	9.07
6	0.8	2.25	5	3.5	3.01	28.84	11.74
7	0.8	2.25	4	5	2.16	29.56	9.45
8	0.8	2.25	5	5	3.80	33.86	12.48
9	0.7	2.25	4.5	3.5	4.55	26.22	21.85
10	0.9	2.25	4.5	3.5	0.99	27.33	10.79
11	0.7	2.25	4.5	5	5.20	31.41	20.16
12	0.9	2.25	4.5	5	1.57	31.56	12.94
13	0.8	1.5	4	4.25	2.28	26.02	10.48
14	0.8	3	4	4.25	1.31	28.83	7.55
15	0.8	1.5	5	4.25	3.94	30.04	13.36
16	0.8	3	5	4.25	2.45	32.80	10.36
17	0.7	2.25	4	4.25	4.34	27.65	20.69
18	0.9	2.25	4	4.25	0.88	27.66	12.43
19	0.7	2.25	5	4.25	5.90	30.82	21.83
20	0.9	2.25	5	4.25	2.27	32.53	11.88

Table 3. *Cont.*

Number	α_p	B_{s0}	B_{s1}	H_m	$T_c(\text{N·m})$	$T_a(\text{N·m})$	$T_r\ (\%)$
21	0.8	1.5	4.5	3.5	2.55	25.37	10.88
22	0.8	3	4.5	3.5	1.47	27.88	8.34
23	0.8	1.5	4.5	5	3.12	29.65	11.69
24	0.8	3	4.5	5	1.89	32.73	8.91
25	0.8	2.25	4.5	4.25	2.50	29.33	10.28

To ensure the applicability of the regression equation corresponding to the three optimal objectives, the coefficient of determination, R^2 (ranging from 0 to 1), was utilized to assess the precision of the regression model fit. The closer to 1 the polynomial regression model fits, the better. R^2 can be calculated using Equation (23):

$$R^2 = 1 - \frac{\sum(y_a - y)^2}{\sum\left(y_a - \bar{y}\right)^2} \tag{23}$$

where y_a is the actual value of the objective function and y is the mean of the actual values of the objective function. The four parameters α_p, B_{s0}, B_{s1} and H_m were fitted as a function of the cogging torque, average torque and torque ripple, respectively, according to Table 4. The R^2 values of all three functions obtained by fitting were greater than 0.95, indicating that the fitted functional equation had good applicability.

Table 4. The range of values for local optimization of structural parameters.

Parameter (Unit)	Initial	RSM+NSGA-II	Local Range of Values
α_p	0.82	0.9	0.86–0.9
$B_{s0}(\text{mm})$	2	3	2.7–3

3.4.2. Multi-Objective Optimized Design Based on NSGA-II

For SPMSM multi-objective optimization issues, optimizing the performance of one objective frequently degrades the performance of the other objectives. They are constrained by one another. It is not possible to find a solution that maximizes the performance across all objectives. However, the optimized method can achieve equilibrium so that each aim can be optimized to the greatest extent possible. NSGA-II is suitable for optimizing complicated structures such as SPMSM due to its enhanced global search capabilities and increased resilience. In this research, NSGA-II was used for the thorough optimization of the cogging torque, torque ripple and average torque of the motor, with the objective function specified as follows:

$$\min : \begin{cases} f_1(x) = T_c \\ f_2(x) = -T_a \\ f_3(x) = T_r \end{cases} \tag{24}$$

where min means to minimize the target. In order to minimize the cogging torque, only one constraint was added to the NSGA-II optimization search process: $f_1(x)$ is less than the initial motor cogging torque ($T_c < 4.77\text{Nm}$).

The RSM-based polynomial equations were integrated with the NSGA-II multi-objective optimization technique to obtain the optimal structural parameter design space. The NSGA-II flowchart is depicted in Figure 4a. Figure 4b depicts the Pareto solution of three optimized objectives obtained after 50 NSGA-II iterations. Each solution presented in the diagram had a greater average torque than the initial motor. As a result, the option with the lowest cogging torque was chosen as the optimal solution. This is represented by the green dot in Figure 4b.

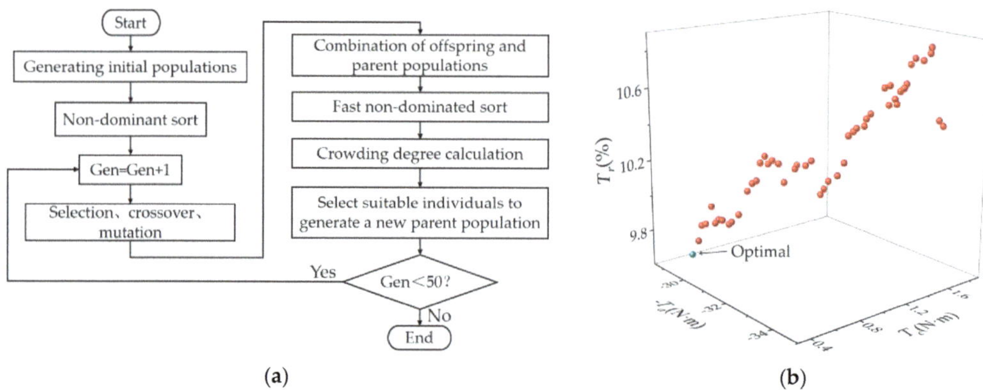

Figure 4. NSGA-II: (**a**) flow chart of NSGA-II; (**b**) Pareto solution set.

To significantly reduce the SPMSM's cogging torque and enhance its torque performance, in this research, we offered RSM+NSGA-II-LR, a method that employs RSM analysis for local optimization after an NSGA-II search for optimal solutions. The procedure for implementation was as follows: After first optimizing SPMSM using NSGA-II, a new FEM was developed. The structural characteristics with the greatest impact on the cogging torque were then chosen. Figure 3 shows that α_p, B_{s0} and B_{s1} had a significant impact on the cogging torque. However, the connection between B_{s1} and the average torque and cogging torque was inverse, and B_{s1}'s effect on the average torque was substantially bigger than its effect on the cogging torque, hence B_{s1} was not considered. The local optimization range of the structural parameters α_p and B_{s0} was redefined, with the new value being 20% of the initial range of the structural parameters. RSM was also utilized to maximize the cogging torque of the motor. Assuming that the initial value range of structural parameters is $[a, c]$ with a value of b, the optimization range of the local optimization can be described as follows:

$$\begin{cases} (c - a) \times 10\% = d \\ [a, a + 2d], & b - a < d \\ [c - 2d, c], & c - b < d \\ [b - d, b + d], & b - a > d, c - b > d \end{cases} \tag{25}$$

The local search ranges of α_p and B_{s0} were computed in accordance with Equation (25) and Table 4, with the results being displayed in Table 4. Figure 5 depicts the construction of the response surface model using the BBD design following the determination of the local search range.

The cogging torque increased and then dropped with the increase in B_{s0} and decreased with the increase in α_p, as shown in Figure 5a. Figure 5b demonstrates that when α_p increased, the average torque likewise increased slightly. The average torque increased as B_{s0} increases, although the increase was not evident, and the average torque varied by less than 1 Nm. Figure 5c demonstrates that the torque ripple increased significantly with increasing B_{s0} and diminished with increasing α_p.

In conclusion, the increase in α_p was advantageous for the reduction in the cogging torque and torque ripple as well as the enhancement of the average torque, so α_p was equal to 0.9. The decrease in B_{s0} was advantageous for the reduction in the cogging torque and torque ripple but not for the rise in the average torque. However, the effect of B_{s0} on the average torque was diminished, and the objective of this optimization was to lower the motor's cogging torque; therefore, B_{s0} was set as 2.7 mm.

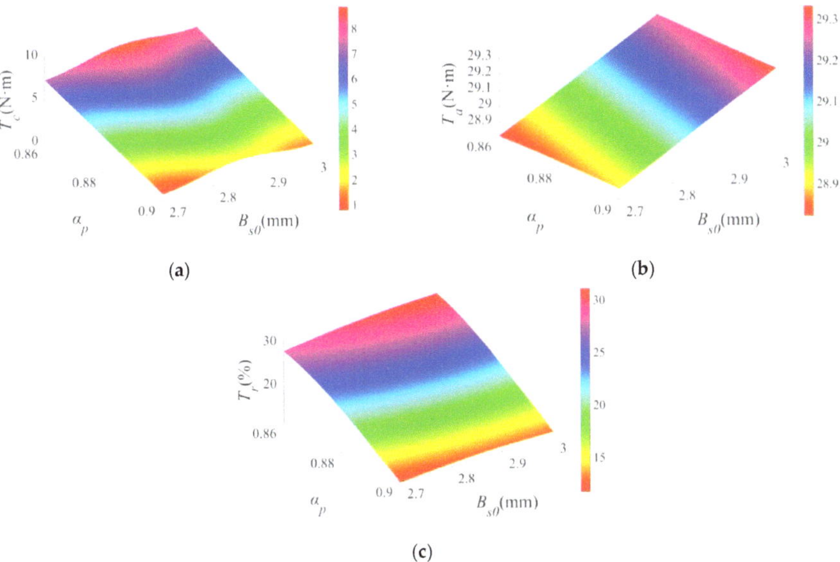

Figure 5. Response surface diagram: (**a**) the response surfaces of α_p, B_{s0} and T_c; (**b**) the response surfaces of α_p, B_{s0} and T_a; (**c**) the response surfaces of α_p, B_{s0} and T_r.

4. Optimized Result Analysis

The improved motor's structural parameters were produced from the aforementioned study, and a numerical simulation model was created to compare optimization objectives before and after optimization, as depicted in Figure 6. Table 5 displays the particular values. After the NSGA-II parameter search, the torque performance of the motor was significantly enhanced compared to the initial motor. In particular, the cogging torque and torque ripple were decreased by 71% and 33%, respectively, while the average torque was raised by 3%. This significantly increased the motor's driving performance. Using the RSM+NSGA-II-LR optimization, the torque performance of the motor was enhanced further. Compared to the RSM+NSGA-II-optimized motor torque performance, the cogging torque and torque ripple were reduced by 43% and 9%, respectively, and the average torque was reduced by 2% but was still higher than the initial motor's average torque, which was consistent with the objectives of this optimization. After optimizing the motor, this solution was chosen as the optimal combination of structural parameters. The motor's optimal structural parameters were obtained, as given in Table 6.

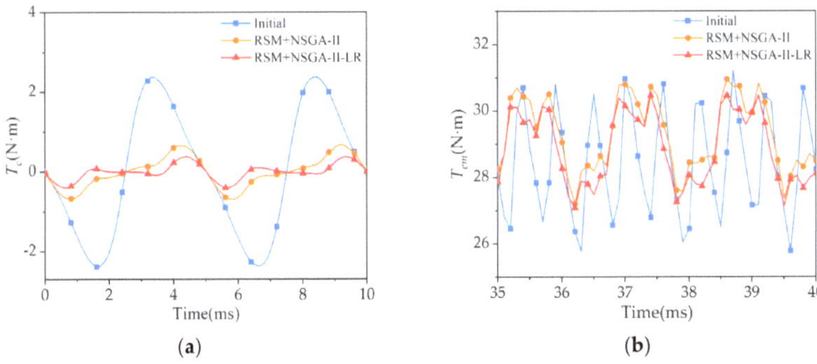

Figure 6. Results of numerical analysis before and after optimization: (**a**) curve diagram of cogging torque; (**b**) curve diagram of electromagnetic torque.

Table 5. Torque performance of the motor before and after optimization.

	$T_c(\mathbf{N \cdot m})$	$T_a(\mathbf{N \cdot m})$	T_r (%)
Initial	4.77	28.42	19.19
RSM+NSGA-II	1.37	29.33	12.81
Taguchi method	2.74	35.17	10.61
RSM	1.2	30.81	12.27
RSM+NSGA-II-LR	0.78	28.89	11.69

Table 6. Parameters before and after SPMSM optimization.

Parameter (Unit)	Initial	RSM+NSGA-II-LR
α_p	0.82	0.9
H_m(mm)	4.5	4.28
B_{s0}(mm)	2	2.7
B_{s1}(mm)	4.2	4.13

The results of optimizing the same four structural parameters using the Taguchi technique and RSM analysis are shown in Table 5. After comparison, it was evident that the RSM+NSGA-II-LR approach described in this study lowered the cogging torque while satisfying the assumption that the average torque must be greater than the original motor's torque.

5. Conclusions

This study combined the sensitivity analysis approach, RSM and NSGA-II and offered a method based on RSM+NSGA-II-LR to optimize the torque performance of a PMSM in order to address the issue of excessive cogging torque. The findings indicated that the method ensured that the average torque of the motor was not less than the average torque before optimization and efficiently minimized the cogging torque and torque ripple of the motor. Consequently, this study presented an efficient method for building an efficient, low-vibration and low-noise SPMSM, which may also be applied to enhance the performance of other motors. In future work, we will study more effective optimization techniques and robust design optimization methods to improve the stability and reliability of motor design and make experimental prototypes.

Author Contributions: Conceptualization, Y.Y. and Y.P.; methodology, Y.Y.; software, Y.P.; writing—original draft, Y.Y. and Y.P. writing—review and editing, Y.Y., Q.C., D.Z., Y.H., H.-H.G., Z.Z. and S.N. All authors have read and agreed to the published version of the manuscript.

Funding: This research was supported in part by the National Natural Science Foundation of China (Grant No. 52067006, U2034211, 52162044), in part by the Foreign Expert Bureau of the Ministry of science and technology of China (Grant No. G2021022002L), in part by long-term project of innovative leading talents in the "Double Thousand Plan" of Jiangxi Province (jxsq2019101027) and by the Key Research Program of Jiangxi Province (Grant No. 20212BBE51014).

Institutional Review Board Statement: Not applicable.

Informed Consent Statement: Not applicable.

Data Availability Statement: Not applicable.

Acknowledgments: The authors would like to thank anonymous reviewers for their helpful comments and suggestions to improve the manuscript.

Conflicts of Interest: The authors declare no conflict of interest.

References

1. Zhang, J.M.; Su, H.; Ren, Q.; Li, W.; Zhou, H.C. Review on development and key technologies of permanent magnet synchronous traction system for rail transit. *J. Traffic Transp. Eng.* **2021**, *21*, 63–77.
2. Zhang, W.; Shi, L.; Liu, K.; Li, L.; Jing, J. Optimization Analysis of Automotive Asymmetric Magnetic Pole Permanent Magnet Motor by Taguchi Method. *Int. J. Rotating Mach.* **2021**, *2021*, 6691574. [CrossRef]
3. Chen, Y.Y.; Zhu, X.Y.; Quan, L.; Han, X.; He, X.J. Parameter Sensitivity Optimization Design and Performance Analysis of Double-Salient Permanent-Magnet Double-Stator Machine. *Trans. China Electrotech. Soc.* **2017**, *32*, 160–168.
4. Wu, Z.; Fan, Y.; Lee, C.H.; Gao, D.; Yu, K. Vibration Optimization of FSCW-IPM Motor Based on Iron-Core Modification for Electric Vehicles. *IEEE Trans. Veh. Technol.* **2020**, *69*, 14834–14845. [CrossRef]
5. Hwang, M.-H.; Lee, H.-S.; Yang, S.-H.; Lee, G.-S.; Han, J.-H.; Kim, D.-H.; Kim, H.-W.; Cha, H.-R. Cogging Torque Reduction and Offset of Dual-Rotor Interior Permanent Magnet Motor in Electric Vehicle Traction Platforms. *Energies* **2019**, *12*, 1761. [CrossRef]
6. Anuja, T.; Doss, M. Reduction of Cogging Torque in Surface Mounted Permanent Magnet Brushless DC Motor by Adapting Rotor Magnetic Displacement. *Energies* **2021**, *14*, 2861. [CrossRef]
7. Gao, J.; Xiang, Z.; Dai, L.; Huang, S.; Ni, D.; Yao, C. Cogging Torque Dynamic Reduction Based on Harmonic Torque Counteract. *IEEE Trans. Magn.* **2021**, *58*, 8103405. [CrossRef]
8. Arias, A.; Caum, J.; Ibarra, E.; Grino, R. Reducing the Cogging Torque Effects in Hybrid Stepper Machines by Means of Resonant Controllers. *IEEE Trans. Ind. Electron.* **2018**, *66*, 2603–2612. [CrossRef]
9. Lei, G.; Zhu, J.; Guo, Y.; Liu, C.; Ma, B. A Review of Design Optimization Methods for Electrical Machines. *Energies* **2017**, *10*, 1962. [CrossRef]
10. Bramerdorfer, G.; Tapia, J.A.; Pyrhonen, J.J.; Cavagnino, A. Modern Electrical Machine Design Optimization: Techniques, Trends, and Best Practices. *IEEE Trans. Ind. Electron.* **2018**, *65*, 7672–7684. [CrossRef]
11. Ueda, Y.; Takahashi, H. Cogging Torque Reduction on Transverse-Flux Motor with Multilevel Skew Configuration of Toothed Cores. *IEEE Trans. Magn.* **2019**, *55*, 8203005. [CrossRef]
12. Moon, J.-H.; Kang, D.-W. Torque Ripple and Cogging Torque Reduction Method of IPMSM Using Asymmetric Shoe of Stator and Notch in Stator. *J. Electr. Eng. Technol.* **2022**, *17*, 3465–3471. [CrossRef]
13. Jo, I.-H.; Lee, H.-W.; Jeong, G.; Ji, W.-Y.; Park, C.-B. A Study on the Reduction of Cogging Torque for the Skew of a Magnetic Geared Synchronous Motor. *IEEE Trans. Magn.* **2018**, *55*, 8100505. [CrossRef]
14. Brahim Ladghem, C.; Kamel, B.; Rachid, I. Permanent magnet shaping for cogging torque and torque ripple reduction of PMSM. *COMPEL-Int. J. Comput. Math. Electr. Electron. Eng.* **2018**, *37*, 2232–2248. [CrossRef]
15. Orosz, T.; Rassõlkin, A.; Kallaste, A.; Arsénio, P.; Pánek, D.; Kaska, J.; Karban, P. Robust Design Optimization and Emerging Technologies for Electrical Machines: Challenges and Open Problems. *Appl. Sci.* **2020**, *10*, 6653. [CrossRef]
16. Lan, Z.Y.; Yang, X.Y.; Wang, F.Y.; Zheng, C.D. Application for Optimal Designing of Sinusoidal Interior Permanent Magnet Synchronous Motors by Using the Taguchi Method. *Trans. China Electrotech. Soc.* **2011**, *26*, 37–42.
17. Wang, X.Y.; Zhang, L.; Xu, W.G. Multi-objective Optimal Design for Interior Permanent Magnet Synchronous Motor Based on Taguchi Method. *Micromotors* **2016**, *49*, 1780–1783.
18. Si, J.K.; Zhang, L.F.; Feng, H.C.; Xu, X.Z.; Zhang, X.L. Multi-objective optimal design of a surface-mounted and interior permanent magnet synchronous motor. *J. China Coal Soc.* **2016**, *41*, 3167–3173.
19. Zou, J.J.; Zhao, S.W.; Yang, X.Y. Optimization of Cogging Torque of External Rotor PMSM Based on Genetic Algorithm. *Small Spec. Electr. Mach.* **2020**, *48*, 1–6.
20. Ilka, R.; Alinejad-Beromi, Y.; Yaghobi, H. Cogging torque reduction of permanent magnet synchronous motor using multi-objective optimization. *Math. Comput. Simul.* **2018**, *153*, 83–95. [CrossRef]
21. Wu, M.Q. *Design and Characteristic Research of Permanent Magnet Synchronous Motor for New Energy Vehicle*; Chongqing Jiaotong University: Chongqing, China, 2021.
22. Krishnan, R. *Permanent Magnet Synchronous and Brushless DC Motor Drives*; CRC Press: Boca Raton, FL, USA, 2017. [CrossRef]
23. Tang, R.Y. *Modern Permanent Magnet Machines: Theory and Design*; China Machine Press: Beijing, China, 2016.
24. Lei, G.; Liu, C.; Zhu, J.; Guo, Y. Techniques for Multilevel Design Optimization of Permanent Magnet Motors. *IEEE Trans. Energy Convers.* **2015**, *30*, 1574–1584. [CrossRef]
25. Shields, M.D.; Zhang, J. The generalization of Latin hypercube sampling. *Reliab. Eng. Syst. Saf.* **2016**, *148*, 96–108. [CrossRef]
26. Cao, Y.J.; Feng, L.L.; Mao, R.; Yu, L.; Jia, H.Y.; Jia, Z. Multi-objective Stratified Optimization Design of Axial-flux Permanent Magnet Memory Motor. *Proc. CSEE* **2021**, *41*, 1983–1992.
27. Gao, F.Y.; Gao, J.N.; Li, M.M.; Yao, P.; Song, Z.X.; Yang, K.W.; Gao, X.Y. Optimization Design of Halbach Interior Permanent Magnet Synchronous Motor Based on Parameter Sensitivity Stratification. *J. Xi'an Jiaotong Univ.* **2022**, *56*, 180–190.

Article

A Non-Permanent Magnet DC-Biased Vernier Reluctance Linear Machine with Non-Uniform Air Gap Structure for Ripple Reduction

Zhenyang Qiao [1], Yunpeng Zhang [1], Jian Luo [1,*], Weinong Fu [2], Dingguo Shao [1] and Haidong Cao [3]

[1] School of Mechatronic Engineering and Automation, Shanghai University, Shanghai 200444, China
[2] Shenzhen Institutes of Advanced Technology, Chinese Academy of Sciences, Shenzhen 518055, China
[3] Shanghai Engineering Research Center of Motor System Energy Saving Co., Ltd., Shanghai 200063, China
* Correspondence: luojian@shu.edu.cn

Abstract: Thrust ripple and density greatly impact the performance of the linear machine and other linear actuators, causing positioning control precision, dynamic performance, and efficiency issues. Generalized pole-pair combinations are difficult to satisfy both the thrust and ripple for double salient reluctance linear machines. In this paper, a DC-Biased vernier reluctance linear machine (DCB-VRLM) is proposed to solve the abovementioned issues. The key to the proposed design is to reduce the ripple and enhance the thrust density with non-uniform teeth by utilizing and optimizing the modulated flux in the air gap. To effectively verify the proposed design, the DCB-VRLMs with different winding pole pairs and secondary poles are compared. The 12-slot/10-pole combination is chosen to adopt a non-uniform air gap structure. Moreover, the energy distribution of AC/DC winding is studied and optimized to further enhance the performance of the proposed DCB-VRLM. The results indicate that the DCB-VRLM with the non-uniform air gap has a lower thrust ripple, better overload capability, and higher thrust density, which confirms its superiority in long-stroke linear rail transit and vertical elevator applications.

Keywords: DC biased current; doubly salient; linear machines; vernier reluctance machines; non-uniform air gap

Citation: Qiao, Z.; Zhang, Y.; Luo, J.; Fu, W.; Shao, D.; Cao, H. A Non-Permanent Magnet DC-Biased Vernier Reluctance Linear Machine with Non-Uniform Air Gap Structure for Ripple Reduction. *Actuators* **2023**, *12*, 7. https://doi.org/10.3390/act12010007

Academic Editor: Ahmad Taher Azar

Received: 25 November 2022
Revised: 14 December 2022
Accepted: 17 December 2022
Published: 22 December 2022

1. Introduction

Non-permanent magnet electric machines have been widely used in industrial applications and are concerned by researchers for they employ no permanent magnets (PMs), which have no risk of demagnetization and a slight cost issue. As one type of non-permanent magnet machine, the switched reluctance machine (SRM) adopts the doubly salient structure, which has neither PMs nor windings on the rotor. A variable flux reluctance machine (VFRM) was proposed, in which DC excitation windings are uniformly placed in the stator slot [1]. The number of rotor poles can be odd, which is no longer limited by the selection of rotor poles of SRMs. Reference [2] proposed a control strategy that integrates VFRM excitation windings and armature windings, saving stator windings and forming DC-biased vernier reluctance machines (DCB-VRMs). The DC-biased machine injects DC current into the armature windings, so the phase current is divided into the positive biased part and negative biased part. Based on the flux modulation principle, the torque generation mechanism of DCB-VRMs was analyzed [3]. It is found that although the torque density of DCB-VRMs is not as high as that of PM machines, the good overload capability can ensure that the output torque is close to PM machines by increasing the current. The slots/poles combinations and winding structure of DCB-VRMs were further studied [4], in which the electromagnetic properties of several slot/pole combinations, such as inductance, back electromotive force (back EMF), torque, loss, etc., are compared by using finite element analysis. According to the principle of magnetic field modulation [5], it is also possible to improve the torque density of DCB-VRMs by installing a few PMs in appropriate positions [6–8]. Short-distance concentrated windings are commonly used in DC-VRMs since DC-VRMs are developed from SRMs, but distributed windings have also been applied to DCB-VRMs [9].

The machines with doubly salient structures have the advantage of simple structure, low cost, and high reliability, which is suitable for long-stroke linear applications. Several machines have been applied in the field of long-stroke linear applications successfully, such as induction machines, SRMs, and doubly salient permanent magnet machines (DSP-MMs) [10]. The doubly salient linear machines with PM on the slots were proposed in [11], which can provide a much higher thrust force than linear variable flux reluctance machines at the same copper loss. A hybrid-excited doubly salient permanent magnet linear machine with DC-biased armature current was studied [12], which exhibits better thrust capability under high power dissipation. However, the existence of PMs will limit the overload capability of DCB-VRLMs, and also increase the manufacturing cost, so the linear machines without PMs also have important research significance.

The torque ripple of rotation electric machines with doubly salient structures is large, which aggravates the vibration, noise, and speed ripple of machines. This problem is more serious in linear machines with doubly salient structures due to the addition of the edge effect. The common methods to reduce the thrust ripple include skewing poles, adjusting the primary length, slot shifting, etc. In addition, a structure with unequal windings is proposed to optimize the non-uniform inductance caused by the non-uniform winding distribution in a dual three-phase permanent magnet linear machine [13]. A non-uniform air gap structure was used to reduce the torque ripple for permanent magnet brushless motors, which also made the motor achieve a wide speed range [14]. This change in rotor shape also can be applied to linear machines.

In this paper, the 12-slot/10-pole DCB-VRLM with non-uniform air gap structure is proposed to reduce the thrust ripple. The magnetic field modulation principle and combinations of 12-slot DCB-VRLMs with different pole pairs and secondary poles are analyzed. Four possible combinations are selected and optimized by the multi-objective optimization algorithm. Then, the 12-slot/10-pole combination is chosen to adopt a non-uniform air gap structure for its highest average thrust. The non-uniform air gap structure changes the flux density distribution in the air gap and reduces the cogging force of DCB-VRLMs, so the thrust ripple can be reduced at the same time. The results show that the DCB-VRLM with non-uniform air gap has a lower thrust ripple than the constant air gap structure. At the same time, the analytical method and 3D FEM are used to verify the thrust result. Hence, the DCB-VRLM with non-uniform air gap has lower thrust ripple and higher thrust density, which can be used as a linear actuator in rail transit, vertical elevators, and other long-stroke applications.

2. Magnetic Field Modulation Principle and Configurations of DCB-VRLMs

2.1. Magnetic Field Modulation Principle

The doubly salient structure of DCB-VRLMs is shown in Figure 1, and concentrated winding is used to reduce the copper loss. Based on the equivalent magnetic circuit method, the magnetic motive force (MMF) of the DC component F_{dc} can be expressed as (1) [15]:

$$F_{dc}(x) = \sum_{j=1,\,3,\,5...}^{+\infty} \frac{4N_{dc}I_{dc}}{j\pi} cos\left(jP_{dc}\frac{2\pi}{L_p}x \right) \tag{1}$$

where N_{dc} and I_{dc} represent the number of coil turns and the current value of DC windings, respectively, P_{dc} is the DC component pole pairs, L_p is the length of the primary, and x is the position of the secondary.

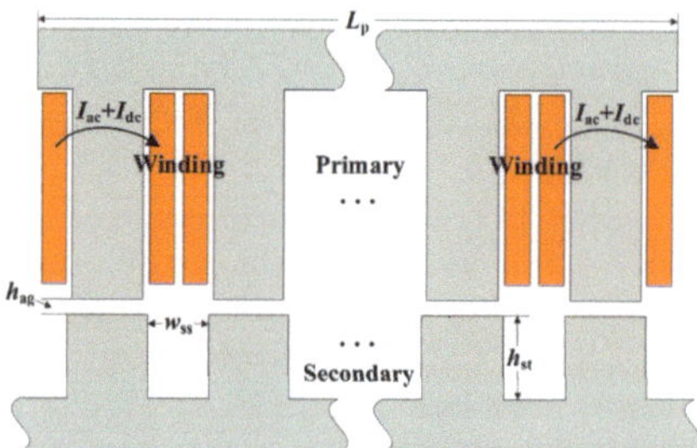

Figure 1. Structure and parameters of DCB-VRLMs.

With the influence of primary slotting ignored, the air gap permeability can be expressed as:

$$\lambda(x,t) = \lambda_0 + \sum_{i=1}^{+\infty} \lambda_i cos\left[iN_{sp}\frac{2\pi}{L_p}(x - vt - x_0)\right] \tag{2}$$

$$\lambda_0 = \frac{N_{sp}}{L_p}\left(\frac{\mu_0}{h_{ag}}w_{st} + \frac{\mu_0}{h_{ag} + h_{st}}w_{ss}\right) \tag{3}$$

$$\lambda_i = \frac{2}{i\pi}\left[\left(\frac{\mu_0}{h_{ag}} - \frac{\mu_0}{h_{ag} + h_{st}}\right)sin\left(iN_{sp}w_{st}\frac{\pi}{L_p}\right)\right] \tag{4}$$

where N_{sp} is the secondary pole number, v the is secondary velocity, x_0 is the initial position of the secondary, μ_0 is vacuum permeability, h_{ag} is the length of the air gap, h_{st} is the height of the secondary tooth, w_{st} is the width of the secondary tooth, and w_{ss} is the width of the secondary slot.

Ignoring the influence of high-order harmonics, the no-load air gap magnetic flux density can be calculated as (5). The second and third components in (5) are generated by the modulation effect of the secondary salient poles on the DC component magnetic field. Due to the large amplitude of the lower harmonic in the traveling wave magnetic field, the pole pairs of the low harmonics are generally taken as the pole pairs of the armature winding to make full use of the harmonic. Therefore, the third component in (5) should be selected as the effective component of the machines to enhance the back EMF.

$$\begin{aligned}
B(x,t) &= F_{dc}(x)\lambda(x,t) \\
&= \frac{4N_{dc}I_{dc}}{\pi}\lambda_0 cos\left(P_{dc}\frac{2\pi}{L_p}x\right) + \frac{2N_{dc}I_{dc}}{\pi}\lambda_1 cos\left[(N_{sp} + P_{dc})\frac{2\pi}{L_p}\left(x - \frac{N_{sp}vt+N_{sp}x_0}{N_{sp}+P_{dc}}\right)\right] \\
&+ \frac{2N_{dc}I_{dc}}{\pi}\lambda_1 cos\left[(N_{sp} - P_{dc})\frac{2\pi}{L_p}\left(x - \frac{N_{sp}vt+N_{sp}x_0}{N_{sp}-P_{dc}}\right)\right]
\end{aligned} \tag{5}$$

Hence, the winding pole pairs can be expressed as:

$$P_a = \left|N_{sp} - P_{dc}\right| \tag{6}$$

$$v_a = \frac{N_{sp}}{\left|N_{sp} - P_{dc}\right|}v = Gv \tag{7}$$

where P_a is the armature winding pole pairs, v_a is the motion velocity of the armature magnetic field, v is the motion velocity of the secondary, G is the transmission ratio.

If the influence of higher harmonics is considered, the winding pole pairs can be expressed as:

$$P_a = \left|iN_{sp} \pm (2j - 1)P_{dc}\right| \tag{8}$$

where i = 0, 1, 2, 3… and j = 1, 2, 3…

According to the formula of magnetic flux density, the phase flux linkage can be calculated, and then the no-load back EMF can be obtained, so the thrust can be expressed as:

$$F = \frac{P_{out}}{v} = \frac{e_a i_a + e_b i_b + e_c i_c}{v} \tag{9}$$

where P_{out} is output power, e_a, e_b, and e_c are no-load back EMF, i_a, i_b, and i_c are armature currents.

2.2. Configurations of Slots and Poles

It is necessary to consider how to reduce the thrust ripple when choosing the combinations of winding pole pairs and secondary poles. According to (6), three variables P_a, N_{sp}, and P_{dc} need to be determined.

The case of 12-slot is taken as an example in this paper; hence, the number of slots N_s is 12. The better choice of P_a in the 12-slot is 4, 5, 7, and 8 since their winding factor is higher than 0.866 [16]. When the number of secondary poles is close to the primary slots, the thrust ripple will be smaller. Therefore, 10, 11, 13, and 14 should be selected for N_{sp} in 12-slot machines. The P_{dc} in 12-slot includes 4 configurations, as shown in Table 1. Therefore, P_{dc} can be equal to 1, 2, 3, and 6, but it has been proved in rotating electrical machines that choosing $P_{dc} = 6$ is helpful to reduce torque ripple [4]. However, the P_a, N_{sp} and P_{dc} can also be chosen for flexibility by different design aims.

Table 1. Configurations of DC Component Pole Pairs.

N_s	$2P_{dc}$	$N_s/2P_{dc}$	Expressed as
	2	6	NNNNNNSSSSSS
12	4	3	NNNSSSNNNSSS
	6	2	NNSSNNSSNNSS
	12	1	NSNSNSNSNSNS

In summary, the possible combinations of $P_a/N_{sp}/P_{dc}$ are 4/10/6, 5/11/6, 7/13/6, and 8/14/6 according to (6).

2.3. Non-Uniform Air Gap Structure

To reduce the thrust ripple, the structure of the air gap is changed from constant to non-uniform, as shown in Figure 2. Here, the unilateral non-uniform air gap is analyzed as a simple example.

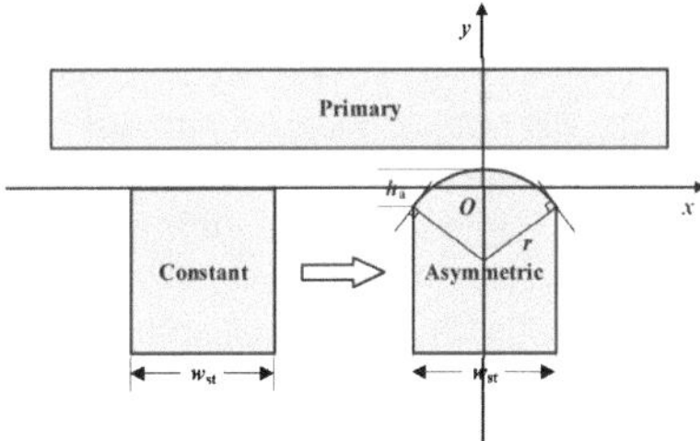

Figure 2. Structure of constant and non-uniform air gap.

The influence of primary slotting is ignored in the following analysis. According to the coordinate system established in Figure 2, the relationship between the length of the air gap and the secondary teeth width is given as:

$$(r - h_a)^2 + \left(\frac{1}{2} w_{st}\right)^2 = r^2 \tag{10}$$

$$x^2 + \left(y + r - \frac{1}{2}h_a\right)^2 = r^2 \tag{11}$$

where r is the radius of an arc in the non-uniform structure, h_a is the height of an arc in the non-uniform structure, x and y are the coordinates of a certain point in the arc.

The air gap length at any point in the secondary teeth is calculated as:

$$y = \sqrt{\left(\frac{1}{2}h_a + \frac{w_{st}^2}{8h_a}\right)^2 - x^2} - \frac{w_{st}^2}{8h_a} \tag{12}$$

Hence, the effective air gap length in the original air gap permeability should be replaced by $h_{ag} - y$, and the transformed formulations can be expressed as:

$$\lambda_0 = \frac{N_{sp}}{L_p}\left(\frac{\mu_0}{h_{ag} - y}w_{st} + \frac{\mu_0}{h_{ag} + h_{st}}w_{ss}\right) \tag{13}$$

$$\lambda_i = \frac{2}{i\pi}\left[\left(\frac{\mu_0}{h_{ag} - y} - \frac{\mu_0}{h_{ag} + h_{st}}\right)sin\left(iN_{sp}w_{st}\frac{\pi}{L_p}\right)\right] \tag{14}$$

The effective air gap length $\delta\,(x, a)$ can be obtained by the position function of primary and secondary. The cogging force of the machine is determined by the magnetic co-energy in the air gap:

$$T_{cog} = -\frac{\partial W}{\partial a} \tag{15}$$

$$W = -\frac{1}{2\mu}\int_V B^2 dV \tag{16}$$

where a is the position of secondary.

3. Optimization of Structural Parameters

3.1. Optimization for Different Slots/Poles Combinations

The machine structure is optimized using the NSGA-II multi-objective optimization algorithm [17], which improves the distribution of search solutions in the decision space based on the NSGA algorithm. In the optimization, the end length is adjusted to reduce the influence of the end effect. The main optimization variables include primary tooth width, secondary tooth width, secondary height, end length, etc. The population size is set as 100, and the evolutionary generation is set as 50. The optimization goal is to increase the average thrust and reduce the thrust ripple.

To reduce the normal magnetic force, the bilateral structure is selected for study in this paper. The performance of the models was analyzed by the finite element method (FEM) using Ansys Maxwell software. The structures of the four machines are optimized under the same constraints, and the non-dominated solution of the final generation is shown in Figure 3.

The selection principle of the optimal point is as follows: arrange the non-dominated solution of final generation (N points) in ascending order, and then subtract the two adjacent solutions to get ($\Delta Thrust$, $\Delta Ripple$) for N-1 points, and select the design point with the largest $\Delta Thrust/\Delta Ripple$ value as the optimal point. Then, the optimal point is determined according to the optimization results. The winding connection of four combinations is shown in Figure 4. The key parameters of DCB-VRMs are listed in Table 2, where the primary length in the table does not include the end length.

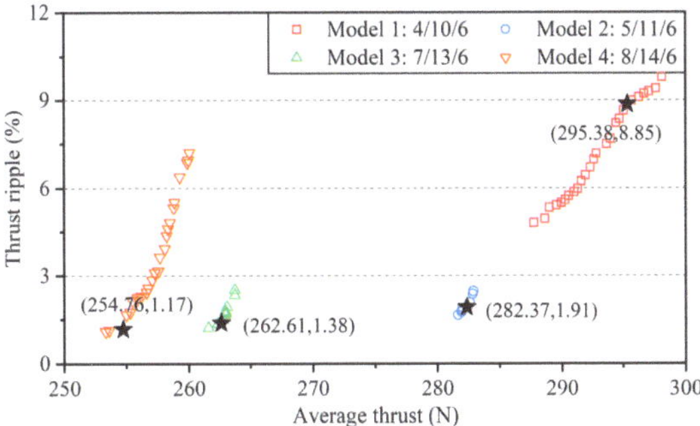

Figure 3. Multi-objective optimization of DCB-VRLMs using NSGA-II algorithm.

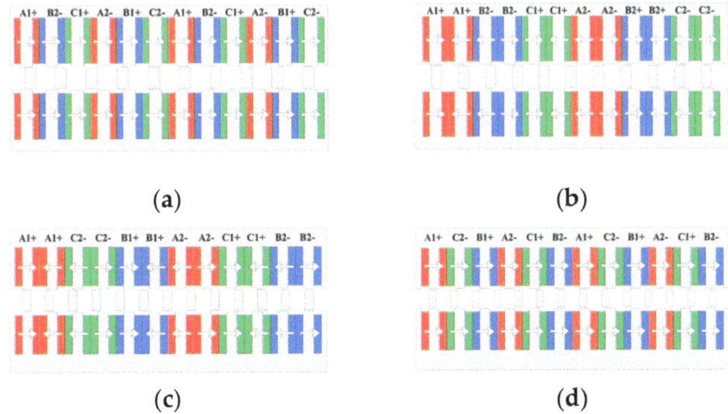

Figure 4. The DCB-VRLMs with different combinations of $P_a/N_{sp}/P_{dc}$. (**a**) Model 1: 4/10/6. (**b**) Model 2: 5/11/6. (**c**) Model 3: 7/13/6. (**d**) Model 4: 8/14/6.

Table 2. Key Parameters of DCB-VRLMs.

Parameters	Model 1	Model 2	Model 3	Model 4
Primary unilateral slot number			12	
Secondary pole number	10	11	13	14
Primary length (mm)			288	
Primary unilateral height (mm)			57	
End length (mm)	3.5	8.0	4.9	5.3
Stack length (mm)			150	
Air gap (mm)			1	
Primary tooth width (mm)	12.0	11.9	10.1	10.0
Primary slot width (mm)	12.0	12.1	13.9	14.0
Secondary tooth width (mm)	11.0	10.2	8.6	7.1
Secondary height (mm)	19.9	19.9	18.9	16.4
Lamination materials			50DW465	
AC current (A)			5	
DC current (A)			5	
Wire diameter (mm)			1.78	
Slot filling factor			0.6	
Number of coil turns			60	

3.2. Optimization for Non-Uniform Air Gap Structure

The air gap structure of 12-slot/10-pole DCB-VRLM can be changed to non-uniform by making both the primary and secondary into an arc. The structure parameters of the machine are optimized by NSGA-II multi-objective optimization algorithm under the same conditions as Section 3.1, and the result is shown in Figure 5. The optimal point can be determined from the optimization result and marked in Figure 5. The configuration of DCB-VRLM with non-uniform air gap structure and its 3D model is shown in Figure 6, the key parameters are listed in Table 3.

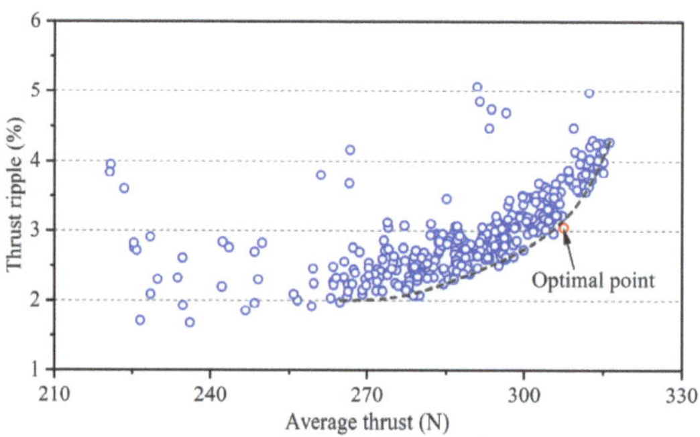

Figure 5. Multi-objective optimization of DCB-VRLM with non-uniform air gap using NSGA-II algorithm.

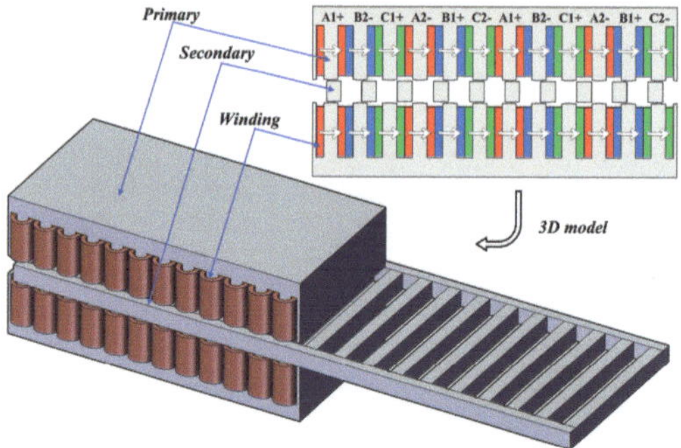

Figure 6. The DCB-VRLM with non-uniform air gap structure.

Table 3. Key Parameters of DCB-VRLM with Non-uniform Air Gap Structure.

Parameters	Constant Air Gap	Non-Uniform Air Gap
Primary unilateral slot number	12	
Secondary pole number	10	
Primary length (mm)	288	
Primary unilateral height (mm)	57	
End length (mm)	3.5	3.1
Stack length (mm)	150	
Air gap (mm)	1	0.8~1.3
Primary tooth width (mm)	12.0	11.4
Primary slot width (mm)	12.0	12.6
Secondary tooth width (mm)	11.0	11.9
Secondary height (mm)	19.9	15.9
Lamination materials	50DW465	
AC current (A)	5	
DC current (A)	5	
Wire diameter (mm)	1.78	
Slot filling factor	0.6	
Number of coil turns	60	

4. Performance Analysis of Different Modulation Combinations

The four models of DCB-VRMs are shown in Figure 4 and Table 2 in Section 3.1. They have the same primary length and winding current density. In addition, the motion velocity is set as 2 m/s to compare under the same application specifications. For DCB-VRMs, the velocity relationship can be expressed as:

$$w_e = N_r w_m = N_r \frac{2\pi n}{60} = 2\pi f \tag{17}$$

$$n = \frac{60f}{N_r} \tag{18}$$

where w_e is the electric angular velocity, w_m is the mechanical angular velocity, and N_r is the number of rotor poles.

Hence, the motion velocity v of DCB-VRLMs can be expressed as:

$$v = N_r \frac{n}{60} \tau = f\tau \tag{19}$$

where f is the current frequency, and τ is the distance between two secondary poles.

The inverter circuit of DCB-VRLMs is shown in Figure 7, which can flexibly control the machine by adjusting the DC/AC ratio. The inverter can be a dual-three-phase or six-phase inverter. Phase currents of DCB-VRLMs can be expressed as:

$$\begin{cases} i_{A\pm} = \sqrt{2}I_{ac}sin(2\pi f + \theta_0) \pm I_{dc} \\ i_{B\pm} = \sqrt{2}I_{ac}sin\left(2\pi f - \frac{2}{3}\pi + \theta_0\right) \pm I_{dc} \\ i_{C\pm} = \sqrt{2}I_{ac}sin\left(2\pi f + \frac{2}{3}\pi + \theta_0\right) \pm I_{dc} \end{cases} \tag{20}$$

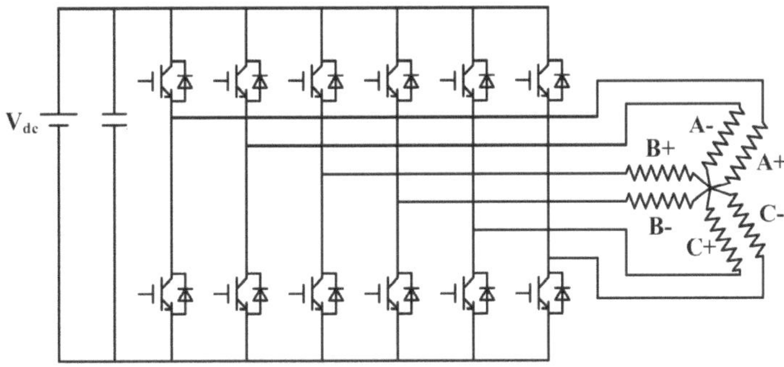

Figure 7. The inverter circuit of DCB-VRLMs.

4.1. No-Load Back-EMF

When only DC current component is applied, the no-load performance of the four machines can be obtained. The no-load back-EMF and harmonic order of phase A are shown in Figure 8.

As shown in Figure 8a, Model 1 and Model 4 are more sinusoidal than Model 2 and Model 3. Meantime, the total harmonic distortions (THD) of the four models can be calculated as 2.33%, 12.02%, 10.86%, and 3.17%. In Figure 8b, the fundamental amplitude decreases with the increase in the number of secondary poles. The third harmonic amplitude of Model 2 and Model 3 is higher than that of Model 1 and Model 4.

Similarly, when only DC current component is applied, the no-load performance of the DCB-VRLMs with constant and non-uniform air gap structures can be obtained. The back-EMF and harmonic order of phase A are shown in Figure 9.

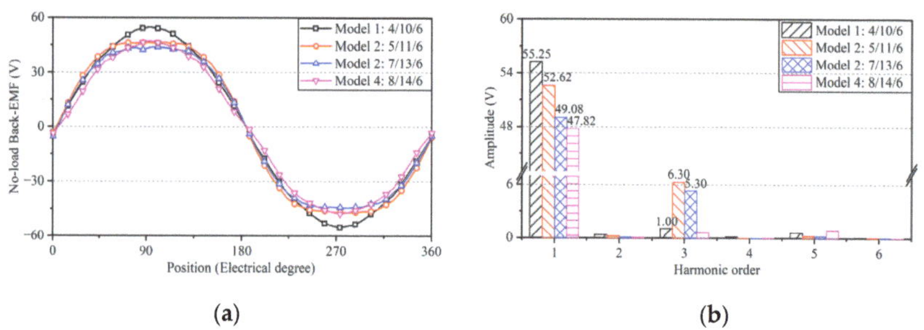

Figure 8. The no-load back-EMF (**a**) and harmonic order (**b**) of DCB-VRLMs with different combinations.

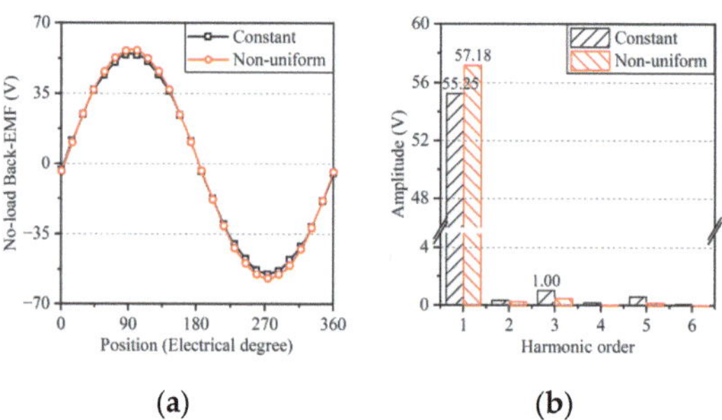

Figure 9. The no-load back-EMF (**a**) and harmonic order (**b**) of DCB-VRLM with the constant and non-uniform air gap.

As can be seen in Figure 9a, the DCB-VRLM with non-uniform air gap is more sinusoidal than the constant ones. It can be calculated that the THD of DCB-VRLM with constant and non-uniform air gap is 2.33% and 1.08%. In Figure 9b, the non-uniform air gap structure increases the fundamental amplitude and decreases the third harmonic amplitude compare to the constant ones.

Figure 10 shows the detent force, the normal magnetic flux of the air gap, and its harmonic order of DCB-VRLM with the constant and non-uniform air gap. As can be seen in Figure 10a, the maximum detent force of the non-uniform air gap structure motor is reduced by 61.4% compared to the constant air gap from 10.1 N to 3.9 N. Figure 10b shows the waveform of normal air gap magnetic density (ignoring the end length), it can be seen that the non-uniform air gap structure improves the air gap magnetic density distribution. Figure 10c shows that the 4th and 7th harmonics are the main working harmonics of the machine, and the 6th, 18th, and 30th harmonics are stationary with no thrust contribution.

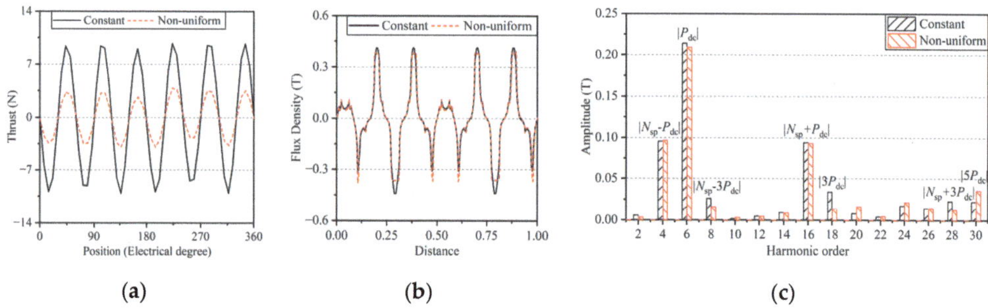

Figure 10. The detent force (**a**), the normal magnetic flux of air gap (**b**), and its harmonic order (**c**) of DCB-VRLM with the constant and non-uniform air gap.

4.2. Rated On-Load Characteristics

Figure 11 shows the average thrust variation with the phase current angle when the AC current and DC current are both 5 A. It is shown that the maximum thrust occurs when the phase current angle is $-3°\sim1°$. Therefore, the current angle is set to maximize the average thrust. Then, the magnetic flux distributions when $t = 0$ of four models are shown in Figure 12, and the thrust comparison of four possible combinations are shown in Figure 13. It can be seen that the average thrust decreases with the increase in the number of secondary poles, and the thrust ripple of Model 2~Model 4 is lower than Model 1. Model 2 with 11-pole has low thrust ripple and high average thrust, which should be a better application choice.

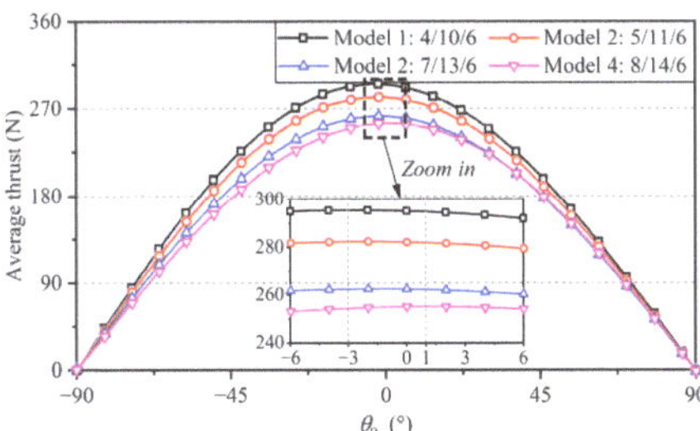

Figure 11. Average thrust variation with phase current angle.

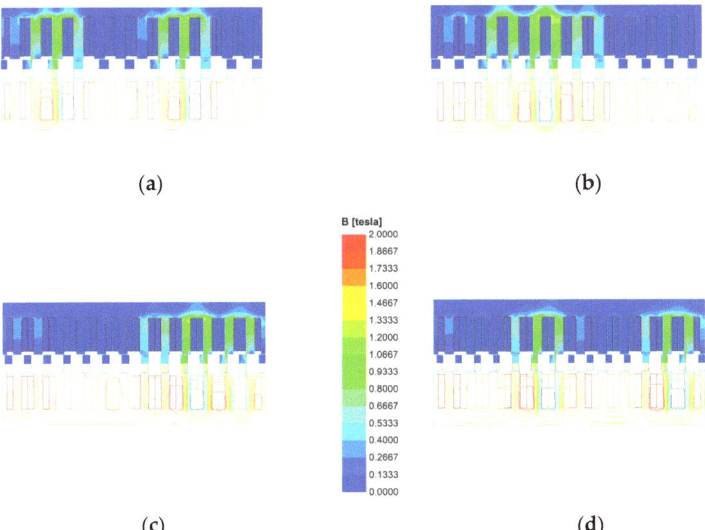

Figure 12. The magnetic flux density distribution of DCB-VRLMs with different combinations of $P_a/N_{sp}/P_{dc}$. (**a**) Model 1: 4/10/6. (**b**) Model 2: 5/11/6. (**c**) Model 3: 7/13/6. (**d**) Model 4: 8/14/6.

The average thrust, thrust ripple, and other performances of the four models are shown in Table 4. The copper loss for DCB-VRLMs can be calculated as (the end winding is considered):

$$P_{\text{Copper}} = 3R_{\text{Phase}}\left(I_{\text{ac}}^2 + I_{\text{dc}}^2\right) \tag{21}$$

where R_{phase} is phase resistance.

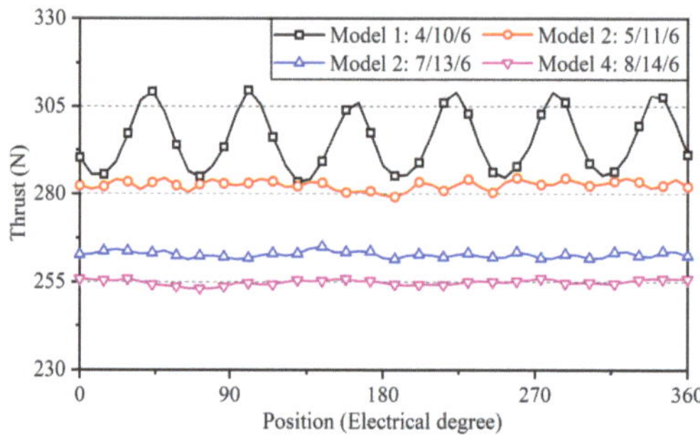

Figure 13. The thrust comparison of four possible combinations.

Table 4. Performance of Four Models.

Parameters	Model 1	Model 2	Model 3	Model 4
Average thrust (N)	295.38	282.37	262.61	254.76
Thrust ripple (%)	8.85	1.91	1.38	1.17
Max normal force (N)	1.74	0.45	0.38	0.71
Copper Loss (W)			200	
Core Loss (W)	17.97	19.69	18.36	18.24
Efficiency (%)	72.00	70.97	69.65	69.05
Weight (kg)	38.07	39.30	38.41	38.25
Thrust density (N/kg)	7.76	7.18	6.84	6.66

The winding temperature of the machine is set as 85 °C at 200 W copper loss and 150 °C at 400 W copper loss approximately [12]. R_{phase} increases with the increase in temperature. $R_{\text{phase}} = R_{20} [1 + \alpha (t - 20)]$, where $\alpha = 0.003862$ is the temperature coefficient. Then, the iron loss can be calculated by the Bertotti iron loss model as:

$$P_{\text{Fe}} = P_{\text{h}} + P_{\text{c}} + P_{\text{e}} = K_{\text{h}} f B_{\text{m}}^{\alpha} + K_{\text{c}} f^2 B_{\text{m}}^2 + K_{\text{e}} f^{1.5} B_{\text{m}}^{1.5} \tag{22}$$

where P_{h} is hysteresis core loss, P_{c} is classical eddy current core loss, P_{e} is hysteresis excess core loss, B_{m} is the amplitude of magnetic flux density, K_{h} is the hysteresis core loss coefficient, K_{c} is the classical eddy-current core loss coefficient, and K_{e} is the excess core loss coefficient.

The excessive loss of the machines is roughly set to 2% of the output power. The performance of the four models in Table 4 is compared at the same motion velocity (2 m/s) and copper loss (200 W). The secondary weight is calculated as twice the number of secondary poles.

It can be found that the average thrust and efficiency decrease with the increase in the number of secondary poles. Among them, Model 1 exhibits the highest average thrust and highest thrust ripple, Model 2~Model 4 have low thrust ripple than Model 1, while the max normal magnetic force of the four models is less than 2 N. Model 1 and Model 2 exhibit a higher thrust density than Model 3 and Model 4.

The efficiency of DCB-VRLMs is low since the low-speed application condition in this paper, but it can increase when motion velocity is high. Among the four models, Model 2 with 11 poles is the better choice from a comprehensive point of view, since it has a higher average thrust, lower thrust ripple, and lower max normal magnetic force.

For the non-uniform air gap structure, when both the DC and AC current are applied, the thrust comparison of DCB-VRLM with the constant and non-uniform air gap is shown in Figure 14. The AC current and DC current are both set as 5 A. It can be seen that the thrust ripple of DCB-VRLM with the non-uniform air gap is lower than the constant ones, and the average thrust is higher than the constant one.

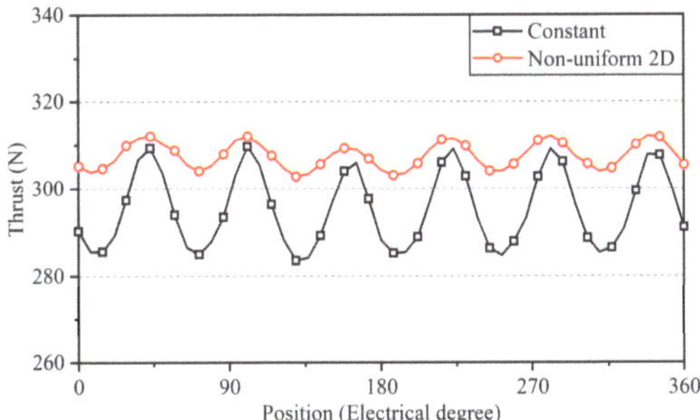

Figure 14. The thrust comparison of DCB-VRLM with constant and non-uniform air gap.

The average thrust, thrust ripple, and other performances comparison of DCB-VRLM with constant and non-uniform air gap are shown in Table 5, which are compared at the same motion velocity (2 m/s) and copper loss (200 W). It can be found that the average thrust of DCB-VRLM with non-uniform air gap is 4.1% higher than the constant ones, while the thrust ripple of DCB-VRLM with non-uniform air gap is 65.6% lower than the constant ones, and the reduction percentage is consistent with the detent force.

Table 5. Key Parameters of DCB-VRLM with Non-uniform Air Gap Structure.

Parameters	Constant Air Gap	Non-Uniform Air Gap
Average thrust (N)	295.38	307.41
Thrust ripple (%)	8.85	3.04
Max normal force (N)	1.74	2.82
Copper Loss (W)	200	
Core Loss (W)	17.97	16.54
Efficiency (%)	72.00	72.88
Weight (kg)	38.07	38.60
Thrust density (N/kg)	7.76	7.96

4.3. Overload Capability Verification

The average thrust performance of DCB-LVRMs under different currents is shown in Figure 15. The AC current is set as equal to the DC current, and they are both set from 0 A to 12.5 A. The results show that the average thrust of Model 1~4 at I = 10 A can reach 898.72 N, 851.83 N, 800.03 N, and 799.73 N, respectively, with good overload capability. Model 4 exhibits higher thrust than Model 3 when the current is greater than 10.4 A.

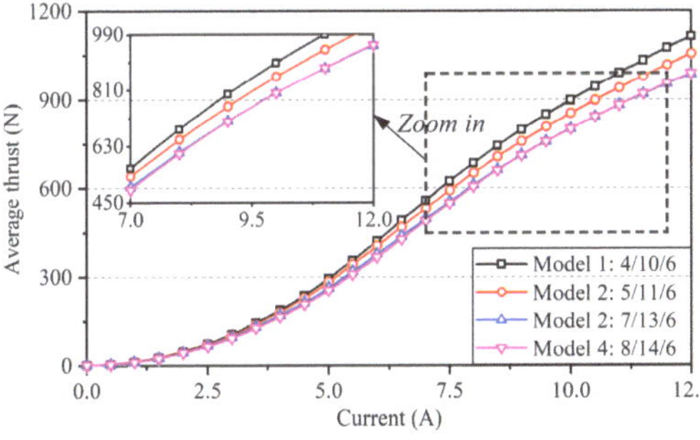

Figure 15. The average thrust performance under different currents.

The overload performance of DCB-LVRMs with constant and non-uniform air gap under different currents is shown in Figure 16. The AC current is set as equal to DC current, and they are both set from 0 A to 12.5 A. As can be seen in Figure 16a, the result shows that DCB-LVRM with the non-uniform air gap has good overload capability too, and the DCB-VRLM with non-uniform air gap exhibits about 3.1% higher thrust than the constant ones. As can be seen in Figure 16b, the thrust ripple of the DCB-VRLM with non-uniform air gap exhibit better performance than the constant ones when the current increases.

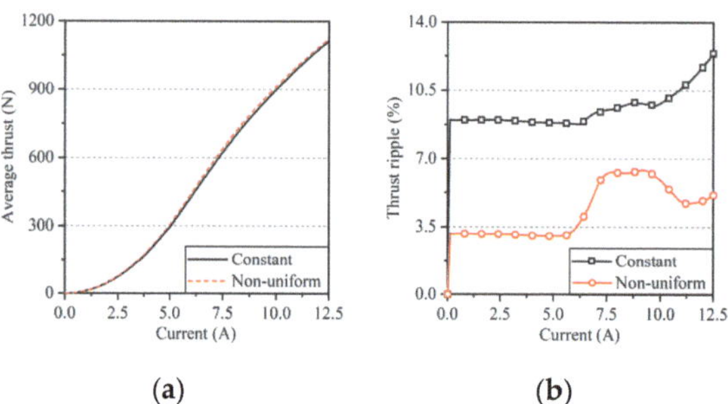

(a)   **(b)**

Figure 16. The overload performance (**a**) average thrust (**b**) thrust ripple of DCB-VRLMs with constant and non-uniform air gap under different currents.

4.4. Energy Distribution of AC/DC Current

Figure 17 shows the average thrust and estimated efficiency under different AC currents and DC currents of DCB-VRLMs with the non-uniform air gap. Here, the change of resistance with temperature has been considered by numerical fitting when the current changes. The results show that the highest efficiency of DCB-VRLMs can reach more than 80%. The DC/AC ratio with the highest efficiency is 1 when the AC current is 0~4 A and decreases gradually when the AC current increase. When the AC current is 12.5 A, the DC/AC ratio with the highest efficiency is reduced to 0.52.

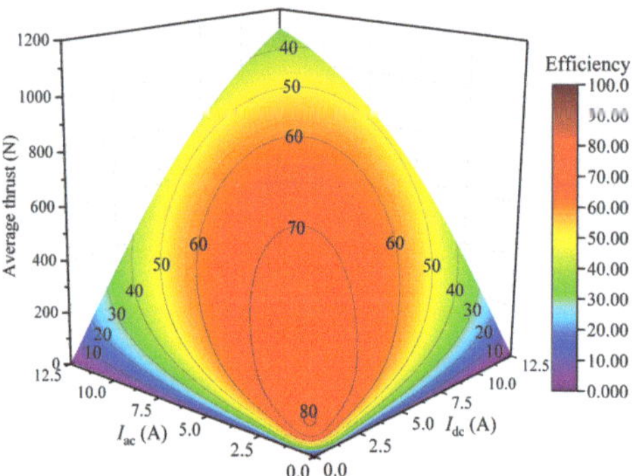

Figure 17. The average thrust and efficiency under different AC currents and DC currents of DCB-VRLM with non-uniform air gap.

4.5. 3D FEM Verification and Comparison

In order to verify the effectiveness of the 2D FEM calculation results, the DCB-VRLM with the non-uniform air gap was calculated using the analytical calculation method and 3D FEM, respectively, as shown in Table 6. The 3D magnetic flux density distribution map at $t = 0$ is shown in Figure 18, and the thrust waveform of different calculation methods is shown in Figure 19.

Table 6. Comparison of different calculation methods.

Parameters	Analytical	2D FEM	3D FEM
Average thrust (N)	303.92	307.41	309.96
Thrust ripple (%)	1.80	3.04	3.94
Element numbers	-	36,303	1,555,803
Calculation time (h)	-	0.0694	58.79

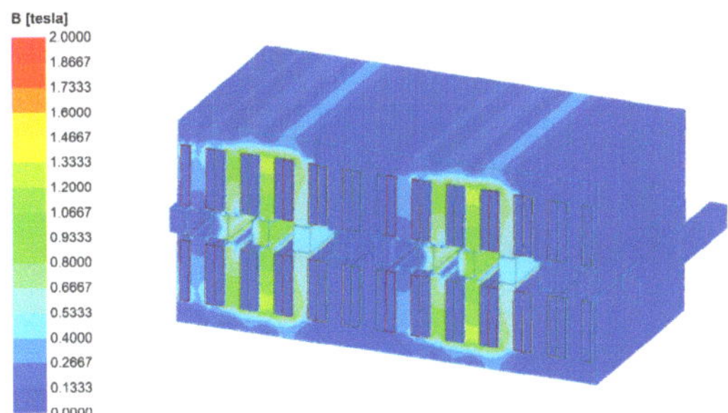

Figure 18. 3D magnetic flux density distribution map of DCB-VRLM with non-uniform air gap.

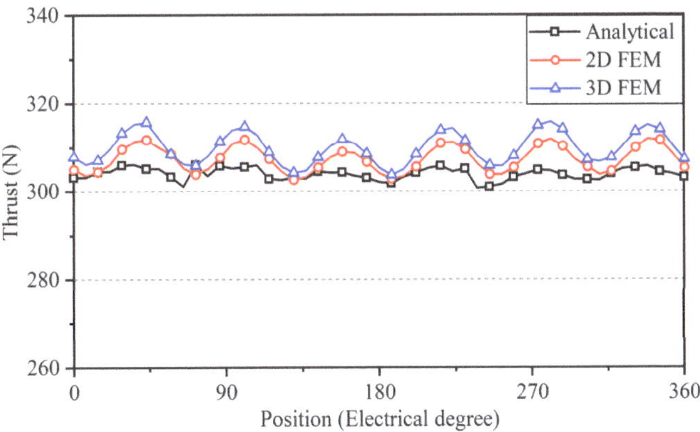

Figure 19. 3D magnetic flux density distribution map of DCB-VRLM with non-uniform air gap.

It can be seen from Figure 19 and Table 6 that the calculation error between 2D FEM and the other two methods is less than 1.2%, but the calculation time of 3D FEM is about 861.5 times that of 2D FEM. Therefore, the analysis results calculated by 2D FEM are effective, and can save a lot of computing resources.

5. Conclusions

This paper proposed a 12-slot/10-pole bilateral non-permanent magnet DCB-VRLM with non-uniform air gap structure. The main contribution of the proposed design is concluded as follows.

The thrust ripple is effectively reduced with the non-uniform air gap structure under 12-slot/10-pole combination. Results indicate the thrust ripple of DCB-VRLM with the non-uniform air gap is reduced by 65.6% lower than the constant ones. At the same time, the average thrust of DCB-VRLM with the non-uniform air gap is improved compared to the constant ones of 4.1%.

A better overload capability is achieved with the proposed design. Under a heavy load condition, the average thrust of the proposed DCB-LVRM can be improved to 914 N with excitation currents of 10 A while that under the rated condition is 307.4 N with excitations of 5 A.

Based on the discussion, the DCB-VRLM with non-uniform air gap structure has better overload capability, low thrust ripple, and high thrust density, which confirm its superiority in long-stroke linear applications.

Author Contributions: Conceptualization, W.F. and J.L.; methodology, Z.Q. and Y.Z.; software, Z.Q. and H.C.; validation, Y.Z., D.S. and J.L.; formal analysis, D.S. and H.C.; writing—original draft preparation, Z.Q.; writing—review and editing, J.L., W.F. and Y.Z. All authors have read and agreed to the published version of the manuscript.

Funding: This work was funded by the Shanghai Sailing Program, grant number 20YF1412600.

Institutional Review Board Statement: Not applicable.

Informed Consent Statement: Not applicable.

Data Availability Statement: Not applicable.

Conflicts of Interest: The authors declare no conflict of interest.

References

1. Liu, X.; Zhu, Z.Q. Electromagnetic performance of novel variable flux reluctance machines with DC-field coil in stator. *IEEE Trans. Magn.* **2013**, *49*, 3020–3028. [CrossRef]
2. Zhu, Z.Q.; Lee, B.; Liu, X. Integrated field and armature current control strategy for variable flux reluctance machine using open winding. *IEEE Trans. Ind. Appl.* **2016**, *52*, 1519–1529.
3. Jia, S.; Qu, R.; Kong, W.; Li, D.; Li, J. Flux modulation principles of DC-biased sinusoidal current vernier reluctance machines. *IEEE Trans. Ind. Appl.* **2018**, *54*, 3187–3196. [CrossRef]
4. Jia, S.; Qu, R.; Kong, W.; Li, D.; Li, J.; Zhang, R. Stator/rotor slot and winding pole pair combinations of DC-biased current vernier reluctance machines. *IEEE Trans. Ind. Appl.* **2018**, *54*, 5967–5977. [CrossRef]
5. Chen, Y.; Fu, W.; Weng, X. A concept of general flux-modulated electric machines based on a unified theory and its application to developing a novel doubly-fed dual-stator motor. *IEEE Trans. Ind. Electron.* **2017**, *64*, 9914–9923. [CrossRef]
6. Jia, S.; Qu, R.; Kong, W.; Li, D.; Li, J.; Yu, Z.; Fang, H. Hybrid excitation stator PM vernier machines with novel DC-biased sinusoidal armature current. *IEEE Trans. Ind. Appl.* **2018**, *54*, 1339–1348. [CrossRef]
7. Jia, S.; Qu, R.; Li, J.; Li, D.; Kong, W. A stator-PM consequent-pole vernier machine with hybrid excitation and DC-biased sinusoidal current. *IEEE Trans. Magn.* **2017**, *53*, 8105404. [CrossRef]
8. Sheng, T.; Niu, S.; Fu, W.; Lin, Q. Topology exploration and torque component analysis of double stator biased flux machines based on magnetic field modulation mechanism. *IEEE Trans. Energy Convers.* **2018**, *33*, 584–593. [CrossRef]
9. Jia, S.; Sun, P.; Liang, D.; Dong, X.; Zhu, Z.; Liu, J. Analysis of DC-biased vernier reluctance machines having distributed windings. *IEEE Trans. Magn.* **2021**, *57*, 8106605. [CrossRef]
10. Eguren, I.; Almandoz, G.; Egea, A.; Ugalde, G.; Escalada, A.J. Linear machines for long stroke applications—A review. *IEEE Access* **2020**, *8*, 3960–3979. [CrossRef]
11. Shen, Y.; Zeng, Z.; Lu, Q.; Wu, L. Design optimization and performance investigation of linear doubly salient slot permanent magnet machines. *IEEE Trans. Ind. Appl.* **2019**, *55*, 1524–1535. [CrossRef]
12. Shen, Y.; Lu, Q.; Shi, T.; Xia, C. Analysis and evaluation of hybrid-excited doubly salient permanent magnet linear machine with DC-biased armature current. *IEEE Trans. Ind. Appl.* **2021**, *57*, 3666–3677. [CrossRef]
13. Jiang, Q.; Lu, Q.; Li, Y. Thrust ripple and depression method of dual three-phase permanent magnet linear synchronous motors. *Trans. China Electrotech. Soc.* **2021**, *36*, 883–892.
14. Chau, K.T.; Cui, W.; Jiang, J.Z.; Wang, Z. Design of permanent magnet brushless motors with asymmetric air gap for electric vehicles. *J. Appl. Phys.* **2006**, *99*, 08R322. [CrossRef]
15. Du, Y.; Cheng, M.; Chau, K.T. Design and analysis of a new linear primary permanent magnet vernier machine. *Trans. China Electrotech. Soc.* **2012**, *27*, 22–30.
16. Carraro, E.; Bianchi, N.; Zhang, S.; Koch, M. Design and performance comparison of fractional slot concentrated winding spoke type synchronous motors with different slot-pole combinations. *IEEE Trans. Ind. Appl.* **2018**, *54*, 2276–2284. [CrossRef]
17. Deb, K.; Pratap, A.; Agarwal, S.; Meyarivan, T.A.M.T. A fast and elitist multiobjective genetic algorithm: NSGA-II. *IEEE Trans. Evol. Comput.* **2002**, *6*, 182–197. [CrossRef]

Article

Characteristic Analysis of a New Structure Eccentric Harmonic Magnetic Gear

Libing Jing [1], Youzhong Wang [1], Dawei Li [2,*] and Ronghai Qu [2]

[1] College of Electrical Engineering and New Energy, Hubei Provincial Engineering Technology Research Center for Microgrid, China Three Gorges University, Yichang 443002, China; jinglibing163@163.com (L.J.); wangyouzhong@ctgu.edu.cn (Y.W.)

[2] Department of Electrical and Electronic Engineering, Huazhong University of Science and Technology, Wuhan 430074, China; ronghaiqu@hust.edu.cn

* Correspondence: daweili@hust.edu.cn

Abstract: An eccentric harmonic magnetic gear (EHMG) is better suited for situations requiring larger transmission ratios than magnetic-field-modulated magnetic gears. In the meantime, to increase the torque density even further, a new structure for EHMGs is presented in this paper. The stator's permanent magnets (PMs) are irregularly distributed, while the rotor's PMs are applied to a fan-shaped structure. Moreover, a Halbach array is adopted in both the rotor and the stator. A two-dimensional finite element (FE) model of the proposed EHMG is developed, and the flux density distribution and torque of the EHMG are calculated and verified via FE analysis. When compared to a conventional EHMG, the presented model's torque increases from 38.04 Nm to 50.41 Nm. In addition, for the sake of avoiding the oscillation and noise caused by resonance, a modal analysis of the proposed model is conducted and the consequences show that it has better antivibration properties. Finally, a prototype is made, a test bench is established, and the correctness and effectiveness of the proposed model are verified.

Keywords: EHMG; electromagnetic torque; magnetic field; modal analysis

Citation: Jing, L.; Wang, Y.; Li, D.; Qu, R. Characteristic Analysis of a New Structure Eccentric Harmonic Magnetic Gear. *Actuators* **2023**, *12*, 248. https://doi.org/10.3390/act12060248

Academic Editor: Leonardo Acho Zuppa

Received: 8 May 2023
Revised: 8 June 2023
Accepted: 13 June 2023
Published: 14 June 2023

1. Introduction

Magnetic gears (MGs), as a technology that replaces mechanical gears, have distinct benefits such as low vibration, no friction, automatic overload protection, and no need for maintenance or lubrication [1–3]. They offer a very wide range of development prospects in the domains of motors, hydropower, wind turbines, aerospace, and the marine industry [4–6].

In 1901, C. G. Armstrong submitted a patent application for a power transmission device for the first time, creating a precedent for MGs. However, the constraints of permanent magnet (PM) materials and overly simple structures resulted in low transmission efficiency and performance in early MGs; as a consequence, they could not be produced and used [7]. As permanent magnetic materials continued to advance, particularly with the discovery of rare-earth PMs, scholars began to focus again on MG research in the 1990s. Until 2001, magnetic-field-modulated coaxial magnetic gears (CMGs) were put forward, which greatly solved the problems of PMs' low use rate and low torque density [8]. Since then, CMGs have gained popularity as a research topic [9]. To date, CMGs are still put into use in numerous fields such as wind turbine applications due to their high torque density and good performance [10]. Researchers have attempted to improve the torque density and reduce the torque ripple of MGs by changing the magnetization method of PMs, and the results show that Halbach arrays have better performance compared to radial magnetization [11–13]. At the same time, scholars are also attempting to change the topology of MGs to improve their performance. Especially in [14], a new MG structure with an eccentric pole and a Halbach array was proposed; moreover, the high-temperature superconducting blocks were added between the magnetic adjusting rings, and the consequences show that the torque density of the magnetic gear was as high as 173 kN·m/m^3. However, the drawback of a low transmission ratio became apparent. In particular, its torque density will drastically drop when its transmission ratio is too large [15].

To address this issue, eccentric harmonic magnetic gears (EHMGs) were introduced, which are appropriate for large transmission ratios and still have a torque density greater than 150 kN·m/m^3 [16]. Likewise, researchers have introduced Halbach arrays into EHMGs, which considerably increases their air gap magnetic field and electromagnetic torque [17]. However, compared to mechanical gears, their torque density is still relatively low, so improving torque density is still a problem that researchers need to solve.

Considering the above research, although EHMGs have good advantages in terms of transmission ratio, there are still few advancements in improving the topology structure to increase torque density and reduce torque ripple. In this paper, a new topology for an EHMG is proposed, the PM structure is changed in the topology, and the PM is redivided and magnetized.

Although noncontact gear structures that transmit energy and speed through magnetic fields generated by PMs can avoid vibration caused by friction, vibration can still be generated by electromagnetic forces acting on the magnetic gear components [18]. In order to avoid vibration or even resonance causing significant damage to the EHMG, it is necessary to conduct modal analysis to obtain modal shapes and natural frequencies.

To further enhance the torque density and lower the torque ripple, a new structure of EHMG is proposed in this paper [19]. Inspired by [20], a fan-shaped structure is applied to the low-speed rotor, while the PMs on the stator are irregularly distributed in a 1:n:1 ratio. In addition, by rearranging and combining PMs, a Halbach array is used. Using the finite element (FE) method, the presented EHMG model with 16 pole pairs on the stator and 15 pole pairs on the rotor is established in Section 2. In Section 3, the flux density and torque are analyzed and contrasted with the conventional radially magnetized EHMG; moreover, to verify that the proposed model has good antivibration performance, a modal analysis of the EHMG is conducted, and the natural frequency and modal shapes of the EHMG are obtained. A prototype is established to verify the proposed model in Section 4. Finally, a conclusion is drawn in Section 5.

2. Topology and Basic Principle

Based on mechanical harmonic gears, an EHMG was proposed. In Figure 1, the topology of EHMG is shown. An EHMG primarily consists of five parts: stator, bearings, low-speed rotor, high-speed rotor, and air gap. The low-speed rotor corresponds to the flexible wheel in a mechanical harmonic gear; likewise, the stator is equivalent to the rigid wheel, and the high-speed rotor corresponds to the harmonic generator.

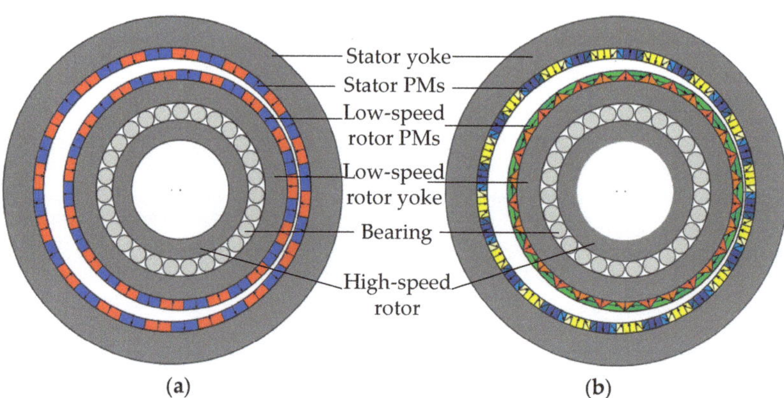

Figure 1. Structure of EHMG. (**a**) Conventional model; (**b**) proposed model.

Since the stator and rotor are not centered on the same axis, the air gap between the rotor and stator is nonuniform. When the EHMG operates normally, the low-speed rotor will be driven to revolve by the high-speed rotor through bearings, causing the air gap to change periodically. When the low-speed rotor is propelled by the high-speed rotor to revolve around the stator axis, similar to orbital motion, the low-speed rotor will revolve around its own axis in the opposing direction at the same time. The time-varying, nonuniform air gap can generate space harmonics of the air gap permeance function, and the air gap permeance function rotates with orbit motion, thereby adjusting the spatial magnetic flux harmonics to promote gear transmission behavior, achieving torque and speed transmission. During this process, the nonuniform air gap's length can be represented [21] as

$$g = \frac{g_{max} + g_{min}}{2} + \left(\frac{g_{max} - g_{min}}{2}\right)\cos(p_a(\theta - \omega_h t)) \tag{1}$$

where g_{min} is the minimum length, g_{max} is the maximum length, ω_h is the high-speed rotor's angular speed, and p_a is the number of sinusoidal periodic changes in the air gap, which depends on the rotor's shape. As the rotor has a circular structure in this paper, the value of p_a is equal to 1. θ is the angle at which a point within the air gap deviates from the x-axis.

On the basis of [16], the pole pair in spatial harmonic flux density distribution produced by the low-speed rotor is

$$p_{m,k} = mp + (-1)^k p_a \tag{2}$$

where k is introduced to distinguish various asynchronous spatial harmonics associated with each harmonic of the magnetic field generated by the PM. m is related to the magnitude of asynchronous spatial harmonics. $k = 1,2$ and $m = 1,3,5 \ldots \infty$.

The maximal air gap magnetic flux density amplitude may be attained at $m = 1$ and $k = 2$, which results in the greatest electromagnetic torque. Hence, the pole pair of the low-speed rotor can be decided as $p_{1,2}$; as a result, the relationship between p_s, p, and p_a can be attained [16] as

$$p_s = p + p_a \tag{3}$$

where p_s denotes the number of pole pairs of the stator, and p represents the number of pole pairs of the low-speed rotor.

The angular speed of the spatial harmonic [17] is

$$\omega_{m,k} = p_a\omega_h + (-1)^k mp\omega_r \tag{4}$$

where ω_r the angular speed of the low-speed rotor.

As can be observed, the angular speed of the space harmonic differs from the rotors' angular velocities. Due to the highest asynchronous spatial harmonic occurring at m = 1, the number of poles on the stator must be equal to $p + p_\omega$ in order to achieve maximum torque transmission between the stator and the low-speed rotor. When the pole pairs of the rotor and stator meet the above formula, it can be inferred that EHMG's transmission ratio [17] is

$$G_r = (-1)^{k+1}\frac{p_a}{p} = -\frac{1}{p} \tag{5}$$

The pole pair of the low-speed rotor attaches great importance to the EHMG transmission ratio. The large pole pair of the low-speed rotor provides a high transmission ratio.

Assuming that the stator is held immobile, the low-speed rotor will revolve $\alpha/15$ in a clockwise direction about its axis when continuously pushed by the high-speed rotor to revolve α counterclockwise about the stator axis. The placement of the rotors relative to each other is depicted in Figure 2, where Figure 2a represents the original position with two black-colored PMs on the low-speed rotor. In Figure 2b, the low-speed rotor revolves 9° around its axis when driven to revolve 135° counterclockwise round the stator axis. Similarly, as shown in Figure 2c, 15° of rotation corresponds to a revolution of 225°. After completing one cycle of revolution in the counterclockwise direction, the low-speed rotor revolves 24° clockwise relative to its original position, which is compatible with the EHMG's transmission ratio. When the pole pairs of the low-speed rotor are n, that is, the transmission ratio is $-1{:}n$, and when the high-speed rotor rotates β, the low-speed rotor will be driven to rotate β/n. The ratio of the rotation angle between the low-speed rotor and the high-speed rotor is related to the pole pairs of the low-speed rotor, which is consistent with the transmission ratio.

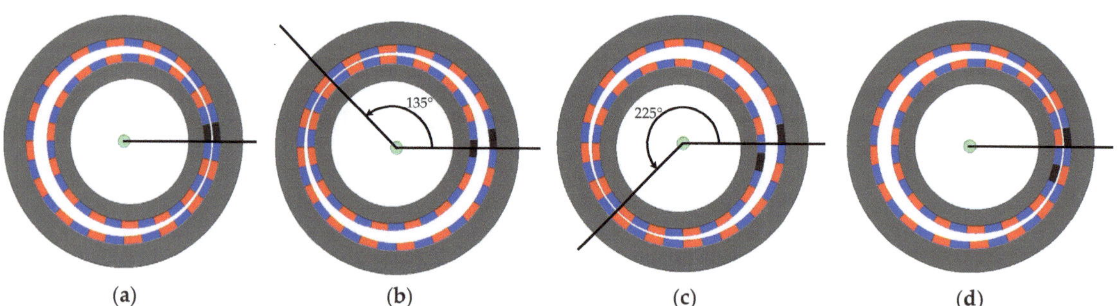

(a)   (b)   (c)   (d)

Figure 2. Relative placement of stator and rotor at different angles. (**a**) 0°; (**b**) 135°; (**c**) 225°; (**d**) 360°.

The material used for the PM in this paper is NdFeB, which has a high magnetic energy product and coercive force. Compared with ferrite materials, it can store more magnetic energy in a smaller volume, provide higher magnetic field strength, and improve equipment efficiency. Although its cost may be higher than that of ordinary permanent magnets, its performance benefit is higher than the cost benefit. PMs have a wide range of magnetization modes. Radial magnetization is the most classic type of magnetization. Halbach arrays and Spoking magnetization have recently gained popularity. Different air gap magnetic field distributions will result from various magnetization modes. The Halbach array, which is close to the ideal PM magnetization mode, has an apparent unilateral effect and enables a sinusoidal distribution of the air gap flux density, both of which dramatically increase the electromagnetic torque and the usage of PM. The magnetization angle and installation order are the two key variables affecting the Halbach array effect. In most cases, magnetization angles of 45°, 60°, and 90° are employed. The smaller the magnetization angle, the higher the sine degree achieved, and the more obvious the unilateral effect. However, a smaller magnetization angle will make PM installation more difficult. Additionally, the unilateral effect's predominant side will depend on the installation order. In this paper, the low-speed rotor uses a fan-shaped structure, and an inhomogeneous PM segment is applied to the stator in the proposed model. Moreover, both the stator and the rotor adopt a Halbach array.

The structure of the two EHMG models is exhibited in Figure 1. As shown in Figure 1a, radial magnetization is used in the conventional model. A Halbach array is adopted in the presented model, as revealed in Figure 1b. The unequal blocks of stator PMs are exhibited in Figure 3 and the fan-shaped structure of PMs is shown in Figure 4, respectively. Figure 3 shows the stator PMs are distributed unequally with a ratio of 1:n:1, and a 60° Halbach array is used. The PMs in Figure 4 are 90°-magnetized. The arrows represent the magnetization direction of PMs.

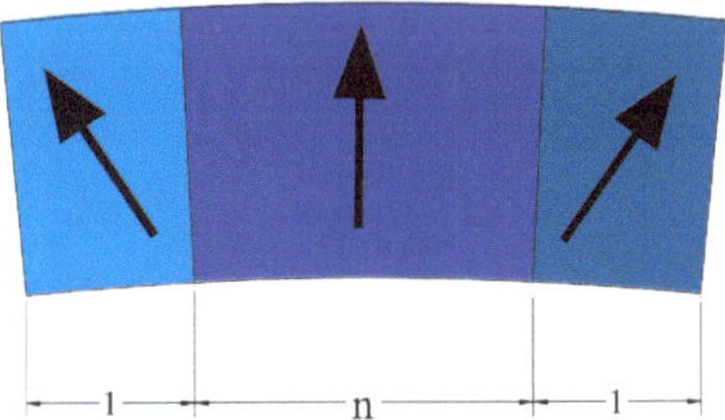

Figure 3. The unequal blocks of PMs.

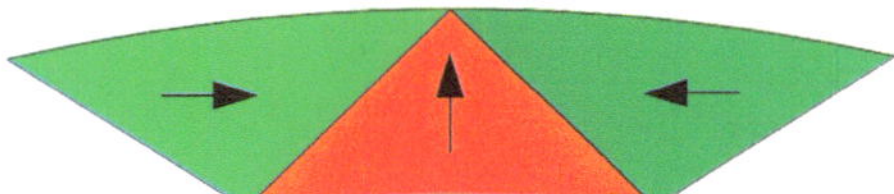

Figure 4. The fan-shaped structure of PMs.

3. Performance Analysis

Table 1 provides the EHMG's details, which were obtained based on experience and reference [16,21]. The performance of the suggested and conventional EHMGs was analyzed and compared using FE analysis.

Table 1. Details of the EHMG.

Parameters	Values
Internal radius of stator yoke R_{sin}	49 mm
External radius of stator yoke R_{sout}	60 mm
Internal radius of rotor yoke R_{rin}	30 mm
External radius of rotor yoke R_{rout}	38 mm
Thickness of stator PMs	3.5 mm
Thickness of rotor PMs	3.5 mm
Maximal length of air gap g_{max}	7 mm
Minimal length of air gap g_{min}	1 mm
Eccentric distance e	3 mm
Stator pole pairs P_s	16
Low-speed rotor pole pairs P_r	15
Axial length L_{ef}	40 mm
Relative permeability	1
PM remanence	1.25 T
Materials of PMs	NdFeB

3.1. Magnetic Field Analysis

One of the most crucial aspects of EHMG performance is its magnetic field characteristics. The main flux, which assists in the transmission of torque, is the flux that intersects with low-speed rotor PMs and stator PMs. The flux that only crosses the low-speed rotor PMs or the stator PMs is referred to as the leakage flux, which is not involved in the transfer of torque. Hence, increasing the main flux while decreasing the leakage flux is required to achieve higher torque.

Figure 5 displays the magnetic flux distribution for the two EHMG models. As can be seen in Figure 5, in the air gap, where the main flux dominates the bulk of space, the magnetic flux lines gather more densely on the suggested EHMG, and the torque mainly relies on the air gap permeance function in the air gap to modulate the harmonic of the air gap flux density, so as to ensure the consistency of the harmonic pole pairs of the inner and outer PMs, and ensure the transmission of torque. Therefore, it is more favorable for torque transmission when the leakage flux predominates in the yokes, resulting in fewer magnetic flux lines distributed within them, which significantly enhances the utilization of PMs. In addition, it corresponds with the unilateral Halbach array effect.

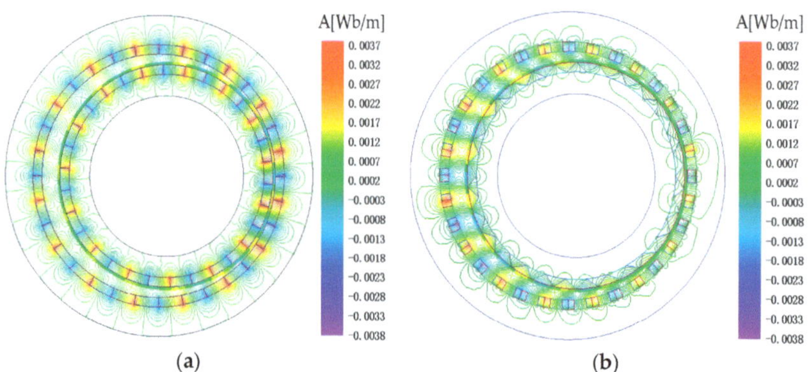

Figure 5. Magnetic flux distribution. (**a**) Conventional; (**b**) proposed.

As shown in Figure 5, we can see that under the same magnetic flux density, as shown by the scale, the magnetic flux density at the junction of the PMs in the traditional model is stronger, with red and blue being more prominent in the figure. In the proposed model, the magnetic flux density between PMs is relatively weak, with fewer red and blue colors in the figure. In the proposed model, it is only due to the effect of the Halbach array that the magnetic flux density on one side (the air gap side) is enhanced, while it is appropriately reduced at the iron yoke and the junction of the PMs. This precisely enhances the magnetic field cross-linking between the inner and outer PMs, and under the modulation of the air gap permeability function, it better transmits torque while also reducing the demagnetization risk of the PMs. In addition, the rare-earth PM NdFeB has the advantages of high magnetic properties, strong machinability, and high cost-effectiveness, and is not easily demagnetized. Its magnetic permeance coefficient is mainly affected by multiple factors such as chemical composition, microstructure, magnetization process, and temperature, so it is not significantly affected by the wide air gap side.

Since flux density is a vector, decomposing it into radial and tangential components can allow a better analysis of its characteristics. By choosing a circular arc in the air gap, the tangential and radial air gap flux density can be analyzed. The consequences are exhibited in Figure 6.

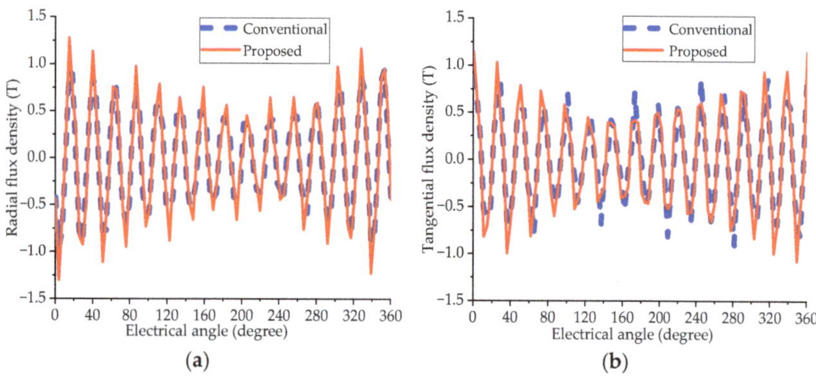

Figure 6. Flux density. (**a**) Radial flux density; (**b**) tangential flux density.

The consequences illustrate that there are 15 troughs and peaks in flux density throughout an electric cycle, which corresponds to the transmission ratio. The radial flux density of the presented EHMG is much higher than that of the conventional EHMG, as shown in Figure 6a. At roughly 30° and 330° electrical angles, the maximal radial flux density exceeds 1.25 T. As shown in Figure 6b, although the tangential flux density of the proposed EHMG is smaller than that of the conventional EHMG at very few angular positions, the overall tangential flux density of the proposed model is still larger than or close to that of the conventional model at most other angular positions. As a consequence, the suggested EHMG model may successfully increase the flux density.

Based on the principle of torque generation, only when two magnetic fields with identical pole pairs interact are they capable of transmitting torque steadily. The modulation effect of harmonics in the air gap permeance function accounts for how regularly the unequal pole pairs of inner and outer PMs may accomplish torque transmission and energy conversion. The coupling of spatial harmonics with the same pole pairs and same speeds may generate the working torque, but the interaction between the spatial harmonics with the same pole pairs while at different speeds will cause torque ripple. Consequently, it is imperative to analyze the space harmonics' pole pairs. MATLAB software was used to carry out a Fourier analysis on the tangential and radial flux density. Figure 7 shows the harmonic pole pair distribution spectrum of flux density. Only 60 harmonic orders were gathered for convenience of analysis because, as harmonic pole pairs grow larger, their amplitudes grow smaller and smaller, reducing their influence on electromagnetic torque.

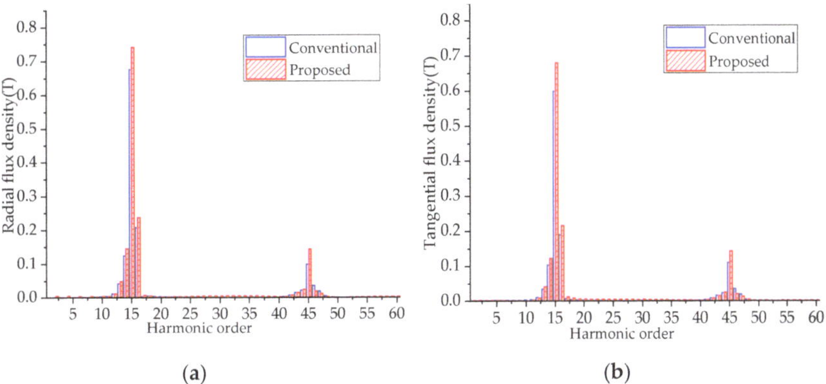

Figure 7. Harmonic pole pair distribution spectrum of flux density. (**a**) Radial component; (**b**) tangential component.

The space harmonics are mostly made up of 14th, 15th, 16th, 44th, 45th, and 46th, as seen in Figure 7. EHMG flux density space harmonics can be decomposed into pole pairs of 15 and 45 harmonics, representing the fundamental and third components, respectively. The working torque is effectively formed by the fundamental components, not the third harmonic components, which is inactive in this process. As can be seen from the above, p_s is 16, p is 15, and p_a is 1. The fundamental component with a harmonic pole pair of 15 is made up of the harmonic components from the low-speed rotor and the stator that are modulated by the air gap. Meanwhile, the harmonic pole pair of 16 comprises two parts: the harmonic component from the stator and the harmonic component from the low-speed rotor that are modulated by the air gap. Lastly, the low-speed rotor modulated by the air gap forms the fundamental component with a harmonic pole pair of 14. Working torque is formed by the three fundamental components. Figure 7 illustrates how the fundamental components of the presented model greatly improved when compared to the conventional EHMG. Similar to the fundamental components, the third components will likewise be produced appropriately but will not be involved in the creation of working torque. The

44th and 46th harmonics diminish, despite the fact that the 45th harmonic also rises. The suggested EHMG is generally useful for increasing working torque.

3.2. Torque Analysis

One of the most significant aspects of EHMG performance is the torque. It is indispensable to analyze the torque.

Torque can be divided into static torque and steady-state torque. The static torque can be determined by keeping the high-speed rotor stationary while the low-speed rotor rotates around the stator's axis for an electric cycle, which is exhibited in Figure 8. The steady-state torque may be attained by recording the torque value every 1.25° while the low-speed rotor revolves from 0° to 22.5° by spinning the high-speed rotor around the stator axis at a speed of 1500 rpm and the low-speed rotor revolves around its axis in the opposing direction at a speed of 100 rpm, as illustrated in Figure 9.

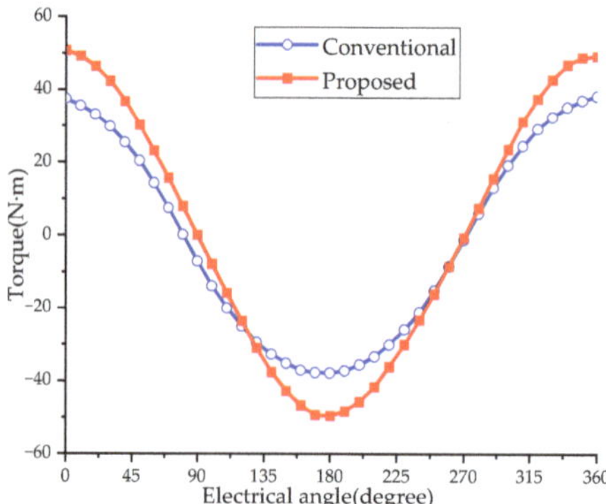

Figure 8. Static torque.

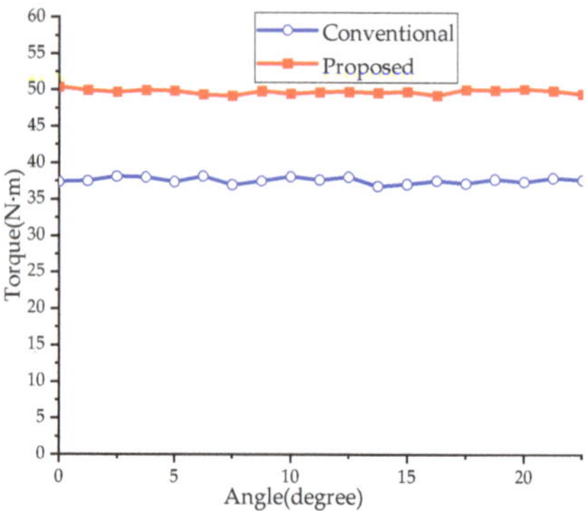

Figure 9. Steady-state torque.

According to Figure 8, the curve of static torque is a typical sine wave. The highest static torque of the conventional model is 38.04 Nm, whereas the maximum static torque of the presented EHMG is 50.41 Nm. It is evident that the suggested EHMG's output torque rose by 32.5%. The torque density was also raised by 32.5% as a result of the identical mass and volume of the conventional and suggested EHMGs. The maximal static torque corresponds to the highest load that the EHMG can bear. The magnetic gear will become unstable over this number. To put it another way, the load capacity of the presented EHMG is higher than that of the conventional EHMG.

As illustrated in Figure 9, it can be seen that the value of the steady-state torque is basically consistent with the amplitude of the static torque. Moreover, another key benefit of EHMG is that the torque ripple is extremely low in the EHMG. Even so, the presented EHMG has a lower torque ripple than the conventional model. Consequently, the proposed EHMG has a better, smoother operation.

Compared with conventional models, the proposed model has both advantages and disadvantages, mainly reflected in the following aspects: due to changes in the shape of PMs and the use of a Halbach array, the cost of PMs inevitably increases. In this paper, the cost increased by approximately USD 400. In terms of the usage of PMs, the proposed model concentrated the magnetic field towards the air gap side, which was more conducive to the modulation effect of the air gap and better transmitted torque, thus improving the usage of PMs; in terms of air gap flux density and output torque, the proposed model had a positive effect, that is, increasing the air gap magnetic density led to an increase in the output torque. In addition, it also reduced the torque ripple.

3.3. Modal Analysis

Although an MG has the merits of no friction and low noise compared with mechanical gear, an EHMG will inevitably produce vibration and noise. When an MG gets close to its natural frequency, resonance will occur, giving rise to great noise and deformation. In order to better study the inherent characteristics of an EHMG and avoid the disastrous impact caused by resonance on EHMGs, it is essential to obtain the natural frequency and modal shape through a modal analysis of the EHMG.

To carry out modal analysis, it is necessary to understand the cause and effect of gear vibration noise. The gear vibration noise mainly consists of the following three parts: air vibration noise, mechanical vibration noise, and electromagnetic vibration noise. Air vibration noise is generally generated by the surrounding air rotating with the gear, which can be almost ignored. Mechanical vibration noise is mainly caused by the manufacturing process, which can be avoided by improving the processing accuracy and process level. Electromagnetic vibration noise is mainly generated by the electromagnetic force in the MG. On the one hand, the electromagnetic force will generate the torque that makes the gear rotate; on the other hand, it can also cause rotor deformation and vibration, thus generating noise and vibration. Hence, the study of MG vibration mainly focuses on electromagnetic vibration noise.

During the operation of an EHMG, tangential electromagnetic force density will produce electromagnetic torque, while radial electromagnetic force density will cause gear deformation, vibration, and noise.

In accordance with the Maxwell tensor method, the tangential and radial electromagnetic force density [17] can be obtained:

$$f_t = \frac{1}{\mu_0} B_r B_t \tag{6}$$

$$f_r = \frac{1}{2\mu_0}\left(B_r{}^2 - B_t{}^2\right) \tag{7}$$

where B_r and B_t represent the radial and tangential magnetic induction intensities, respectively; μ_0 is the vacuum magnetic permeability.

The FE method was used to calculate the space distribution of the radial electromagnetic force density at no-load, as shown in Figure 10. Since the radial electromagnetic force density is a vector, the positive and negative only represent the opposite direction. It can be seen that in a cycle, the radial electromagnetic force density of the proposed EHMG was reduced by 60.06% compared with the conventional model, and it means that it will be more difficult to cause deformation, vibration, and noise using the presented EHMG.

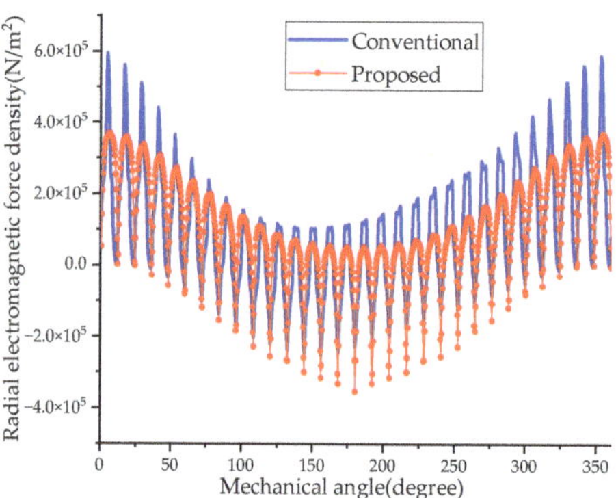

Figure 10. Space distribution of the radial electromagnetic force density.

To conduct a modal analysis, firstly, import the 3D model of the EHMG into the Ansys Workbench software, and then assign the material properties attributes to the EHMG. Finally, by imposing fixed constraints on the stator, the modal shapes and the natural frequencies of the rotor at different orders can be obtained, which are shown in Figure 11. The material properties of different orders are shown in Table 2.

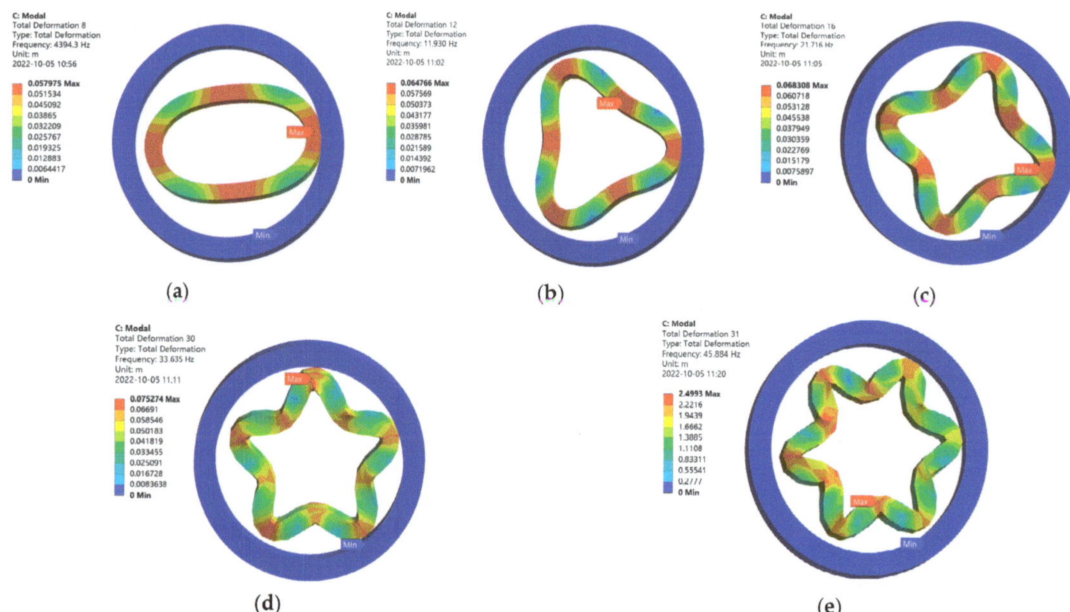

Figure 11. Modal shapes of rotor. (**a**) Second order; (**b**) third order; (**c**) fourth order; (**d**) fifth order; (**e**) sixth order.

Table 2. Material properties.

Material Properties	Stator and Rotor
Young's modulus (N/m^2)	1.738×10^{11}
Density (g/m^3)	7262.5
Poisson's ratio	0.3

As can be seen in Figure 11, with the order increasing, the natural frequency of the rotor became larger, making it more difficult to excite the vibration mode, and the modal shapes of the higher natural frequencies became more complex. As is exhibited in Figure 11a, the modal shape of the rotor was similar to an oval, the natural frequency was 4394.3 Hz, and the corresponding maximum deformation was 0.057975 m; in Figure 11b, the modal shape of the rotor was similar to a triangle, the natural frequency was 11,930 Hz, and the corresponding maximum deformation was 0.064766 m. The number of angles of different modal shapes was related to their respective orders. In Figure 11c, the modal shape had four angles, which was consistent with the fourth order; the corresponding deformation was 0.068308 m; and, similarly, the fifth-order modal shape had five angles, the sixth-order modal shape had six angles, and the corresponding maximum deformation also increased, as is exhibited in Figure 11d, e. The natural frequencies of the fourth to sixth orders were 21,716 Hz, 33,635 Hz, and 45,884 Hz, respectively. The natural frequencies of different orders are shown in Table 3. It is obvious that with the increase in natural frequency, the corresponding maximum deformation also increased. Although the deformation caused by the vibration of high-order natural frequency was large, most of it disappeared in a very short time; hence, the modal shapes and natural frequencies of the low-order modes are the most important. Among the first two orders of vibration, the minimum natural frequency of the rotor was 4394.3 Hz for the second order and 11,930 Hz for the third order. Hence, it is necessary to avoid making the frequency of the gear reach this value as much as possible to reduce the possibility of resonance. In addition, the radial electromagnetic force density of the proposed EHMG was smaller than that of the conventional model, making the proposed model have better antivibration performance compared to the conventional model.

Table 3. Natural frequencies of rotor.

Order	Natural Frequency of Rotor
2	4394.3 Hz
3	11,930 Hz
4	21,716 Hz
5	33,635 Hz
6	45,884 Hz

4. Experiments

To verify the analysis and simulation results of the proposed model, a prototype was manufactured and experiments were carried out, as shown in Figure 12.

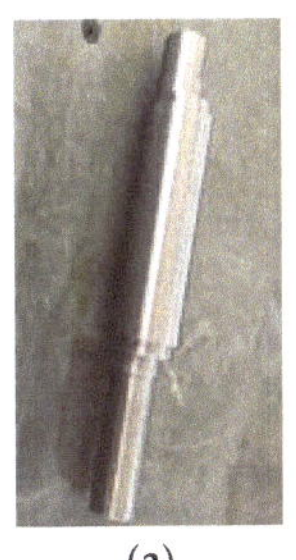

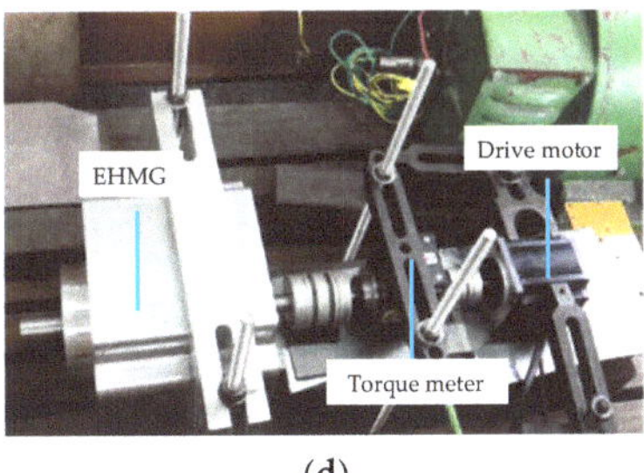

(a) (b) (c) (d)

Figure 12. Prototype parts and test bench. (**a**) High-speed rotor; (**b**) low-speed rotor; (**c**) stator; (**d**) test bench.

The prototype test bench consisted of a DC drive motor, a torque meter, and an EHMG. As a driving motor, a DC motor was connected to the high-speed rotor of the EHMG through a torque sensor, which could measure the speed and torque of the EHMG. The sensitivity of the torque meter (torque meter FTE) was 1.5 mV/V; as for the accuracy, there were multiple ranges to choose from, mainly including the following: 0–0.2 Nm, 0–0.5 Nm, 0–1 Nm, 0–2 Nm, 0–3 Nm, 0–5 Nm, 0–10 Nm, 0–20 Nm, 0–30 Nm, 0–50 Nm, and 0–100 Nm. In this paper, we mainly used the 0–100 Nm range for testing. The DC drive motor had excellent starting characteristics, and after starting the equipment, they could drag the high-speed rotor to rotate, thus starting the EHMG. By recording the torque values at different speeds and comparing them with the torque values obtained with the FE method in this paper, the results are shown in Figure 13.

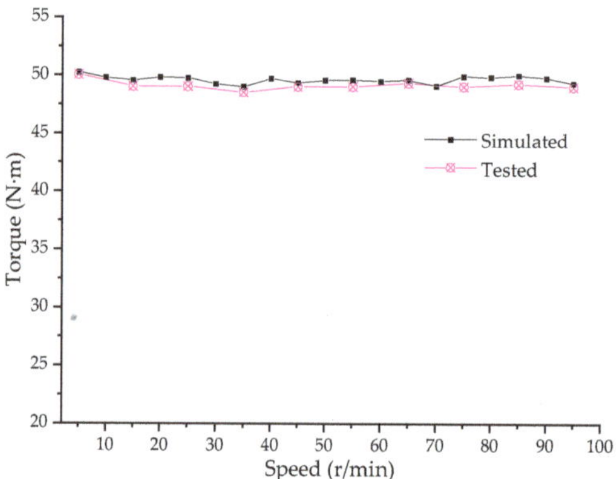

Figure 13. Output torque.

As can be seen in Figure 13, the torque obtained from the FE simulation was about 50.4 Nm, while the torque measured from the experiments was about 48.8 Nm; the data obtained from the experiments were generally smaller than those obtained from the FE simulation. This is because the experimental platform can inevitably generate errors—while the FE simulation is conducted under idealized parameters—but overall, it is not significantly different from the FE method in terms of numerical values within the error range; in other words, the test results are basically consistent with the simulation results, indicating the effectiveness of this experiment.

In order to verify the efficiency of the EHMG, load tests were conducted in this paper, recording the input torque and output torque at different speeds. The efficiency is the ratio of output torque to input torque. Figure 14 shows the transmission efficiency curve of the proposed EHMG. It can be seen that the operating efficiency of the proposed EHMG was stable, and the efficiency reached 92%.

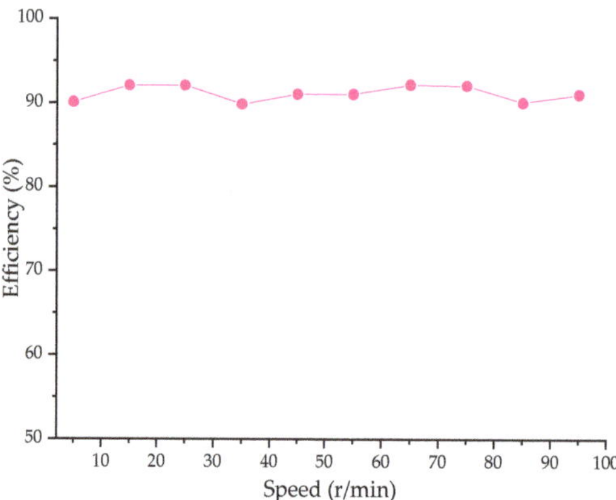

Figure 14. Measured efficiency.

5. Conclusions

In order to further improve the torque density of MGs in the field of high transmission ratios, a novel EHMG with a fan-shaped structure and Halbach array was proposed. It had a better unilateral effect in magnetic flux lines, which greatly improved the utilization rate of PMs. Meanwhile, there was also an improvement in the air gap flux density and output torque, indicating that it had better load capacity and smoother operation. In addition, in order to better verify its vibration characteristics, a modal analysis was conducted to obtain its natural frequency and radial electromagnetic force density, indicating that the proposed model had better antivibration characteristics.

An experimental test bench was established, and a prototype was constructed. The results were consistent with the FE method and the transmission efficiency was superior, indicating that the proposed model was an effective attempt.

Author Contributions: Conceptualization, L.J. and D.L.; methodology, L.J. and Y.W.; software, Y.W.; validation, L.J., Y.W., D.L. and R.Q.; resources, L.J., D.L. and R.Q.; writing—original draft, L.J. and Y.W.; writing—review and editing, Y.W.; supervision, L.J. All authors have read and agreed to the published version of the manuscript.

Funding: This research received no external funding.

Data Availability Statement: All data generated or analyzed during this study are included in this published article.

Conflicts of Interest: The authors declare no conflict of interest.

References

1. Jing, L.; Su, Z.; Wang, T.; Wang, Y.; Qu, R. Multi-objective optimization analysis of magnetic gear with HTS bulks and uneven Halbach arrays. *IEEE Trans. Appl. Supercond.* **2023**, *33*, 5202705. [CrossRef]
2. Jing, L.; Pan, Y.; Wang, T.; Qu, R.; Cheng, P. Transient analysis and verification of a magnetic gear integrated permanent magnet brushless machine with Halbach arrays. *IEEE J. Emerg. Sel. Top. Power Electron.* **2022**, *10*, 1881–1890. [CrossRef]
3. Ding, J.; Yao, L.; Xie, Z.; Wang, Z.; Chen, G. A novel 3-D mathematical modeling method on the magnetic field in nutation magnetic gear. *IEEE Trans. Magn.* **2022**, *58*, 8000910. [CrossRef]
4. Dai, B.; Nakamura, K.; Suzuki, Y.; Tachiya, Y.; Kuritani, K. Cogging torque reduction of integer gear ratio axial-flux magnetic gear for wind-power generation application by using two new types of pole pieces. *IEEE Trans. Magn.* **2022**, *58*, 8002205. [CrossRef]
5. Jing, L.; Tang, W.; Wang, T.; Ben, T.; Qu, R. Performance Analysis of Magnetically Geared Permanent Magnet Brushless Motor for Hybrid Electric Vehicles. *IEEE Trans. Transp. Electrif.* **2022**, *8*, 2874–2883. [CrossRef]
6. Fang, L.; Li, D.; Qu, R. Torque improvement of vernier permanent magnet machine with larger rotor pole pairs than stator teeth number. *IEEE Trans. Ind. Electro.* **2023**, 1–11. [CrossRef]
7. Mezani, S.; Atallah, K.; Howe, D. A high-performance axial-field magnetic gear. *J. Appl. Phys.* **2006**, *99*, 08R303. [CrossRef]
8. Atallah, K.; Howe, D. A novel high-performance magnetic gear. *IEEE Trans. Magn.* **2001**, *37*, 2844–2846. [CrossRef]
9. Okano, M.; Tsurumoto, K.; Togo, S.; Tamada, N.; Fuchino, S. Characteristics of the magnetic gear using a bulk high-Tc superconductor. *IEEE Trans. Appl. Supercond.* **2002**, *12*, 979–983. [CrossRef]
10. Jing, L.; Liu, W.; Tang, W.; Qu, R. Design and optimization of coaxial magnetic gear with double-layer PMs and spoke structure for tidal power generation. *IEEE/ASME Trans. Mechatron.* **2023**, 1–9. [CrossRef]
11. Jian, L.; Chau, K.T. A coaxial magnetic gear with Halbach permanent-magnet arrays. *IEEE Trans. Energy Convers.* **2010**, *25*, 319–328. [CrossRef]
12. Jenne, K.; Pakdelian, S. Magnetic design aspects of the trans-rotary magnetic gear using quasi-Halbach arrays. *IEEE Trans. Ind. Electron.* **2020**, *67*, 9582–9592. [CrossRef]
13. Liu, X.; Qiu, J.; Chen, H.J.; Xu, X.Y.; Wen, Y.M.; Li, P. Design and optimization of an electromagnetic vibration energy harvester using dual Halbach arrays. *IEEE Trans. Magn.* **2015**, *51*, 8204204. [CrossRef]
14. Jing, L.B.; Zhang, T.; Gao, Y.T.; Qu, R.H.; Huang, Y.H.; Ben, T. A novel HTS modulated coaxial magnetic gear with eccentric structure and Halbach arrays. *IEEE Trans. Appl. Supercond.* **2019**, *29*, 5000705. [CrossRef]
15. Rens, J.; Clark, R.; Calverley, S.; Atallah, K.; Howe, D. Design, Analysis and Realization of a Novel Magnetic Harmonic Magnetic Gear. In Proceedings of the 2008 18th International Conference on Electrical Machines, Vilamoura, Portugal, 6–9 September 2008.
16. Rens, J.; Atallah, K.; Calverley, S.D.; Howe, D. A novel magnetic harmonic gear. *IEEE Trans. Ind. Appl.* **2010**, *46*, 206–212. [CrossRef]
17. Jing, L.; Gong, J. Research on eccentric magnetic harmonic gear with Halbach array. *Prog. Electromagn. Res Lett.* **2020**, *89*, 37–44. [CrossRef]
18. Lee, J.; Chang, J. Analysis of the vibration characteristics of coaxial magnetic gear. *IEEE Trans. Magn.* **2017**, *53*, 8105704. [CrossRef]

19. Wang, Y.; Jing, L. A Novel Eccentric Harmonic Magnetic Gear with Inhomogeneous Structure and Halbach Array. In Proceedings of the 2022 IEEE 3rd China International Youth Conference on Electrical Engineering (CIYCEE), Wuhan, China, 3–5 November 2022.
20. Baninajar, H.; Modaresahmadi, S.; Wong, H.Y.; Bird, J.; Williams, W.; Dechant, B. Designing a Halbach rotor magnetic gear for a marine hydrokinetic generator. *IEEE Trans. Ind. Appl.* **2022**, *58*, 6069–6080. [CrossRef]
21. Zhang, Y.; Zhang, J.; Liu, R. Magnetic field analytical model for magnetic harmonic gears using the fractional linear transformation method. *Chin. J. Electr. Eng.* **2019**, *5*, 47–52. [CrossRef]

Article

An Efficient and High-Precision Electromagnetic–Thermal Bidirectional Coupling Reduced-Order Solution Model for Permanent Magnet Synchronous Motors

Yinquan Yu [1,2,3,*], Pan Zhao [1,2,3], HuiHwang Goh [4], Giuseppe Carbone [5,*], Shuangxia Niu [6], Junling Ding [1,2,3], Shengrong Shu [1,2,3] and Zhao Zhao [7]

1   School of Mechatronics and Vehicle Engineering, East China Jiaotong University, Nanchang 330013, China
2   Key Laboratory of Conveyance and Equipment of Ministry of Education, East China Jiaotong University, Nanchang 330013, China
3   Institute of Precision Machining and Intelligent Equipment Manufacturing, East China Jiaotong University, Nanchang 330013, China
4   School Electrical Engineering, Guangxi University, Nanning 530004, China
5   Department of Mechanical, Energy, and Management Engineering, University of Calabria, I-87036 Rende, Italy
6   Department of Electronic and Information Engineering, The Hong Kong Polytechnic University, Hong Kong, China
7   Faculty of Electrical Engineering and Information Technology, Otto-Von-Guericke University Magdeburg, 39106 Magdeburg, Germany
*   Correspondence: yu_yinquan@ecjtu.edu.cn (Y.Y.); giuseppe.carbone@unical.it (G.C.)

Citation: Yu, Y.; Zhao, P.; Goh, H.; Carbone, G.; Niu, S.; Ding, J.; Shu, S.; Zhao, Z. An Efficient and High-Precision Electromagnetic–Thermal Bidirectional Coupling Reduced-Order Solution Model for Permanent Magnet Synchronous Motors. *Actuators* **2023**, *12*, 336. https://doi.org/10.3390/act12080336

Academic Editor: Dong Jiang

Received: 19 July 2023
Revised: 7 August 2023
Accepted: 11 August 2023
Published: 21 August 2023

Abstract: The traditional electromagnetic–thermal bidirectional coupling model (EMTBCM) of permanent magnet synchronous motors (PMSMs) requires a long time to solve, and the temperature-induced torque change is not accounted for in the finite element (FE) numerical calculation of the EM field. This paper presents a precise and efficient EMTBC reduced-order solution model. The specific methods are as follows: First, a torque control technology based on the current injection method is proposed for determining the effect of temperature on the properties of EM materials and EM torque in an EM field, and the accuracy of the FE numerical calculation model is improved. Second, we use the improved EM field finite element numerical calculation model (FEMNCM) to analyze the correlation between the EM loss, the temperature, and the load, and we replace the FEMNCM with the EM field reduction model using the least-squares method. Then, we analyze the law of the PMSM's internal temperature distribution. We choose the GA-BP algorithm with as few samples as possible and a high accuracy and stability to build the regression prediction model of the temperature field. We use this regression prediction model to replace the complex temperature field calculation. After analyzing the EMTBCM solution strategy, the original complex EMTBC numerical calculation model is substituted with iterations of the magnetic field reduction model and the temperature field regression prediction model. The FE numerical calculation is then used to validate the reduced-order model. The proposed model is validated through numerical simulations. The numerical results indicate that the proposed reduced-order EMTBC model in this paper is accurate and computationally efficient.

Keywords: PMSM; electromagnetic–thermal bidirectional coupling; least-squares method; GA-BP algorithm; solution strategy; reduced-order EMTBC model

1. Introduction

As an electric drive permanent magnet synchronous motor (PMSM) has a small size, large torque, wide speed regulation range, high power density, and high efficiency, it has been extensively utilized [1,2] in a variety of applications. In the design of PMSMs, thermodynamic optimization is equally as essential as electromagnetic (EM) optimization, because the performance of the motor is not only determined by the EM performance but also strongly influenced by the temperature field [3]. Temperature increase is a significant factor that impedes motor miniaturization and high-power density, so it is essential to have temperature data for the motor's overall design and online monitoring. Due to the aforementioned factors, it is crucial to develop an accurate and efficient thermal calculation model for the multi-physics design of PMSMs.

Various thermal analysis methodologies for electric motors have been proposed in recent years. The lumped parameter thermal network method concentrates the source of heat loss on each discrete node, connects the nodes through thermal resistance, and then establishes the network topology relationship according to the direction and path of heat transfer within the motor, which significantly reduces the computation time [4,5]. The adopted lumped parameter thermal network method analyzes the temperature field of the motor [6], and the lumped parameter thermal network method has a high calculation efficiency. However, it can only determine the average temperature of the motor and cannot determine the temperatures of individual components. Moreover, owing to the equivalence and approximation of the thermal resistance network construction procedure, the calculation results of the lumped parameter thermal loop lack precision, so the lumped parameter thermal network method is not considered in this paper.

For the calculation of motor temperature rise, there are two primary methods: the unidirectional coupling calculation method [7] and the bidirectional coupling calculation method [8]. The numerical calculation for unidirectional coupling directly couples the EM field loss to the temperature field. Since the electrical resistivity of the copper wire of the motor increases as the motor temperature rises and the remanence of the PM decreases as the temperature rises, the unidirectional coupling calculation method in the EM field does not account for the temperature characteristics of the EM material. Therefore, this motor temperature calculation procedure is inaccurate [9]. The EMTBC numerical calculation incorporates EM field loss data into the temperature field, while temperature field data are transferred to the EM field. This calculation method considers the effect of temperature on the properties of EM materials and enhances the precision of temperature rise calculations [10,11]. Comparing the temperature rise calculation results of the EM-thermal unidirectional coupling method and the EMTBC method, numerical calculations and experimental results demonstrate that the EMTBC method provides more accurate calculations [12].

Even though the PMSM method for solving the temperature field with EMTBC has a high level of accuracy, the iterative process of the EM field and temperature field requires a great deal of calculation time, which is clearly insufficient to meet the demand for high thermal calculation efficiency. Alternatively, during the actual operation of the motor, when the PM is demagnetized, the controller autonomously adjusts the power angle to maintain torque [13]. Nonetheless, in the process of the EMTBC numerical calculation, the increase in temperature reduces the remanence of the PM, but the power angle of the EM field does not change automatically, resulting in a reduction in accuracy; therefore, it is necessary to propose a new methodology to further consider the adaptive control of torque.

In this paper, a high-precision reduced-order EMTBC solution model is proposed to address the computational issue of the traditional EMTBC model. Our strategies include the following: First, the effect of temperature variations on the EM torque is considered. Then, the stability of the EM torque is regulated by temperature feedback, as well as control methods. (In this way, the temperature accuracy of the solution through the EMTBC will be greatly improved). Secondly, the EM field and temperature field are calculated, and their respective reduced-order calculation models are then established based on the results of the calculations. (The aim is to replace the calculation of EM and temperature fields with proxy modeling). Thirdly, the two established one-dimensional models are coupled iteratively in accordance with the bidirectional coupling solution strategy to simulate the bidirectional EM-thermal coupling solution process. (With the progress of the bidirectional coupling iteration, the obtained temperature value will be closer to the real value). Finally, a finite element analysis (FEA) is used to validate the results of our proposed method.

2. Theoretical Analysis

2.1. Theoretical Analysis of Demagnetization of PM

Since PMs have a high temperature coefficient, their remanence (B) and coercivity (Hci) vary as the temperature changes. The temperature coefficient α_{Br} quantifies the extent to which the residual magnetic induction intensity of PM materials can be reversibly altered by temperature, and the unit is K^{-1} [14].

$$\alpha_{Br} = \frac{B_1 - B_0}{B_0(t_1 - t_0)} \times 100 \tag{1}$$

Similarly, α_{Hci} is also commonly used to represent the degree to which the intrinsic coercivity of PM materials changes reversibly with temperature, in K^{-1} [14].

$$\alpha_{Hci} = \frac{Hci_1 - Hci_0}{Hci_0(t_1 - t_0)} \times 100 \tag{2}$$

where B_0, B_1 and Hci_0, Hci_1 are the values at temperatures t_0 and t_1, respectively.

Equations (1) and (2) demonstrate that both the remanence and coercivity of the PM decrease as the temperature rises. Consequently, the influence of temperature rise on the properties of EM materials must be accounted for in the calculation of the motor temperature rise via the mutual coupling iteration of the EM field and temperature field.

As demonstrated in (3), for the irreversible magnetization loss of the PMSM, because the remanence (B) decreases, this leads to a decrease in B_r and B_θ, which further leads to a decrease in the motor drive torque [15].

$$T_{em} = \frac{L_{ef}}{\mu_0} \oint r^2 B_r B_\theta d\theta \tag{3}$$

where r is the radius of any circle located in the air gap; B_r and B_θ are the radial and tangential components of the magnetic density of the air gap at radius r. L_{ef} calculates the length of the armature.

When magnetic pole i loses magnetic K_i (K_i denotes the percentage of the ith pole demagnetization), the winding back electromotive force (Back-EMF) of the motor also changes [16].

$$e_{si,A1}(t) = E_s(1 - \frac{K_i}{2p})\sin(2\pi f_e t - \frac{\pi}{6}) - E_s K_i \sum_{k=1}^{\infty} \frac{1}{k\pi}\sin(\frac{k\pi}{2p}) \cdot$$
$$\sin[2\pi f_e t(1 \pm \frac{k}{p}) \mp \frac{(3i-1)k\pi}{3p} - \frac{\pi}{6}] \tag{4}$$

where e_{si} is the Back-EMF of a single-slot winding group during no-load operation at the rated motor speed, E_s is the no-load fundamental wave Back-EMF amplitude of a single slot, 2p is the number of poles, and $A1$ represents the first slot of the A-phase winding.

Equation (4) demonstrates that, when the magnetic pole is demagnetized, the value of the Back-EMF decreases, which causes the winding current to increase.

In summary, a reduction in the PM's remanence further diminishes the motor's torque. In order to maintain a constant EM torque, the closed-loop controller increases the winding current. Therefore, if the PM is demagnetized during actual motor operation, the winding current increases to compensate for the magnetic density amplitude of the PM. In the numerical calculation of the conventional EMTBC, however, only the influence of temperature on the EM material is considered; however, the torque variation due to the influence of temperature increase must also be considered.

2.2. Loss Analysis

2.2.1. Core Loss

Iron loss accounts for the majority of PMSM losses. The formula for calculating the iron core loss is as follows [17]:

$$p_{fe} = p_h + p_c + p_a = k_h f B_m^{\alpha} + k_c f^2 B_m^2 + k_a f^{1.5} B_m^{1.5} \tag{5}$$

Since the resistivity of ferromagnetic materials is affected by temperature and increases as the temperature rises, the resistivity of ferromagnetic materials increases [18,19].

$$\rho(T) = \rho(20)(1 + k_t \cdot (T - 20)) \tag{6}$$

where T is the temperature, $\rho(20)$ is the resistivity at room temperature, T = 20 °C, and, finally, k_t denotes the temperature coefficient.

Therefore, a temperature coefficient, k_t, is introduced to characterize the effect of temperature on the eddy current loss term. By substituting Equation (6) into the expression for the eddy current loss coefficient, it can be deduced [18,19]:

$$p_{fe} = p_h + p_c + p_a = k_h f B_m^{\alpha} + \frac{k_c}{1 + k_t(T - 20)} f^2 B_m^2 + k_a f^{1.5} B_m^{1.5} \tag{7}$$

where p_h is the hysteresis loss. p_c is the eddy current loss. p_a is the abnormal eddy current loss. B_m is the magnetic density amplitude of the iron core. f is the frequency. k_h is the hysteresis loss coefficient. k_c is the abnormal eddy current loss coefficient, k_a is the additional loss coefficient, and α is the Steinmetz coefficient. The standard electric steel loss factors (k_h, k_c, k_a) are used in this paper.

2.2.2. Copper Loss

The copper loss p_{Cu} of the PMSM can be expressed as follows [20]:

$$P_{cu} = mI^2 R \tag{8}$$

where m is the number of phases, I is the effective value of the phase current, and R is the phase resistance.

The resistance of copper is shown in Equation (7):

$$R = \rho \frac{L}{S} \tag{9}$$

where ρ is the resistivity of copper.

Since the resistivity of copper varies with temperature, the following expression is valid [20]:

$$\rho = \rho_0[1 + \alpha(t - t_0)] \tag{10}$$

where α is the resistance temperature coefficient of the material.

Because a low-speed motor is utilized in this study, the AC loss and mechanical loss are not taken into account in this work.

3. Methodologies

A high-power-density and compactly designed PMSM is extremely demanding, but these characteristics make motor heat dissipation difficult to design. An improper thermal design of a motor can result in an excessive internal temperature rise and failure. Obtaining accurate temperature field calculation results for motor thermal design, nevertheless, is a time-consuming process. Therefore, we proposed a method for EMTBC dimensionality reduction in order to develop an effective and high-precision model for the solution. The strategy is depicted in Figure 1.

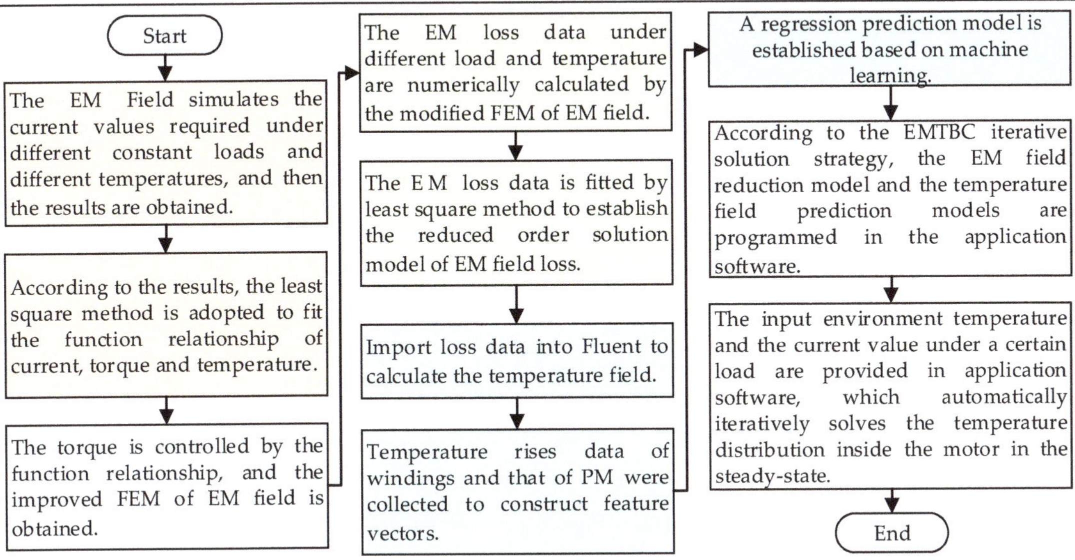

Figure 1. The steps of constructing a reduced-order solution model.

3.1. Finite Element Model (FEM)

This article utilizes a 1.5 kW surface-mounted PMSM as its research subject. The motor's specifications are listed in Table 1. The motor's PM type is N38H, its winding insulation is class B, its maximum operating temperature is 130 °C, and its cooling structure is air-cooled. Because the PMSM is a symmetrical structure, the 1/4 model is chosen to determine the transient magnetic field in order to reduce the calculation time. Figure 2a depicts the mesh partitioning for the simulation investigation. Figure 2b depicts the motor architecture.

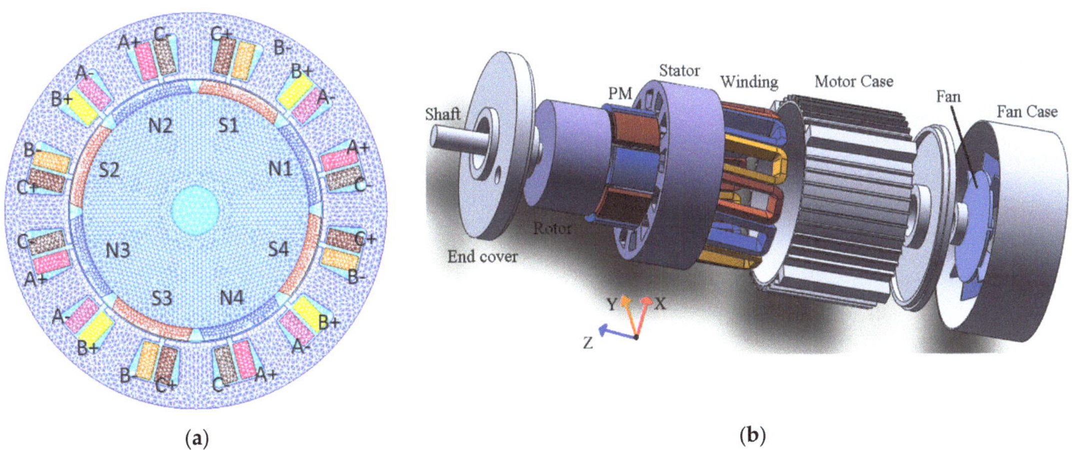

(a)   (b)

Figure 2. Model of PMSM: (a) FEM; (b) Geometric Model.

Table 1. Specifications of the adopted motor.

Parameter	Value	Parameter	Value
Rated output power	1.5 KW	Rated speed	3000 r/min
Length	40 mm	Number of poles (2p)	8
Stator outer diameter	120 mm	Number of slots (Q)	12
Stator inner diameter	80 mm	Magnet thickness	4 mm
Rotor outer diameter	79 mm	air gap	0.5 mm
Rated frequency	50 Hz	PM material	N38H

3.2. Enhancement of EM Field FEM

Using the traditional EMTBC method to solve the temperature field, the remanence of the PM diminishes as the temperature increases, and the torque also decreases. Figure 3 displays the calculated motor torque at various temperatures. In the numerical calculation, one can see that an increase in temperature decreases the EM torque of the motor, and the higher the temperature, the greater the torque decrease.

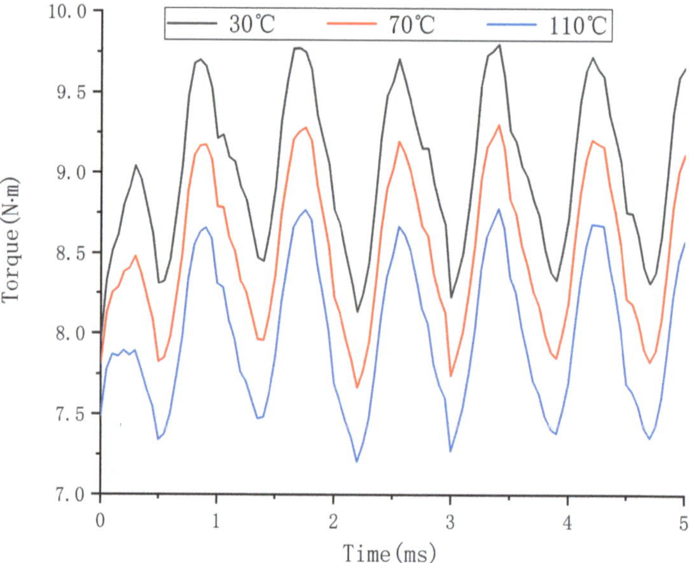

Figure 3. The influence of temperature on torque under the traditional model is solved.

To maintain a constant EM torque under the EMTBC solution model, a torque control method based on the current injection technique is proposed. The fundamental concept is to increase the input current as the temperature rises and to maintain a constant torque by increasing the magnetic field intensity. Consequently, this paper employs a numerical calculation method to ascertain the temperature and current change under a constant torque, and then it fits the corresponding relationship between the temperature and current under a varying torque. By controlling the change in the current value during the EMTBC procedure, it is intended to prevent the EM torque from varying with the temperature rise. Table 2 contains the numerical results. It is important to note that, as the output torque increases, the temperature of each motor component also increases. We determined the solution method of the intermediate gradient change in Table 2 by calculating the maximum temperature rise under various torques.

Table 2. Corresponding table of change in temperature and current under constant torque.

Temperature (°C)	Current at Different Temperatures at Rated Torque (A)	Current at Different Temperatures at 1.25 Times Torque (A)	Current at Different Temperatures at 1.5 Times Torque (A)	Current at Different Temperatures at 1.75 Times Torque (A)	Current at Different Temperatures at 2 Times Torque (A)
20	10	13.55	17.16	20.68	24.32
30	10.32	13.89	17.55	21.07	24.72
40	10.62	14.23	17.9	21.46	25.12
50	10.93	14.56	18.26	21.85	25.54
60	11.26	14.89	18.62	22.25	25.96
70	11.58	15.24	18.99	22.65	26.39
80		15.58	19.38	23.06	26.82
90			19.79	23.49	27.26
100				23.92	27.71
110					28.17

The data in Table 3 are fitted using the least-squares method, with the current serving as the dependent variable and the temperature and torque serving as the independent variables. Figure 4 depicts the results, and the following equation can be used to calculate the current value under arbitrary torque and temperature conditions:

$$z = z0 + ax + by + cx^2 + dy^2 + fxy \tag{11}$$

Table 3. Effect of load on EM loss.

Current (A)	Iron Loss (W)	Eddy Current Loss of PM (W)	Copper Loss (W)
13.89	64.19	6.30	8.27
17.55	65.53	6.62	23.94
21.07	65.71	7.05	34.47
24.72	66.15	7.63	47.45

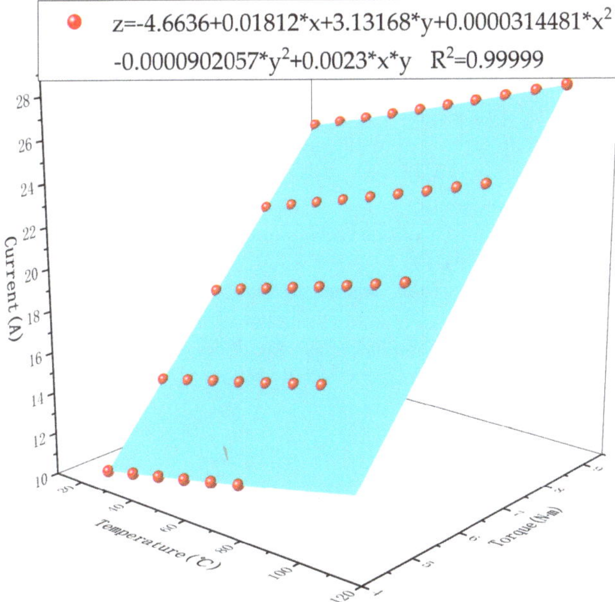

Figure 4. Temperature−torque−current fitting.

In the equation, z represents the current, x represents the temperature, y represents the torque, $z0 = -4.6636$, $a = 0.01812$, $b = 3.13168$, $c = 0.0000314481$, $d = -0.0000902057$, and $f = 0.0023$. When $R^2 = 0.99999$, this indicates that the fitting effect is good.

If the output torque is known, the feedback temperature value can be used to modify the current value during the EMTBC to maintain a constant torque. To verify the efficacy of the proposed control method, the FE numerical calculation was performed, as depicted in Figure 5.

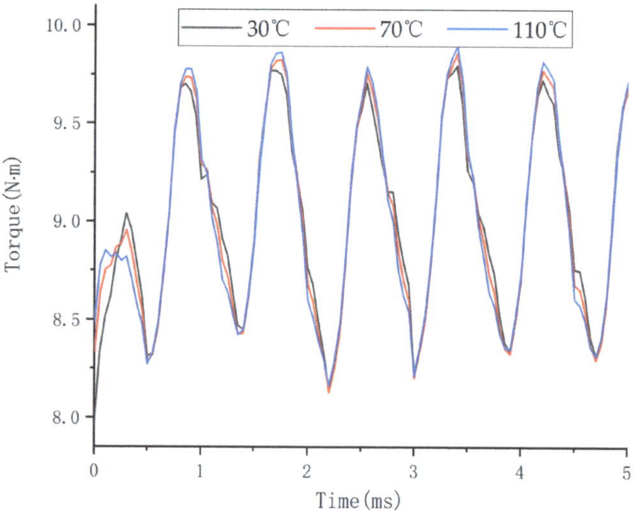

Figure 5. Torque comparison based on the improved solution model.

Figure 5 is a comparison of torque simulations at various temperatures. Based on the comparison of the results, it is evident that the EMTBC temperature rise calculation after the control is implemented can keep the torque essentially constant (improving the accuracy of the temperature rise calculations), which further demonstrates the efficacy of the employed control method.

3.3. Solution of Reduced-Order Model of EM Loss

3.3.1. Influence of Load and Temperature on EM Loss

In the previous section, we proposed a new model to considerably improve the calculation efficiency of FEM in the EM field. This section employs this model for further analysis. PMSM has a wide operating load range, and the EM loss varies with varied loads. Alternatively, the temperature affects the remanence of the PM, which in turn modifies the load angle of the motor's operation; consequently, the temperature also affects the loss. To numerically analyze the influence of the load and temperature rise on the EM loss of the motor, each parameter was sequentially altered within the predetermined range of the load and temperature. The effect of each parameter on the EM loss of the motor is illustrated in Table 3, Table 4, and Figure 6.

Table 4. Effect of temperature on EM loss.

Temperature (°C)	Iron Loss (W)	Eddy Current Loss of PM (W)	Copper Loss (W)
30	65.60	7.05	34.47
60	61.25	6.52	45.78
90	57.84	6.19	59.22
120	51.26	6.15	74.95

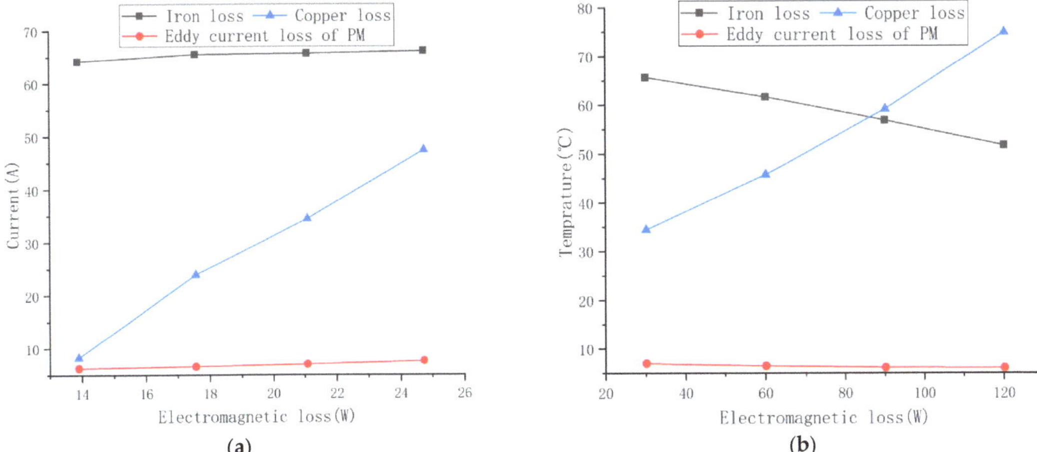

Figure 6. EM loss with load and temperature rise change relationship. (**a**) Effect of load on EM loss. (**b**) Effect of temperature on EM loss.

As shown in Figure 6, the results show the following:

- The iron loss increases with the increasing load current and decreases with the increasing temperature. This is because the increase in the current enhances the induced magnetic field of the stator, thus increasing the stator iron loss; however, the increase in the temperature reduces the remanent magnetic field of the PM, as well as the loss coefficient of the silicon steel sheet, thus decreasing the iron loss.
- The copper loss increases with the increasing load current and temperature. However, the increasing temperature increasing the resistance of the copper wire leads to an increasing copper loss.
- The eddy current loss of the PM increases with the increasing load current and decreases with the increasing temperature. The reason for this is that the increase in the load current enhances the working magnetic field of the PM, while the increase in the temperature reduces the remanent magnetic field of the PM. Therefore, the EM loss has a certain correlation with the load current and temperature.

3.3.2. The Specific Law of Load and Temperature Effects on EM Loss

In order to analyze the relationship between the EM loss, load current, and temperature, this section uses temperature and various load torques as independent simulation variables, and the iron loss and eddy current loss of the PM as dependent variables. Combined with the data in Table 2, the results of the FE simulation are shown in Table 5. In addition, since there is an error in the calculation of copper consumption in Maxwell (Maxwell calculates the resistance value from the cross-sectional area of the stator slots and the length of the stator, and it does not take into account the real winding structure), we manually compute the copper consumption of the windings.

In order to facilitate observation and acquire specific rules, polynomial fitting is performed using the least-squares method on Table 5 data. The effect of the fitting is illustrated in Figure 7. Equations (10)–(12) illustrate the specific principles of the load current, temperature, and iron consumption; the eddy current loss of the PM; and copper consumption. Consequently, the specific value of the EM loss can be calculated using the load current and temperature.

$$P_{iron\_loss} = z0 + ax + by + cx^2 + dy^2 + fxy \tag{12}$$

Table 5. Influence of load and temperature on iron loss and eddy current loss of PM.

Temperature (°C)	Loss at Rated Torque (W)		Loss at 1.25 Times the Torque (W)		Loss at 1.5 Times the Torque (W)		Loss at 1.75 Times the Torque (W)		Loss at 2 Times the Torque (W)	
	Iron Loss	PM Eddy Current Loss	Iron Loss	PM Eddy Current Loss	Iron Loss	PM Eddy Current	Iron Loss	PM Eddy Current Loss	Iron Loss	PM Eddy Current Loss
20	66.19	6.22	66.33	6.47	66.55	6.83	66.87	7.25	67.45	7.80
30	64.78	5.95	65.17	6.22	65.38	6.62	65.60	7.05	66.04	7.63
40	63.39	5.69	63.92	5.97	63.98	6.39	64.25	6.85	64.59	7.47
50	61.89	5.44	62.61	5.74	62.35	6.19	62.74	6.68	63.01	7.34
60	60.21	5.20	61.09	5.52	60.76	6.00	61.25	6.52	61.42	7.23
70	58.53	4.97	59.49	5.31	59.10	5.83	59.52	6.38	59.66	7.14
80			57.58	5.11	57.42	5.67	57.84	6.19	57.96	7.06
90					55.86	5.55	56.22	6.19	56.40	7.06
100							54.43	6.14	54.63	7.07
110									53.20	7.14

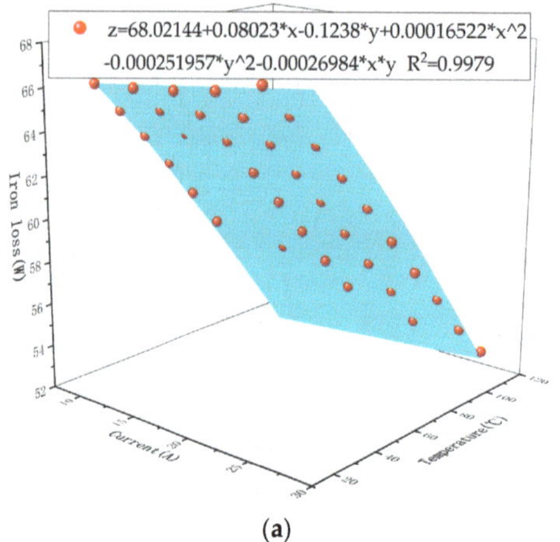

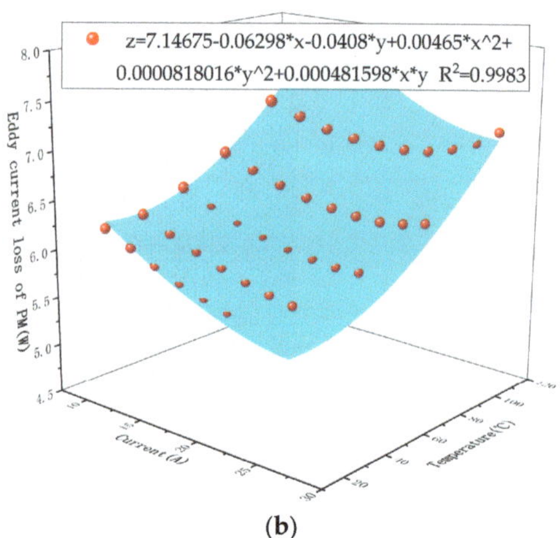

(a)   (b)

Figure 7. The influence of load and temperature on EM loss: (**a**) the influence of load and temperature on iron loss; (**b**) the influence of load and temperature on eddy current loss of PM.

In the equation, x represents the current value, y represents the temperature, z0 = 68.02144, a = 0.08023, b = −0.1238, c = 0.00016522, d = −0.000251957, and f = −0.00026984. When R^2 = 0.9979, this indicates that the fitting effect is good.

$$P_{PM\_Loss} = z0 + ax + by + cx^2 + dy^2 + fxy \tag{13}$$

In the equation, x represents the current value, y represents the temperature, z0 = 7.14675, a = −0.06298, b = −0.0408, c = 0.00465, d = −0.0000818016, and f = 0.000481598. When R^2 = 0.9983, this indicates that the fitting effect is good.

Due to the inaccuracy of the Maxwell solution for winding copper consumption and the fact that the motor is a low-speed motor, only the DC winding loss is considered. The equation for copper winding consumption (14) can be derived from Equations (8)–(10):

$$P_{cu} = 3 \times I^2 \times R \times [1 + 0.0044 \times (T - 20)] \tag{14}$$

In the Equation (14), R is the phase resistance. According to the structure of the winding, the resistance value of the winding is calculated as 0.03303 Ohm.

3.4. Regression Prediction Model of Temperature Field

3.4.1. Temperature Distribution Rule

Due to the different thermal conductivities of the materials within the motor, the temperature distribution within the motor is not uniform. In this paper, Fluent software is used to calculate the temperature of the motor, and the thermal conductivity of the materials used in this paper for each part of the motor is shown in Table 6 [20–22]. Figure 8 depicts the results of the EMTBC numerical calculation of the PMSM used to determine the temperature distribution field inside the motor.

Table 6. Physical properties of the motor material.

Motor Components	Density (Kg/m^3)	Thermal Conductivity (W/(m·K))	Specific Heat Capacity (J/(Kg·K))
Equivalent winding	8396	378	368
Equivalent insulation layer	804	0.15	885.5
PM	7700	8.95	465
Equivalent silicon steel sheet	7440	x–38.01 y–38.01 z–3.65	518
Shaft	7900	43	440
Equivalent air gap	1.225	0.024	1006
Shell	2719	202.4	871

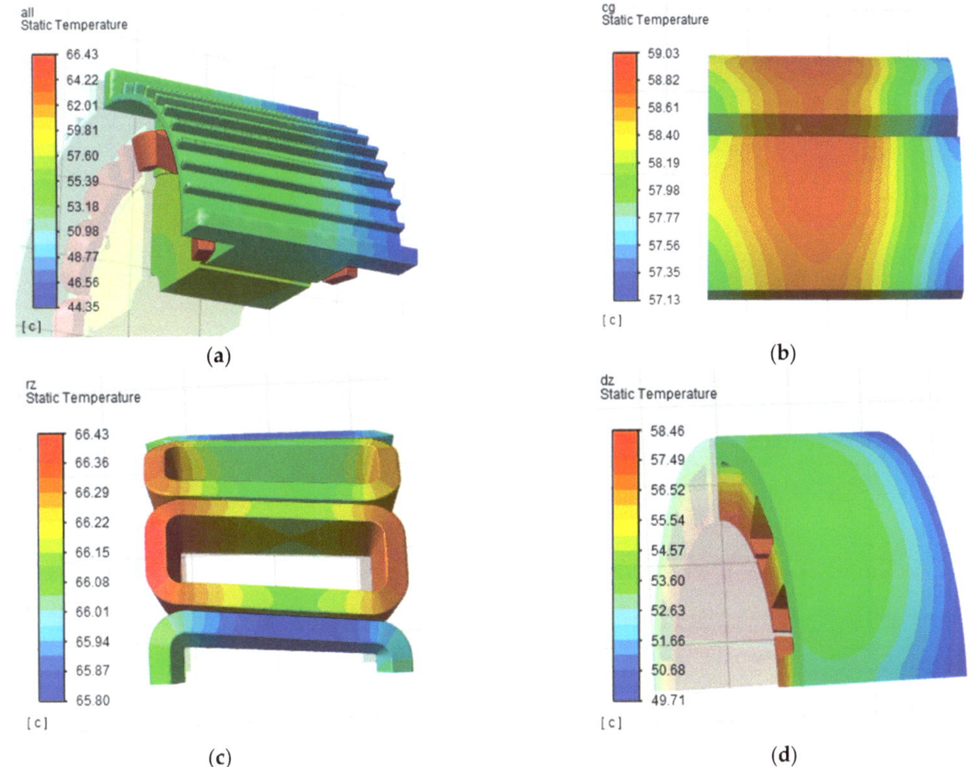

Figure 8. The temperature distribution diagram of each part of the motor: (**a**) the temperature distribution diagram of the whole motor; (**b**) the temperature distribution diagram of the PM; (**c**) the temperature distribution diagram of the winding; (**d**) the temperature distribution diagram of the stator.

To derive the specific temperature distribution rule within the motor, the temperatures of various components in Figure 8 were extracted, and the temperature gradients of the various components within the motor are depicted in Figure 9a. Because the high temperature of the permanent magnet (PM) causes demagnetization and the high insulation temperature of the winding causes a short-circuit defect, the temperature rise distribution of the PM and winding should be analyzed within a suitable range, as depicted in Figure 9b–d

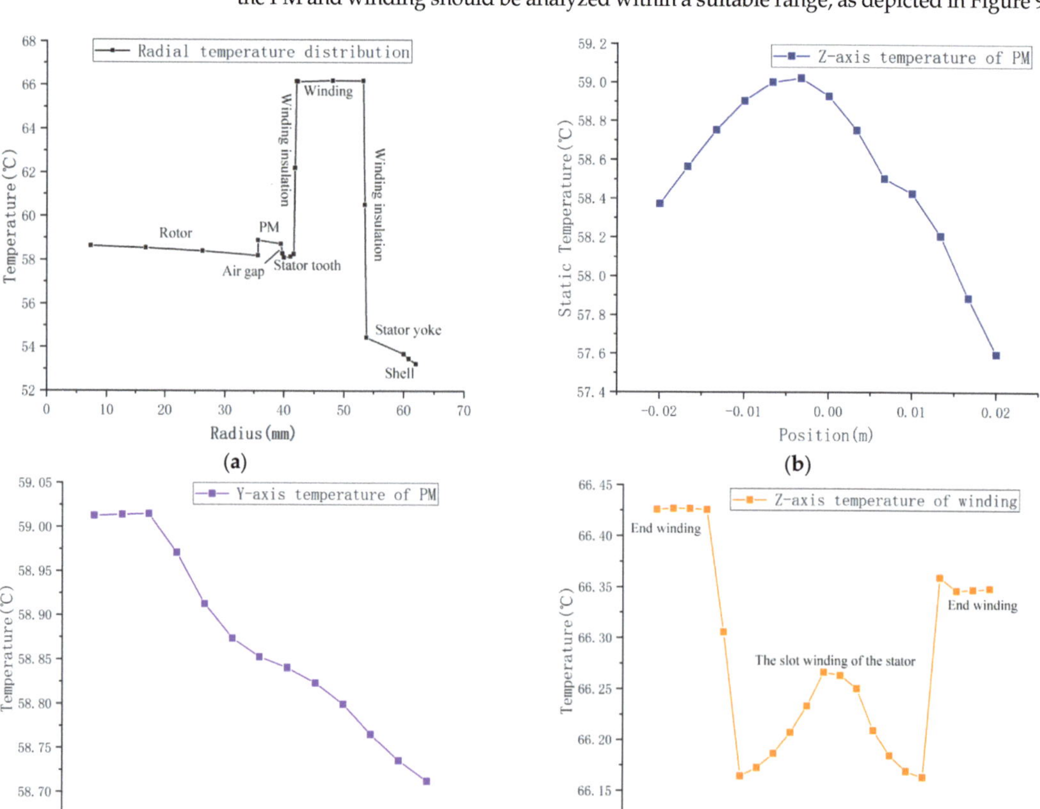

Figure 9. Temperature gradients of various parts inside the motor; (**a**) radial temperature distribution of the motor; (**b**) axial temperature distribution of the PM; (**c**) lateral temperature distribution diagram of PM; (**d**) winding axial temperature distribution.

Figure 9a depicts the temperature gradient changes of the various motor components (where the orientation of the specific coordinate system is shown in Figure 2b). It indicates that the winding temperature is the highest, and the motor case temperature is the lowest. There are two protrusions: the temperature distribution of the PM and the temperature distribution of the winding. Since both the PM and the winding are heat sources, their temperatures are slightly higher than those of the surrounding contact portions. Figure 9b depicts the PM's axial temperature distribution diagram. The PM temperature rise is higher in the middle and lower at both extremities, with the temperature on the right being lower than the temperature on the left. The following are the factors for this change:

1.   The axial middle section of the PM has poor heat dissipation;
2.   The motor is an axial air-cooled structure, and because the right side of the PM is near to the fan, its temperature is lower than that of the left side; Figure 9c is a diagram of the PM's transverse temperature distribution.

The temperature near the middle border of the PM is the highest. The highest temperature of the PM is located in the axial middle of the PM and marginally near the edge of the middle of the two PMs, whereas the lowest temperature is located near the fan's axial end. Figure 9d depicts the distribution of the winding's axial temperature rise. Due to inadequate heat dissipation at the end of the winding, the temperature is the highest at both ends of the winding, and the temperature is the lowest near the fan and connected to the end of the winding.

3.4.2. Establishment of Prediction of Regression Temperature Field Model

In the process of solving the internal temperature field of the PMSM using FEA, on the one hand, a large number of grids are required to simulate the real heat transfer process inside the PMSM; on the other hand, to account for the influence of temperature on the properties of EM materials, the temperature field and the EM field must constantly reverse iterate in order to achieve a certain level of calculation accuracy. Therefore, the traditional EMTBC method for solving temperature fields requires a great deal of calculation effort. To address this issue, we use the EM field loss data as the input to the temperature field and the highest and lowest winding and PM temperatures as the output. By constructing a high-precision proxy model to fit the mapping relationship between the input and output, it is possible to obtain a high-precision regression prediction model without an excessive amount of data samples, which significantly improves the computational efficiency of the temperature field.

To establish a high-precision regression prediction model, the sample point selection must satisfy the following criteria:

1. The iron loss, eddy current loss of the PM loss, and copper loss input data have a certain correlation and cannot be combined at random.
2. The PMSM load torque and operating temperature are two factors that influence the EM loss; consequently, the operating point should be evenly selected within the torque and temperature range of the motor.
3. Since the solution of the temperature field is time-consuming, a regression prediction model with a higher accuracy should be obtained with fewer sample points than those of a traditional FEM.

In conclusion, based on the numerical calculation data of the EM field in Table 5, we selected 20 °C, 40 °C, 60 °C, 70 °C, 80 °C, 90 °C, 100 °C, and 110 °C and the rated EM torque, 1.25 times EM torque, 1.5 times EM torque, 1.75 times EM torque, and 2 times EM torque, and we conducted a numerical calculation of the temperature field at a total of 24 operating points.

This paper compares five common regression prediction models, which are as follows: (1) BP (Back Propagation), (2) SVR (support vector regression), (3) CNN (Convolutional Neural Network), (4) RF (Random Forest), and (5) LSTM (Long Short-Term Memory). First, the five regression prediction algorithms were used to build a proxy model for the input variable (loss) and the output target (temperature). Then, the accuracy of the proxy model was compared by calculating the evaluation index of the regression model. Finally, the one with the best evaluation index was selected as the regression prediction model in this paper. Commonly used regression model evaluation indexes include R^2 (the coefficient of determination), among which R^2 is the evaluation index that can best reflect the degree of fitting, and the closer the coefficient is to 1, the better the fitting degree. The calculation of this regression evaluation index can be expressed as follows [23]:

$$R^2 = 1 - \frac{\sum_{i=1}^{n} \left(y_i - \hat{y}_i\right)^2}{\sum_{i=1}^{n} \left(y_i - \overline{y}_i\right)^2} \tag{15}$$

where n is the number of samples of the test set, y_i is the actual value of the test set sample, $\hat{y}_i$ is the predicted value of the test set sample, and $\overline{y}$ is the average value of the actual value of the test set sample. The specific evaluation indicators of each regression model after training are shown in Table 7.

Table 7. Each agent model evaluation index.

Regression Prediction Method	Evaluating Indicator	Output Result			
		Tmax-Winding (°C)	Tmin-Winding (°C)	Tmax-PM (°C)	Tmin-PM (°C)
BP	R^2	**0.996**	**0.996**	**0.994**	**0.996**
SVR	R^2	0.989	0.992	0.965	0.992
CNN	R^2	0.944	0.954	0.960	0.922
RF	R^2	0.731	0.762	0.727	0.823
LSTM	R^2	0.996	0.986	0.998	0.991

In Table 6, it can be seen that both the BP neural network and LSTM models have a high level of prediction accuracy for the four output targets, but the BP neural network is superior. For each of the four output targets, the model developed by RF has a weak predictive accuracy. For all four output targets, the SVR- and CNN-constructed models provide more precise predictions. In conclusion, we propose the use of a BP neural network to build a regression prediction model.

Since the initial weight and threshold of a BP neural network are arbitrary, the network's output is unstable [24]. If the initial weight and threshold are inadequate, the network will reach a local optimal state, resulting in a subpar prediction effect. To enhance the stability of the BP neural network's prediction effect, we use a genetic algorithm to determine the optimal weight and threshold. Figure 10 depicts the iterative convergence process, while Table 6 depicts the final model effect.

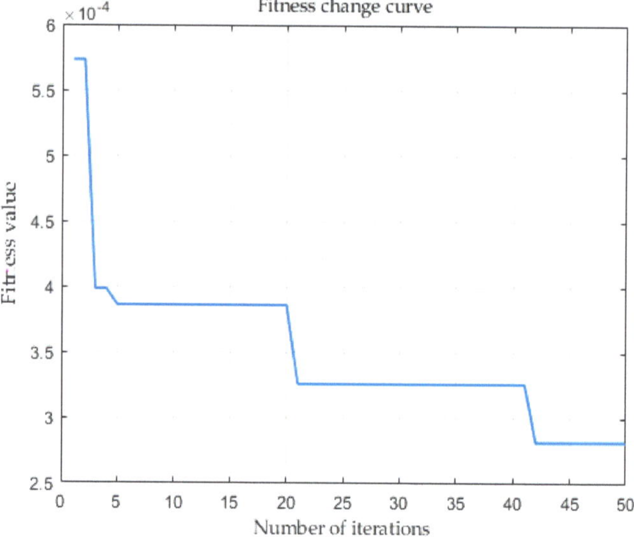

Figure 10. Iterative process of genetic algorithm.

Table 8 demonstrates that the genetic algorithm is used to determine the optimal weight and bias for BP, and that the enhanced GA-BP algorithm is more accurate and stable. Therefore, we propose the GA-BP model as the temperature field regression prediction model.

Table 8. Model comparison before and after optimization.

Regression Prediction Method	Evaluating Indicator	Output Result			
		Tmax-Winding (°C)	Tmin-Winding (°C)	Tmax-PM (°C)	Tmin-PM (°C)
BP	R^2	0.996	0.996	0.994	0.996
GA-BP	R^2	0.999	0.999	0.999	0.999

3.5. Bidirectional Coupling Solution Strategy Based on Reduced-Order Model

In Sections 3.3 and 3.4, we establish the reduced-order solving models for the EM field and temperature, respectively. To simulate the iterative process of the EMTBC while taking into account the effect of the temperature field on EM materials, two reduced-order models are coupled and iterated to simulate the solution process of the EMTBC. The specific procedure is reported in the flow-chart in Figure 11.

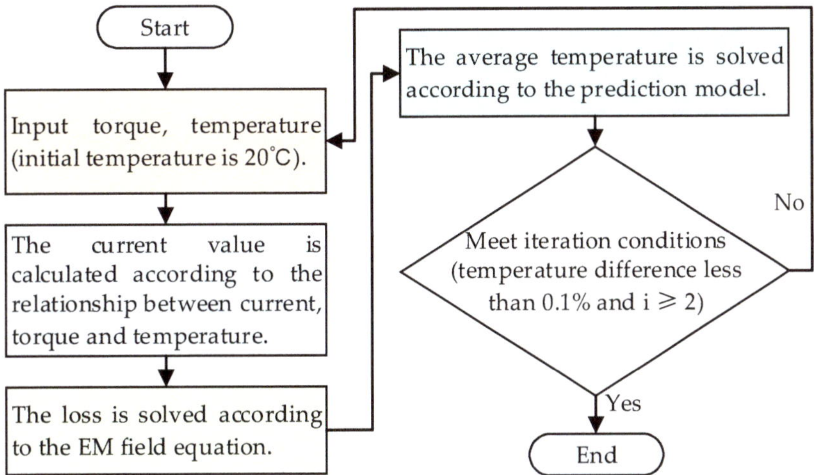

Figure 11. Bidirectional coupling solution strategy based on reduced-order model.

This section is pre-programmed with two reduced-order models, so we only need to provide the initial load torque, and the model will solve the temperature distribution of the winding and PM automatically.

4. Results and Analysis

To verify the correctness of the efficient reduced-order EMTBC solution model and solution strategy proposed in this paper, the traditional EMTBC numerical temperature rise calculation, the traditional unidirectional EM thermal numerical temperature rise calculation, and the reduced-order EMTBC model temperature rise calculation were performed, and the three solution results were compared and analyzed as shown in Table 9.

Compared to the unidirectional EM-thermal numerical temperature rise calculation, the maximum temperature rise of the winding in the EMTBC numerical temperature rise calculation increases by 5.56 percent; the minimum temperature rise of the winding increases by 5.58 percent; and the maximum and minimum temperature rises of the PM increase by 0.21 percent and 0.27 percent, respectively. This also demonstrates that the EMTBC numerical temperature rise calculation is more precise. In the process of the EMTBC iteration, however, the solution time also increases significantly; the solution time is approximately 3 h, which is 2.9 times longer than the calculation time for the unidirectional EM-thermal numerical temperature rise. Moreover, by comparing the numerical temperature rise calculation results and solution time of our proposed reduced-order EMTBC model and the traditional EMTBC, it can be seen that there is a small difference in the numerical temperature rise calculation results between these two strategies, but the solving time is reduced by 159 times. Our proposed EMTBC dimensionality reduction model is therefore precise and effective.

Table 9. Comparison of three calculation methods of temperature rise.

	Tmax-Winding (°C)	Tmin-Winding (°C)	Tmax-PM (°C)	Tmin-PM (°C)	Solution Time (Minute)
Traditional EMTBC numerical temperature rise calculation	61.85	61.27	56.17	54.75	160
Traditional unidirectional EM-thermal numer-ical temperature rise calculation	58.40	57.85	56.05	54.60	41
Difference (%)	5.56	5.58	0.21	0.27	
Reduced-order EMTBC numerical temperature rise calculation	**61.90**	**61.30**	**56.25**	**54.77**	**1**
Difference (%)	0.081	0.049	0.14	0.036	

5. Conclusions

This paper analyzes the effect of temperature on EM materials and concludes that temperature can reduce the EM performance of PMs in conventional electromagnetic FEA, resulting in a reduction in the EM torque. To keep the EM torque constant, the EMTBC injects an additional control current to enhance the precision of the results. However, the adopted EMTBC's numerical calculation is time-consuming. To increase the computational efficiency, we propose a reduced-order EMTBC. The proposed model has been validated numerically, demonstrating high precision and efficiency. Nonetheless, the model proposed in this paper is intended to compute the temperature distribution under steady-state conditions. In future work, we will investigate and analyze the rise in temperature of the temperature field in real time under conditions of flux variation.

Author Contributions: Conceptualization, Y.Y. and P.Z.; methodology, Y.Y. and P.Z.; software, P.Z.; writing—original draft, Y.Y. and P.Z. writing—review and editing, Y.Y., H.G., G.C., S.N., J.D., S.S. and Z.Z. All authors have read and agreed to the published version of the manuscript.

Funding: This research was supported in part by the National Natural Science Foundation of China (Funder: (Yinquan Yu), Grant No. 52067006; Funder: (Shengrong Shu), Grant No. 52165058). We would like to express our sincere thanks for the funding.

Institutional Review Board Statement: Not applicable.

Informed Consent Statement: Not applicable.

Data Availability Statement: Not applicable.

Conflicts of Interest: The authors declare no conflict of interest.

References

1. Zheng, J.Q.; Zhao, W.X.; Ji, J.H.; Li, H.T. Review on Design Methods of Low Harmonics of Fractional-slot Concentrated-windings Permanent-magnet Machine. *Proc. CSEE* **2020**, *40*, 272–280.
2. Cai, X.; Cheng, M.; Zhu, S.; Zhang, J. Thermal modeling of flux-switching permanent-magnet machines considering anisotropic conductivity and thermal contact resistance. *IEEE Trans. Ind. Electron.* **2016**, *63*, 3355–3365. [CrossRef]
3. Wang, Y.F.; Zhao, J.; Huang, J.B.; Wang, F. Impedence Parameters Analysis of Permanent Magnet Synchronous Motor under Inter-Turn Short Circuit Fault. *Electr. Mach. Control Appl.* **2017**, *44*, 105–109.
4. Ding, S.Y.; Jiang, X.; Zhu, M.; Liu, W. Starting and steady temperature rise investigation for permanent magnet synchronous motor based on lumped-parameter thermal-network. *Electr. Mach. Control* **2020**, *24*, 143–150.
5. Mo, L.; Zhu, X.; Zhang, T.; Quan, L.; Wang, Y.; Huang, J. Temperature Rise Calculation of a Flux-Switching Permanent-Magnet Double-Rotor Machine Using Electromagnetic-Thermal Coupling Analysis. *IEEE Trans. Magn.* **2018**, *54*, 1–4. [CrossRef]

6.	Yu, W.; Hua, W.; Qi, J.; Zhang, H.; Zhang, G.; Xiao, H.; Xu, S.; Ma, G. Coupled Magnetic Field-Thermal Network Analysis of Modular-Spoke-Type Permanent-Magnet Machine for Electric Motorcycle. *IEEE Trans. Energy Convers.* **2020**, *99*, 120–130. [CrossRef]
7.	Wang, Z.P.; Wang, M.; Ding, S. Analysis on Correlated Sensitivity Factors of Temperature Rise for Surface Permanent Magnet Synchronous Motors. *Proc. CSEE* **2014**, *34*, 5888–5894.
8.	Alberti, L.; Bianchi, N. A Coupled Thermal–Electromagnetic Analysis for a Rapid and Accurate Prediction of IM Performance. *IEEE Trans. Ind. Electron.* **2008**, *55*, 3575–3582. [CrossRef]
9.	Ding, S.Y.; Shen, S.F.; Yang, Z.; Chen, S.X.; Dai, Y. Fluid-solid coupling simulation and performance analysis of high-speed permanent magnet synchronous motor. *Electr. Mach. Control* **2021**, *25*, 112–121.
10.	Zhang, G.; Hua, W.; Cheng, M.; Zhang, B.; Guo, X. Coupled Magnetic-Thermal Fields Analysis of Water-Cooling Flux-Switching Permanent Magnet Motors by an Axially Segmented Model. *IEEE Trans. Magn.* **2017**, *53*, 1–4. [CrossRef]
11.	Li, Y.; Zhang, C.; Xu, X. Bidirectional electromagnetic–thermal coupling analysis for permanent magnet traction motors under complex operating conditions. *Trans. Can. Soc. Mech. Eng.* **2022**, *46*, 541–560. [CrossRef]
12.	Jia, M.; Hu, J.; Xiao, F.; Yang, Y.; Deng, C. Modeling and Analysis of Electromagnetic Field and Temperature Field of Permanent-Magnet Synchronous Motor for Automobiles. *Electronics* **2021**, *10*, 2173. [CrossRef]
13.	Zhao, P.; Yu, Y. Analysis of Influence of Permanent Magnet Demagnetization on Motor Temperature Rise. *J. Phys. Conf. Ser.* **2022**, *2395*, 012022. [CrossRef]
14.	Tang, R.Y. *Modern Permanent Magnet Machines Theory and Design*; China Machine Press: Beijing, China, 2015.
15.	Qi, G. Research on Dual Three-Phase Permanent Magnet Brushless AC Motors Accounting for Cross-Coupling Magnetic Saturation. Ph.D. Thesis, Huazhong University of Science and Technology, Wuhan, China, 2010.
16.	Zhong, Q.; Ma, H.Z.; Liang, W.M.; Chen, C. Preliminary Analysis of Demagnetization Fault Mathmatical Model for Permanent Magnet Synchronous Motor of Electric Vehicles. *Micromotors* **2013**, *46*, 9–12+18.
17.	Zhang, Y.; McLoone, S.; Cao, W. Electromagnetic loss modeling and demagnetization analysis for high-speed permanent magnet machine. *IEEE Trans. Magn.* **2017**, *54*, 1–5. [CrossRef]
18.	Liu, M.Y. Research on Iron Loss Model of High-Speed Permanent Magnet Motor Considering Multiphysics Factors. Master's Thesis, Shenyang University of Technology, Shenyang, China, 2019.
19.	Kuczmann, M.; Orosz, T. Temperature-Dependent Ferromagnetic Loss Approximation of an Induction Machine Stator Core Material Based on Laboratory Test Measurements. *Energies* **2023**, *16*, 1116. [CrossRef]
20.	Shao, H. Research on Vibration and Temperature Rise of In-Wheel Motor For Electric Vehicle. Master's Thesis, East China Jiaotong University, Nanchang, China, 2022.
21.	Yao, H.Y. Simulation Study on Temperature Rise and Cooling Characteristics of Electric Vehicle In-Wheel Motor. Master's Thesis, Jilin University, Changchun, China, 2019.
22.	Wu, Y.H.; Liu, X.M. Research on Thermal Conductivity of Small Motor Windings. *Micromotors* **2015**, *48*, 27–29+37.
23.	Yu, Y.; Pan, Y.; Chen, Q.; Hu, Y.; Gao, J.; Zhao, Z.; Niu, S.; Zhou, S. Multi-Objective Optimization Strategy for Permanent Magnet Synchronous Motor Based on Combined Surrogate Model and Optimization Algorithm. *Energies* **2023**, *16*, 1630. [CrossRef]
24.	Li, W.H.; Niu, G.B.; Liu, Y.J. Hydraulic System Fault Diagnosis Based on Genetic Algorithm Optimized BP Neural Network. *Mach. Tool Hydraul.* **2023**, *51*, 159–164.

Article

Parameter Identification of Permanent Magnet Synchronous Motor with Dynamic Forgetting Factor Based on H∞ Filtering Algorithm

Tianqing Yuan [1], Jiu Chang [2,*] and Yupeng Zhang [3]

[1]   Key Laboratory of Modern Power System Simulation and Control & Renewable Energy Technology, Ministry of Education, Northeast Electric Power University, Jilin 132012, China; 20192929@neepu.edu.cn
[2]   Department of Electrical Engineering, Northeast Electric Power University, Jilin 132012, China
[3]   School of Information Science and Engineering, Northeastern University, Shenyang 110819, China; 2310263@stu.neu.edu.cn
*   Correspondence: 2202100007@neepu.edu.cn; Tel.: +86-157-6554-5764

Abstract: To address system parameter changes during permanent magnet synchronous motor (PMSM) operation, an H∞ filtering algorithm with a dynamic forgetting factor is proposed for online identification of motor resistance and inductance. First, a standard linear discrete PMSM parameter identification model is established; then, the discrete H∞ filtering algorithm is derived using game theory reducing state and measurement noise influence. A cost function is defined, solving extremes values of different terms. A dynamic forgetting factor is introduced to the weighted combination of initial and current measurement noise covariance matrices, eliminating identification issues from different initial values. On this basis, a dynamic forgetting factor is added to weigh the combination of the initial measurement noise covariance matrix and the current measurement noise covariance matrix, which eliminates the influence of the discrimination error caused by the different initial values. Finally, the identification model is built in MATLAB/Simulink for simulation analysis to verify the feasibility of the proposed algorithm. The simulation results show the proposed H∞ filtering algorithm rapidly and accurately identifies resistance and inductance values with significantly improved robustness. The forgetting factor enables quick stable recognition even with poor initial values, enhancing PMSM control performance.

Keywords: PMSM; H∞ filtering algorithm; parameter identification analysis; dynamic forgetting factor

Citation: Yuan, T.; Chang, J.; Zhang, Y. Parameter Identification of Permanent Magnet Synchronous Motor with Dynamic Forgetting Factor Based on H∞ Filtering Algorithm. *Actuators* **2023**, *12*, 453. https://doi.org/10.3390/act12120453

Academic Editor: Dong Jiang

Received: 27 October 2023
Revised: 1 December 2023
Accepted: 6 December 2023
Published: 7 December 2023

1. Introduction

The structure of a permanent magnet synchronous motor and other factors can result in differences in motor parameters under different working conditions. This can reduce the accuracy of the overall control system and potentially impact system stability [1]. Accurate identification of motor parameters is therefore crucial.

Compared to offline identification methods, ideal online methods can accurately estimate motor parameters in real time, enhancing system control performance [2–4]. Current online parameter identification methods for permanent magnet synchronous motors primarily utilize least squares [5–7], model reference adaptive [8–10], and Kalman filter algorithms [11–13]. Uddin et al. [14] realizes online alternating axis inductance by identification via a model-referenced adaptive algorithm, assuming known, constant stator resistance and permanent magnet magnetic chain. Gao et al. [15] proposed a model-referenced adaptive system based on disturbance compensation, designed a real-time disturbance estimator, and updated the adaptive rate according to the disturbance. This reduces system uncertainty and disturbance effects and increases model-referenced adaptive algorithm application scenarios. Based on the permanent magnet synchronous motor mechanical and electromagnetic models, Tang et al. [16] designed an adaptive rate improvement model reference adaptive algorithm. This estimates rotor position and load torque while designing a load torque feedforward compensator controller for quick load response. Kalman filtering for linear systems can be updated in real-time to optimally estimate parameters but often fails with unknown noise and large modeling errors. The motor system is nonlinear, so many scholars improved Kalman filtering, such as via extended Kalman filtering [17,18], traceless Kalman filtering [19], unscented Kalman filtering [20], etc. Kalman filtering assumes Gaussian distributed measurement noise, but system noise statistics are often unknown or time-varying in reality. The adaptive Kalman filtering algorithm in reference [21] addresses this to some extent by selecting the appropriate covariance distribution to estimate the covariance matrix, which does not vary significantly. However, it is limited by the linear Gaussian state model. Researchers have achieved many achievements in engine parameter identification, but existing methods often only consider Gaussian noise, while practice engines are often disturbed by non-Gaussian noise and other factors. For Gaussian noise with unknown covariance, the traditional H∞ filtering algorithm is used. Chen et al. [22] proposes an adaptive H∞ filtering algorithm for parameter identification, based on the traditional extended H∞ filtering algorithm, and investigates the online identification of motor stator inductance and resistance. However, the covariance of the traditional H∞ filtering algorithm is set by humans, which affects the accuracy of the H∞ filtering algorithm. As inductance parameters are influenced by the motor's operating state, Liu et al. [23] establishes a motor model based on the motor's transient voltage equation and introduces a forgetting factor to improve the least squares identification method, successfully enhancing algorithm tracking performance. However, setting the forgetting factor to a specific value makes ensuring the least squares method's robustness during identification difficult. Fang et al. [24] takes the error between theoretical and actual output as a variable, dynamically adjusting the forgetting factor, to accelerate algorithm convergence and ensure robustness.

Therefore, this paper proposes a discrete H∞ filtering algorithm containing dynamic forgetting factor based on minimizing maximum estimation error requiring no assumptions about system or observation noise characteristics. The remaining paper is organized as follows: First, a PMSM parameter identification model is established from the PMSM's d-q mathematical model. Second, the cost function is defined per H∞ filtering, with the extreme point solved. Then, a dynamic forgetting factor is introduced to reduce abnormal initial value influence on the algorithm. Finally, this H∞ filtering with dynamic forgetting factor identifies PMSM parameters, verifying effectiveness via simulation.

2. Modeling of PMSM Parameter Identification

To facilitate the study, the mathematical model under the synchronous rotation coordinate system d-q of the permanent magnet synchronous motor is usually selected, so that the stator voltage equation can be expressed as

$$
\begin{cases}
u_\mathrm{d} = R_\mathrm{s} i_\mathrm{d} + \frac{d\psi_\mathrm{d}}{dt} - \omega_\mathrm{e}\psi_\mathrm{q} \\
u_\mathrm{q} = R_\mathrm{s} i_\mathrm{q} + \frac{d\psi_\mathrm{q}}{dt} + \omega_\mathrm{e}\psi_\mathrm{d}
\end{cases}.
\tag{1}
$$

The equation for the stator chain can be expressed as follows:

$$
\begin{cases}
\psi_\mathrm{d} = L_\mathrm{d} i_\mathrm{d} + \psi_\mathrm{f} \\
\psi_\mathrm{q} = L_\mathrm{q} i_\mathrm{q}
\end{cases}.
\tag{2}
$$

Substituting Equation (2) into Equation (1) gives the following Equation:

$$
\begin{cases}
u_\mathrm{d} = R_\mathrm{s} i_\mathrm{d} + L_\mathrm{d}\frac{d}{dt} i_\mathrm{d} - \omega_\mathrm{e} L_\mathrm{q} i_\mathrm{q} \\
u_\mathrm{q} = R_\mathrm{s} i_\mathrm{q} + L_\mathrm{q}\frac{d}{dt} i_\mathrm{q} + \omega_\mathrm{e}(L_\mathrm{d} i_\mathrm{d} + \psi_\mathrm{f})
\end{cases},
\tag{3}
$$

where u_d and u_q are the voltages of the d and q axes, respectively; i_d and i_q are the currents of the d and q axes, respectively; R_s is the stator resistance; Ψ_d and Ψ_q are the stator chain components, respectively; ω_e is the electric angular velocity of the rotor; L_d and L_q are the inductances of the d and q axes, respectively; and Ψ_f is the magnetic chain of the permanent magnet.

The state space equations for the permanent magnet synchronous motor are established by selecting the i_d and i_q of the current in the d and q axes and identifying the parameters L_d and L_q, and R_s as state variables. This paper focuses on the surface-mounted motor, which satisfies $L_d = L_q = L_s$. Thus, Equation (3) can be rearranged as follows:

$$\frac{d}{dt}\begin{bmatrix} i_d \\ i_q \\ R_s \\ L_s \end{bmatrix} = \begin{bmatrix} -\frac{R_s}{L_s} & \omega_e & 0 & 0 \\ -\omega_e & -\frac{R_s}{L_s} & 0 & 0 \\ 0 & 0 & 0 & 0 \\ 0 & 0 & 0 & 0 \end{bmatrix} \begin{bmatrix} i_d \\ i_q \\ R_s \\ L_s \end{bmatrix} + \begin{bmatrix} \frac{1}{L_s} & 0 \\ 0 & \frac{1}{L_s} \\ 0 & 0 \\ 0 & 0 \end{bmatrix} \begin{bmatrix} u_d \\ u_q - \omega_e \psi_f \end{bmatrix}. \tag{4}$$

Since the coefficient matrix of Equation (4) contains coupling terms, the direct identification of R_s and L_s becomes more complex. Therefore, intermediate variables a and b are introduced to simplify the identification equations. Let $a = R_s/L_s$, $b = 1/L_s$. Equation (4) can be rearranged as follows:

$$\frac{d}{dt}\begin{bmatrix} i_d \\ i_q \\ a \\ b \end{bmatrix} = \begin{bmatrix} 0 & \omega_e & -i_d & u_d \\ -\omega_e & 0 & -i_q & u_q - \omega_e \psi_f \\ 0 & 0 & 0 & 0 \\ 0 & 0 & 0 & 0 \end{bmatrix} \begin{bmatrix} i_d \\ i_q \\ a \\ b \end{bmatrix} + w. \tag{5}$$

The output y can be expressed as

$$y = \begin{bmatrix} 1 & 0 & 0 & 0 \\ 0 & 1 & 0 & 0 \end{bmatrix} \begin{bmatrix} i_d \\ i_q \\ a \\ b \end{bmatrix} + v, \tag{6}$$

where w represents the process noise of the system and v represents the measurement noise of the system. The state variable matrix of the system is denoted by $x = [i_d\ i_q\ a\ b]^T$, and the output variable matrix is denoted by $y = [i_d\ i_q]^T$. By discretizing Equations (5) and (6) using the sampling period T_s, the standard linear discrete system form presented below is obtained as

$$\begin{cases} x_{k+1} = F_k x_k + w_k \\ y_k = H_k x_k + v_k \end{cases}, \tag{7}$$

where w_k and v_k are noise terms that are random and of unknown statistical properties; F_k and H_k represent the coefficient matrices as follows:

$$\begin{cases} F_k = \begin{bmatrix} 1 & \omega_e T_s & -i_d T_s & u_d T_s \\ -\omega_e T_s & 1 & -i_q T_s & (u_q - \omega_e \psi_f) T_s \\ 0 & 0 & 1 & 0 \\ 0 & 0 & 0 & 1 \end{bmatrix}. \\ \\ H_k = \begin{bmatrix} 1 & 0 & 0 & 0 \\ 0 & 1 & 0 & 0 \end{bmatrix} \end{cases} \tag{8}$$

Thus, the algorithm in this study uses the H∞ filtering algorithm to detect parameters a and b using the above model. Next, the values of resistance R_s and inductance L_s are obtained by relating a and b and R_s and L_s.

3. H∞ Filtering Algorithm Based on Game Theory

The estimation of x_k is denoted as $\hat{x}_k$, and the estimation of the initial state is denoted as $\hat{x}_0$. In the game-theoretic approach to derive $H\infty$ filtering and estimate x_k, including $N - 1$ moments and $N - 1$ moments before the measurement condition, the cost function [22] must be defined as shown in Equation (9):

$$J_1 = \frac{\sum\limits_{k=0}^{N-1} \|x_k - \hat{x}_k\|^2_{S_k}}{\|x_0 - \hat{x}_0\|^2_{P_0^{-1}} + \sum\limits_{k=0}^{N-1}\left(\|w_k\|^2_{Q_k^{-1}} + \|v_k\|^2_{R_k^{-1}}\right)}. \tag{9}$$

The presence of disturbances, such as natural noise, produces w_k, v_k, and x_0, that maximizes J_1. Thus, the cost function places w_k, v_k and x_0 in the denominator. To minimize J_1, we must estimate x_k in the cost function and find the appropriate solution. Equation (9) employs symmetric positive definite matrices P_0, Q_k, R_k and S_k, chosen based on the specific problem.

Minimizing J_1 directly is challenging; thus, we choose a performance limit that ensures the J_1 cost function meets the following condition:

$$J_1 < \frac{1}{\theta}. \tag{10}$$

θ is the performance boundary. We set the following:

$$J = J_1 - \frac{1}{\theta} = \frac{-1}{\theta}\|x_0 - \hat{x}_0\|^2_{P_0^{-1}} + \sum_{k=0}^{N-1}\left[\|x_k - \hat{x}_k\|^2_{S_k} - \frac{1}{\theta}\left(\|w_k\|^2_{Q_k^{-1}} + \|v_k\|^2_{R_k^{-1}}\right)\right] < 0. \tag{11}$$

Therefore, from Equation (11), it can be seen that J can be minimized by choosing an appropriate w_k, v_k, and x_0, while w_k, v_k, and x_0 generated by the noise effect can maximize J. The noise effect can be expressed by substituting v_k into Equation (11). From $y_k = H_k x_k + v_k$ in Equation (7), it can be seen that $v_k = y_k - H_k x_k$, and substituting v_k into Equation (11) can be expressed as

$$\begin{aligned} J &= -\tfrac{1}{\theta}\|x_0 - \hat{x}_0\|^2_{P_0^{-1}} + \sum_{k=0}^{N-1}\left[\|x_k - \hat{x}_k\|^2_{S_k} - \tfrac{1}{\theta}\left(\|w_k\|^2_{Q_k^{-1}} + \|y_k - H_k x_k\|^2_{R_k^{-1}}\right)\right] \\ &= \psi(x_0) + \sum_{k=0}^{N-1} L_k \end{aligned} \tag{12}$$

where $\Psi(x_0)$ and L_k can be expressed by Equation (13), and to solve the extreme-problem that exists in Equation (12), the extreme points of J with respect to w_k and x_0 can be found first, and then the extreme points of J with respect to $\hat{x}_k$ and y_k.

$$\begin{cases} \psi(x_0) = -\tfrac{1}{\theta}\|x_0 - \hat{x}_0\|^2_{P_0^{-1}} \\[2mm] L_k = \|x_k - \hat{x}_k\|^2_{S_k} - \tfrac{1}{\theta}(\|w_k\|^2_{Q_k^{-1}} + \|y_k - H_k x_k\|^2_{R_k^{-1}}) \end{cases} \tag{13}$$

3.1. Extreme Solutions for w_k and x_0

To obtain the maximum value with respect to J, we define the Hamiltonian function as follows:

$$H = L_k + \frac{2\lambda_{k+1}^{\mathrm{T}}}{\theta}(F_k x_k + w_k). \tag{14}$$

The $2\lambda^{\mathrm{T}}_{k+1}/\theta$ is the time-varying Lagrange multiplier to be computed ($k = 0,\ldots, N-1$). It is clear from the theory of dynamic constrained optimization that we can solve the constrained optimization problem of J with respect to w_k and x_0 via the following Equation:

$$\begin{cases} \dfrac{2\lambda_0^{\mathrm{T}}}{\theta} + \dfrac{\partial \psi_0}{\partial x_0} = 0 \\[2mm] \dfrac{2\lambda_N^{\mathrm{T}}}{\theta} = 0 \\[2mm] \dfrac{\partial H}{\partial w_k} = 0 \\[2mm] \dfrac{2\lambda_k^{\mathrm{T}}}{\theta} = \dfrac{\partial H}{\partial x_k} \end{cases} \tag{15}$$

Simplified Equation (15):

$$\begin{cases} \dfrac{2\lambda_0}{\theta} - \dfrac{2}{\theta} P_0^{-1}(x_0 - \hat{x}_0) = 0 \\[2mm] x_0 = \hat{x}_0 + P_0\lambda_0 \\[2mm] \lambda_N = 0 \\[2mm] w_k = Q_k\lambda_{k+1} \\[2mm] \dfrac{2\lambda_k}{\theta} = 2S_k(x_k - \hat{x}_k) + \dfrac{2}{\theta} H_k^{\mathrm{T}} R_k^{-1}(y_k - H_k x_k) + \dfrac{2}{\theta} F_k^{\mathrm{T}}\lambda_{k+1} \\[2mm] \lambda_k = F_k^{\mathrm{T}}\lambda_{k+1} + \theta S_k(x_k - \hat{x}_k) + H_k^{\mathrm{T}} R_k^{-1}(y_k - H_k x_k) \end{cases} \tag{16}$$

This can be obtained by substituting $w_k = Q_k\lambda_{k+1}$ in Equation (16) into Equation (7):

$$x_{k+1} = F_k x_k + Q_k\lambda_{k+1}. \tag{17}$$

From Equation (16), we obtain $x_0 = \hat{x}_0 + P_0\lambda_0$, so we can set that

$$x_k = \mu_k + P_k\lambda_k. \tag{18}$$

Equation (18) holds for all k. μ_k and P_k are functions to be determined, P_0 is given, and the initial value $\mu_0 = \hat{x}_0$. Assume that x_k is an affine function of λ_k, if the final result is correct, our assumption is correct. Substituting Equation (18) into Equation (17), we obtain

$$\mu_{k+1} + P_{k+1}\lambda_{k+1} = F_k\mu_k + F_k P_k\lambda_k + Q_k\lambda_{k+1}. \tag{19}$$

Substituting $\lambda_k = F_k^{\mathrm{T}}\lambda_{k+1} + \theta S_k(x_k - \hat{x}_k) + H_k^{\mathrm{T}} R_k^{-1}(y_k - H_k x_k)$ in Equation (16) into Equation (18), we obtain

$$\begin{aligned} &\lambda_k - \theta S_k P_k\lambda_k + H_k^{\mathrm{T}} R_k^{-1} H_k P_k\lambda_k = \\ &F_k^{\mathrm{T}}\lambda_{k+1} + \theta S_k(\mu_k - \hat{x}_k) + H_k^{\mathrm{T}} R_k^{-1}(y_k - H_k\mu_k) \end{aligned} \tag{20}$$

Shifting the terms gives λ_k as follows:

$$\begin{aligned} \lambda_k &= \left[I - \theta S_k P_k + H_k^{\mathrm{T}} R_k^{-1} H_k P_k \right]^{-1} \times \\ &\quad \left[F_k^{\mathrm{T}}\lambda_{k+1} + \theta S_k(\mu_k - \hat{x}_k) + H_k^{\mathrm{T}} R_k^{-1}(y_k - H_k\mu_k) \right] \end{aligned} \tag{21}$$

Substituting the expression of Equation (21) into Equation (19) gives the following:

$$
\begin{aligned}
&\mu_{k+1} - F_k\mu_k - F_kP_k\left[I - \theta S_kP_k + H_k^TR_k^{-1}H_kP_k\right]^{-1} \times \\
&\left[\theta S_k(\mu_k - \hat{x}_k) + H_k^TR_k^{-1}(y_k - H_k\mu_k)\right] = \\
&\left[-P_{k+1} + F_kP_k\left[I - \theta S_kP_k + H_k^TR_k^{-1}H_kP_k\right]^{-1}F_k^T + Q_k\right]\lambda_{k+1}
\end{aligned}
\tag{22}
$$

This equation holds when both sides of the above equation are zero at the same time. Setting the left side to zero gives:

$$
\begin{aligned}
\mu_{k+1} = &\, F_k\mu_k + F_kP_k\left[I - \theta S_kP_k + H_k^TR_k^{-1}H_kP_k\right]^{-1} \times \\
&\left[\theta S_k(\mu_k - \hat{x}_k) + H_k^TR_k^{-1}(y_k - H_k\mu_k)\right]
\end{aligned}
\tag{23}
$$

Let the right side of Equation (22) be zero to obtain

$$
\begin{aligned}
P_{k+1} &= F_kP_k\left[I - \theta S_kP_k + H_k^TR_k^{-1}H_kP_k\right]^{-1}F_k^T + Q_k \\
&= F_k\tilde{P}_kF_k^T + Q_k
\end{aligned}
\tag{24}
$$

Define $\tilde{P}_k$ as

$$
\tilde{P}_k = P_k\left[I - \theta S_kP_k + H_k^TR_k^{-1}H_kP_k\right]^{-1} = \left[P_k^{-1} - \theta S_k + H_k^TR_k^{-1}H_k\right]^{-1}.
\tag{25}
$$

It follows from Equation (25) that if P_k, S_k, and R_k are symmetric, then they will also be positive definite; and it follows from Equation (24) that if Q_k is positive definite, then P_{k+1} will also be positive definite; so, for all k, P_0, S_k, Q_k, and R_k, if they are all symmetric, then $\tilde{P}_k$ and P_k will be symmetric at some point.

It turns out that we are able to find the extreme points of J, so the above assumption is correct. Using the values of x_0 and w_k already obtained, we can again find the extreme points of the function J with respect to $\hat{x}_k$ and y_k.

3.2. Extreme Solutions for $\hat{x}_k$ and y_k

Based on the solution of the problem of the extreme points of x_0 and w_k, we also need to find the extreme points of the function J with respect to $\hat{x}_k$ and y_k. From the initial condition of μ_k in Equation (18), we can see that

$$
\begin{cases}
\lambda_k = P_k^{-1}(x_k - \mu_k) \\
\lambda_0 = P_0^{-1}(x_0 - \hat{x}_0)
\end{cases}
\tag{26}
$$

The following can be obtained from Equation (26):

$$
\|\lambda_0\|_{P_0}^2 = \lambda_0^TP_0\lambda_0 = \|x_0 - \hat{x}_0\|_{P_0^{-1}}^2.
\tag{27}
$$

In this case, Equation (12) can be rewritten as

$$
J = -\frac{1}{\theta}\|\lambda_0\|_{P_0}^2 + \sum_{k=0}^{N-1}\left[\|x_k - \hat{x}_k\|_{S_k}^2 - \frac{1}{\theta}\left(\|w_k\|_{Q_k^{-1}}^2 + \|y_k - H_kx_k\|_{R_k^{-1}}^2\right)\right].
\tag{28}
$$

Substituting the expression for x_k, the Equation (28) can be rewritten as

$$J = -\tfrac{1}{\theta}\|\lambda_0\|_{P_0}^2 +$$
$$\sum_{k=0}^{N-1}\left[\|\mu_k + P_k\lambda_k - \hat{x}_k\|_{S_k}^2 - \tfrac{1}{\theta}\left(\|w_k\|_{Q_k^{-1}}^2 + \|y_k - H_k(\mu_k + P_k\lambda_k)\|_{R_k^{-1}}^2\right)\right] \tag{29}$$

Substituting the expression for w_k in Equation (16) into this position of Equation (29) gives the new equation:

$$\|w_k\|_{Q_k^{-1}}^2 = w_k^T Q_k^{-1} w_k = \lambda_{k+1}^T Q_k \lambda_{k+1}. \tag{30}$$

As Q_k is a symmetric matrix, Equation (30) can be written as

$$J = -\tfrac{1}{\theta}\|\lambda_0\|_{P_0}^2 +$$
$$\sum_{k=0}^{N-1}\left[\|\mu_k + P_k\lambda_k - \hat{x}_k\|_{S_k}^2 - \tfrac{1}{\theta}\|y_k - H_k(\mu_k + P_k\lambda_k)\|_{R_k^{-1}}^2\right] - \tfrac{1}{\theta}\sum_{k=0}^{N-1}\|\lambda_{k+1}\|_{Q_k}^2 \tag{31}$$

It follows from Equation (16) that $\lambda_N = 0$; hence,

$$\sum_{k=0}^{N}\lambda_k^T P_k \lambda_k - \sum_{k=0}^{N-1}\lambda_k^T P_k \lambda_k = 0. \tag{32}$$

Equation (32) can be written as

$$0 = -\frac{1}{\theta}\|\lambda_0\|_{P_0}^2 - \frac{1}{\theta}\sum_{k=0}^{N-1}\left(\lambda_{k+1}^T P_{k+1}\lambda_{k+1} - \lambda_k^T P_k \lambda_k\right). \tag{33}$$

Equation (31) is obtained by subtracting Equation (33) and simplifying the following:

$$J = \sum_{k=0}^{N-1}\left[(\mu_k - \hat{x}_k)^T S_k(\mu_k - \hat{x}_k) + 2(\mu_k - \hat{x}_k)^T S_k P_k \lambda_k + \lambda_k^T P_k S_k P_k \lambda_k +\right.$$
$$\tfrac{1}{\theta}\lambda_{k+1}^T(P_{k+1} - Q_k)\lambda_{k+1} - \tfrac{1}{\theta}\lambda_k^T P_k \lambda_k - \tfrac{1}{\theta}(y_k - H_k\mu_k)^T R_k^{-1}(y_k - H_k\mu_k) +$$
$$\tfrac{2}{\theta}(y_k - H_k\mu_k)^T R_k^{-1} H_k P_k \lambda_k - \tfrac{1}{\theta}\lambda_k^T P_k H_k^T R_k^{-1} H_k P_k \lambda_k \tag{34}$$

Taking Equation (24) into Equation (34) and organizing it gives the following:

$$\lambda_{k+1}^T(P_{k+1} - Q_k)\lambda_{k+1}$$
$$= \lambda_k^T P_k \lambda_k - \theta \lambda_k^T P_k S_k P_k \lambda_k + \lambda_k^T P_k H_k^T R_k^{-1} H_k P_k \lambda_k -$$
$$2\theta(\mu_k - \hat{x}_k)^T S_k P_k \lambda_k - 2(y_k - H_k\mu_k)^T R_k^{-1} H_k P_k \lambda_k +$$
$$\theta^2(\mu_k - \hat{x}_k)^T S_k \widetilde{P}_k S_k(\mu_k - \hat{x}_k) + 2\theta(\mu_k - \hat{x}_k)^T S_k \widetilde{P}_k H_k^T R_k^{-1}(y_k - H_k\mu_k) +$$
$$(y_k - H_k\mu_k)^T R_k^{-1} H_k \widetilde{P}_k H_k^T R_k^{-1}(y_k - H_k\mu_k) \tag{35}$$

Taking Equation (35) into Equation (34) and organizing it gives the following:

$$J = \sum_{k=0}^{N-1} \Big[(\mu_k - \hat{x}_k)^{\mathrm{T}} \Big(S_k + \theta S_k \widetilde{P}_k S_k \Big)(\mu_k - \hat{x}_k) +$$

$$2(\mu_k - \hat{x}_k)^{\mathrm{T}} S_k \widetilde{P}_k H_k^{\mathrm{T}} R_k^{-1}(y_k - H_k \mu_k) +$$

$$\tfrac{1}{\theta}(y_k - H_k \mu_k)^{\mathrm{T}} \Big(R_k^{-1} H_k \widetilde{P}_k H_k^{\mathrm{T}} R_k^{-1} - R_k^{-1} \Big)(y_k - H_k \mu_k) \Big] \tag{36}$$

The goal is to find the solutions to the extreme value problem of J with respect to $\hat{x}_k$ and y_k. Therefore, the J of Equation (36) is made to take partial derivatives with respect to $\hat{x}_k$ and y_k, respectively, and the partial derivatives are made to be zero. We obtain the following:

$$\begin{cases} \frac{\partial J}{\partial \hat{x}_k} = 2\Big(S_k + \theta S_k \widetilde{P}_k S_k \Big)(\hat{x}_k - \mu_k) + 2 S_k \widetilde{P}_k H_k^{\mathrm{T}} R_k^{-1}(H_k \mu_k - y_k) = 0 \\[2mm] \frac{\partial J}{\partial y_k} = \frac{2}{\theta} \Big(R_k^{-1} H_k \widetilde{P}_k H_k^{\mathrm{T}} R_k^{-1} - R_k^{-1} \Big)(y_k - H_k \mu_k) + 2 R_k^{-1} H_k \widetilde{P}_k S_k(\mu_k - \hat{x}_k) = 0 \end{cases} \tag{37}$$

The solutions to Equation (37) are as follows:

$$\begin{cases} \hat{x}_k = \mu_k \\[2mm] y_k = H_k \mu_k \end{cases} \tag{38}$$

$\hat{x}_k$ and y_k in Equation (38) are the extreme points of Equation (37). If the second-order partial derivatives of J are positive definite, it means that the extreme point is the minimum point. The second-order partial derivatives of J with respect to $\hat{x}_k$ are as follows:

$$\frac{\partial^2 J}{\partial \hat{x}_k^2} = 2\Big(S_k + \theta S_k \widetilde{P}_k S_k \Big), \tag{39}$$

If $S_k + \theta S_k \widetilde{P}_k S_k$ is positive definite, $\hat{x}_k$ will be the point where J is minimized. The choice of S_k in Equation (9) is always positive definite, so $\hat{x}_k$ will be the point of minimum of J as long as $\widetilde{P}_k$ is positive definite.

From Equations (23), (24) and (38), a filtering method as shown in Equation (40) can be derived such that the cost function J_1 can be smaller than $1/\theta$.

$$K_k = P_k \Big[I - \theta S_k P_k + H_k^{\mathrm{T}} R_k^{-1} H_k P_k \Big]^{-1} H_k^{\mathrm{T}} R_k^{-1}$$

$$\hat{x}_{k+1} = F_k \hat{x}_k + F_k K_k(y_k - H_k \hat{x}_k)$$

$$P_{k+1} = F_k P_k \Big[I - \theta S_k P_k + H_k^{\mathrm{T}} R_k^{-1} H_k P_k \Big]^{-1} F_k^{\mathrm{T}} + Q_k \tag{40}$$

where K_k is the gain matrix.

In order to have a solution to the problem of the observer of Equation (40), the following conditions must always be satisfied:

$$P_k^{-1} - \theta S_k + H_k^{\mathrm{T}} R_k^{-1} H_k > 0. \tag{41}$$

4. Forgetting Factor H∞ Filtering Algorithm

In the H∞ filtering algorithm, the noise covariance matrix is artificially set based on experience, and its initial value affects the accuracy and convergence of the algorithm [25–27]. In this chapter, assuming a fixed process noise covariance matrix, we design a dynamic forgetting factor to weight the combination of the initial and current measurement noise covariance matrices. The initial matrix is gradually forgotten to minimize the effect of anomalous initial values on the algorithm.

Define the best estimate of the measurement noise via the following:

$$V_k = y_k - H_k \hat{x}_k. \tag{42}$$

Combined with Equation (7), these yield the following:

$$v_k = V_k - H_k(x_k - \hat{x}_k). \tag{43}$$

At this point, the measurement noise covariance matrix is as follows:

$$R_k = \text{cov}(v_k) = V_k V_k^T - H_k P_k H_k^T. \tag{44}$$

Define the dynamic forgetting factor β_k:

$$\beta_k = \frac{1-\alpha}{1-\alpha^k}, \tag{45}$$

where α is a constant, usually taken as 0.96~0.99.

Weighting the measurement noise covariance matrix with a dynamic forgetting factor strengthens the role of the measurement noise covariance matrix in the estimation at that moment and gradually forgets the initial measurement noise covariance matrix:

$$R_{k+1} = \beta_k(V_k V_k^T - H_k P_k H_k^T) + (1 - \beta_k)R_k. \tag{46}$$

Combining Equation (46) with the filtering method shown in Equation (40) can lead to the H∞ filtering algorithm with dynamic forgetting factor shown in Equation (47):

$$\begin{aligned}
K_k &= P_k\left[I - \theta S_k P_k + H_k^T R_k^{-1} H_k P_k\right]^{-1} H_k^T R_k^{-1} \\
R_{k+1} &= \beta_k(V_k V_k^T - H_k P_k H_k^T) + (1 - \beta_k)R_k \\
\hat{x}_{k+1} &= F_k \hat{x}_k + F_k K_k(y_k - H_k \hat{x}_k) \\
P_{k+1} &= F_k P_k\left[I - \theta S_k P_k + H_k^T R_k^{-1} H_k P_k\right]^{-1} F_k^T + Q_k
\end{aligned} \tag{47}$$

5. Experimental Analysis and Comparison

In order to verify the feasibility of the parameter identification algorithm proposed in this paper, relevant simulations are carried out in this chapter to verify the simulation flowchart, and the motor parameters used are shown in Figure 1 and Table 1.

Table 1. Parameters of the PMSM control system.

Parameter	Value	Unit
DC voltage	24	V
Stator resistance	0.48	Ω
d-axis inductance	2	mH
q-axis inductance	2	mH
Flux linkage	0.01	Wb
Number of pole pairs	4	-

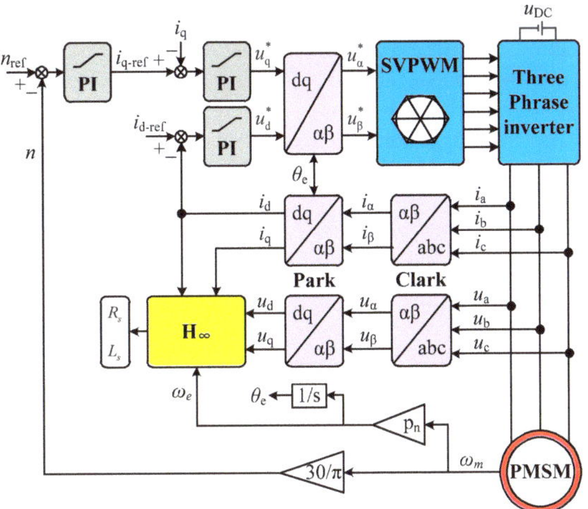

Figure 1. Permanent magnet synchronous motor system model.

In this section, the parameter identification simulation under a steady-state condition is carried out first to verify the effectiveness of the proposed identification algorithm then to verify the robustness of the proposed parameter identification algorithm, the simulation analysis is carried out for three conditions of motor load change, stator resistance change, and stator inductance change in turn. At the end, the effectiveness of the parameter identification algorithm with the addition of a forgetting factor is verified.

5.1. Steady-State Performance

The motor is operating in a steady-state condition. The motor load is set to 0.3 N·m and the motor speed is set to 600 rpm. The parameters of the recognition algorithm are set as follows: $x_0 = [0.01\ 5\ 280\ 550]$, $P_0 = \mathrm{diag}([0.01\ 0.1\ 1\ 1])$, $S_k = \mathrm{diag}([0.18\ 0.06\ 0\ 0])$, $Q_k = \mathrm{diag}([0\ 0\ 0.9\ 1.18])$, and $R_k = [1\ 1]$, $T_s = 0.0001$ s. The simulation results are shown in Figure 2.

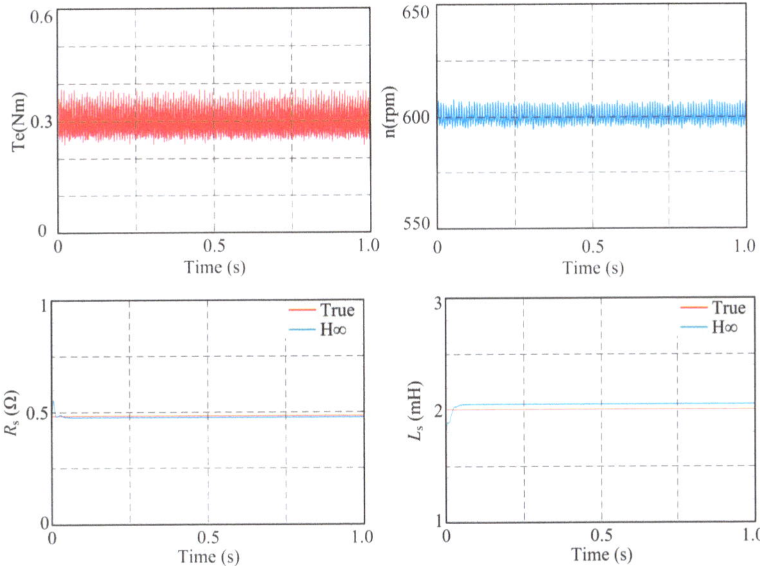

Figure 2. Parameter identification of R_s and L_s under steady-state conditions.

From Figure 2, it can be seen that the proposed parameter identification algorithm can achieve the identification of resistance and inductance in a short time. The difference between the final identification result and the actual value of the resistance is almost 0. The actual value of the inductance is 2 mH, the final identification is 2.1 mH, and the difference between the final identification result and the actual value of inductance is within 5%, which proves the effectiveness of the proposed parameter identification algorithm.

5.2. Robustness Verification

5.2.1. Load Torque

The motor speed is set to 900 rpm and the torque changes from 0.2 N·m to 0.4 N·m at 0.5 s. The simulation results are shown in Figure 3.

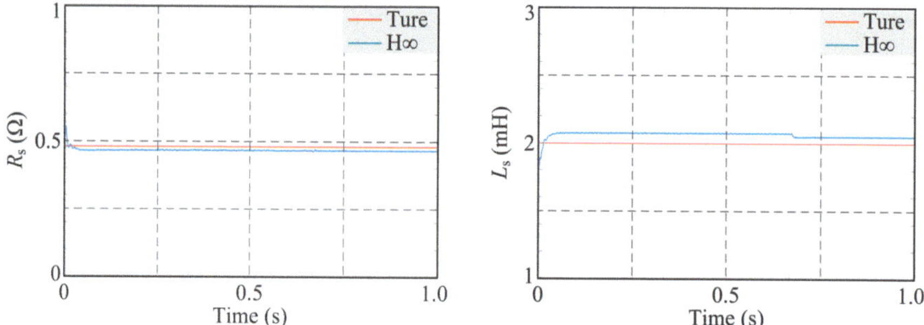

Figure 3. Parameter identification of R_s and L_s under load torque variation.

From Figure 3, it can be seen that the proposed parameter identification algorithm can guarantee the identification of the parameters when the torque is changed (twice). The resistance parameter identification results remain almost unchanged when the torque is changed. The inductance parameter identification results are 2.2 mH and 2.1 mH, respectively, and the difference between the changed identification result and the previous one is within 2%, which proves that the proposed parameter identification algorithm is robust to the torque-change condition.

5.2.2. Stator Resistance

In this subsection, the simulation simulates two operating conditions: sudden change in resistance due to motor failure and slow increase in resistance due to temperature rise and other factors. The stator resistance increased stepwise from 0.48 Ω to 0.8 Ω and gradually to 0.8 Ω, respectively. The motor speed is set to 900 rpm, the motor load is set to 0.3 N·m, and the simulation results are shown in Figure 4.

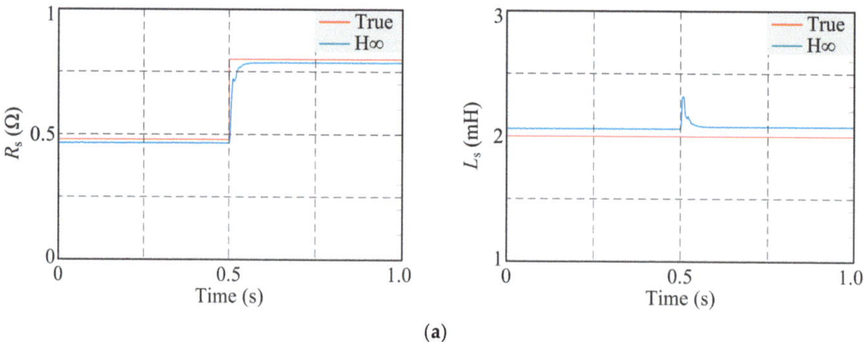

Figure 4. *Cont.*

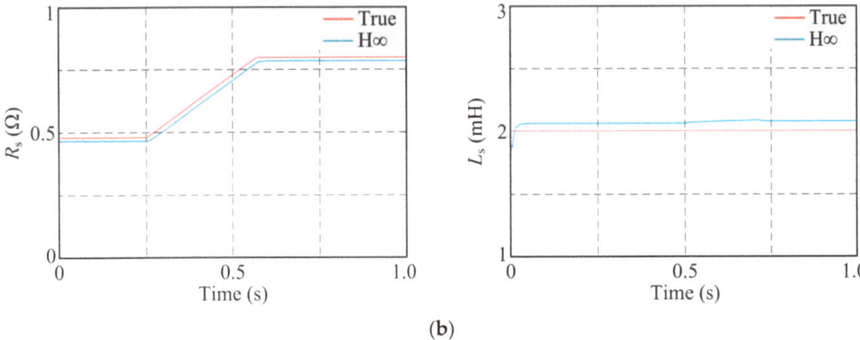

(b)

Figure 4. Parameter identification of R_s and L_s under varying stator resistance: (**a**) sudden change in resistance due to motor failure; (**b**) slow increase in resistance due to temperature rise and other factors.

From Figure 4, it can be seen that the proposed parameter identification algorithm can guarantee the identification of the parameters when the resistance is changed and guarantees the qualified response speed when the resistance is changed abruptly, which proves that the proposed parameter identification algorithm connects the robustness to the working condition of the resistance change.

5.2.3. Stator Inductance

In this section, the simulation simulates the inductance change condition corresponding to the previous section. The stator inductance is abruptly changed from 2 mH to 4 mH and gradually increases from 2 mH to 4 mH. The motor speed is set to 900 rpm, the motor load is set to 0.3 N·m, and the simulation results are shown in Figure 5.

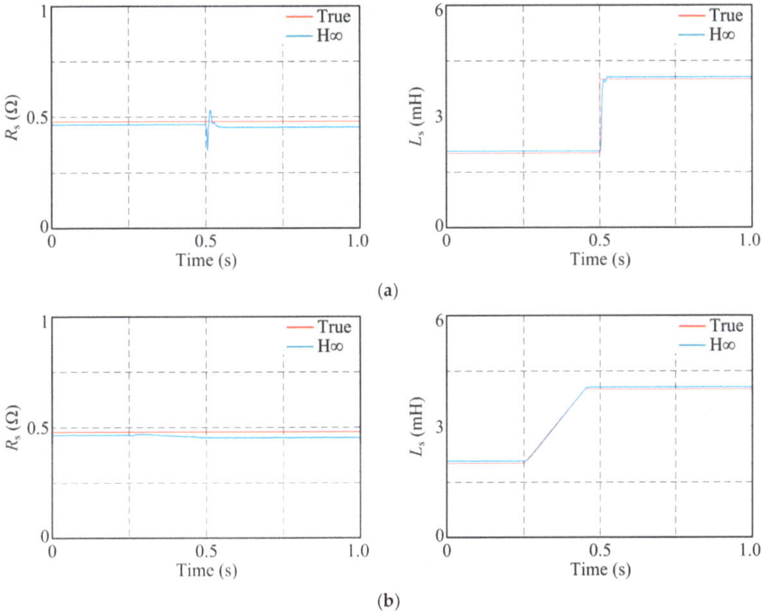

(a)

(b)

Figure 5. Parameter identification of R_s and L_s under varying stator inductance. (**a**) The stator inductance changed abruptly from 2 mH to 4 mH. (**b**) The stator inductance changed gradually from 2 mH to 4 mH.

From Figure 5, it can be seen that the proposed parameter identification algorithm can guarantee the identification of the parameters when the inductance is changed and guarantees the qualified response speed when the inductance is changed abruptly, which proves that the proposed parameter identification algorithm connects the robustness to the working condition of the inductance change.

5.3. Validation of the Forgetting Factor

In this section, the steady-state condition of Section 5.1 is re-simulated for the identification algorithm before and after adding the forgetting factor. Then, the R matrix is changed to [10 10] to verify the effectiveness of the forgetting factor proposed in this paper. The comparative simulation results are shown in Figure 6.

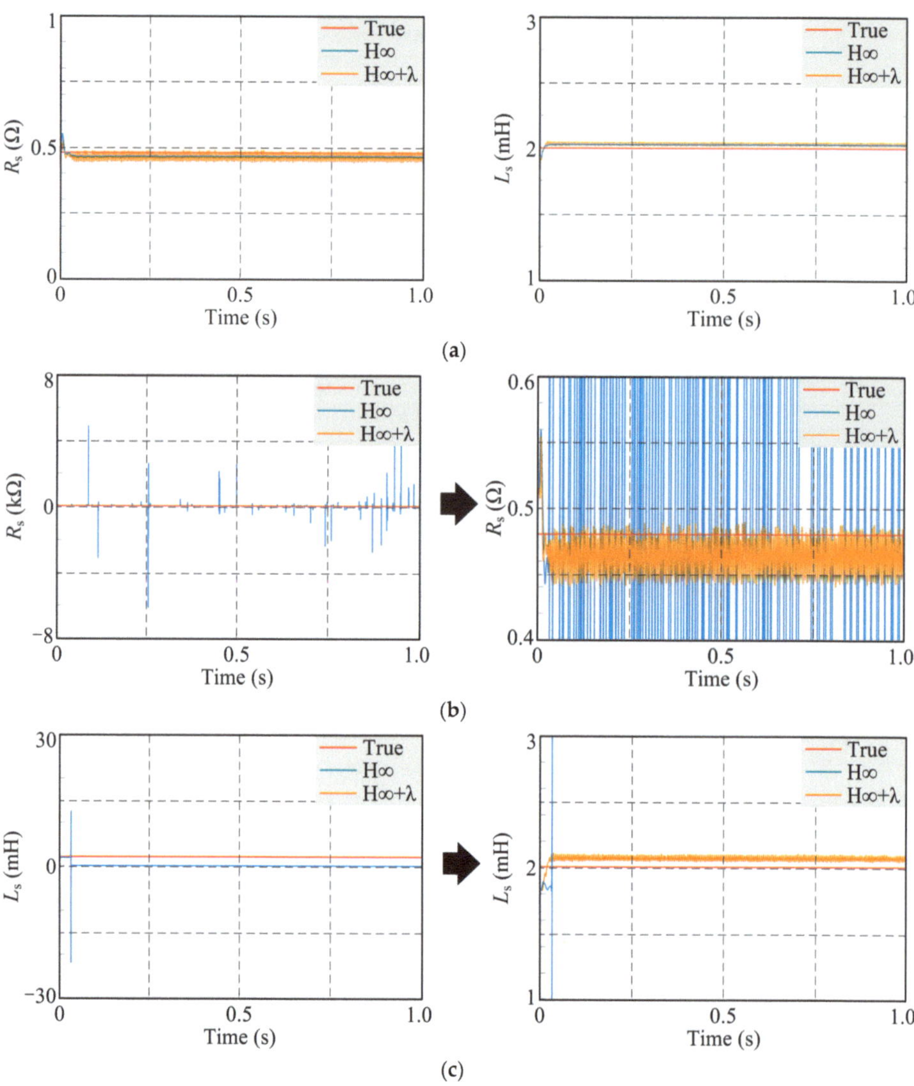

Figure 6. Comparison of the effect of parameter identification after adding forgetting factor. (**a**) The condition of steady-state. (**b**) Identification of parameter R_s when the initial value is abnormal. (**c**) Identification of parameter L_s when the initial value is abnormal.

From Figure 6, it can be seen that the recognition algorithm with the added forgetting factor is not much different from the previous algorithm when the initial parameters are normal. However, when the initial parameters are abnormal, the observation results of the recognition algorithm without the added forgetting factor are abnormal, while the proposed forgetting factor is able to correct the error and recognize the parameters in time.

6. Discussion

With the development of modern power electronics technology, PMSMs are increasingly used in CNC machine tools, robots, and new energy vehicles due to their simple structure, high efficiency, and high functionality. However, there are modeling errors and noise uncertainties in PMSM systems. To meet the system's requirements for robustness, we adopt the H∞ filtering algorithm. However, the noise covariance matrix and the upper performance limit of the H∞ filtering algorithm are set empirically, which may affect the accuracy of the algorithm. If they are not set appropriately, it may lead to a decrease in system accuracy and even to filtering divergence.

The application of the H∞ filtering algorithm to real PMSMs requires online identification of several parameters, such as motor speed, rotor position, and magnetic chain. These parameters will be collected by measuring instruments in the motor system and processed by the H∞ filtering algorithm, which reduces the influence of noise and other external disturbances, achieving high-accuracy online parameter identification of the motor and improving system robustness. However, implementing the H∞ filtering algorithm in a real PMSM system faces many challenges.

(1) Sensor noise: The PMSM control system uses sensors to obtain measured values of the motor state, which may contain sensor noise. The H∞ filtering algorithm must consider the influence of sensor noise when dealing with external noise interference. If the statistical characteristics of the sensor noise change, the accuracy of the H∞ filtering algorithm may be affected.

(2) High sampling rate and data processing requirements: Servo motors are generally divided into three control rings-current, speed, and position. The frequency of each ring determines its position, with higher frequencies corresponding to inner rings. PMSM control systems require high sampling rates for accurate measurement and control, increasing hardware and real-time performance requirements. Additionally, H∞ filtering may need to process large amounts of data, which is challenging for devices with limited data processing capabilities.

7. Conclusions

Through theoretical analysis and simulation verification, it can be concluded that the H∞ filtering algorithm based on game theory can obtain the recognition results quickly and accurately without making any assumptions about the noise, and its robustness has been significantly improved. The H∞ filtering algorithm after adding the improved forgetting factor can quickly and stably obtain the recognition result under the situation of poor initial value, which compensates the recognition error caused by human setting.

The algorithm proposed in this paper improves the estimation accuracy and robust performance of the original algorithm to some extent, but there are still deficiencies to be improved:

(1) This paper improves the H∞ filtering algorithm by adding a dynamic forgetting factor, achieving weighted estimation of the initial and current measurement noise covariances. Although the accuracy of the algorithm is improved, it takes more time due to multiple iterations per time step.

(2) The motor in the simulation ran at low speed, and the algorithm is inadequate for high-speed operation. The subsequent work can focus on identifying motor parameters during high-speed operation.

Author Contributions: Conceptualization, T.Y., J.C. and Y.Z.; methodology, J.C.; software, J.C. and Y.Z.; validation, T.Y., J.C. and Y.Z.; formal analysis, T.Y., J.C. and Y.Z.; investigation, T.Y.; resources, T.Y.; data curation, Y.Z.; writing—original draft preparation, J.C. and Y.Z.; writing—review and editing, J.C. All authors have read and agreed to the published version of the manuscript.

Funding: This research was funded by the project is supported by Scientific Research Project of Education Department of Jilin Province (No. JJKH20230121KJ).

Data Availability Statement: Data are contained with the article.

Conflicts of Interest: The authors declare no conflict of interest.

References

1. Ali, N.; Alam, W.; Pervaiz, M.; Iqbal, J. Nonlinear adaptive backsteepping control of permanent magnet synchronous motor. *RRST-EE* **2023**, *66*, 15–20.
2. Zhang, X.H.; Zhao, J.W.; Wang, L.J.; Hu, D.B.; Wang, L. High Precision Anti-interference Online Multiparameter Estimation of PMSLM With Adaptive Interconnected Extend Kalman Observer. *Proc. CSEE* **2022**, *12*, 4571–4581.
3. Ma, Y.L.; Yuan, H.; Yin, W.; Yang, H. An on-line parameter identification method for PMSM DC signal injection considering equivalent electromagnetic loss resistance offset. *Trans. China Electrotech. Soc.* **2023**, 6015–6026.
4. Zhang, Q.S.; Fan, Y. The Online Parameter Identification Method of Permanent Magnet Synchronous Machine under Low-Speed Region Considering the Inverter Nonlinearity. *Energies* **2022**, *15*, 4314. [CrossRef]
5. Yu, J.W. *Research on Online Parameter Identification of Permanent Magnet Synchronous Motor Considering the Variation of Operating Condition*; Harbin Institute of Technology: Harbin, China, 2021.
6. Lian, C.Q.; Xiao, F.; Liu, J.L.; Gao, S. Parameter and VSI Nonlinearity Hybrid Estimation for PMSM Drives Based on Recursive Least Square. *IEEE Trans. Transp. Electrif.* **2023**, *9*, 2195–2206. [CrossRef]
7. Purbowaskito, W.; Wu, P.Y.; Lan, C.Y. Permanent Magnet Synchronous Motor Driving Mechanical Transmission Fault Detection and Identification: A Model-Based Diagnosis Approach. *Electronics* **2022**, *11*, 1356. [CrossRef]
8. Liu, H.; Zhang, H.X.; Zhang, H.; Chen, G.Q. Model Reference Adaptive Speed Observer Control of Permanent Magnet Synchronous Motor Based on Single Neuron PID. *J. Phys. Conf. Ser.* **2022**, *2258*, 012052. [CrossRef]
9. Miao, J.L.; Li, X.; Dong, B. Vector Control Strategy of Permanent Magnet Synchronous Motor Based on Model Reference Adaptive. *Mech. Eng. Autom.* **2020**, *6*, 16–18.
10. Niu, F.; Chen, X.; Huang, S.P.; Huang, X.Y.; Wu, L.J.; Li, K.; Fang, Y.T. Model predictive current control with adaptive-adjusting timescales for PMSMs. *CES Trans. Electr. Mach. Syst.* **2021**, *5*, 108–117. [CrossRef]
11. Wang, L.; Li, J.W.; Ma, Y.; Kan, H.Y.; Sun, B.B.; Gao, S.; Wang, D. Estimation of Rotor Position of PMSM Based on Strong Tracking Cubature Kalman Filter. *Micromotors* **2020**, *53*, 61–65.
12. Niedermayr, P.; Alberti, L.; Bolognani, S.; Abl, R. Implementation and Experimental Validation of Ultrahigh-Speed PMSM Sensorless Control by Means of Extended Kalman Filter. *IEEE J. Emerg. Sel. Top. Power Electron.* **2022**, *10*, 3337–3344. [CrossRef]
13. Dini, P.; Saponara, S. Design of an Observer-Based Architecture andNon-Linear Control Algorithm for Cogging Torque Reduction in Synchronous Motors. *Energies* **2020**, *13*, 2077. [CrossRef]
14. Uddin, M.N.; Chy, M.M.I. Online Parameter-Estimation-Based Speed Control of PM AC Motor Drive in Flux-Weakening Region. *IEEE Trans. Ind. Appl.* **2008**, *44*, 1486–1494. [CrossRef]
15. Gao, D.X.; Zhou, L.; Chen, J.; Pan, C.Z. Design of Derivative-free Model-reference Adaptive Control for a Class of Uncertain Systems Based on Disturbance Compensation. *Control Theory Appl.* **2023**, *40*, 735–743.
16. Tang, Y.R.; Xu, W.; Liu, Y.; Dong, D.H. Dynamic Performance Enhancement Method Based on Improved Model Reference Adaptive System for SPMSM Sensorless Drives. *IEEE Access* **2021**, *9*, 135012–135023. [CrossRef]
17. Dang, D.Q.; Rafaq, M.S.; Choi, H.H.; Jung, J.-W. Online Parameter Estimation Technique for Adaptive Control Applications of Interior PM Synchronous Motor Drives. *IEEE Trans. Ind. Electron.* **2016**, *63*, 1438–1449. [CrossRef]
18. Liu, X. *Research on Torque Ripple Suppression of Brushless DC Motor Based on Buck-Boost Topology and Kalman Filter*; Jiangsu University: Zhenjiang, China, 2018.
19. Khazraj, H.; Silva, F.F.D.; Bak, C.L. A performance comparison between extended Kalman Filter and unscented Kalman Filter in power system dynamic state estimation. In Proceedings of the 51st International Universities Power Engineering Conference (UPEC), Coimbra, Portugal, 6–9 September 2016.
20. Wasim, M.; Ali, A.; Choudhry, M.A.; Saleem, F.; Shaikh, I.U.H.; Iqbal, J. Unscented Kalman filter for airship model uncertainties and wind disturbance estimation. *PLoS ONE* **2021**, *16*, e0257849. [CrossRef] [PubMed]
21. Huang, Y.L.; Zhang, Y.G.; Wu, Z.M.; Li, N.; Chambers, J. A Novel Adaptive Kalman Filter With Inaccurate Process and Measurement Noise Covariance Matrices. *IEEE Trans. Autom. Control* **2018**, *63*, 594–601. [CrossRef]
22. Chen, L.J. *Parameter Estimations for Brushless DC Motor Based on Adaptive H-Inifnity Filter*; Nanjing University of Posts and Telecommunications: Nanjing, China, 2022.
23. Liu, X.; Wang, X.P.; Wang, S.H.; Liang, L.B.; Huang, J.W. Inductance Identification of Permanent Magnet Synchronous Motor Based on Least Square Method. *Electr. Mach. Control Appl.* **2020**, *47*, 1–5+32.

24. Fang, G.H.; Wang, H.C.; Gao, X. Parameter Identification Algorithm of Permanent Magnet Synchronous Motor Based on Dynamic Forgetting Factor Recursive Least Square Method. *Comput. Appl. Softw.* **2021**, *38*, 280–283.
25. Wang, R.T.; Wang, S.Q. Research on split-source two-stage matrix converter and its modulation strategy. *J. Northeast. Electr. Power Univ.* **2023**, *43*, 47–54.
26. Chen, J.K.; Dong, J.Q.; Li, H.R.; Zhu, S.Q. STATCOM harmonic circulation and sub-module voltage imbalance analysis and suppression method. *J. Northeast. Electr. Power Univ.* **2023**, *43*, 8–17.
27. Yan, G.G.; Jia, X.H.; Wang, Y.P.; Cui, C.; Cui, Y.S. Control parameter identification method of wind turbine grid-connected converter driven by dynamic response error. *J. Northeast. Electr. Power Univ.* **2023**, *43*, 1–7.

Optimal Design of a Surface Permanent Magnet Machine for Electric Power Steering Systems in Electric Vehicle Applications Using a Gaussian Process-Based Approach

Gilsu Choi [1], Gwan-Hui Jang [1], Mingyu Choi [1,†], Jungmoon Kang [1], Ye Gu Kang [2,*] and Sehwan Kim [3]

[1] Department of Electrical and Computer Engineering, Inha University, Incheon 22212, Republic of Korea; gchoi@inha.ac.kr (G.C.); jkh9452@inha.edu (G.-H.J.); mingyu5832@inha.edu (M.C.); jmkang@inha.edu (J.K.)
[2] School of Electrical, Electronics & Communication Engineering, Korea University of Technology and Education, Cheonan 31253, Republic of Korea
[3] Korea Institute of Machinery and Materials, Pusan 46744, Republic of Korea; sehwan@kimm.re.kr
* Correspondence: kang@koreatech.ac.kr
† Graduated.

Abstract: The efficient design optimization of electric machines for electric power steering (EPS) applications poses challenges in meeting demanding performance criteria, including high power density, efficiency, and low vibration. Traditional optimization approaches often fail to find a global solution or suffer from excessive computation time. In response to the limitations of traditional approaches, this paper introduces a novel methodology by incorporating a Gaussian process-based adaptive sampling technique into a surrogate-assisted optimization process using a metaheuristic algorithm. Validation on a 72-slot/8-pole interior permanent magnet (IPM) machine demonstrates the superiority of the proposed approach, showcasing improved exploitation–exploration balance, faster convergence, and enhanced repeatability compared to conventional optimization methods. The proposed design process is then applied to two surface PM (SPM) machine configurations with 9-slot/6-pole and 12-slot/10-pole combinations for EPS applications. The results indicate that the 12-slot/10-pole SPM design surpasses the alternative design in torque density, efficiency, cogging torque, torque ripple, and manufacturability.

Keywords: design optimization; electric machines; electric power steering systems; electric vehicles; gaussian process; adaptive sampling

Citation: Choi, G.; Jang, G.-H.; Choi, M.; Kang, J.; Kang, Y.G.; Kim, S. Optimal Design of a Surface Permanent Magnet Machine for Electric Power Steering Systems in Electric Vehicle Applications Using a Gaussian Process-Based Approach. *Actuators* **2024**, *13*, 13. https://doi.org/10.3390/act13010013

Academic Editor: Dong Jiang

Received: 6 December 2023
Revised: 26 December 2023
Accepted: 28 December 2023
Published: 29 December 2023

1. Introduction

In the era of electric mobility, the pursuit of efficient and high-performance electric power steering (EPS) systems has become central to the evolution of vehicular dynamics. The electric machine, at the heart of these systems, plays a pivotal role in converting electrical energy into precise mechanical assistance. Electric machines for EPS systems are required to meet demanding performance requirements, including high power density, efficiency, low noise and vibration, and fault tolerance [1]. Additionally, compactness and lightweight characteristics are essential for seamless integration into vehicles without compromising space or adding excessive weight.

Addressing specific design goals, such as the reduction of torque ripple and cogging torque, becomes crucial as the rotating movement is converted into the linear movement of the steering rack [2,3]. These factors directly impact the vibration and overall driving comfort of the vehicle. Ensuring fault tolerance is equally important to guarantee continued system functionality after a failure [4]. As electric vehicles (EVs) become mainstream, optimizing the performance characteristics of EPS machines has become critical. However, the limitations of traditional machine design approaches are becoming apparent, necessitating more powerful and efficient design methodologies.

Rotating electrical machines, due to their complex geometry and nonlinear magnetic saturation characteristics, are often designed using numerical methods like Finite Element Analysis (FEA) rather than analytical approaches. While computer simulations such as FEA aid in the design process, the optimization of electrical machines for varying performance characteristics poses challenges. These challenges include multi-variable and multi-objective optimization problems requiring exploration of the entire design space. However, the computational expense associated with FEA makes full design space exploration impractical, particularly as the number of design variables increases [5–9].

In response to the limitations of conventional optimization approaches, recent years have seen active research in metaheuristic, algorithm-based machine design optimization. Techniques utilizing genetic algorithms, particle swarm optimization, and similar methods based on FEA have been explored. Although these metaheuristic algorithms offer optimal solutions, they demand substantial computations, resulting in significant computation time. To address this, recent studies have investigated surrogate-assisted metaheuristic algorithms [10,11]. The approach exemplified in [11] proposes a systematic design optimization process for internal permanent magnet synchronous machines (IPMSMs), utilizing surrogate models (SMs), such as Kriging, artificial neural networks (ANNs), and support vector regression (SVM). While these approaches significantly reduce computation time without compromising accuracy, the selection of effective samples for SM construction and evolutionary search algorithms has not yet been fully discussed.

The accuracy and efficiency of SM is closely related to the quality and quantity of the dataset used for training [12–14]. For optimal design, the training dataset must be representative of the entire input space to avoid bias and allow the surrogate model to generalize well [15–17]. Despite the critical role of dataset selection, comprehensive discussions on optimizing the data selection strategy for efficiency and fast convergence are lacking in past literature on electric machine design. This paper addresses this imperative challenge by introducing a cutting-edge approach: application of Gaussian process-based algorithms for the optimal design of EPS machines.

The Gaussian process (GP), known for its ability to model complex and nonlinear relationships, provides a promising avenue for systematically exploring the design space. Its objective, in the context of electric machine design, is to identify design configurations that maximize power density while enhancing overall performance. This paper provides a systematic and detailed description of how the GP can be integrated into surrogate-assisted optimization techniques using metaheuristic algorithms to develop an efficient design optimization process. The major contributions of this paper are outlined as follows:

- Development of a design optimization process utilizing adaptive sampling that blends exploitation and exploration to simultaneously improve model accuracy and convergence speed.
- Validation of the developed optimization process through its application to a 72-slot/8-pole IPMSM for traction applications.
- Comprehensive design and analysis of surface permanent magnet (SPM) machines to address design challenges for EPS applications.
- Comparative analysis and design optimization of two promising PM machine topologies: SPM machines equipped with fractional-slot concentrated windings (FSCW) with 9-slot/6-pole and 12-slot/10-pole.
- Experimental verification of the optimal design through the construction and testing of a prototype machine.

The subsequent sections of this paper are organized as follows: Section 2 provides an overview of the Gaussian process-based adaptive sampling algorithm proposed in this paper and presents validation results on the performance of the proposed algorithm through a case study. Details on the chosen baseline machine topologies for an EPS application and their basic electromagnetic performance are presented in Section 3. Section 4 describes a design optimization process utilizing the proposed adaptive sampling technique to find the global optimal solution for an EPS motor. Section 4 also presents the results of the comparative study in which the proposed optimization process is applied to two promising machine configurations. Finally, experimental verification results for the final optimized design are presented in Section 5 by comparing the measured machine performance with FE predictions.

2. Gaussian Process-Based Adaptive Sampling Algorithm

The Gaussian Process (GP), also known as the Kriging method, serves as an interpolation technique for predicting data in a high-dimensional space based on input and output data. Widely applied in the design optimization of electrical machines, the GP algorithm significantly reduces computational time by constructing surrogate models from pre-computed simulation results, especially beneficial for extensive nonlinear numerical calculations. Recent research has concentrated on building surrogate models that ensure required accuracy with minimal computation, emphasizing the efficacy of surrogate-based optimization (SBO) with adaptive sampling for efficient and accurate design exploration.

Adaptive sampling, a pivotal technique in optimization processes, iteratively refines the surrogate model by strategically introducing new samples into crucial regions of the design space. This refinement, guided by response surface information from the existing surrogate model, facilitates the efficient construction of surrogate models, enabling improved accuracy with reduced sample size and calculation time. Two primary techniques for sample selection criteria, exploitation and exploration, play essential roles in optimizing the efficiency of the process [18].

Exploitation involves generating the next sample by identifying the point predicted as the best value of the function based on given information, employing techniques like K-fold cross-validation [19,20] and leave-one-out cross validation (LOOCV) [21]. Figure 1a shows an example of a case where additional sampling using an exploit (yellow dots) is applied. The advantage of using this technique is that it allows for efficient design space exploration to improve sample distribution and reduce the time spent searching for the optimal point. [22]. However, caution is necessary to avoid overlooking the space of interest.

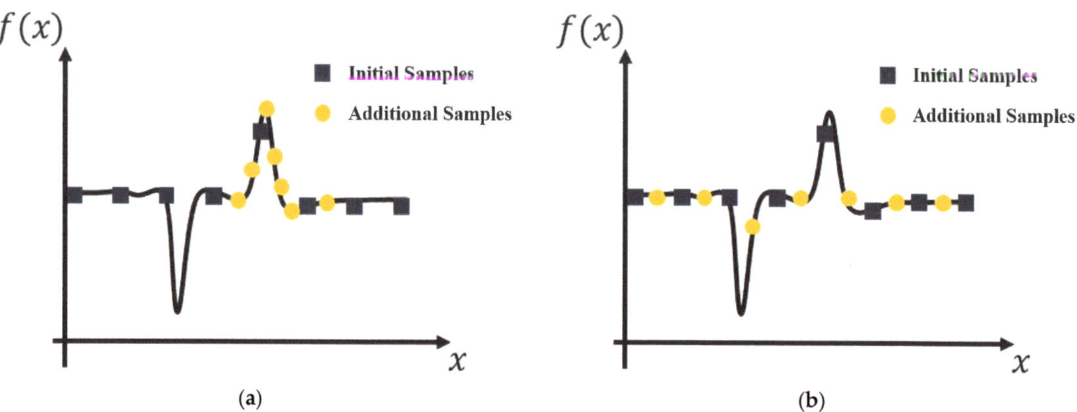

Figure 1. Comparison of additional sampling cases. (**a**) Case 1: Application of the exploitation, (**b**) Case 2: Application of the exploration.

On the other hand, exploration generates subsequent samples to gather information from the design space with large variance or empty space that may lead to better results. Figure 1b shows an example of additional sampling using exploration. Techniques based on the distance between samples [23,24], the variance of samples [25–27], and space-filling [28] are widely used in exploration due to their advantages in reducing uncertainty and improving the prediction accuracy of surrogate models. However, the computational burden may increase if samples are created in unnecessary space.

The paper addresses the exploitation–exploration tradeoff, a significant challenge in multi-variable, multi-objective optimization problems. Various adaptive sampling algorithms have been proposed to solve this problem, including expected improvement (EI), probability of improvement (PI), and upper confidence bound (UCB). Among them, we chose the Gaussian process-based UCB (GP-UCB) sampling method, which employs an efficient sampling approach with flexible parameter tuning capabilities to enhance the accuracy of surrogate models and improve the optimization efficiency.

The GP-UCB-based adaptive sampling proposed in this paper generates the next sample from data with the largest sum of mean and variance, as shown by the yellow line in Figure 2. The GP-UCB function $U(x)$ for an input variable x can be expressed as follows [29]:

$$U(x) = \mu(x) + \kappa\sigma(x) \tag{1}$$

where $\mu(x)$ is the mean value calculated by GP regression, $\sigma(x)$ is the variance, and κ is a hyperparameter to control the characteristics of the confidence bounds. The larger the hyperparameter, the larger the upper bound, and the algorithm favors solutions that explore currently unexplored regions of the design space. On the other hand, when κ is small, the algorithm focuses more on finding high-performance solutions.

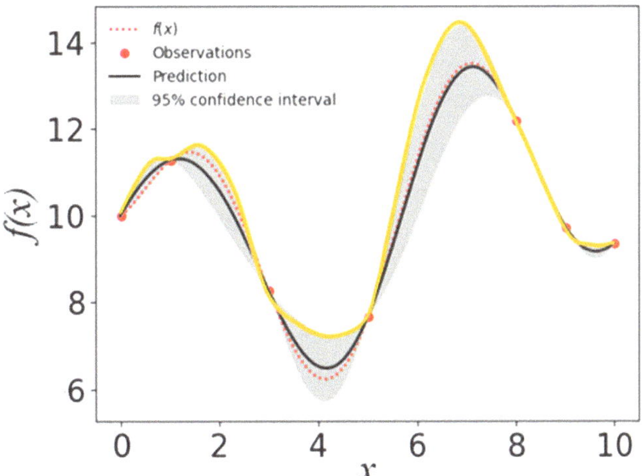

Figure 2. Illustration of the input–output relationship of the Gaussian process and GP-UCB (yellow line).

When tackling problems involving one-dimensional design variables, the sampling process is straightforward, involving the computation of responses across the entire design space. However, in electric machine design, the abundance of design variables gives rise to the "curse of dimensionality", rendering exhaustive exploration of the entire design space excessively computationally time-consuming. Additionally, electric motors designed for e-mobility applications commonly involve multiple objective functions, including torque, mass, cost, and efficiency. In the following sections, the application of the proposed GP-UCB-based design optimization process to two case studies will be presented, effectively delivering a comprehensive optimal solution to multi-objective, multi-variable problems.

2.1. Proposed Optimization Process

Figure 3 presents a comparative analysis between the conventional SM-based optimization process and the proposed optimization process employing the GP-UCB-based adaptive sampling technique. Both methods start with initial samples generated via a design of experiment (DOE) technique to construct an initial surrogate model, approximating the performance response of the electric machine under consideration. In the traditional approach depicted in Figure 3a, the prediction results of the SM model are iteratively compared with FE calculation results, and the calculated FEA results are incorporated into the model training set to enhance accuracy until convergence criteria are met. In contrast, as illustrated in Figure 3b, the proposed technique leverages GP-UCB and a metaheuristic algorithm to identify the Pareto front for the given objective function. This information is then utilized to generate the subsequent dataset for SM training. However, it is acknowledged that GP-UCB predictions may encounter local minima due to approximation errors. To avoid this challenge, the proposed optimization process integrates a space-filling technique to generate additional samples from undiscovered regions. This has a similar effect to mutation in a genetic algorithm.

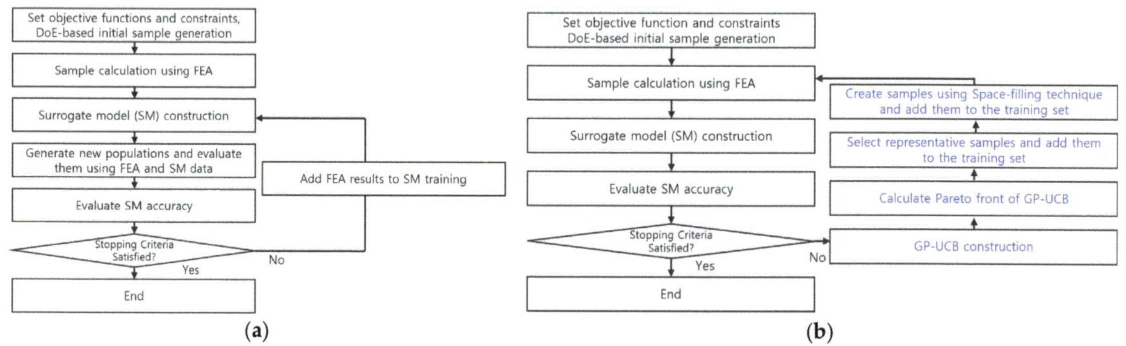

Figure 3. Comparison of the flowcharts. (**a**) Conventional optimization process, (**b**) proposed GP-UCB-based optimization process.

2.2. Case Study: 72-Slot/8-Pole IPM Traction Machine

This section presents the results of the case study conducted to validate the performance of the proposed adaptive sampling algorithm. The chosen reference model is an interior permanent magnet synchronous machine (IPMSM) originally designed in [30], featuring 72 slots, 8 poles, and hairpin windings on the stator. The choice of this reference model was based on the recognition that the proposed optimization process may not perform well when faced with very complex datasets. IPMSMs are known to be challenging to model, with significant nonlinearities due to deep magnetic saturation. Figure 4 shows a cross-section of the baseline IPMSM with key design parameters, and Table 1 provides information on the key parameters of the reference motor.

The two objective functions for this case study are the torque density of the machine and the total active material cost, which are expressed as follows:

$$\text{minimize:} \quad \begin{array}{l} 1.\ -(Tpk/m_{total})\ [\text{Nm/kg}] \\ 2.\ \text{Active material cost}\ [\$] \end{array}$$

where Tpk is the peak torque and m_{total} is the total mass of active materials. The assumed material costs are \$2.36/kg for the iron core, \$118/kg for magnets, and \$9.44/kg for copper.

Table 1. Main parameters of the 72-slot/8-pole IPMSM motor.

Parameter	Value
Slot/Pole	72/8
Peak current density	25 Arms/mm^2
Maximum current	400 Arms
Airgap length	0.75 mm
Rotor outer diameter	150 mm
Stator outer diameter	mm
Stack length	90 mm

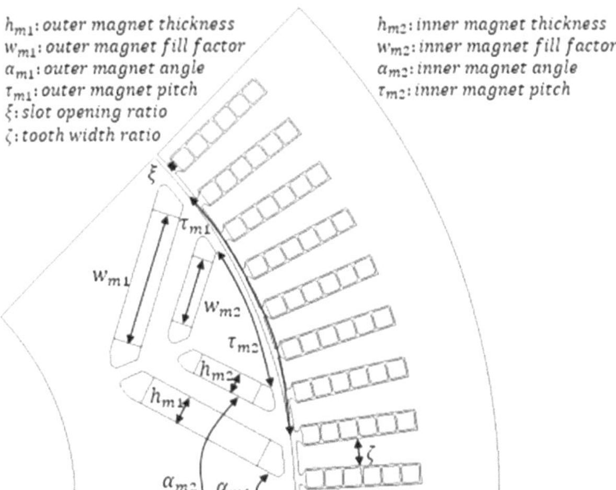

Figure 4. Cross section of the 72-slot/8-pole IPMSM baseline model.

To assess the efficacy of the proposed algorithm, critical performance metrics are examined, including exploitation–exploration balance, convergence speed, sensitivity to initial sampling, and repeatability. The SM is established using the GP-based algorithm detailed in [11], with NSGA-II employed for metaheuristic optimization. NSGA-II parameters include a crossover probability of 0.9 and a mutation probability of 0.05. The hyperpameter for GP-UCB is set to 1, and root-mean-square-error (*RMSE*) serves as a key metric, comparing SM and FEA results and acting as a convergence criterion. The mathematical expression of *RMSE* can be written as:

$$RMSE = \sqrt{\frac{1}{n}\sum_{i=1}^{n}(y_i - \hat{y}_i)} \times 100\% \tag{2}$$

where y_i is the ith data calculated using FEA, and $\hat{y}_i$ is the predicted value by SM.

Convergence occurs when the *RMSE* value is less than 0.5% of the average of the results of all accumulated FEA calculations. This can be expressed as follows:

$$RMSE \leq \frac{1}{n}\sum_{i=1}^{n} y_i \times 0.5\% \tag{3}$$

The adaptive sampling technique aims to optimize a balance between exploitation and exploration, promoting comprehensive exploration while exploiting regions likely to contain optimal solutions. Figure 5 compares exploitation–exploration balance among three sampling techniques: Latin hypercube sampling (LHS), NSGA-II, and the proposed GP-UCB-based adaptive sampling. Figure 5d demonstrates superior performance of the GP-UCB-based method, enhancing the Pareto front and concentrating sample distribution near it, with the same number of the initial 200 samples. Alternatively, if fewer samples are used in the GP-UCB-based method, similar performance can still be achieved compared to the other two techniques.

Furthermore, the proposed GP-UCB-based method exhibits accelerated convergence, as shown in Figure 6. During the first iteration with 50 samples, *RMSE* values for the GP-UCB-based method (0.57%) outperform those for the other methods (1.42% and 2.25%). In fact, the *RMSE* value of the GP-UCB method almost satisfies the predefined convergence condition in (2) in one iteration. This improved convergence speed is attributed to the adaptive sampling technique prioritizing regions with high uncertainty or sensitivity, refining the surrogate model more rapidly. Consequently, the GP-UCB-based adaptive sampling demonstrates robustness across multiple optimization runs by reducing the sensitivity to the quality of the initial sample generated by random sampling techniques such as LHS.

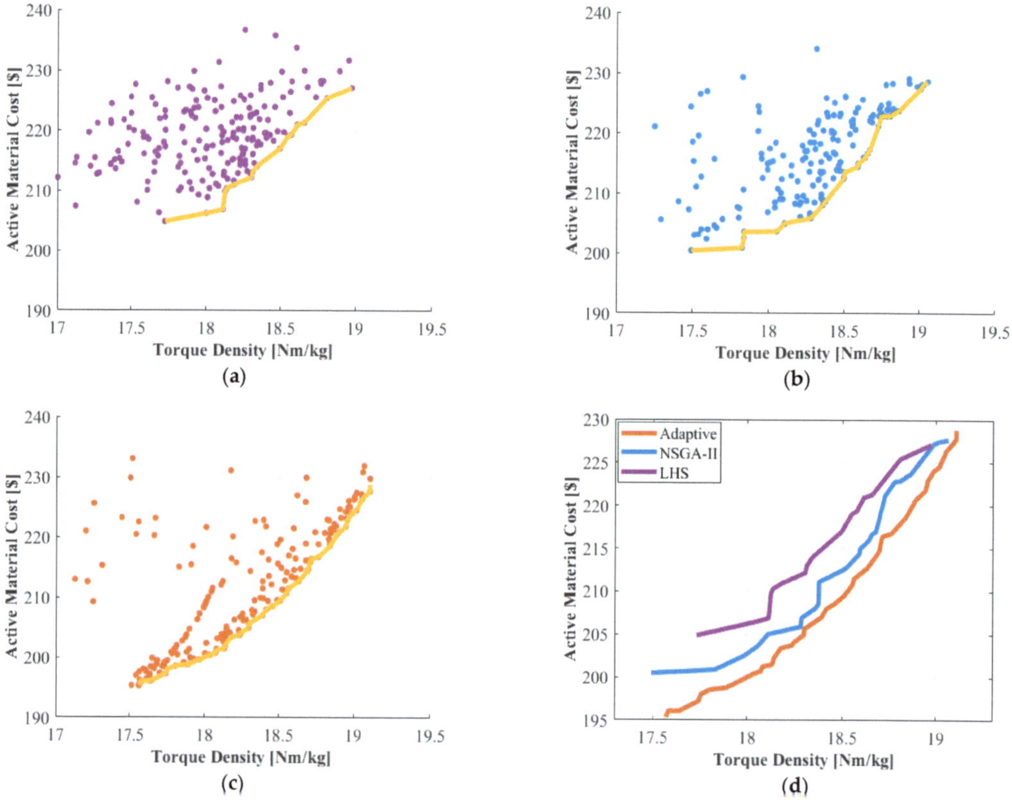

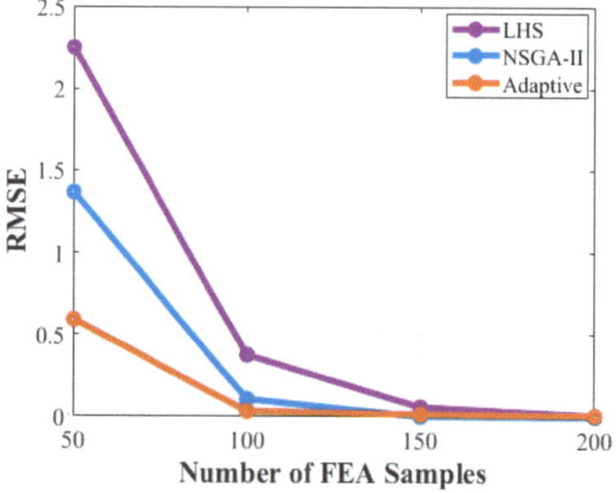

Figure 5. Comparison of exploitation–exploration balance. (**a**) Sample distribution using LHS, (**b**) sample distribution using NSGA-II, (**c**) sample distribution using GP-UCB, (**d**) comparison of Pareto fronts. (dots: Pareto dominated solutions, lines: Pareto fronts).

Figure 6. Comparison of convergence speed.

The performance of the optimization process can vary depending on the initial sample distribution [31]. Figure 7 illustrates the distribution and initial Pareto front of samples generated from five independent runs of LHS. Performing optimizations for each dataset generated by the random sampling technique can lead to repeatability issues. The proposed GP-UCB-based optimization method helps to mitigate this repeatability problem by striking an exploration–development balance to rapidly improve model accuracy and convergence speed. Figure 8 shows that the GP-UCB-based adaptive sampling proved to be robust, with nearly identical Pareto front results across the five optimization runs.

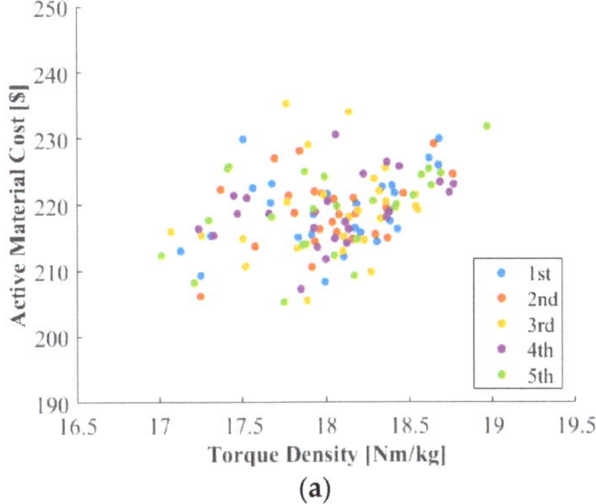

(a)

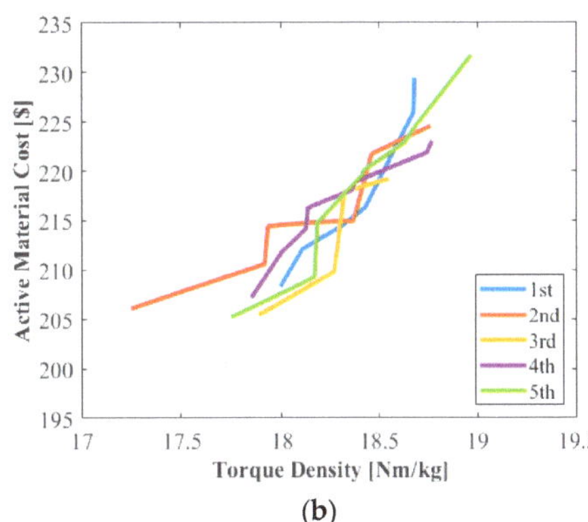

(b)

Figure 7. Sensitivity to initial sampling. (**a**) Data distribution after 5 initial sampling runs (50 sample generation), (**b**) Pareto fronts comparison.

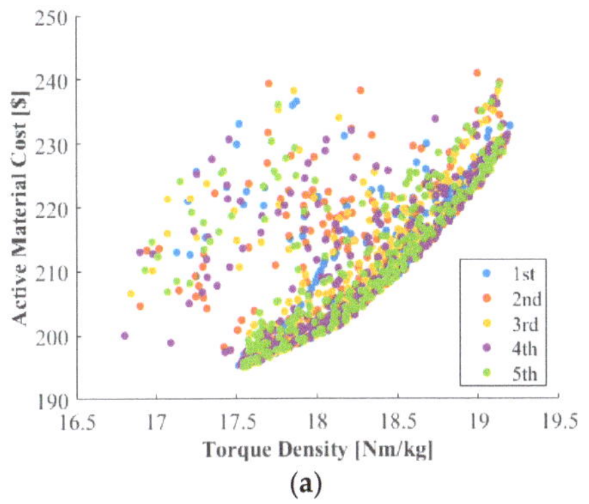

(a)

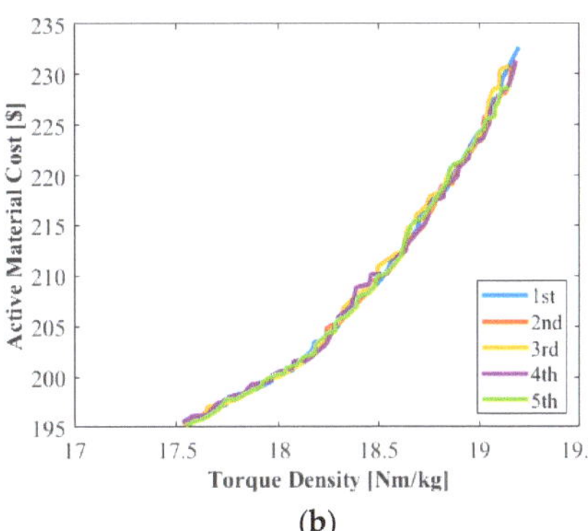

(b)

Figure 8. Repeatability comparison between five simulation runs. (**a**) Sample distribution, (**b**) Pareto fronts comparison (the number of generations is 5, with 50 samples per generation).

In summary, the proposed GP-UCB-based adaptive sampling technique outperforms traditional methods, offering enhanced exploration of the design space, improved exploitation–exploration balance, faster convergence, reduced sensitivity to initial samples, and improved repeatability that is evident in nearly identical Pareto fronts across multiple optimization runs.

3. EPS Motor Design

Permanent magnet synchronous machines (PMSMs) with fractional-slot concentrated windings (FSCWs) have gained significant attention over the past two decades, driven by their advantages in power density, efficiency, and fault tolerance [32–34]. This section explores two distinct FSCW-PMSM design families characterized by slot-per-pole-per-phase ratios of 1/2 and 2/5 for EPS applications. The 1/2 family is favored for its low harmonic contents in stator MMFs, resulting in minimal rotor losses. However, it exhibits a relatively low fundamental winding factor of 0.866. Conversely, the 2/5 family, with a fundamental winding factor (k_{w1}) of 0.933 and a least common multiple (LCM) value of 60, offers high torque density and low cogging torque. In particular, the 2/5 family further increases the winding factor to 0.966 when applying single-layer windings, and provides fault tolerance capability due to zero mutual coupling between phases. For balanced torque performance and machine losses, the 2/5 family with a double-layer winding configuration is selected for comparison. The two baseline designs with surface-mounted magnets (SPMs) on a rotor, featuring 9-slot/6-pole and 12-slot/10-pole combinations, are illustrated in Figure 9, with key parameters summarized in Table 2.

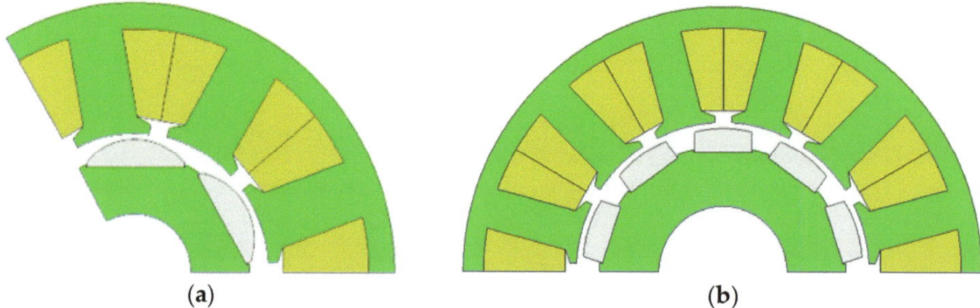

(a) **(b)**

Figure 9. Cross-sections of the baseline designs. (**a**) Design 1: 9-slot/6-pole, (**b**) Design 2: 12-slot/10-pole.

Table 2. Key machine parameters for the two baseline designs.

Parameter	Design 1	Design 2
Slot/Pole	9/6	12/10
DC bus voltage	48 V	
Rated current	7.57 Arms	
Stator diameter	86 mm	85 mm
Rotor diameter	44 mm	47 mm
Stack length	37 mm	36 mm
Series turns	105	100
# of parallel circuit	3	2
Rotor skew	Yes	No
Current density	6.54 Arms/mm^2	6.36 Arms/mm^2

The stator design variables to be optimized consist of slot width ratio bs, slot opening ratio bo, and tooth-tip thickness htt, as shown in Figure 10a. Among these, the slot width ratio and slot height were determined, considering a given number of winding turns and current density level. The current density level was set to around 6.5 Arms/mm^2 at rated conditions, assuming the motor is air-cooled. The rotor design variables shown in Figure 10a, such as magnet arc eccentricity radius rec, magnet thickness hpm, and magnet width ratio τpm, were also optimized for optimal machine performance. The protruding rotor caps between the magnets prevent magnet displacement during rotor rotation. The cap structure can be categorized as rectangular or round, as shown in Figure 10b. The rotor magnets were designed in a breadloaf shape to produce near-sinusoidal back-emf waveforms to reduce cogging torque and torque ripple. The key operating points considered for optimization are shown in Table 3.

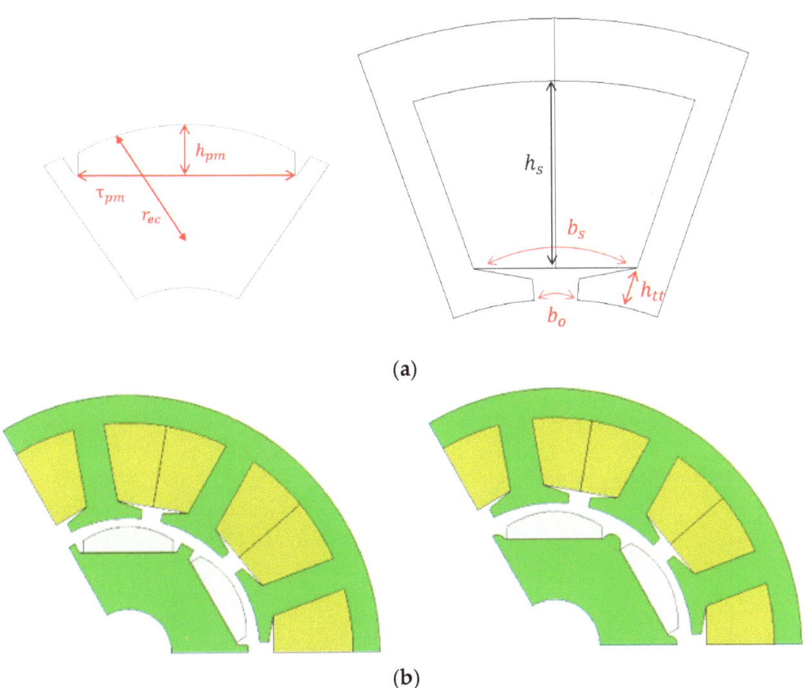

Figure 10. Key design variables. (**a**) Key stator and rotor variables, (**b**) rectangular cap (**left**) and round cap (**right**).

Table 3. Key operating points considered for optimization.

Operating Point	Torque [Nm]	Speed [r/min]	Power [W]
Point 1	1.9	790	157
Point 2	1.1	1750	202

As discussed extensively in previous literature, mitigating cogging torque and torque ripple in EPS motors is essential to prevent degradation of driving performance and ride quality. In this section, we conduct a comprehensive comparison of fundamental electromagnetic performance metrics, including back-emf voltage, cogging torque, and torque ripple, for the 9-slot/6-pole machine (Design 1) illustrated in Figure 9a. Figure 11 compares the back-emf waveforms and FFT spectrum for Design 1 with and without rotor skewing and rotor caps. Rotor skewing is observed to diminish the amplitude of back-emf voltage while concurrently reducing harmonic components, as shown in Figure 11b. While this reduction has a negative impact on the average torque, it is more desirable to minimize back-emf harmonics, which are a major source of torque ripple in EPS motor applications. Figure 12a further compares torque waveforms of Design 1 with and without rotor skewing and rotor cap structure under the rated load condition. Consistent with the cogging torque results in Figure 12b, rotor skewing demonstrates a reduction in both average torque and torque ripple.

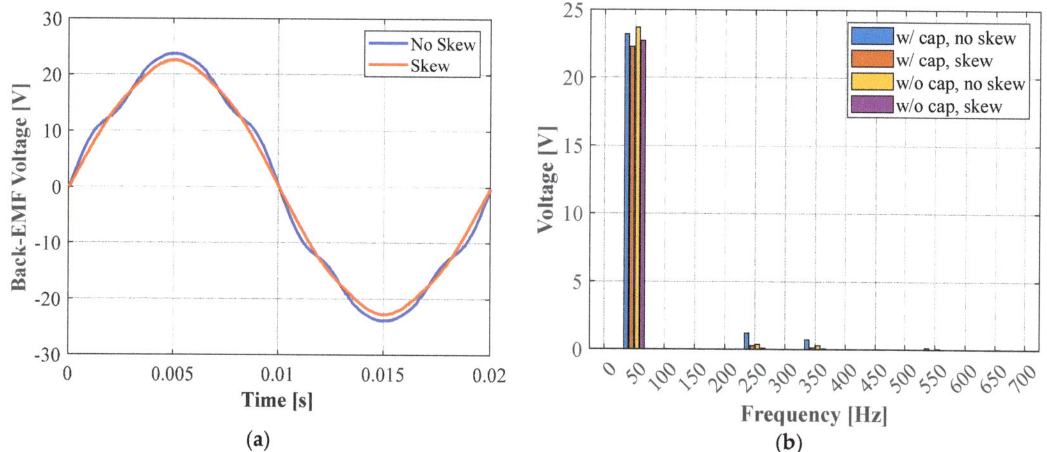

Figure 11. Back-emf voltage waveforms of 9-slot/6-pole design at 1000 r/min. (**a**) Back-emf waveforms comparison with and without rotor skewing, (**b**) FFT spectrum.

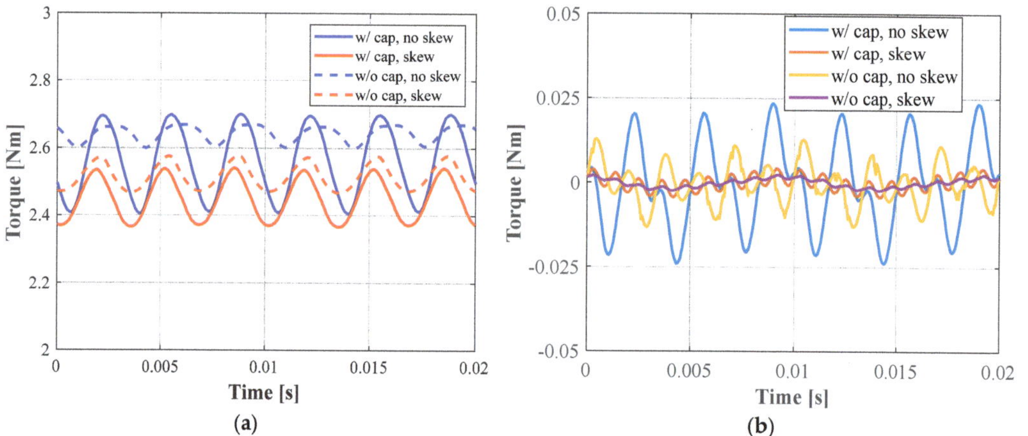

Figure 12. Cogging torque comparison of 9-slot/6-pole design with and without skewing and rotor cap at 790 r/min. (**a**) Instantaneous torque waveforms at rated load, (**b**) cogging torque waveforms comparison.

Figure 12 compares the cogging torque waveform with the torque waveform under the rated load condition at 790 r/min. The results in Figure 12b show that rotor skewing significantly reduces the cogging torque amplitude by a factor of 5, regardless of the presence of the rotor cap. The rotor cap, which acts as a magnet stopper, contributes to increasing the cogging torque amplitude. To balance structural and electromagnetic considerations, the height of the rotor cap is optimized to 0.5 mm. To optimize the efficacy of rotor skewing, we chose a skew step number of 3 and a 40/3° skew angle to minimize the fundamental component of the cogging torque.

4. Design Optimization

This section provides a detailed analysis of the results of applying the multi-objective, multi-variable design optimization process with GP-UCB-based adaptive sampling described in Section 2 to find the global optimal solution for the EPS motor.

The optimization process can be expressed as follows:

$$\text{minimize:} \qquad \begin{array}{l} 1.\ -(Tpk/m_{total})\ [\text{Nm/kg}] \\ 2.\ \text{Ploss [W]} \\ 3.\ \text{Tcog [Nm]} \end{array}$$

$$\text{subject to:} \qquad \begin{array}{l} 1.\ \text{Active material cost} \leq \$20 \\ 2.\ Tripple \leq 4\% \end{array}$$

where Ploss is the sum of the machine losses calculated at Point 1 and 2 (see Table 3), *Tcog* is the cogging torque, and *Tripple* is the torque ripple. The definition of torque ripple is as follows:

$$Torque\ Ripple\ (\%) = \frac{T_{max} - T_{min}}{T_{avg}} * 100\% \tag{4}$$

where T_{max} is the maximum value of the instantaneous torque waveform, T_{min} is the minimum value, and T_{avg} is the average value. Table 4 shows the range of the input design variables for optimization (see Figure 10a).

Table 4. Design variables used in the optimal design and their ranges.

Design Parameter	Symbol	Range	
		Min	Max
Magnet arc eccentricity radius	r_{ec}	22.5 mm	15.5 mm
Magnet width ratio	τ_{pm}	0.85	0.6
Magnet thickness	h_{pm}	4 mm	3 mm
Slot opening ratio	b_o	0.55	0.35
Slot width ratio	b_s	0.55	0.35
Tooth-tip thickness	h_{tt}	2.4 mm	1.2 mm

Figure 13 shows the proposed optimization process, divided into three steps: (1) Initial setup to define objective functions and constraints, followed by parent sample generation and evaluation through FEA; (2) Construction of the SM employing FE-calculated data until convergence criteria are met; and (3) Surrogate-based optimization utilizing the NSGA-II algorithm. The flowchart in Figure 13 is developed by incorporating the proposed GP-UCB-based adaptive sampling shown in Figure 3b into the surrogate-assisted optimization process introduced in [11]. As shown in the figure, the proposed adaptive sampling is applied to both step 2 and step 3, leveraging the advantages of being applied to each phase.

Figure 14 shows a side-by-side comparison of the initial geometry and optimized design for the 9-slot/6-pole model, showing the changes in tooth width, magnet geometry, and slot openings. Table 5 compares the machine performance before and after optimization, showing improvements across all aspects except torque ripple. Despite the inverse relationship between torque density and losses, the optimized design exhibits a 13% increase in torque density with a slight decrease in losses, and an 11% reduction in cogging torque, from ±7.4 mNm to ±6.6 mNm. Although torque ripple increased from 2.17% to 2.94%, it is still within the targeted 3% constraint. To successfully suppress cogging torque and torque ripple, the step skew technique mentioned earlier was applied to Design 1.

Table 5. Performance comparison of the 9-slot/6-pole design before and after optimization.

	Torque Density [Nm/kg]	Cogging Torque [mNm]	Torque Ripple [%]	Losses [W]	Cost [$]
Before opt.	1.31	7.41	2.17	64.94	20.83
After opt.	1.48	6.62	2.94	63.94	19.20
Difference [%]	+12.98	−10.66	+35.48	−1.54	−8.83

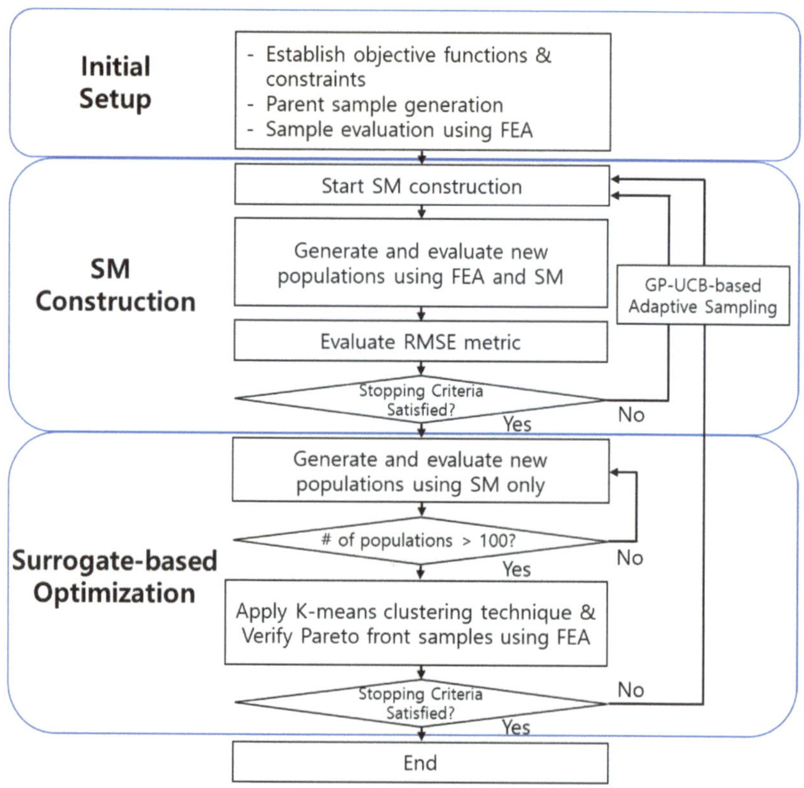

Figure 13. Multi-objective design optimization process using GP-UCB-based adaptive sampling.

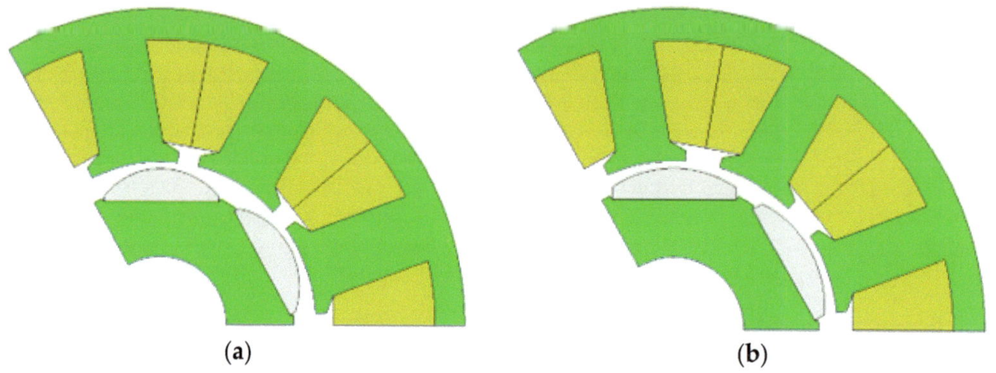

Figure 14. Cross sections of Design 1. (**a**) Before optimization, (**b**) after optimization.

Figure 15 shows a comparison of the initial and optimized geometry for the 12/10 model. The optimized design shows a slight change in tooth width, a wider arc angle, and reduced magnet height. Table 6 shows the difference in performance before and after the optimization, showing an overall improvement in performance. Losses increased slightly from 54.5 W to 55.7 W, but torque density increased by 5.2%, and cogging torque plummeted from ±40.1 mNm to ±6.3 mNm without rotor skew, a reduction of 84%. Importantly, for the same operating conditions outlined in Table 3, the 12/10 model had 15% lower losses than the 9/6 model, and achieved similar levels of cogging torque and torque ripple without applying step skew to the rotor. Torque ripple increased from 2.1% to 2.7%, staying within the targeted 3% constraint. Table 7 provides the change in design parameters before and after optimization for Design 1 and Design 2, respectively.

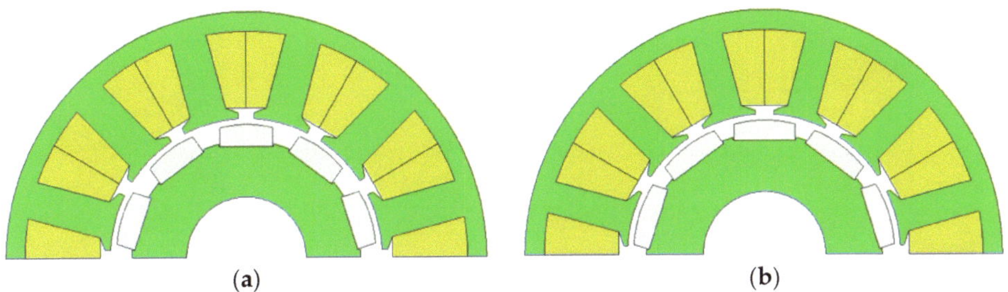

Figure 15. Cross sections of Design 2. (**a**) Before optimization, (**b**) after optimization.

Table 6. Performance comparison of the 12-slot/10-pole design before and after optimization.

	Torque Density [Nm/kg]	Cogging Torque [mNm]	Torque Ripple [%]	Losses [W]	Cost [$]
Before opt.	1.71	40.12	2.08	54.58	17.48
After opt.	1.80	6.33	2.69	55.66	16.75
Difference [%]	+5.23	−84.19	+29.44	+1.98	−4.21

Table 7. Parameter variations before and after optimization.

Design Parameter	Symbol	Design 1		Design 2	
		Before	After	Before	After
Magnet arc eccentricity radius [mm]	r_{ec}	16	16	16.5	21.4
Magnet width ratio [-]	τ_{pm}	0.85	0.85	0.85	0.85
Magnet thickness [mm]	h_{pm}	4.00	4.13	4.00	3.53
Slot opening ratio [-]	b_o	0.35	0.41	0.40	0.41
Slot width ratio [-]	b_s	0.40	0.48	0.40	0.40
Tooth-tip thickness [mm]	h_{tt}	1.70	1.41	0.90	0.67

The NSGA-II algorithm uses a population of 50 and 40 evolutionary generations, with a crossover probability of 0.9 and a mutation probability of 0.05. Figure 16 shows the 3D projections of the Pareto non-dominated designs and the Pareto front for Design 1 and Design 2, illustrating optimal solution sets among conflicting objective functions. Conventional solution-finding methods based on weighting factors are susceptible to biases, elevating the likelihood of obtaining locally optimal solutions. Hence, in modern optimization practices, machine designers often make selections based on technical requirements, strategically navigating tradeoffs among various objective functions. Indeed, considering the nature of muti-objective optimization, achieving a single global solution optimizing all three objective functions simultaneously is unattainable. Our approach involves prioritizing the reduction of cogging torque while balancing other objectives and constraints, aligning with the characteristics of EPS applications. It is worth noting that the displayed samples represent a subset of the total tested, considering the scale of the plot axes.

Table 8 provides the design optimization results for Design 1 and Design 2. In particular, the 12/0 model (Design 2) shows excellent performance characteristics across torque density, efficiency, cost, and manufacturability. Both models exhibit excellent cogging torque characteristics and maintain torque ripple within the targeted 3%, which is consistent with the desirable characteristics for EPS applications.

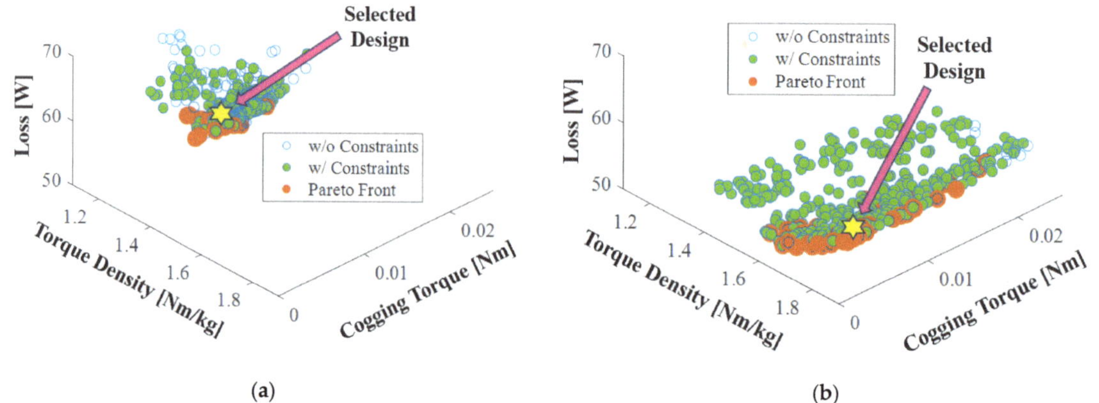

Figure 16. 3D objective space projections of the Pareto non-dominated designs and Pareto front. (**a**) Design 1 (9-slot/6-pole), (**b**) Design 2 (12-slot/10-pole).

Table 8. Performance comparison between the two baseline designs.

	Torque Density [Nm/kg]	Cogging Torque [mNm]	Torque Ripple [%]	Losses [W]	Cost [$]	Rotor Skew
Design 1 (9-slot/6-pole)	1.48	6.62	2.94	63.94	19.20	O
Design 2 (12-slot/10-pole)	1.80 (+22%)	6.33 (−4%)	2.69 (−9%)	55.66 (−13%)	16.75 (−13%)	X

Figure 17 presents instantaneous torque waveforms for Design 1 and Design 2 under no load (i.e., cogging torque) and at rated load. The implementation of rotor step skew effectively controls torque ripple to under 3% at the rated condition for Design 1, with cogging torque amplitude sufficiently suppressed to approximately 6 mNm. Design 2, even without rotor step skew, displays comparable torque ripple and cogging torque, suggesting easier and more cost-effective manufacturing.

Finally, efficiency performance is compared between Design 1 and Design 2. Figure 18 illustrates the efficiency maps for the two baseline designs under the conditions outlined in Table 2. Design 2 has a 7.7% higher fundamental winding factor than Design 1, resulting in noticeably lower losses for the same torque. Looking at the operating points (see Table 3) overlaid on the efficiency maps in Figure 18a,b, we can see that Design 2 has a relatively higher operating efficiency.

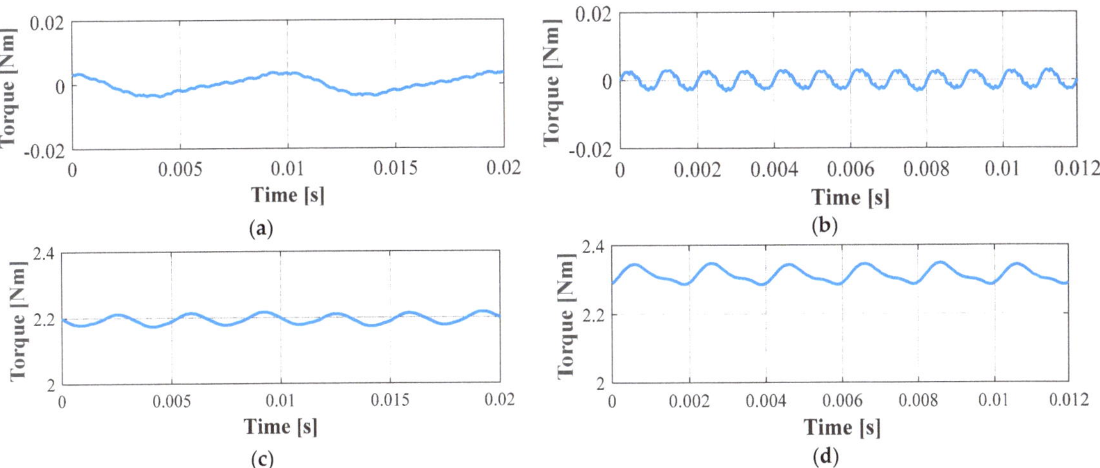

Figure 17. Torque waveforms of Design 1 and Design 2. (**a**) Cogging torque waveform for Design 1, (**b**) cogging torque waveform for Design 2, (**c**) torque ripple waveform for Design 1, (**d**) torque ripple waveform for Design 2.

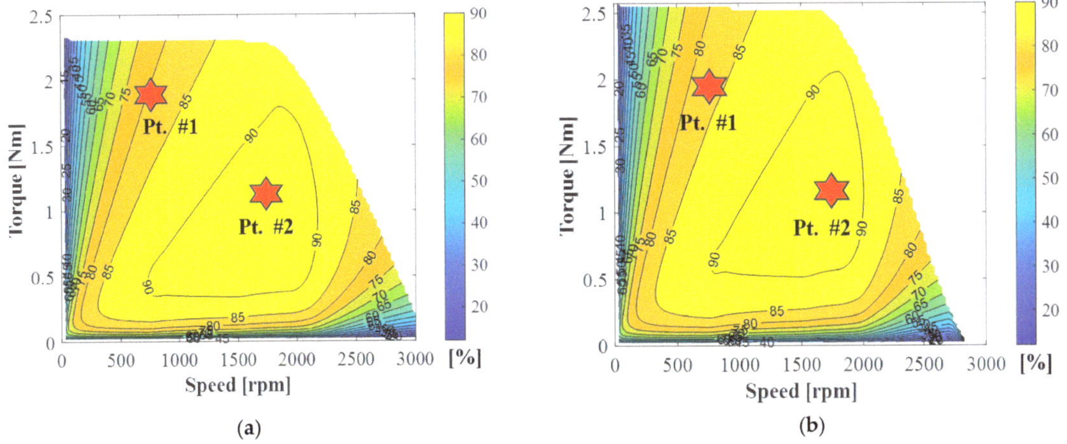

Figure 18. Comparison of efficiency maps. (**a**) Design 1, (**b**) Design 2.

5. Prototype Machine and Experimental Results

To experimentally validate the performance of Design 2, a prototype machine was built and tested. Figure 19 shows a 200 W (rated) 12-slot/10-pole prototype FSCW-SPM machine specifically designed for EPS application. The rotor in this prototype FSCW-SPM machine features N42SH sintered NdFeB magnets, which have excellent thermal resistance up to 120 °C.

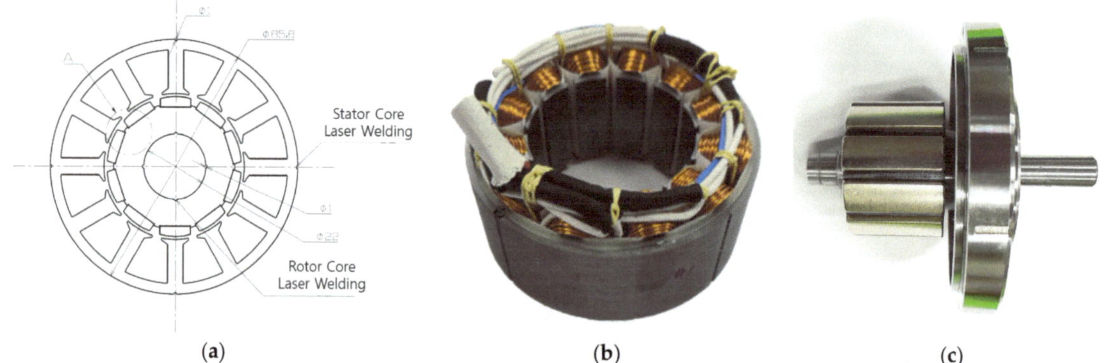

Figure 19. Prototype EPS motor: (**a**) Machine drawing, (**b**) stator core and windings; (**c**) rotor core and magnets with the bearing and front cover attached.

Figure 20a shows the back-to-back dynamometer setup used for testing. The prototype machine was mounted on the dynamometer and an industrial SPM machine with a maximum torque of 4.8 Nm, and a maximum speed of 5000 rpm was used as a prime mover to perform back-emf voltage testing at no-load. This dynamometer setup is controlled by a custom dual-inverter motor drive, as shown in Figure 20b.

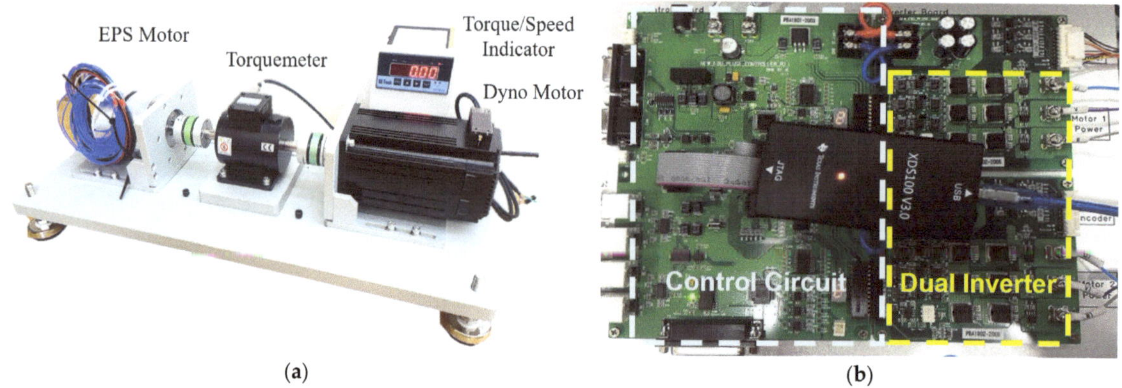

Figure 20. Experimental dynamometer setup. (**a**) Dynamometer jig, (**b**) inverter hardware.

Table 9 summarizes the experimental reverse electromotive force voltage measurement results. Figure 21 provides a comparison between the measured back-emf voltage waveform and the FE-predicted back-emf waveform. As evident in the figure, the measured waveform closely matches the FE predicted waveform, indicating excellent agreement. Figure 22 shows a comparison of the FFT spectra of the FE-predicted waveform and measured back-emf waveform. The resulting calculated THD values are 1.94% for the simulated result and 1.38% for the experimental result, indicating that both waveforms are very close to the ideal sinusoidal waveform.

Table 9. A comparison of measured and FE-predicted phase back-emf voltages at 1000 RPM.

Test Type	Phase-*a*	Phase-*b*	Phase-*c*	Average
Experiment	11.24 Vrms	11.24 Vrms	11.21 Vrms	11.23 Vrms
FEA	11.20 Vrms	11.20 Vrms	11.20 Vrms	11.20 Vrms

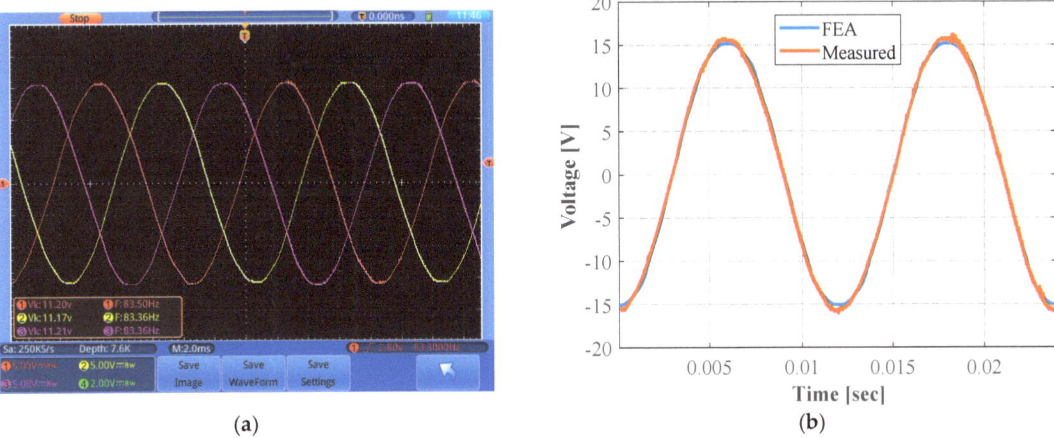

(**a**)

(**b**)

Figure 21. Back-emf waveforms of the prototype FSCW-SPM machine. (**a**) Measured back-emf voltage waveforms, (**b**) measured vs. FE-predicted back-emf waveforms.

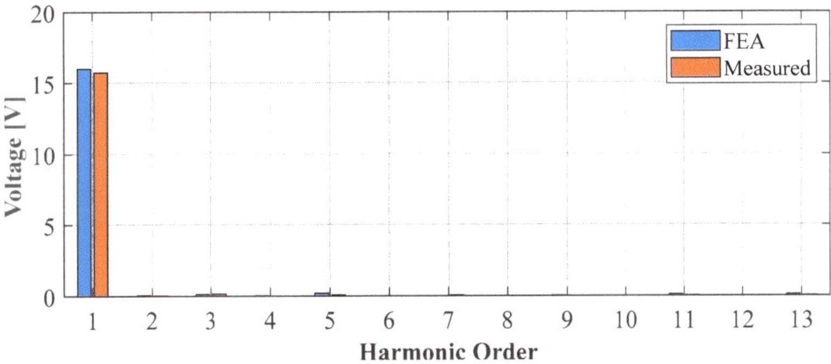

Figure 22. FFT spectrum comparison: FEA vs. measured back-emf waveforms.

6. Conclusions

This paper presented a novel design optimization process utilizing GP-UCB-based adaptive sampling for electric machine design. Through two case studies involving an IPMSM for EV applications and an FSCW-SPMSM for EPS applications, the proposed approach demonstrated superior performance compared to conventional optimization methods lacking adaptive sampling. Specifically, the adaptive sampling technique significantly improved key optimization performance measures, including exploitation–exploration balance, convergence speed, sensitivity to initial sampling, and repeatability.

Subsequently, GP-UCB-based adaptive sampling was employed in the optimization process to identify the optimal design for EPS applications with demanding performance requirements. The results highlighted that an FSCW-SPM design featuring a 12-slot/10-pole configuration exhibited exceptional torque density, high efficiency, low cost, and enhanced manufacturability—aligning well with the desired performance characteristics for EPS applications. The optimization results were further validated through a dynamometer test, revealing an error of only 0.3% between the amplitude of measured and simulated back-emf voltages, indicating excellent agreement.

Finally, it was shown that the proposed approach is applicable to various stator and rotor configurations with minimal modifications. Categorized as a black-box approach, the proposed method exhibits certain limitations compared to a model-based approach. These include a lack of physical insights, limited control over the algorithm's decision-making process, and the need for intensive computational resources. It is up to the judgment of the machine designer to strike a balance between different approaches to improve the efficiency and physical significance of the design process. Planned future work includes further experimental validation under different loading conditions and evaluation of a broader range of optimization algorithms and sampling techniques for different types of electric machines.

Author Contributions: Methodology, software, formal analysis, validation, data curation, writing—original draft preparation, writing—review and editing, supervision, G.C.; software, formal analysis, G.-H.J.; methodology, software, formal analysis, writing—review, M.C.; validation, data curation, J.K.; formal analysis, writing—review, Y.G.K.; conceptualization, funding acquisition, writing—review and editing, S.K. All authors have read and agreed to the published version of the manuscript.

Funding: This work was supported in part by the NRF (National Research Foundation) grant funded by the Korean government (MSIT: Ministry of Science and ICT) (grant number: 2021R1F1A1048754), and in part by the Education and Research Promotion Program of KOREATECH in 2023.

Data Availability Statement: Data are contained within the article.

Conflicts of Interest: The authors declare no conflicts of interest.

References

1. Bianchi, N.; Bolognani, S. Design techniques for reducing the cogging torque in surface-mounted PM motors. *IEEE Trans. Ind. Appl.* **2002**, *38*, 1259–1265. [CrossRef]
2. Bianchi, N.; Pre, M.D.; Bolognani, S. Design of a fault-tolerant IPM motor for electric power steering. *IEEE Trans. Veh. Technol.* **2006**, *55*, 1102–1111. [CrossRef]
3. Jang, J.; Cho, S.G.; Lee, S.J.; Kim, K.S.; Kim, J.M.; Hong, J.P.; Lee, T.H. Reliability-Based Robust Design Optimization with Kernel Density Estimation for Electric Power Steering Motor Considering Manufacturing Uncertainties. *IEEE Trans. Magn.* **2015**, *51*, 8001904. [CrossRef]
4. Naidu, M.; Gopalakrishnan, S.; Nehl, T.W. Fault-Tolerant Permanent Magnet Motor Drive Topologies for Automotive X-By-Wire Systems. *IEEE Trans. Ind. Appl.* **2010**, *46*, 841–848. [CrossRef]
5. Uler, G.F.; Mohammed, O.A.; Koh, C.-S. Design optimization of electrical machines using genetic algorithms. *IEEE Trans. Magn.* **1995**, *31*, 2008–2011. [CrossRef]
6. Gao, J.; Sun, H.; He, L.; Dong, Y.; Zheng, Y. Optimization design of Switched Reluctance Motor based on Particle Swarm Optimization. In Proceedings of the 2011 International Conference on Electrical Machines and Systems, Beijing, China, 20–23 August 2011; pp. 1–5. [CrossRef]
7. Mutluer, M.; Bilgin, O. Design optimization of PMSM by particle swarm optimization and genetic algorithm. In Proceedings of the 2012 International Symposium on Innovations in Intelligent Systems and Applications, Trabzon, Turkey, 2–4 July 2012; pp. 1–4. [CrossRef]
8. Duan, Y.; Harley, R.G.; Habetler, T.G. Comparison of Particle Swarm Optimization and Genetic Algorithm in the design of permanent magnet motors. In Proceedings of the 2009 IEEE 6th International Power Electronics and Motion Control Conference, Wuhan, China, 17–20 May 2009; pp. 822–825. [CrossRef]
9. Murata, T.; Ishibuchi, H. MOGA: Multi-objective genetic algorithms. In Proceedings of the 1995 IEEE International Conference on Evolutionary Computation, Perth, WA, Australia, 29 November–1 December 1995; pp. 289–294. [CrossRef]
10. Li, Y.; Lei, G.; Bramerdorfer, G.; Peng, S.; Sun, X.; Zhu, J. Machine Learning for Design Optimization of Electromagnetic Devices: Recent Developments and Future Directions. *Appl. Sci.* **2021**, *11*, 1627. [CrossRef]
11. Choi, M.; Choi, G.; Bramerdorfer, G.; Marth, E. Systematic Development of a Multi-Objective Design Optimization Process Based on a Surrogate-Assisted Evolutionary Algorithm for Electric Machine Applications. *Energies* **2023**, *16*, 392. [CrossRef]
12. Taran, N.; Ionel, D.M.; Dorrell, D.G. Two-Level Surrogate-Assisted Differential Evolution Multi-Objective Optimization of Electric Machines Using 3-D FEA. *IEEE Trans. Magn.* **2018**, *54*, 8107605. [CrossRef]
13. Gao, Y.; Yang, T.; Bozhko, S.; Wheeler, P.; Dragičević, T. Filter Design and Optimization of Electromechanical Actuation Systems Using Search and Surrogate Algorithms for More-Electric Aircraft Applications. *IEEE Trans. Transp. Electrif.* **2020**, *6*, 1434–1447. [CrossRef]
14. Gu, J.; Hua, W.; Yu, W.; Zhang, Z.; Zhang, H. Surrogate Model-Based Multiobjective Optimization of High-Speed PM Synchronous Machine: Construction and Comparison. *IEEE Trans. Transp. Electrif.* **2023**, *9*, 678–688. [CrossRef]
15. Yoo, C.-H. A New Multi-Modal Optimization Approach and Its Application to the Design of Electric Machines. *IEEE Trans. Magn.* **2018**, *54*, 8202004. [CrossRef]
16. Jackson, D.; Belakaria, S.; Cao, Y.; Doppa, J.R.; Lu, X. Machine Learning Enabled Design Automation and Multi-Objective Optimization for Electric Transportation Power Systems. *IEEE Trans. Transp. Electrif.* **2022**, *8*, 1467–1481. [CrossRef]

17. Gupta, S.; Biswas, P.K.; Debnath, S.; Ghosh, A.; Babu, T.S.; Zawbaa, H.M.; Kamel, S. Metaheuristic Optimization Techniques Used in Controlling of an Active Magnetic Bearing System for High-Speed Machining Application. *IEEE Access* **2023**, *11*, 12100–12118. [CrossRef]
18. Eason, J.; Cremaschi, S. Adaptive sequential sampling for surrogate model generation with artificial neural networks. *Comput. Chem. Eng.* **2014**, *68*, 220–232. [CrossRef]
19. Fushiki, T. Estimation of prediction error by using K-fold cross-validation. *Stat. Comput.* **2011**, *21*, 137–146. [CrossRef]
20. Meckesheimer, M.; Booker, A.J.; Barton, R.R.; Simpson, T.W. Computationally inexpensive metamodel assessment strategies. *AIAA J.* **2002**, *40*, 2053–2060. [CrossRef]
21. Jiang, C.; Cai, X.; Qiu, H.; Gao, L.; Li, P. A two-stage support vector regression assisted sequential sampling approach for global metamodeling. *Struct. Multidiscip. Optim.* **2018**, *58*, 1657–1672. [CrossRef]
22. Fuhg, J.N.; Fau, A.; Nackenhorst, U. State-of-the-Art and Comparative Review of Adaptive Sampling Methods for Kriging. *Arch. Comput. Methods Eng.* **2021**, *28*, 2689–2747. [CrossRef]
23. Liu, H.; Xu, S.; Wang, X.; Wu, J.; Song, Y. A global optimization algorithm for simulation-based problems via the extended DIRECT scheme. *Eng. Optim.* **2015**, *47*, 1441–1458. [CrossRef]
24. Burke, J.V.; Curtis, F.E.; Lewis, A.S.; Overton, M.L.; Simões, L.E.A. Gradient Sampling Methods for Nonsmooth Optimization. *arXiv* **2018**. [CrossRef]
25. Sóbester, A.; Leary, S.J.; Keane, A.J. On the design of optimization strategies based on global response surface approximation models. *J. Glob. Optim.* **2005**, *33*, 31–59. [CrossRef]
26. Kleijnen, J.P.C.; Van Beers, W.C.M. Application-driven sequential designs for simulation experiments: Kriging metamodeling. *J. Oper. Res. Soc.* **2004**, *55*, 876–883. [CrossRef]
27. Liu, H.; Xu, S.; Ma, Y.; Chen, X.; Wang, X. An Adaptive Bayesian Sequential Sampling Approach for Global Metamodeling. *J. Mech. Des.* **2016**, *138*, 011404. [CrossRef]
28. Nuchitprasittichai, A.; Cremaschi, S. An algorithm to determine sample sizes for optimization with artificial neural networks. *AIChE J.* **2013**, *59*, 805–812. [CrossRef]
29. Makondo, N.; Folarin, A.L.; Zitha, S.N.; Remy, S.L. An Analysis of Reinforcement Learning for Malaria Control. *arXiv* **2021**. [CrossRef]
30. Choi, M.; Choi, G. Investigation, and Mitigation of AC Losses in IPM Machines with Hairpin Windings for EV Applications. *Energies* **2021**, *14*, 8034. [CrossRef]
31. Agushaka, J.O.; Ezugwu, A.E. Initialisation Approaches for Population-Based Metaheuristic Algorithms: A Comprehensive Review. *Appl. Sci.* **2022**, *12*, 896. [CrossRef]
32. EL-Refaie, A.M. Fractional-Slot Concentrated-Windings Synchronous Permanent Magnet Machines: Opportunities and Challenges. *IEEE Trans. Ind. Electron.* **2010**, *57*, 107–121. [CrossRef]
33. Cros, J.; Viarouge, P. Synthesis of high performance PM motors with concentrated windings. *IEEE Trans. Energy Convers.* **2002**, *17*, 248–253. [CrossRef]
34. Choi, G.; Jahns, T.M. Analysis and Design Recommendations to Mitigate Demagnetization Vulnerability in Surface PM Synchronous Machines. *IEEE Trans. Ind. Appl.* **2018**, *54*, 1292–1301. [CrossRef]

Article

Model Predictive Control Strategy Based on Loss Equalization for Three-Level ANPC Inverters

Shaoqi Wan [1]**, Bo Wang** [2]**, Jingbo Chen** [2]**, Haiying Dong** [1,]*** **and Congxin Lv** [1]

1 College of New Energy and Power Engineering, Lanzhou Jiaotong University, Lanzhou 730070, China; 12212033@stu.lzjtu.edu.cn (S.W.); 11210380@stu.lzjtu.edu.cn (C.L.)
2 Lanzhou Wanli Airlines Electromechanical Limited Liability Company, Lanzhou 730070, China; wangb18306782.135@mail.avic (B.W.); chenjb18708803.135@mail.avic (J.C.)
* Correspondence: hydong@mail.lzjtu.cn

Abstract: Targeting the issue of high losses of individual switching tubes in Neutral-Point Clamped (NPC) three-level inverters, an Active Neutral-Point Clamped (ANPC) three-level inverter is used, and a model predictive control strategy using the loss equalization of the inverter is proposed. This method organizes and analyzes multiple zero-state current pathway commutation modes and adds mode three under the original two commonly used zero-state commutation modes. On this basis, the three modes are flexibly switched by model predictive control, and the output is optimized according to the value function for the space vector in each operation, while the midpoint voltage control is added to the value function. The simulation results suggest that the recommended strategy in this study may effectively realize the loss equalization control and midpoint voltage control of the ANPC inverter, which improves the operation efficiency of the electromechanical actuator.

Keywords: three-level inverter; active midpoint clamp; loss equalization control; commutation mode; model predictive control; midpoint voltage

---

1. Introduction

Electromechanical servo systems are increasingly used in aerospace [1,2], and electromechanical actuators are the actuators of electromechanical servo systems, which are more efficient, more integrated, and easier to maintain than traditional hydraulic actuators. With the development of aerospace electromechanical actuators, permanent magnet synchronous motors (PMSM) have become an important part of them because of their high efficiency, high power density, and precise control [3,4]. The control effect of PMSM plays a crucial role in the performance and reliability of electromechanical actuator systems [5–7], and most of the existing researches improve the performance of PMSM by adopting multilevel inverters with advanced control strategies and the loss control should be considered by adopting multilevel inverters.

Among several multilevel structures, the NPC inverter circuit is currently one of the more mature structures, and its circuit structure and control method are straightforward and have the benefits of a high voltage withstand level and a low rate of output voltage distortion [8,9]. However, due to its loss, the problem is difficult to solve, resulting in some switching tubes' junction temperatures being too high and reducing the system efficiency, so the redundant zero-level path more ANPC inverter is used to solve the problem.

In the existing problem of ANPC inverter loss, soft-switching technology is used, which can effectively reduce the voltage and current stresses, switching losses, and electromagnetic interference of switching devices [10,11]. However, multiple switching tubes, as well as capacitors and inductors, need to be added to the inverter to generate resonance, which increases the device size and reduces the power density. Due to the high requirement of device power density in electromechanical actuators [1,2], it does not meet the requirements. Part of it combines the zero-level characteristics of the ANPC inverter,

Citation: Wan, S.; Wang, B.; Chen, J.; Dong, H.; Lv, C. Model Predictive Control Strategy Based on Loss Equalization for Three-Level ANPC Inverters. *Actuators* **2024**, *13*, 111. https://doi.org/10.3390/act13030111

Academic Editor: Giorgio Olmi

Received: 6 February 2024
Revised: 7 March 2024
Accepted: 11 March 2024
Published: 12 March 2024

---

replacing the high-frequency part of the switching tube Si(Silicon) IGBT with SiC(Silicon Carbide) MOSFETs to reduce the loss by using the characteristics of SiC devices [12–14]. This method reduces the loss, but the SiC device increases the cost, and the di/dt and dv/dt of the two devices with different switching speeds in the switching process will lead to electromagnetic interference and switching overvoltage [12], so it does not meet the requirements; The remaining part of the improvement of the control strategy, through the reasonable allocation of the redundant zero-level path to achieve the balanced control of the switching tube loss.

Literature [15] based on hardware PWM configuration outputs asymmetric driving waveforms, which makes the turn-on and turn-off losses separate, but its calculation is complicated and does not make full use of the redundant zero level of the ANPC inverter. Literature [16] controls the ANPC inverter two redundant zero level allocation mode one and mode two alternately; for the different allocation modes to bear the switching loss device, different characteristics of the two flexible switching are needed to achieve the switching tube loss equalization control. Literature [17] combines the two methods to turn a single zero-level clamp circuit into two clamp circuits conducting at the same time, which reduces the conduction loss and increases the efficiency. Literature [18] regulates the frequency cycle, changing the duration ratio of mode one and mode 2 in each frequency cycle to realize the balanced adjustment of switching tube losses, but the ratio selection lacks a theoretical basis, and frequent switching between different modes will produce excessive switching losses, affecting the system efficiency [19]. Literature [16,20,21] used a model prediction method of multi-objective optimization for loss equalization control by setting the value function loss minimization but did not classify and make full use of the ANPC inverter redundancy zero level in a manner, whereas literature [20,21] used only manner two and its variants.

Based on the above problems, this paper makes complete use of the redundant zero level of the ANPC inverter and adds mode three on the basis of the two commonly used zero-level path modes. It summarizes the switching tubes and the loss law driven by each mode, designs the loss principle between the modes, and uses the model predictive control to realize the flexible switching on the basis of which the model predictive control is implemented [7,22]. Thus, the switching tube loss equalization control is accomplished, and the solution to the problem of uneven inverter loss in electromechanical actuators is realized, and, at the same time, it realizes inverter midpoint voltage equalization control. Finally, relevant simulations are carried out in Matlab to verify the theoretical analysis's correctness, as well as the feasibility and effectiveness of the recommended plan.

2. ANPC Inverter with Its Zero-Level Commutation Method

2.1. ANPC Inverter with Space Voltage Vectors

The three-level ANPC inverter topology is shown in Figure 1. In the figure, C_1 and C_2 are the upper and lower dc-side capacitors, respectively, and point o is the midpoint of the dc-side bus. S_{a1}, S_{a2}, S_{a3}, and S_{a4} are the four IGBT switching tubes on the bridge arms of each phase, and the clamp diode in the original NPC inverter is replaced by S_{a5} and S_{a6}, and there are antiparallel diodes D_{a1}–D_{a6} connected to each switching tube. U_{dc} is the dc-side bus voltage; io is the current, and i_{c1} and i_{c2} are the currents that flow through the dc-side capacitors C_1 and C_2. io is the midpoint current of the dc side, and i_{c1} and i_{c2} are the currents that flow through the dc side capacitors C_1 and C_2.

Same as the NPC inverter, each phase of the bridge arm has three modes, p, o, and n, which represent the three voltage amplitudes $U_{dc}/2$, 0, and $-U_{dc}/2$ of the output voltage U_{dc}. There are a total of $3^3 = 27$ switching states, which correspond to the space voltage vector, as shown in Figure 2.

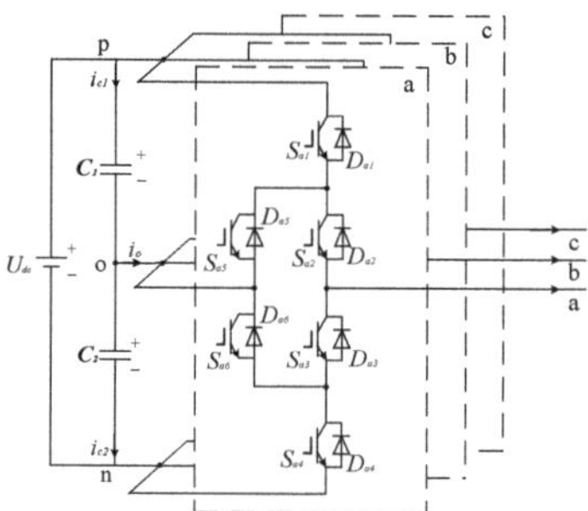

Figure 1. Typical circuit of the three-shunt sensing inverter.

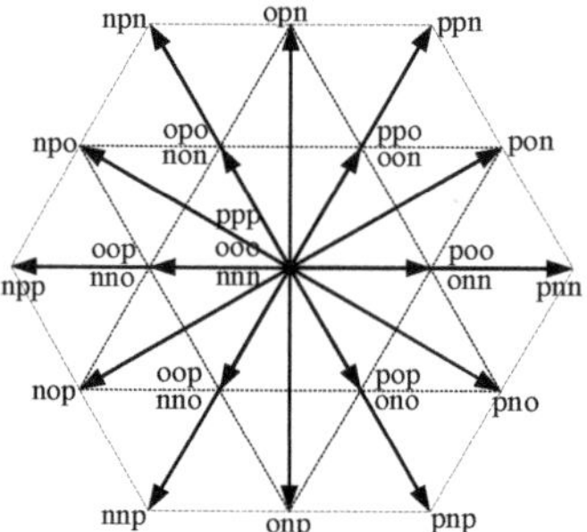

Figure 2. Space vector diagram.

Table 1 shows the space vectors corresponding to the vector types, where the small vectors appear in pairs and have an opposite effect on the midpoint voltage, the medium vectors cause the midpoint voltage to increase, while the zero and large vectors have no impact on the midpoint voltage.

Table 1. Table of vector types.

	Switching State	Midpoint Current
Zero vector	ppp, ooo, nnn	0
Small vector	poo, onn	$-i_a, i_a$
	ppo, oon	$i_c, -i_c$
	opo, non	$-i_b, i_b$
	opp, noo	$i_a, -i_a$
	oop, nno	$-i_c, i_c$
	pop, ono	$i_b, -i_b$
Medium vector	opn, onp	i_a
	pon, nop	i_b
	pno, npo	i_c
Large vector	pnn, ppn, npn npp, nnp, pnp	0

2.2. Space Voltage Vector Modeling

Assuming that the power electronic devices in the inverter are all ideal devices and equating the single-phase bridge arm to a single-pole, three-throw switch, with the three modes p, o, and n corresponding to the three switching states of each bridge arm, the switching function S_x can be expressed in the following way:

$$S_x(x \in a,b,c) = \begin{cases} 1, & \text{p-level} \\ 0, & \text{o-level} \\ -1, & \text{n-level} \end{cases} \tag{1}$$

The phase voltage at the output of the three-level inverter can be expressed as:

$$\begin{cases} u_a = \frac{S_a}{2} U_{dc} \\ u_b = \frac{S_b}{2} U_{dc} \\ u_c = \frac{S_c}{2} U_{dc} \end{cases} \tag{2}$$

where u_a, u_b, u_c are the phase voltages of the three phases of ABC, respectively; U_{dc} is the DC bus voltage; S_a, S_b, and S_c are the switching states of the three-phase bridge arms of ABC.

After the relationship between the line voltage as well as the three-phase current and the three-phase voltage at the space angle of $120°$ to each other, the three-phase output voltage synthesized vector is expressed as follows:

$$U = \frac{U_{dc}}{6} \times \begin{bmatrix} 2 & \frac{1}{2} - j\frac{\sqrt{3}}{2} & \frac{1}{2} + j\frac{\sqrt{3}}{2} \\ -1 & -1 + j\sqrt{3} & \frac{1}{2} + j\frac{\sqrt{3}}{2} \\ -1 & \frac{1}{2} - j\frac{\sqrt{3}}{2} & -1 - j\sqrt{3} \end{bmatrix} \times \begin{bmatrix} S_a \\ S_b \\ S_c \end{bmatrix} \tag{3}$$

The equivalent mathematical model of all 27 space voltage vectors can be obtained from the above equation.

2.3. ANPC Inverter Zero-Level Commutation Method and Corresponding Losses

The current paths in the p-level and n-level of the ANPC inverter are the same as in the NPC topology; while the o-level has more options, as shown in Table 2, the current in the o-level can circulate through S_2 and S_5 or through S_3 and S_6. Therefore, switching, depending on the switching tube turned on, can be categorized into six forms. Where the switching tube S_{a1} with its anti-parallel diode D_{a1} becomes S_1, and the other switching tubes are the same.

Table 2. Switch status table.

	S_1	S_2	S_3	S_4	S_5	S_6
p	1	1	0	0	1	1
n	0	0	1	1	1	0
OUL1	0	1	0	0	1	1
OUL2	0	1	1	0	0	1
OUL3	0	1	1	0	1	0
OUL4	0	0	1	0	1	1
OL1	1	0	1	0	0	1
OL2	0	1	0	1	1	0

Since the bridge arm voltage will switch back and forth between p, o and n states during the inversion process, and because each switching process requires different switching tubes to be controlled, the six forms are categorized into three modes, mode one, mode two and mode three.

As demonstrated in Figure 3, mode 1 combines the OUL1 and OUL4 zero level mode, in the p-level switching o-level process, only S_1 and S_5 switch adjust, S_2 and S_6 switch states remain unchanged; similarly, from the n-level switching to the o-level process, only S_4 and S_6 switch to adjust, S_3 and S_5 switch state remains unchanged. (The arrows in the figure point to the switching tube that needs to change the state, where the dotted line indicates the switching tube that needs to be turned off and the other that needs to be turned on).

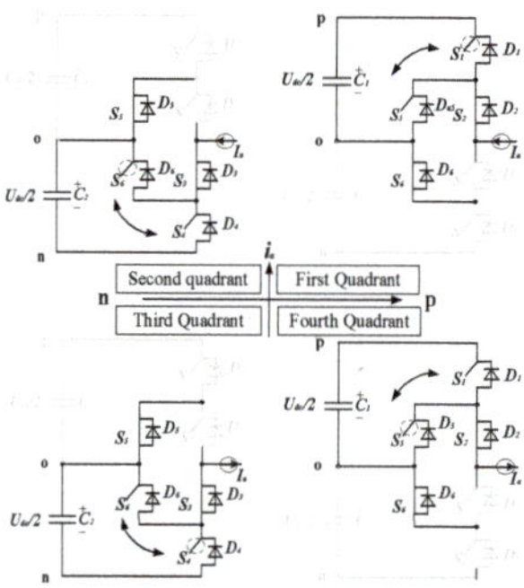

Figure 3. Zero-level path switching mode 1.

Therefore, in current path mode 1, the switching losses are concentrated in the S_1, S_4, S_5, and S_6 tubes, which reduces the losses in the S_2 and S_3 tubes, as shown in Figure 4 for the switching tube losses under mode 1 in the simulation, with higher overall losses S_1 and S_4. The loss collection background is: The test background is exactly the same as that in Section 5, which is the loss generated by the PMSM under model predictive control when running at steady state for 0.1 s at 3000 rpm and 5 N-m load conditions. Then the sampling frequency is set to 100 kHz, and the IGBT adopts Infineon's IKZ75N65EH5, whose turn-on and turn-off losses are 0.68 mj and 0.43 mj, respectively, at 25 °C; the initial voltage drop V_{ceo} of the device is checked by the switching tube datasheet, which is 1.65 V, and the internal resistance r is 10 mΩ in the on-state.

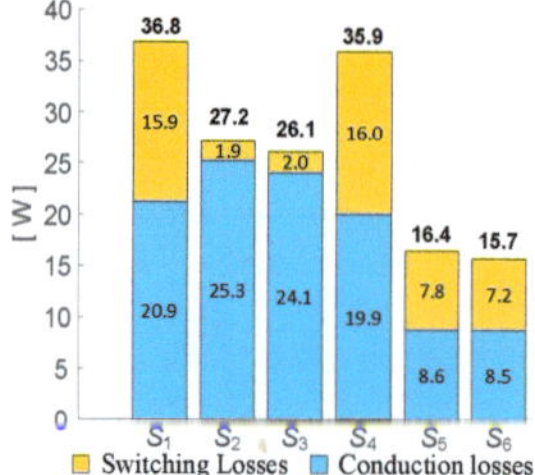

Figure 4. Mode 1 switch tube loss.

As demonstrated in Figure 5, the mode 2 current path method combines the OUL2 and OUL3 o-level methods, in the process of p-level switching o-level only S_2 and S_3 switches are adjusted, and the S_1 and S_6 switch status remains unchanged; similarly, the process of switching from the n-level to the o-level is only adjusted for the S_2 and S_3 switches, and the S_4 and S_5 switches status remains unchanged.

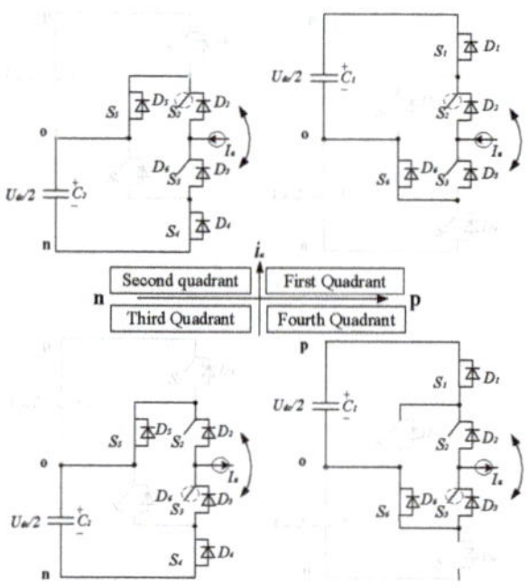

Figure 5. Zero-level path switching mode 2.

Therefore, in the current path switching mode of mode 2, the switching losses are concentrated in the S_2 and S_3 tubes, which reduces the losses of the rest of the switching tubes, as shown in Figure 6 for the switching tube losses under mode 2 in the simulation, and the overall losses S_2 and S_3 are higher.

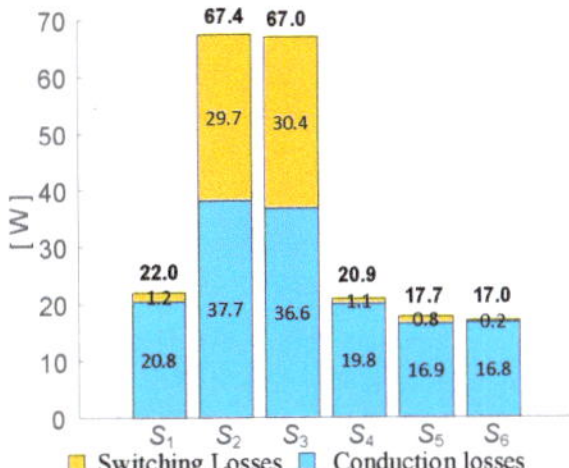

Figure 6. Mode 2 switch tube loss.

As shown in Figure 7 for mode three current path switching mode, which combines the OL1 and OU1 o-level mode, in the process of p-level switching o-level only S_1 and S_3 switch to adjust, S_2 and S_6 switch state remain unchanged; similarly, from the n-level switching to the o-level process is only adjusted to the S_2 and S_4 switch, S_3 and S_5 switch state remains unchanged.

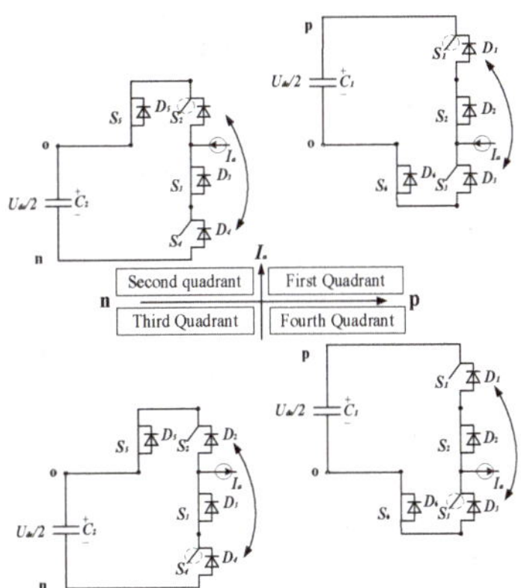

Figure 7. Zero-level path switching mode 3.

Therefore, in the current path switching mode of mode 3, the switching loss is concentrated in the S_1, S_2, S_3 and S_4 tubes, which reduces the loss of the S_5 and S_6 tubes, as shown in Figure 8 for the switching loss under mode 3 in the simulation, the loss of the S_1–S_4 tubes is more average, and the loss of the S_5 and S_6 tubes is lower compared to the previous two modes.

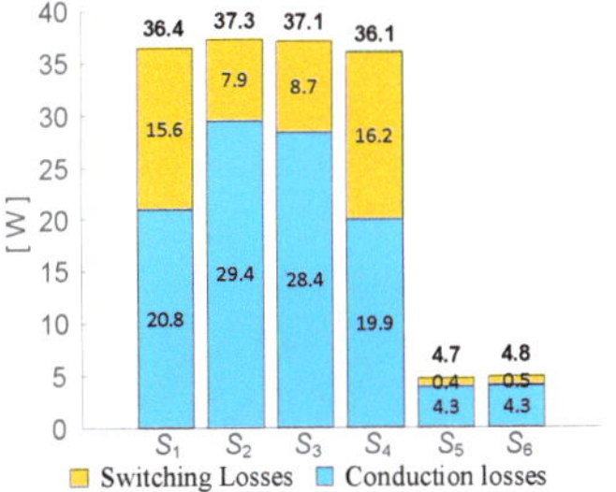

Figure 8. Mode 3 switch tube loss.

3. Switching Tube Loss Evaluation

Most of the inverter power switching tube are used with anti-parallel diode IGBT device composition, it is generally used fast recovery diode, because of its negligible turn-on loss and turn-off loss is much smaller than the IGBT, this paper focuses on the discussion of IGBT loss.

Losses are mainly switching losses and conduction losses of two types, of which switching losses include turn-on losses and turn-off losses.

3.1. ANPC Inverter Switching Loss Analysis

The device during the turn-on and turn-off transient loss is referred to as switching loss, the size of the device is mostly determined by the size and direction of the voltage and current on both sides of the device, and the effect is usually a nonlinear relationship, the representation is generally approximated by constructing a linear relationship.

Assume that the IGBT's switching frequency is f_s, in a switching frequency of switching change process in the role of time $[t_1, t_2]$, then the average switching loss in the role of time is roughly equal to:

$$P_{sw} = \int_{t_1}^{t_2} (E_{on} + E_{off}) \times f_S \times i(t) \, dt \tag{4}$$

where E_{on} and E_{off} are the turn-on loss energy and turn-off loss energy of the IGBT under the actual peak operating current, respectively (which can be found in the switching tube datasheet), f_s is the switching frequency, and $i(t)$ is the real-time current.

3.2. ANPC Inverter Conduction Loss Analysis

Conduction loss is the loss generated by the IGBT during the conduction period. The size of the loss is determined by the length of its own conduction time and the size of the current flowing during conduction during an action time, and its expression can be approximated as:

$$P_{on} = V_{ceo} \times i(t) + i(t)^2 \times r \tag{5}$$

where r is the on-state internal resistance, V_{ceo} is the initial voltage drop of the device (which can be found in the switching tube datasheet); $i(t)$ is the actual on-state current.

Therefore, let the switching frequency of the IGBT be f_s, and the action time during the switching process at one switching frequency be $[t_1, t_2]$, then the average conduction loss during the action time can be expressed as:

$$P_{on\_T} = \int_{t_1}^{t_2} (V_{ceo} + i(t)r) \times i(t) dt \tag{6}$$

4. Model Predictive Control Loss Equalization Method

The basic principle of the model prediction algorithm is to forecast the response of the future era based on the current state, take the first item of the control sequence as the control quantity at the next moment, and then make the predicted response close to the set target through rolling optimization and feedback correction. This is accomplished by bringing the relevant variables into the value function and then comparing the variables with the smallest error in the finite set so that the control value gradually approaches the reference value.

4.1. PMSM Current Prediction Model

The model predictive control performance is heavily dependent on the mathematical model of the controlled object, and a mismatch of the controlled object parameters or other unmodeled dynamics can affect the performance [23]. In this section, an electromechanical actuator in normal operation is targeted to be modeled under the following assumptions: the influence of external factors such as temperature and frequency on the motor parameters is not taken into account, space harmonics and the higher harmonic components of the magnetic field are neglected, rotor losses are not taken into account, and the stator windings are perfectly uniformly distributed in space. The final continuous domain mathematical model of the permanent magnet synchronous motor in the synchronous rotating d-q coordinate system can be expressed as:

$$\begin{cases} u_d = Ri_d + \frac{d}{dt}(L_d i_d + \psi_f) - \omega_1(L_q i_q) \\ u_q = Ri_q + \frac{d}{dt}(L_q i_q) - \omega_1(L_d i_d + \psi_f) \end{cases} \tag{7}$$

where u_d, u_q are the d-axis and q-axis output voltages, i_d and i_q are the d-axis and q-axis output currents, ψ_f is the permanent magnet chain, R is the stator resistance, and ω_1 is the electrical angular frequency.

The output currents i_d, i_q can be approximated using the forward Euler discretization method as:

$$\begin{cases} \frac{di_d}{dt} \approx \frac{i_d(k+1) - i_d(k)}{T_c} \\ \frac{di_q}{dt} \approx \frac{i_q(k+1) - i_q(k)}{T_c} \end{cases} \tag{8}$$

where $i_d(k+1)$ and $i_q(k+1)$ are the d-axis and q-axis output currents at the time of $k+1$, $i_d(k)$, $i_q(k)$ are the d-axis and q-axis output currents at the time of k, respectively, and T_c signifies the output current's control period.

Substitute (8) into (7), it is possible to obtain the compilation:

$$\begin{cases} i_d(k+1) = (1 - \frac{RT_c}{L_d})i_d(k) + \frac{\omega_1 L_q T_c}{L_d}i_q(k) + \frac{T_c}{L_d}u_d(k) \\ i_q(k+1) = -\frac{\omega_1 L_d T_c}{L_q}i_d(k) + (1 - \frac{RT_c}{L_q})i_q(k) + \frac{T_c}{L_q}u_q(k) - \frac{\omega_1 \psi_f T_c}{L_q} \end{cases} \tag{9}$$

where $u_d(k)$ and $u_q(k)$ are the components of the space voltage vector on the d-axis and q-axis at time of k, respectively. Through (9), The projected value of output current at $k+1$ moments under the operation of any space voltage vector can be calculated using k moments.

4.2. Current Control Value Function

After collecting the current $i_d(k)$ and $i_q(k)$ values, the mathematical model synthesized with the 27 voltage vectors is sequentially substituted into the prediction model. The expected values $i_d(k+1)$, $i_q(k+1)$ for the following moment of the 27 voltage vectors can then be derived, and then the value function is set to compare and find the optimization with the reference currents $i_d*(k+1)$, $i_q*(k+1)$. Specifically, it is to find the voltage vector whose predicted value is closest to the reference value, so the predicted values of all the voltage vectors differ from the reference value and squared. Then, optimization is done to find the optimal voltage vector. The value function expression is as follows:

$$\Delta i_{dq} = (i_d(k+1) - i_d*(k+1))^2 + (i_q(k+1) - i_q*(k+1))^2 \tag{10}$$

4.3. Midpoint Voltage Control Value Function

The expression for the midpoint current of the three-level ANPC inverter at moment k is:

$$i_{np}(k) = -(S_a(k)i_a(k) + S_b(k)i_b(k) + S_c(k)i_c(k)) \tag{11}$$

where

$$S_x = \begin{cases} 1, & \text{The } x^{th} \text{ phase output p} \\ 0, & \text{The } x^{th} \text{ phase output o} \quad x = a, b, c \\ -1, & \text{The } x^{th} \text{ phase output n} \end{cases} \tag{12}$$

The expression for the upper and lower capacitance currents and capacitance voltages on the dc side at a time t is given by:

$$\begin{cases} i_{C1}(t) = C_1 \frac{dU_{C1}(t)}{dt} \\ i_{C2}(t) = C_2 \frac{dU_{C1}(t)}{dt} \end{cases} \tag{13}$$

Discretizing (13) yields:

$$\begin{cases} U_{C1}(k+1) = \frac{1}{C}i_{C1}(k)T_s + U_{C1}(k) \\ U_{C2}(k+1) = \frac{1}{C}i_{C2}(k)T_s + U_{C2}(k) \end{cases} \tag{14}$$

The offset of the upper and lower capacitor voltages on the DC side at $k+1$ is then:

$$\Delta U_{np}(k+1) = U_{C1}(k+1) - U_{C2}(k+1) = \frac{1}{C}i_{np}(k)T_s + \Delta U_{np}(k) \tag{15}$$

In the specific calculation process, this paper to the fastest inverter can change the switching frequency time as the role of time, for example, this paper to 100 kHz as the control algorithm sampling time, so $[t_1, t_2]$ is actually $[0, 1 \times 10^{-5}]$, that is, in every 1×10^{-5} s to the loss generated in the previous moment of time for a calculation.

4.4. Loss Equalization Control Logic

The loss equalization control logic is shown in Figure 9. Firstly, the current value passed in each IGBT of the inverter is collected and entered into the loss model to calculate the current loss value, and the cumulative loss of each switching tube is calculated.

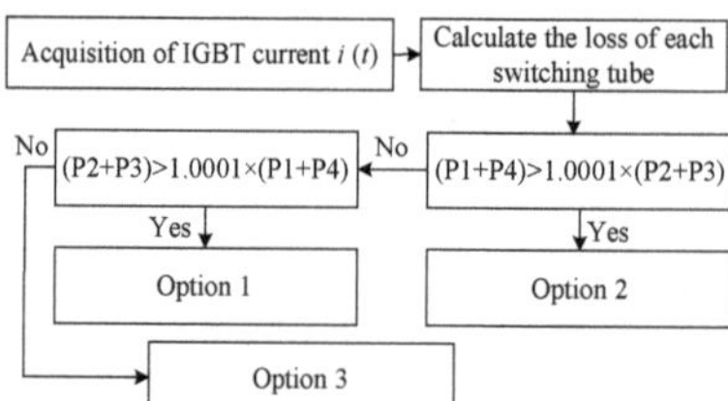

Figure 9. Loss equalization control logic block diagram.

Then, the loss judgment is performed, and the loss judgment time is based on multiple verifications to finally take the value of the electrical frequency, which is carried out once every 0.005 s, with the goal of equalizing the loss of S_1–S_4 tubes and making S_5 and S_6 as small as possible. If the sum of the losses of S_1 and S_4 tubes is greater than the sum of the losses of S_2 and S_3 tubes within a certain range, then the zero level is set to the mode two pathway; conversely, the zero level is set to the mode one pathway; if the losses of S_1 and S_4 tubes are approximately equal, then the zero level is set to the mode three pathway, and the judgment is set to judge the range as large as possible, to decrease the switching loss caused by toggling the three modes. After selecting the zero-level access path, the next judgment is made.

4.5. Finite Set Model Predictive Control

The finite set model prediction algorithm first performs rolling optimization on all voltage vectors in the finite set and then selects the voltage vector with the smallest value function for output. Each rolling is computed for the finite set's 27 voltage vectors.

Figure 10 depicts the total system block diagram, which includes model predictive control for the current inner loop and PI control for the speed outer loop. Firstly, the reference rotational speed ω^* is given, and then the difference with the collected actual rotational speed ω is made and brought into the PI regulator. Next, the reference value of stator q-axis current $i_q{}^*(k+1)$ is obtained by the PI regulator and brought into the value function with the predicted value for calculation, and then the optimal vector is selected by searching for the optimal vector, and the corresponding switching sequence is substituted into the inverter to output the three-phase currents to be supplied to the PMSM, in which the model predictive control controls the currents and the mid-point voltage of the inverter at the same time, that is, (10) and (15) are added together to finally value function is obtained as follows:

$$T = \Delta i_{dq} + n \times \Delta U_{np}(k+1) \tag{16}$$

where n is the midpoint voltage control weight coefficient, and the control of current and midpoint voltage balance is realized by adjusting the n value, In this paper, the value of n is selected as 3500.

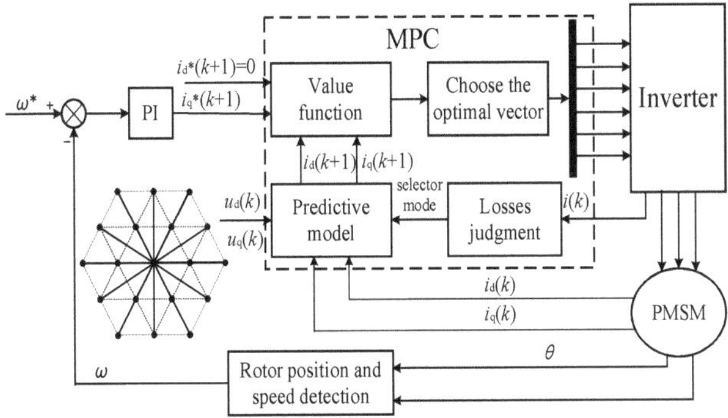

Figure 10. Model predictive control system block diagram.

Where ω and ω^* are the actual value and the given value of the motor speed, respectively, $i_d*(k+1)$, $i_q*(k+1)$ are the given values of stator d-axis and q-axis currents at the moment $k+1$.

The predictive control technique is implemented as follows:

Step 1: The current $i(k)$, the inverter upper and lower capacitor voltages $U_{up}(k)$, $U_{down}(k)$ with the stator dq-axis currents $i_d(k)$, $i_q(k)$ in the PMSM flowing in each switching tube of the inverter at moment k are collected;

Step 2: Bring the current $i(k)$ into the inverter loss model (4) and (6) for loss calculation, and select the zero-level path for the next moment by judging the switching tube loss.

Step 3: After selecting the mode, the collected stator dq-axis currents $i_d(k)$, $i_q(k)$ with the 27 voltage vector mathematical models are brought into the prediction model (9) to obtain the expected currents $i_d(k+1)$, $i_q(k+1)$ for each voltage vector's $k+1$ instant;

Step 4: Substitute the predicted currents with the reference currents $i_d*(k+1)$, $i_q*(k+1)$, and the collected midpoint voltages into the value function (16), calculate and select the voltage vector corresponding to the minimum value of the value function as the optimal vector, and finally act the optimal vector corresponding to the switching sequence on the inverter.

5. Simulation Verification

To test the validity and efficacy of the algorithm suggested in this work, based on the model prediction loss equalization control, Matlab simulation was carried out to compare and analyze the ANPC inverter and permanent magnet synchronous motor. The three phase PMSM is controlled using $i_d = 0$. Table 3 displays the simulation parameters. All of the following loss control backgrounds are the same as in Section 2.3, with only improvements to the control algorithms.

Table 3. Simulation Parameter.

Parameters	Value	Unit
Control frequency	100	[kHz]
DC side voltage	270	[V]
DC side capacitance	4700	[μF]
d-axis inductance	0.395	[mH]
q-axis inductance	0.395	[mH]
Stator resistance	0.0485	[Ω]
Permanent magnet flux	0.1194	[Wb]
Number of pole-pairs	4	[pair]
Maximum motor speed	6000	[r/min]
Locked-rotor torque	33	[N·m]
Rated power	15	[kW]

Figure 11 shows the loss of each switching tube in phase A of the ANPC inverter when only mode one and mode two are used under model control with the parameters of Table 3; Figure 12 shows the loss of each switching tube in phase A of the inverter with the addition of mode 3. In contrast, the improved algorithm has a better control effect, and the overall loss at the same time is reduced by 5.6%, of which the loss of the clamped switch tubes S_5 and S_6 is reduced by 24.6%, which proves the superiority of the control effect under the improved algorithm.

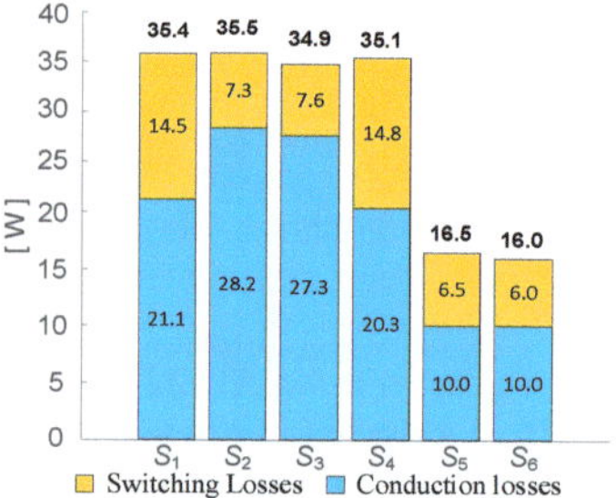

Figure 11. The traditional way of switching tube loss.

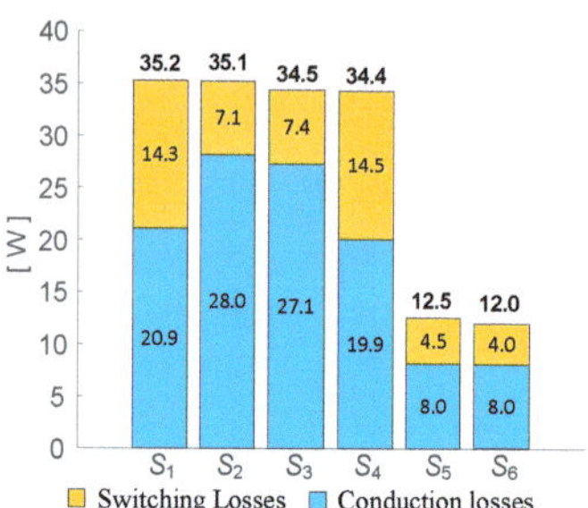

Figure 12. Improved switching tube loss.

Figures 13–18 give the waveforms of the motor operation at the motor setting of 5 N-m and 3000 rpm under the model predictive control with loss equalization. Under this condition, we collect and calculate the average switching frequency of the ANPC inverter, which is 18.7 kHz.

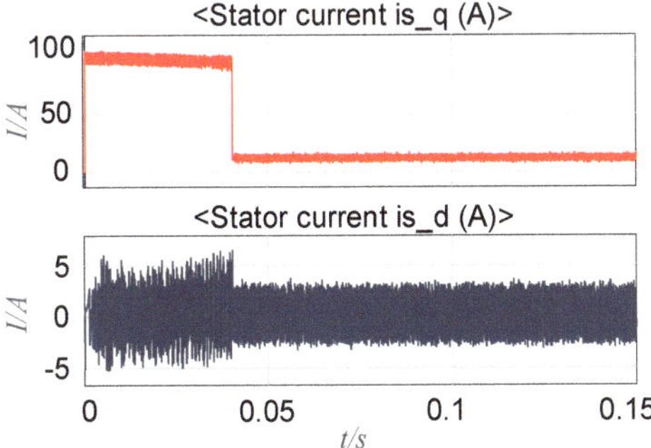

Figure 13. Motor stator dq axis current waveform.

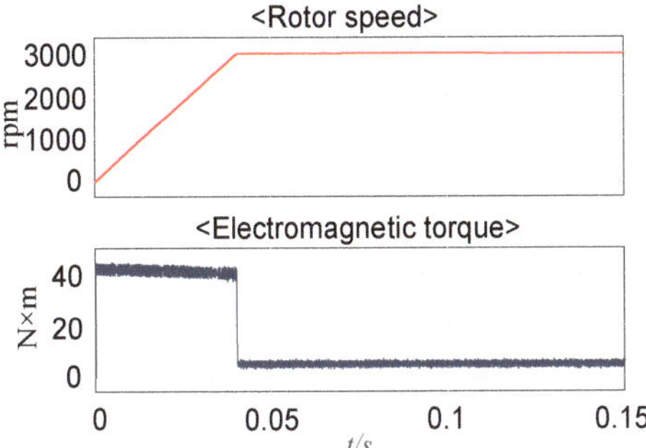

Figure 14. Motor speed and torque waveform.

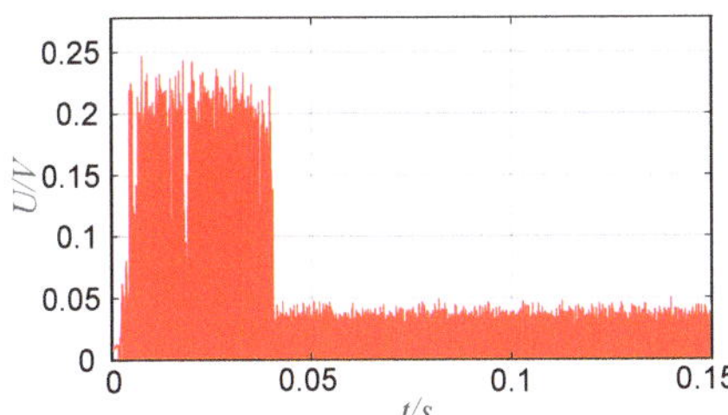

Figure 15. The waveform of the absolute value of midpoint voltage.

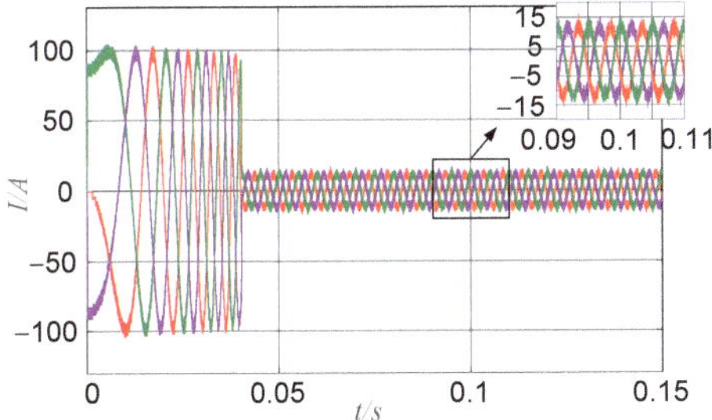

Figure 16. Three-phase current output waveform.

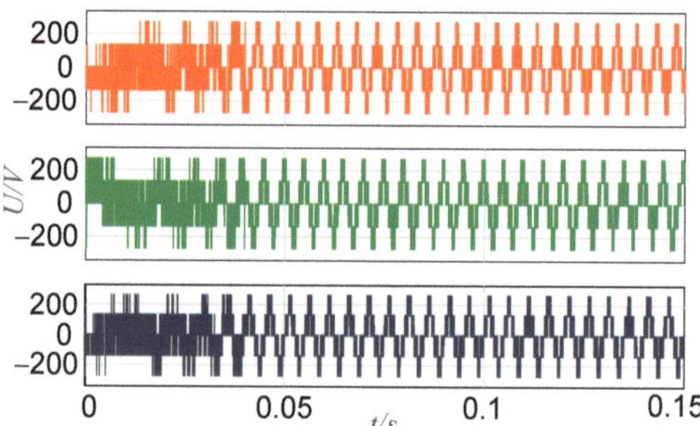

Figure 17. Three-phase voltage output waveform.

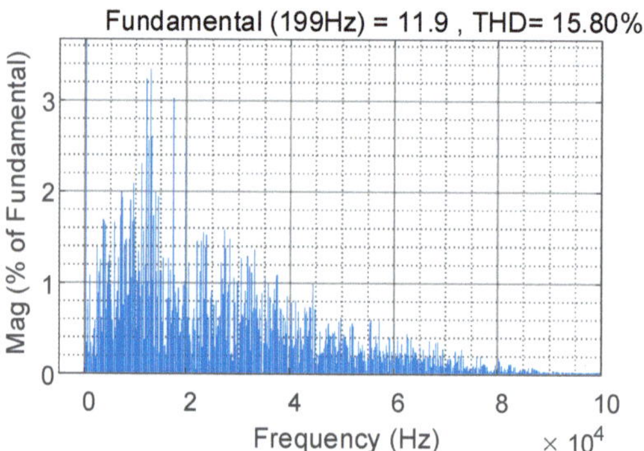

Figure 18. Current total harmonic distortion.

Where: In Figures 16 and 17, the red waveform represents phase a, the green waveform represents phase b, the blue waveform represents phase c.

From Figures 13 and 14, it can be seen that the motor reaches 3000 rpm in 0.04 s and starts to run steadily, at which time the motor stator dq-axis current and torque remain stable.

Figure 15 gives the midpoint voltage control effect of the ANPC inverter under the model predictive control of loss equalization, and it is clear that after the motor speed reaches 3000 rpm and stabilizes, the absolute value of the midpoint upper and lower capacitance voltages is kept within 0.05, which has a better control effect.

Figures 16 and 17 give the three-phase current and three-phase voltage output waveforms of the inverter under the predictive control of the model with loss equalization. It can be seen that the three-phase current and three-phase voltage remain stable after the motor speed reaches 3000 rpm.

Figure 18 shows the total harmonic distortion of phase A current in Figure 16 when the motor is stabilized at 3000 rpm. The three-phase current frequency is measured to be 199 Hz, and it can be seen that at a switching frequency of 100 kHz, the total harmonic distortion of the current is 15.8% and the current is slightly harmonic but generally stable.

Figures 19 and 20 give the waveforms of motor speed and torque during load up and load down, in which the motor speed is set to 3000 rpm, and the torque is switched between 0 N·m and 5 N·m. It can be seen that the rotational speed remains stable during load up and load down, the torque follows the target quickly and then remains stable, and there is no large torque pulsation in the waveforms.

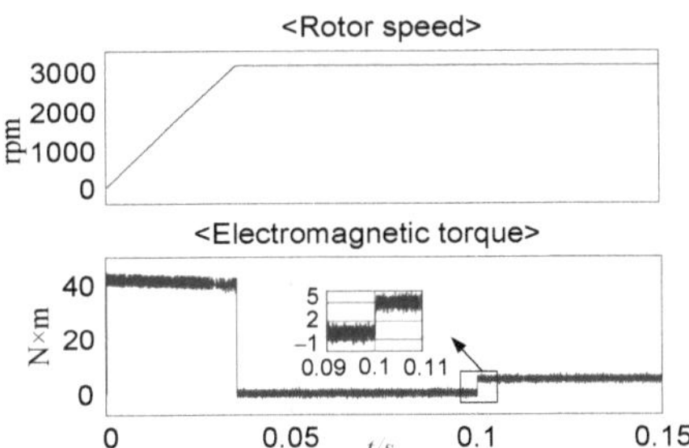

Figure 19. Speed and torque waveform in load lifting.

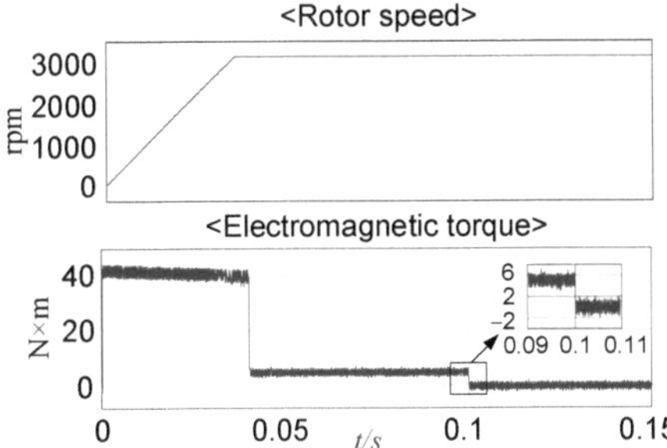

Figure 20. Speed and torque waveform in load shedding.

Figures 21 and 22 give the waveforms of motor speed and torque during acceleration and deceleration, in which the motor torque is set to 5 N·m, and the speed is switched between 1000 and 3000 rpm.

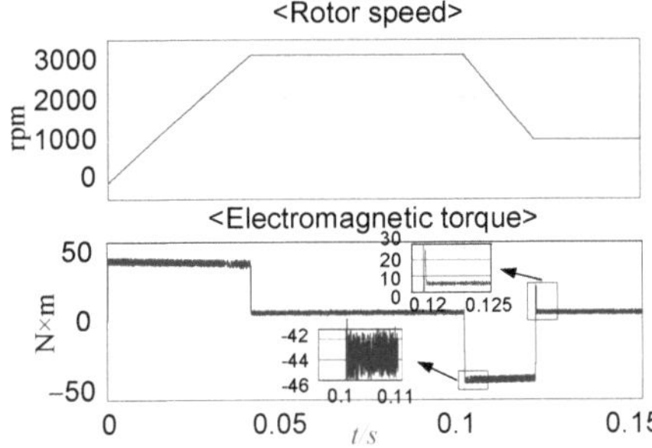

Figure 21. Speed and torque waveform in deceleration.

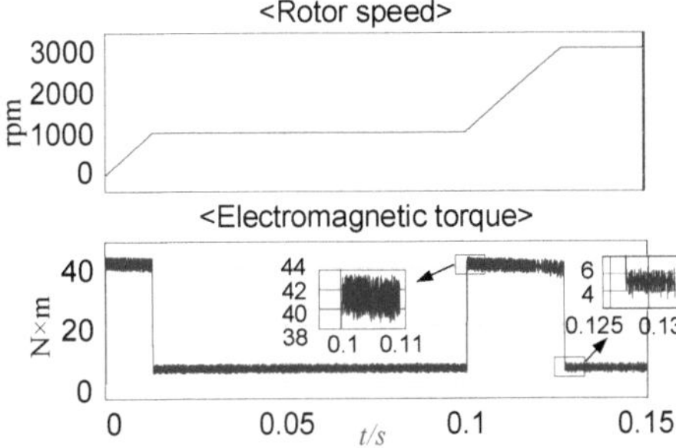

Figure 22. Speed and torque waveform in acceleration.

As can be seen in Figure 21, the motor can be decelerated from 3000 rpm to 1000 rpm in 0.02 s, the torque decreases rapidly and remains stable at the moment of deceleration, and the torque quickly follows and remains stable after the momentary torque pulsation when the deceleration is completed; as can be seen in Figure 19, the motor can be accelerated from 1000 rpm to 3000 rpm in 0.027 s, and the torque rapidly decreases and remains stable at the moment of acceleration, and the torque does not have large torque pulsation and quickly follows and remains stable at the moment of acceleration. When the acceleration is completed, the torque does not have a large torque pulsation, follows quickly, and remains stable.

Figure 23 gives the servo response waveform of the motor during the sudden change of load, in which the displacement of the motor-driven ball screw is set to change by 16 mm in 0.05 s, and the load is set to increase from 5 N·m to 8 N·m in 0.6 s. As shown in the figure, the red line represents the set reference position, and the green line represents the actual position. It can be seen from the figure that the ball screw reaches the preset position in 0.35 s, and at the same time, it remains stable and the position fluctuation is kept within ±0.05 mm in 0.4 s; After the sudden change of load in 0.6 s, the position of the ball screw remains basically unchanged, slightly shifted downward, but the position fluctuation is still kept within ±0.05 mm. mm; in 0.6 s after a sudden change in load, the position of the ball screw remains basically unchanged, slightly shifted downward, but the position fluctuation is still maintained within ±0.05 mm.

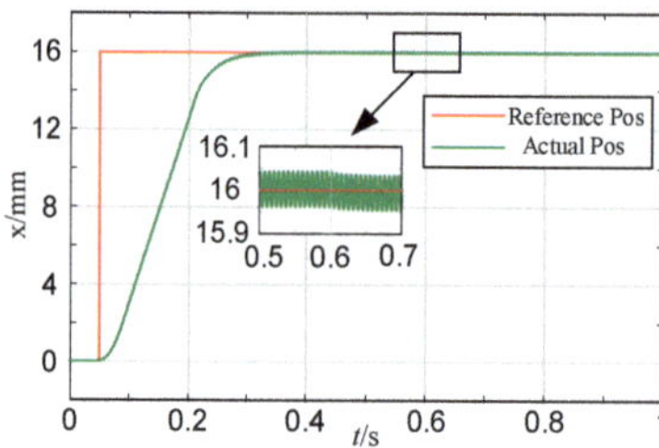

Figure 23. Load sudden change servo response waveform.

Figure 24 gives the servo response waveform of the motor in the displacement change process, in which the ball screw displacement change is set to 16 mm in 0 s, the displacement change is restored to 0 mm in 0.5 s, and the load is set to 5 N·m. From the figure, it can be seen that the ball screw position can be guaranteed to complete the servo response within 0.35 s in both the forward and backward movements and the position fluctuations are kept within ±0.05 mm.

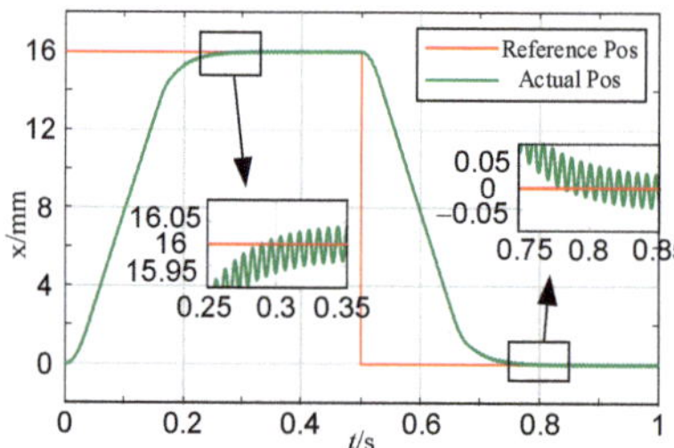

Figure 24. Displacement transform servo response waveform.

6. Conclusions

In this research, a model predictive control technique according to inverter loss equalization is presented to add pathway mode three under the original two commonly used zero-level pathway modes of ANPC inverters, and the three modes are flexibly switched using model predictive control, and the following conclusions can be drawn through the theoretical analysis and simulation verification.

1. The method in this research only improves the driving strategy to avoid the problem of power density reduction in electromechanical actuators and ensures the stability and rapidity of PMSM operation.
2. Under the model predictive control, the improved algorithm realizes the ANPC inverter switching tube loss equalization, and at the same time reduces the overall loss of switching tubes, in which the loss of clamped switching tubes is greatly reduced.
3. The method in this paper uses a model predictive control strategy to simultaneously achieve inverter switching tube loss equalization and midpoint voltage control to improve the operating efficiency of the electromechanical actuator.
4. The method in this paper only changes the driving strategy, which reduces the overall inverter loss as well as the clamped switching tube loss. It can be generalized to the Si and SiC hybrid method for its characteristics.
5. In this paper, the value of its time in the loss judgment is initially selected as 0.005 s each time, but whether this value is also applicable to other rotational speeds is not further verified in this paper, to be followed by further research.

Author Contributions: Conceptualization, S.W. and C.L.; methodology, S.W. and H.D.; software, S.W. and C.L.; validation, H.D. and C.L.; investigation; resources, B.W. and J.C.; data curation, S.W.; writing—original draft preparation, S.W.; writing—review and editing, H.D.; supervision, B.W. and H.D.; project administration; funding acquisition, B.W. and H.D. All authors have read and agreed to the published version of the manuscript.

Funding: This study was supported by the Major Science and Technology Program of Gansu Province, China (21ZD4GA005). Sponsor: Haiying Dong.

Data Availability Statement: The data that has been used are confidential.

Acknowledgments: The completion of this study is due to the collaborative efforts of several co-authors.

Conflicts of Interest: Author Shaoqi Wan was employed by College of New Energy and Power Engineering, Lanzhou Jiaotong University, The other authors declare no conflicts of interest.

References

1. Chunqiang, L.; Guangzhao, L.; Wencong, X. Survey on active disturbance rejection control of permanent magnet synchronous motor for aviation electro-mechanical actuator. *J. Electr. Eng.* **2021**, *16*, 12–24.
2. Shixiao, L.; Jinghua, D.; Yun, L. Fault diagnosis of electromechanical actuators based on one-dimensional convolutional neural network. *Trans. China Electrotech. Soc.* **2022**, *37*, 62–73.
3. Zhuoran, Z.; Li, Y.; Jincai, L. Key Technologies of advanced aircraft electronic machine tools systems for aviation electrification. *J. Nanjing Univ. Aeronaut. Astronaut.* **2017**, *49*, 622–634.
4. Demir, Y.; Aydin, M. A novel dual three-phase permanent magnet synchronous motor with asymmetric stator winding. *IEEE Trans. Magn.* **2016**, *52*, 8105005. [CrossRef]
5. Xinghe, F.; Rui, C.; Ting, D. Review of MTPA control of permanent magnet synchronous motor considering parameter uncertainties. *Proc. CSEE* **2022**, *42*, 796–808.
6. Xin, G.; Jinyue, L.; Zhiqiang, W. Harmonic current suppression strategy for permanent magnet synchronous Motor based on deadbeat current prediction control. *Trans. China Electrotech. Soc.* **2022**, *37*, 6345–6356.
7. Fan, Y.; Ximei, Z.; Hongyan, J. Parameter-free adaptive finite control set model predictive control for PMSM. *Proc. CSEE* **2023**, *43*, 8935–8943.
8. Chunxi, L.; Baoqi, T.; Zhile, L. Adaptive model predictive control for NPC-type three-level grid-connected inverter. *Proc. CSU-EPSA* **2023**, *35*, 143–151.
9. Xinglai, G.; Xiaohua, Z.; Yan, Y. Comparative study on synchronized space vector PWM for three-level neutral point clamped VSI under low carrier ratio. *Proc. CSEE* **2018**, *22*, 24–32.
10. Yenan, C.; Dehong, X. Review of soft-switching topologies for single-phase photovoltaic inverters. *IEEE Trans. Power Electron.* **2022**, *37*, 1926–1944.
11. Yinglai, X.; Ayyanar, R. Naturally adaptive Low-loss zero-voltage-transition circuit for high-frequency full-bridge inverters with hybrid PWM. *IEEE Trans. Power Electron.* **2018**, *33*, 4916–4933.

12. Jing, L.; Enshuai, D.; Yushun, F. A hybrid three-level active-neutral-point-clamped zero-voltage transition soft-switching converter with siliconcarbide and silicon devices. In *Transactions of China Electrotechnical Society*; Machine Press: Beijing, China, 2023; pp. 1–15. [CrossRef]

13. Li, Z.; Xiutao, L.; Chushan, L. Evaluation of different Si/SiC hybrid three-level active NPC inverters for high power density. *IEEE Trans. Power Electron.* **2020**, *35*, 8224–8236.

14. Zhijian, F.; Xing, Z.; Shaolin, Y. Comparative study of 2SiC&4Si hybrid configuration schemes in ANPC inverter. *IEEE Access* **2020**, *8*, 33934–33943.

15. Xiuzhen, L.; Li, Z.; Yongwei, C. A dedicate modulation scheme for 4-SiC 3L-ANPC inverter with loss balanced distribution and efficiency improvement. *Proc. CSEE* **2022**, *42*, 1925–1933.

16. Yi, D.; Jun, L.; Shin, K.H. Improved modulation scheme for loss balancing of three-level active NPC converters. *IEEE Trans. Power Electron.* **2017**, *32*, 2521–2532.

17. Wenchao, W.; Shanxu, D.; Tianbao, Y. Loss equalization and efficiency optimization strategy of active neutral point clamped inverter. *Trans. China Electrotech. Soc.* **2022**, *37*, 4872–4882.

18. Ma, L.; Kerekes, T.; Rodriguez, P. A new PWM strategy for grid-connected half-bridge active NPC converters with losses distribution balancing mechanism. *IEEE Trans. Power Electron.* **2015**, *30*, 5331–5340. [CrossRef]

19. Mengxing, C.; Donghua, P.; Huai, W. Investigation of switching oscillations for silicon carbide MOSFETs in three-level active neutral-point-clamped inverters. *IEEE J. Emerg. Sel. Top. Power Electron.* **2021**, *9*, 4839–4853.

20. Qiang, W.; Daohong, L.; Chuangang, C. ANPC grid-connected inverter based on multi-objective optimal model predictive control. *Power Electron.* **2021**, *55*, 93–97.

21. Cungang, H.; Dajun, M.; Qunjing, W. Control strategy of loss distribution balancing for three-level active neutral-point-clamped inverter. *Trans. China Electrotech. Soc.* **2017**, *32*, 129–138.

22. Shulin, Z.; Jingsong, K.; Yezhe, S. A low common mode voltage FCS-MPC scheme for three-level NPC inverter without weighting factor. *Proc. CSEE* **2023**, *43*, 7614–7626.

23. Brosch, A.; Hanke, S.; Wallscheid, O. Data-driven recursive least squares estimation for model predictive current control of permanent magnet synchronous motors. *IEEE Trans. Power Electron.* **2021**, *36*, 2179–2190. [CrossRef]

Article

Practical Comparison of Two- and Three-Phase Bearingless Permanent Magnet Slice Motors for Blood Pumps

Jonathan E. M. Lawley [1], Giselle C. Matlis [2], Amy L. Throckmorton [2,3] and Steven W. Day [1,*]

[1]   Departments of Biomedical and Mechanical Engineering, Kate Gleason College of Engineering, Rochester Institute of Technology, Rochester, NY 14623, USA; jel9067@rit.edu

[2]   BioCirc Research Laboratory, School of Biomedical Engineering, Science, and Health Systems, Drexel University, Philadelphia, PA 19104, USA; gcm52@drexel.edu (G.C.M.); alt82@drexel.edu (A.L.T.)

[3]   Department of Pediatrics, College of Medicine, St. Christopher's Hospital for Children, Drexel University, Philadelphia, PA 19104, USA

*   Correspondence: steven.day@rit.edu; Tel.: +1-585-475-4738

Abstract: The majority of bearingless permanent magnet slice motors (BPMSMs) used in commercially available rotary blood pumps use a two-phase configuration, but it is unclear as to whether or not a comparable three-phase configuration would offer a better performance. This study compares the performance of two-phase and three-phase BPMSM configurations. Initially, two nominal designs were manufactured and empirically tested for their performance characteristics, namely, the axial stiffness, radial stiffness, and current force. Subsequently, finite element analysis (FEA) models were developed based on these nominal devices and validated against the empirical results. Simulations were then employed to assess the sensitivity of performance characteristics to variations in seven different geometric features of the models for both configurations. Our findings indicate that the nominal three-phase design had a higher axial stiffness and radial stiffness, but resulted in a lower axial-to-radial-stiffness ratio when compared to the nominal two-phase design. Additionally, while the nominal two-phase design shows a higher current force, the nominal three-phase design proves to be slightly superior when the force generated is considered relative to the power usage. Notably, the three-phase configuration demonstrates a greater sensitivity to dimensional changes in the geometric features. We observed that alterations in the air gap and rotor length lead to the most significant variations in performance characteristics. Although most changes in specific geometric features entail equal tradeoffs, increasing the head protrusion positively influences the overall performance. Moreover, we illustrated the interdependent nature of the head height and rotor height on the performance characteristics. Overall, this study delineates the strengths and weaknesses of each configuration, while also providing general insights into the relationship between specific geometric features and performance characteristics of BPMSMs.

Keywords: bearingless; bearingless permanent magnet slice motor; FEA; two-phase; three-phase; rotary blood pump; mechanical circulatory support

Citation: Lawley, J.E.M.; Matlis, G.C.; Throckmorton, A.L.; Day, S.W. Practical Comparison of Two- and Three-Phase Bearingless Permanent Magnet Slice Motors for Blood Pumps. *Actuators* **2024**, *13*, 179. https://doi.org/10.3390/act13050179

Academic Editor: Dong Jiang

Received: 4 March 2024
Revised: 12 April 2024
Accepted: 1 May 2024
Published: 8 May 2024

1. Introduction

Bearingless permanent magnet slice motors (BPMSMs) offer a distinct advantage over systems with conventional bearings by eliminating the physical contact between the rotor and stator. This absence of contact between moving parts results in minimal wear from friction, contributing to the growing popularity of BPMSMs in fluid pump applications. Across various industries, from chemical to biomedical, BPMSM pumps have gained prominence. In the biomedical sector, the BPMSM has found success in medical applications including mechanical circulatory support devices and rotary blood pumps [1–7].

More than 60 million patients globally suffer from heart failure (HF), a debilitating severity of heart disease that renders the heart muscle unable to effectively drive blood to the vital and end organs in the body [8–10]. The current treatment paradigm involves the administration of pharmacologic agents; this results in symptomatic improvement, but does not halt the progression to HF [11,12]. The shortage of donor organs and the further difficulty of donor–recipient size matching create hurdles for cardiac transplantation and extend patients' waiting periods. To address these challenges, alternative treatment strategies are employed to provide bridge-to-transplant circulatory support in the form of a blood pump [13–15].

Blood pumps are designed to supplement the output of the native left ventricle. These pumps generally operate in parallel with the beating diseased ventricle to provide adequate blood flow to the body. The design evolution of these medical devices has concentrated on the bearing support and motor drive systems [2–7]. First-generation blood pumps consist of pulsatile devices with pusher-plate or flexing diaphragms and valve configurations. Second-generation devices comprise continuous-flow or rotary pumps that require mechanical bearings and seals that are in contact with the fluid; and the latest generation of blood pumps, third-generation blood pumps, include axial and radial rotary pumps with no mechanical bearings in contact with the fluid medium, usually integrating magnetic or non-contacting hydrodynamic bearings [13,16]. Figure 1 illustrates a unique blood pump technology that integrates both an axial and a centrifugal pump into one medical device [3].

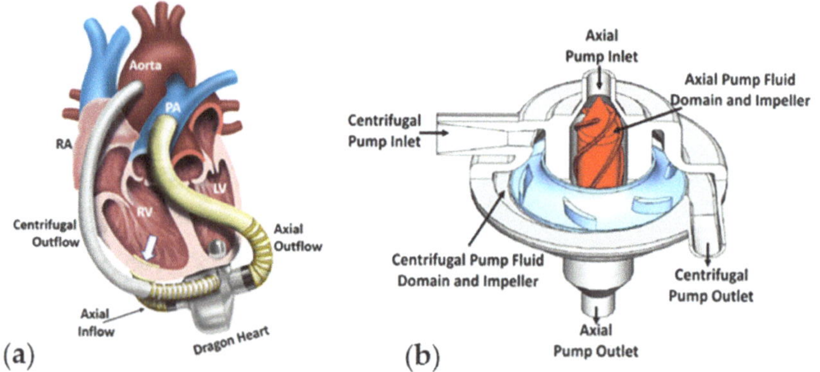

Figure 1. Continuous-flow, magnetically levitated Dragon Heart. (**a**) Implantation of the Dragon Heart medical device. The centrifugal blood pump is designed to support the systemic circulation and the left ventricle, and the axial flow blood pump is designed to support the pulmonary circulation and the right ventricle. RA: right atrium; RV: right ventricle; LV: left ventricle; PA: pulmonary artery. (**b**) TAH design details of the integrated axial and centrifugal pumps into a single device having only two moving parts, the levitated impellers [3].

Notably, the two blood pumps most frequently used clinically, CentrigMag (Abbott, Abbott Park, IL, USA) and HeartMate III (Abbott, Abbott Park, IL, USA), share a two-phase configuration [17,18]. In contrast, BPMSMs in other industries conventionally employ three-phase motors due to the abundance of empirical and developmental data that are readily available. It has not been determined whether there are inherent benefits to one configuration over the other for medical applications such as blood pumps.

Three of the criteria that a designer should consider when choosing either a two-phase or three-phase configuration for a BPMSM blood pump are as follows: the size, air gap, and availability of off-the-shelf controller electronics. In the specific case of implantable blood pumps, there is the desire to reduce the size of the device in an effort to minimize the obtrusion to the user. For most BPMSMs, the quantity of phases will dictate the number of stator arms that are required, which will also affect the space between each arm. The size of this space is further limited by the sensors and coils that are present. Ultimately, there is a direct correlation between how small a device can be made and the quantity of phases used. Another important aspect of BPMSM blood pumps is that there are specific aspects that cannot be arbitrarily changed such as the air gap and blood gap. The air gap, also known as the magnetic gap, is the distance between the rotor and stator. The larger the air gap, the less influence the stator has on the rotor and, thus, the impeller. In a blood pump, a subset of the air gap is known as the blood gap which is defined as the distance between a wall of the impeller and a wall of the inner pump housing. It is advantageous for the blood gap to be large when compared to fluid gaps seen in more traditional pumps. The reason for this is because a small blood gap results in high shear within that region, which, in turn, could damage the blood [19]. The allowed blood gap is generally limited by the magnetic gap, which is, in turn, limited by the minimum allowable pump housing wall thickness. Thus, if either configuration could maintain sufficient control of the rotor with a larger air gap, then the allowable blood gap could be increased. The last thing a designer may consider is that off-the-shelf motor controllers are more prevalent for some phase configurations than others. The unique aspect of the BPMSM is that its rotor is magnetically levitated, but, aside from that, it simply functions as a traditional electric motor. Thus, the vast amounts of motor drive controllers made for traditional electric motors could be used with BPMSMs as long as they have the appropriate amount of phases.

Studies in the field of BPMSMs generally concentrate on developing novel topologies and control schemes [20–23]. Limited research has been carried out to determine the general relationship between specific geometric features and performance characteristics unique to BPMSMs. A notable exception is the study by Zhang et al. where they demonstrate the impact of three different geometric features on a configuration's performance characteristics [24]. This study, however, only focuses on a two-phase configuration and does not offer a comparison to those with different amounts of phases. In studies where the number of phases is considered [25,26], the focus is predominantly on motor-driving characteristics such as the following: the acceleration, torque, and driving efficiency.

To address this knowledge gap, we investigated any innate difference between the two-phase and three-phase configurations and, thus, their appropriateness for use within blood pumps. We accomplished this by initially manufacturing nominal devices for the two-phase and three-phase configurations. The performance characteristics of axial stiffness, radial stiffness, and current force were empirically determined for these two manufactured devices. To extend the study beyond the comparison of only two designs, finite element analysis (FEA) was performed and allowed for dimensional changes for seven different geometric features. These simulations not only facilitated the comparison of how dimensional changes in geometric features affected the two configurations, but also, more broadly, enabled us to quantify the effect of critical geometric features on BPMSM performance. This provided valuable insights into the general nature of BPMSMs. Overall, the findings of this study provide a perspective on the advantages and disadvantages of two-phase and three-phase BPMSMs for use in blood pumps.

1.1. BPMSM Characteristics

We evaluated the three critical performance characteristics that are unique to BPMSMs: the axial stiffness, radial stiffness, and current force. Figure 2 illustrates the nominal designs for the two-phase and three-phase configurations.

The axial and radial stiffness are passive characteristics that are a result of the interaction between the permanent magnet rotor and the ferrous stator. The passive characteristics are defined by the forces resulting from deviations from the rotor's neutral position. We define the neutral position to be the location where the rotor is centered inside of the stator; in this neutral position, as shown in Figure 2, the axial and radial forces are zero. The axial stiffness (N/mm) is the force pulling the rotor towards the neutral position per axial distance displaced from the neutral position. Conversely, the radial stiffness (N/mm) is the force that pulls the rotor away from the neutral position per radial distance. In the majority of BPMSMs, it is operationally optimal to maintain the rotor's position as close as possible to the neutral position. Another passive characteristic is the tilting stiffness which is the restoring torque per degree of the rotor tilt. The tilting stiffness is directly dependent on the axial stiffness for small tilt angles as a characteristic feature of BPMSMs is the large ratio of the rotor diameter to rotor height. This dependence is derived in Supplementary Materials. The axial stiffness and, as a result, tilting stiffness are both desirable due to their capability to return the rotor to its neutral position after an axial or tilt displacement, whereas the radial stiffness is undesirable because it can only displace the rotor from the neutral position. Then, there is the performance characteristic of the current force (N/A), which is the force acting on the rotor per current applied to the coils. The current force is an active characteristic because the user actively controls the force direction and magnitude. The purpose of the current force is to counteract the radial stiffness and center the rotor. Here, we define the current force as the radial force on the rotor per current. It is important to note that the radial stiffness and current force both depend on the rotor angle. To capture this, we analyzed and reported the 0° case as this corresponds with the maximum radial stiffness and current force. Lastly, another important active characteristic for BPMSMs is the motor drive performance, but this has been analyzed in prior work [25,26].

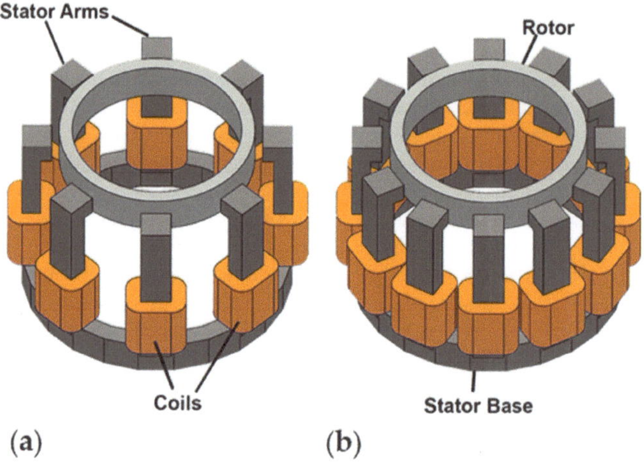

Figure 2. 3D models of the two nominal designs with the pertinent components labelled. (**a**) The two-phase configuration and (**b**) the three-phase configuration. Note that, here, the rotors are displayed in their neutral position.

1.2. Suspension Principle

While there are a wide variety of different suspension principles for two-phase and three-phase configurations, as reported by [27,28], we chose the standard $N_{ps} = N_{pr} + 1$, where N_{ps} represents the number of suspension pole pairs and N_{pr} signifies the number of rotor pole pairs. The aforementioned suspension principle is commonly used [21,22] and corresponds to the configurations of both the CentriMag and HeartMate III [17,18]. Furthermore, the equal, conventional dipole rotor [29,30] was used in this study, meaning that $N_{pr} = 1$, which then required that $N_{ps} = 2$. As a result, the two-phase and three-phase configurations require 8 and 12 arms, respectively. Aside from these differences in the number of arms, the motors also have differing phase configurations, as seen in Figure 3.

As a result of different phase configurations, the current distributions for the active suspension, according to the rotor angle, are shown in Equations (1) and (2):

$$I_U = \frac{I_{in}\cos(\theta)}{2}, \ I_V = \frac{I_{in}\cos(\theta + 120°)}{2}, \ I_W = \frac{I_{in}\cos(\theta - 120°)}{2} \tag{1}$$

and

$$I_U = I_{in}\cos(\theta), \ I_V = I_{in}\cos(\theta + 45°) \tag{2}$$

where I_{in} reflects the current input and θ is the rotor's angle in degrees. These equations are defined such that the total amount of current in either equation would be identical at the same rotor angle.

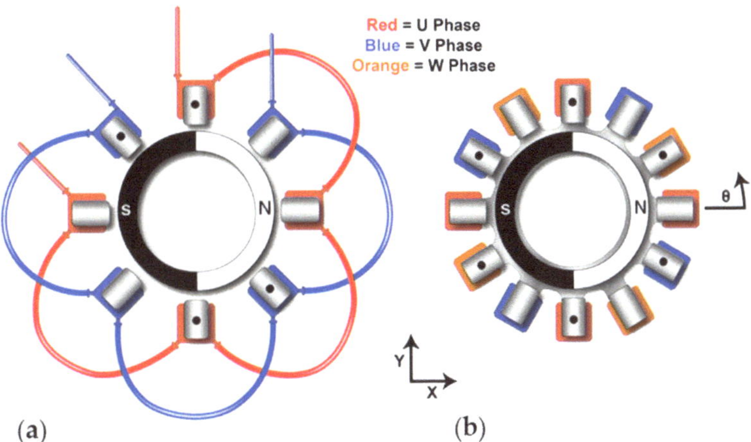

Figure 3. The two-phase and three-phase suspension wiring schemes when viewed from above. (**a**) Two-phase design with the wiring of the phases in series. (**b**) Three-phase design where the phases were wired similarly. The use of a black dot indicates that the coil was specifically wound counterclockwise; otherwise, the coil was wound in the clockwise direction.

2. Materials and Methods

To begin our analysis of two-phase and three-phase configurations, we started by selecting a single design for each configuration, which we refer to as the nominal designs. The manner in which the nominal designs were selected was by first selecting a general design and size that was able to facilitate either 8 or 12 arms. The models of these designs can be seen in Figure 2. The topology was a standard temple design commonly present in BPMSMs [29]. We opted to use a common base size and shape for the stator in both designs, varying only the number of grooves for the arms. Additionally, the arms themselves were identical between the two-phase and three-phase designs with the only difference being the number of arms.

The stator was constructed in-house and made from low-carbon steel 1018. The base and the arms were constructed separately, and then attached via bolts. Each arm received its own coil, which consisted of 150 windings of 24 AWG enamel wire. For the rotor, we employed a custom-grade N50 NdFeB permanent ring magnet with diametric magnetization (SM Magnetics, Pelham, AL, USA). Once the two nominal devices were manufactured, they were then tested to empirically evaluate their performance characteristics.

Afterwards, FEA models of the two nominal designs were created using COMSOL Multiphysics (v. 5.2, COMSOL AB, Stockholm, Sweden), which were informed by results of the empirical tests. The models replicated the manufactured device in both dimensions and materials to the best of our ability. FEA was employed to facilitate a more complete comparison of the two-phase and three-phase configurations by allowing the dimensions to be varied for seven geometric features without requiring different designs to be manufactured.

2.1. Experiments

Figure 4 illustrates a custom test rig that was used to measure the BPMSMs' axial stiffness, radial stiffness, and current force. A schematic representation of the test rig's components is shown in Supplementary Materials Figure S3. The forces were measured in each nominal design by attaching the stator to a three-component force sensor (Kistler Instrument Corp, 9251A, Amherst, NY, USA) via a custom aluminum fixture. The rotor was attached to an aluminum rod via a custom 3D-printed piece. The rod was then rigidly affixed to two separate three-axis translation stages that allowed for its precise positioning for the aforementioned rotor within three-dimensional space. Each stator phase coil grouping was wired to an individual operational amplifier (Apex Microtechnology Inc., PA02A, Tucson, AZ, USA) which allowed for exact allocation of current. Control of the linear stages, current distribution, and data acquisition from the load cell were accomplished using a custom LabVIEW (v. 18.0.1f4) code. The initial rotor position for each experiment was defined as the point where the rotor is centered axially and radially with a 0° rotor angle. To determine the axial stiffness, the rotor was shifted to five evenly spaced axial positions from the points −2 mm to +2 mm and the force was recorded at each location. For radial stiffness, the rotor was moved to five evenly spaced radial positions along the x-axis from the points −1 mm to +1 mm. Finally, to measure the current force, the amperages of 1A, 2 A, 3A, 4A, and 5A were applied to the specific phases, in accordance with Equations (1) and (2).

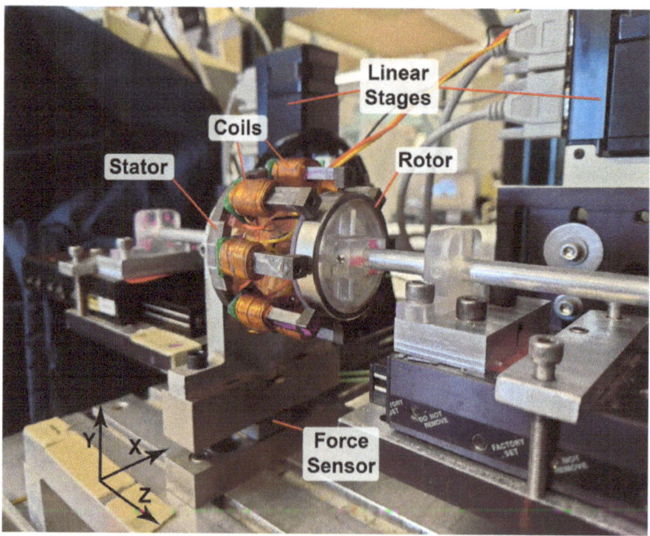

Figure 4. Two-phase nominal design positioned on the force testing rig. Rotor at the centered, neutral position. The rotor and linear stages are moveable, whereas the coils, stator, and force sensor remain stationary.

2.2. Simulations

After empirical testing was completed, we created FEA models of the nominal designs using COMSOL Multiphysics. As mentioned previously, these models replicated the materials and design of the manufactured nominal devices. We then determined the axial stiffness, radial stiffness, and current force by using displacements and currents identical to the ones used in the empirical studies. Validation of the models was assessed by determining the difference between the simulated and empirical results. Our models allowed for the varying of dimensions of seven different geometric features and enabled the evaluation of their effect on the performance of the two different configurations. For this study, we chose to investigate the seven geometric features as follows: air gap, head height, head protrusion, head width, rotor height, rotor length, and rotor outer radius; they are detailed in Figure 5. There are other geometric features including, but not limited to, the stator arm height and baseplate height, that may have an effect on the BPMSMs' performance characteristics. We prioritize the seven geometric features that we selected based on their proximity to the air gap.

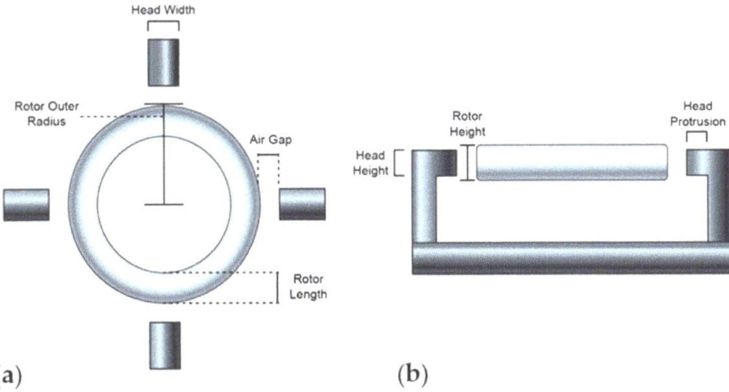

Figure 5. Illustration of the seven key geometric features of the two-phase and three-phase BPMSMs. (a) Rotor outer radius, head width, air gap, and rotor length are shown; top view. (b) Head height, rotor height, and head protrusion are shown; side view.

Along with the nominal design, two other dimensions were tested for each geometric feature as illustrated by Table 1. For this study, only the dimension of a single geometric feature was varied from the nominal design at a time. The two additional dimensions that were selected for each geometric feature were based upon realistic design extremes such as the concept that geometric features cannot be arbitrarily reduced and maintain structural integrity. The values of these dimensions are further limited by physical constraints of the nominal designs. The considered dimensions enabled the sensitivity to be determined for the performance characteristics in response to changes in geometric features. Comparisons were made by calculating the percent difference between the two nominal simulated results. Lastly, preliminary results suggested an interdependent relationship between the geometric features of head height and rotor height. Thus, we simulated all combinations of the dimensions listed in Table 1 for the head height and rotor height.

Table 1. The key geometric features evaluated in the simulations. The bolded numbers indicate the dimensions for the nominal designs. All dimensions are in millimeters.

Air Gap	Head Height	Head Protrusion	Head Width	Rotor Height	Rotor Length	Rotor Outer Radius
1.5	2	1	4	5	**3.5**	**26**
2	**5**	**3**	6	**10**	5.5	29
2.5	8	5	**8**	15	7.5	32

3. Results and Discussion

3.1. Numerical and Empirical Agreement

Table 2 details the performance characteristics for the empirical studies and simulations of the nominal designs. Supplementary Materials Figure S4 demonstrates all of the comparisons between the numerical and empirical tests that were used to derive the aforementioned performance characteristics. The discrepancy between the empirical studies and simulations ranged between 10.3–21.0% for all cases. One source of discrepancy is the fact that the materials used in simulation are the idealized versions whereas inconsistencies in the manufacturing process of these same materials can lead to differing magnetic properties. Additionally, the engineering tolerances of the manufactured device can affect the difference between the numerical and empirical results. This is especially true for tolerances that would affect the air gap because, as we will discuss in later sections, small changes in the air gap have a large impact on the performance characteristics, particularly with regard to the radial stiffness.

Table 2. Comparison between numerical and empirical data.

Characteristics	Two-Phase Numerical	Two-Phase Empirical	Two-Phase Percent Difference	Three-Phase Numerical	Three-Phase Empirical	Three-Phase Percent Difference
Axial Stiffness (N/mm)	4.15	4.60	10.3	5.43	6.08	11.3
Radial Stiffness (N/mm)	10.60	13.44	20.4	14.05	17.34	21.0
Current Force (N/A)	1.97	2.26	13.5	1.24	1.49	18.1

There has been little precedent to determine what defines an adequate agreement between the numerical and empirical data specifically in the case of BPMSMs. In [31], they studied a design similar to our three-phase configuration and they reported 13–38% discrepancies for the performance characteristics of the current force and radial stiffness. With that said, it is important to note that, despite the discrepancies, the trends for both empirical and numerical data as seen in Supplementary Materials Figure S4 are similar in the sense that they are both linear. The reason this aspect of the data is highlighted is because it demonstrates that the underlying mechanisms of the FEA model are in line with those of the physical devices. This is corroborated by the fact that the percent difference of the numerical and empirical results for the three performance characteristics are consistent between the two-phase and three-phase nominal designs. What this means is that, while there are discrepancies within the absolute values, changes in the dimensions of geometric features in the simulations would consistently be 10.3–21.0% different from their empirical counterpart. This then bolsters our FEA model's usage in the simulations of this study, the purpose of the simulation being to show the effect of dimensional changes in geometric features between the two-phase and three-phase configurations.

3.2. Nominal Comparisons

Figure 6 illustrates the experimental results for the axial stiffness, radial stiffness, and current force for the nominal designs. This figure demonstrates the linearity of these three functions within the range of independent values that we explored. Table 3 succinctly summarizes these performance characteristics and provides a quantitative comparison between the two nominal designs.

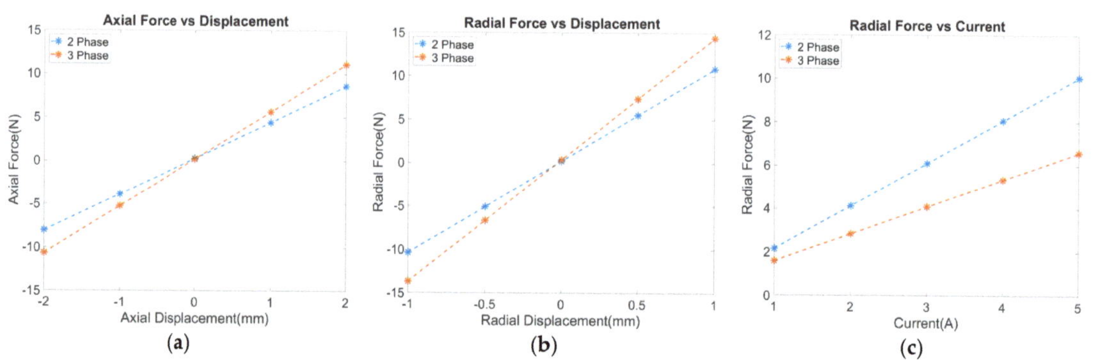

Figure 6. Comparison of the two-phase and three-phase nominal designs: (**a**) the axial force as a function of axial displacement (slope is axial stiffness (N/mm)); (**b**) radial force as a function of radial displacement (slope is radial stiffness (N/mm)); and (**c**) the radial force as a function of current (slope is current force (N/A)).

Table 3. Performance characteristics of the two-phase and three-phase nominal designs.

Characteristics	Two-Phase	Three-Phase	Percent Difference
Axial Stiffness (N/mm)	4.15	5.43	26.7
Radial Stiffness (N/mm)	10.60	14.05	27.9
Current Force (N/A)	1.97	1.24	45.4
Axial-to-Radial Stiffness	0.392	0.386	6.4

The three-phase nominal design demonstrated higher axial and radial stiffnesses, which can be attributed to the fact that the magnetic force is proportional to the surface area [28]; the three-phase design has more arms and, therefore, a greater surface area for a given arm design than the two-phase design. We propose that the ratio of the advantageous axial stiffness to the detrimental radial stiffness can be used to evaluate the general performance of any BPMSM design. In this case, we find that the three-phase nominal design has a lower axial-to-radial-stiffness ratio, 0.386 as compared to 0.392.

The two-phase motor produces a substantially higher (45.4%) current force than the three-phase designs. This is due to the differing distribution of the total current as per Equations (1) and (2). Thus, the total power usage of the three-phase design will be lower than the two-phase design. Figure 7 displays the current force as a function of power rather than current.

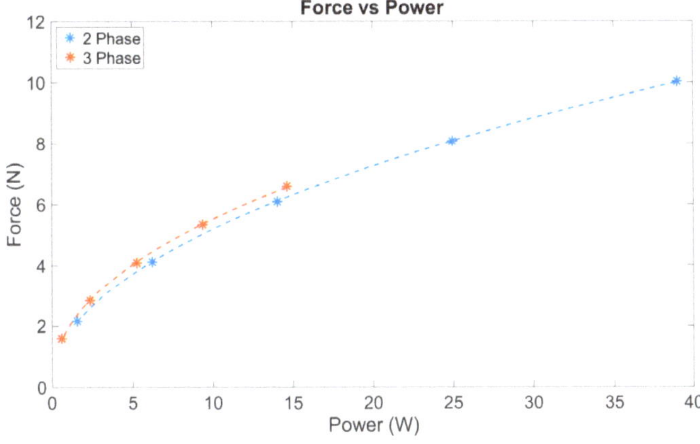

Figure 7. Force generation as a function of power for the nominal two-phase and three-phase designs.

Per Figure 7, more similar efficiencies between the two-phase and three-phase configurations were observed when the force was represented as a function of power, rather than a function of current; Figure 6c. The relationship demonstrated in Figure 7 can then be represented as the following function:

$$F = k\sqrt{P} \tag{3}$$

where F is the force generated (N), P is the power usage (W), and k is the proportionality constant (N/W). k for the two-phase design is 1.57 N/W and that for the three-phase design is 1.62 N/W, which is a 2.7% difference. At this point, it is important to note that the purpose of the current force is to counteract the force resulting from the radial displacement. Thus, while similar force per power values were found, a radial displacement for the three-phase design requires more power than the two-phase design. As an example, a radial displacement of 0.5 mm requires 11.3 W and 18.8 W for the two-phase and three-phase nominal designs, respectively.

3.3. Sensitivity

The purpose of using simulations to vary the dimensions of specific geometric features was twofold. The first reason was to find the sensitivity of the characteristics of each design to dimensional changes in these geometric features for both phase configurations. More generally, it allowed us to determine the relationships between specific geometric features and the active and passive characteristics of a BPMSM with this topology regardless of the configuration. For the ranges of dimensions that we chose to simulate, the relationship between the majority of the characteristics and changes in geometric features were linear, which allows us to represent the sensitivity for these geometric features as a single value as shown in Figure 8.

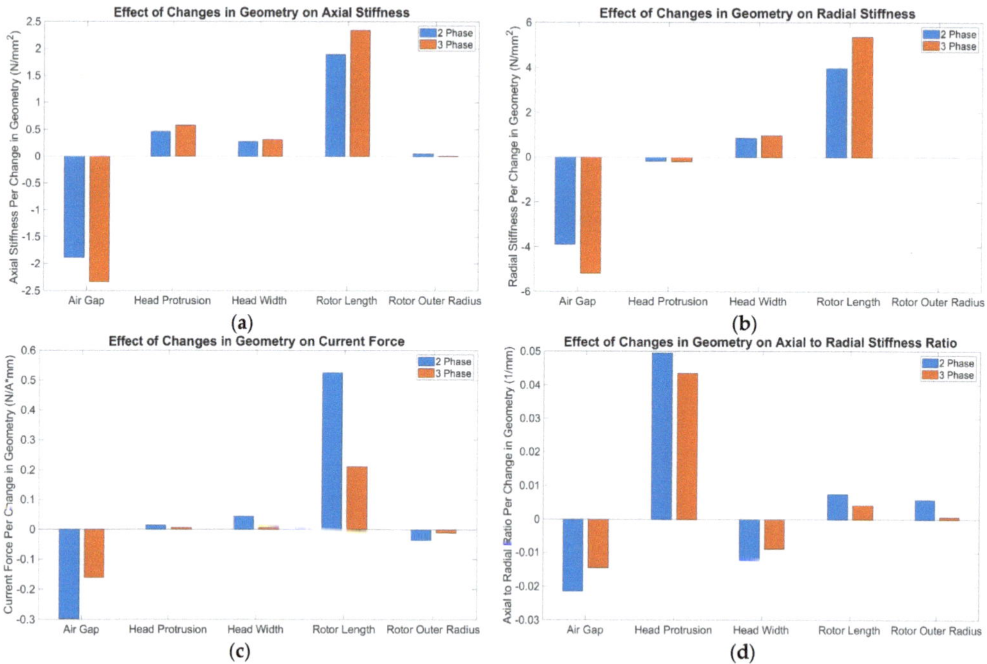

Figure 8. Sensitivity of (**a**) axial stiffness, (**b**) radial stiffness, (**c**) current force, and (**d**) axial-to-radial-stiffness ratio to changes in specified geometric features. Additionally, the comparison between the two-phase and three-phase configurations.

The results in Figure 8 represent how the characteristics of the nominal design would change per dimensional change of the specified geometric feature. Furthermore, a positive value means that increasing the dimension of that geometric feature would increase that performance characteristic, whereas a negative value reflects the opposite. Values of the sensitivities displayed in Figure 8, as well as the percent difference between the two-phase and three-phase configurations, are shown in Supplementary Materials Table S1. Additionally, the graphs from which Figure 8 is derived are shown in Supplementary Materials Figure S5.

In line with the performance characteristics of the nominal designs, we observe that the three-phase design has a higher sensitivity than the two-phase design for both passive characteristics as seen in Figure 8a,b. For example, the radial stiffness sensitivity for the rotor length was 5.4 N/mm^2 versus 4.0 N/mm^2 for three-phase design and two-phase design, respectively. Conversely, Figure 8d demonstrates a consistently higher axial to radial sensitivity for the two-phase configuration as can be seen by the air gap sensitivity being $-0.021\ 1/\text{mm}$ and $-0.014\ 1/\text{mm}$ for the two-phase and three-phase configurations, respectively. Lastly, the three-phase configuration had a lower sensitivity for the active characteristic as exemplified by its $0.21\ \text{NA}/\text{mm}^2$ current force sensitivity to the rotor length as compared to the two-phase configuration's $0.52\ \text{NA}/\text{mm}^2$

Regardless of the configuration, we determine and present the general trends of the BPMSMs. It is shown that the rotor outer radius has virtually no effect on any of the characteristics. For example, the rotor outer radius' axial stiffness sensitivities were only $0.05\ \text{N}/\text{mm}^2$ and $0.09\ \text{N}/\text{mm}^2$ for the two-phase and three-phase configurations, respectively. While most geometric features have the same sign for all of the sensitivities, the head protrusion does not; this suggests that increasing the head protrusion is always beneficial. Additionally, we found that all performance characteristics are sensitive to changes in the air gap and rotor length when compared to the other geometric features. As an example of this, the axial stiffness sensitivity for the two-phase configuration is $-1.88\ \text{N}/\text{mm}^2$ and $1.89\ \text{N}/\text{mm}^2$ for the air gap and magnet length, respectively, but are only $0.46\ \text{N}/\text{mm}^2$ and $0.27\ \text{N}/\text{mm}^2$ for the head protrusion and head width, respectively.

3.4. Interdependent Geometric Features

There are two geometric features whose sensitivities are non-linear and, thus, excluded from Figure 8: the head height and rotor height. These sensitivities also exhibit non-monotonic behavior, and this suggests an interdependent relationship between them. As mentioned previously, the relationship between the two geometric features was explored by conducting simulations for the six additional combinations of the head height and rotor height. The results of these studies are seen in Figure 9.

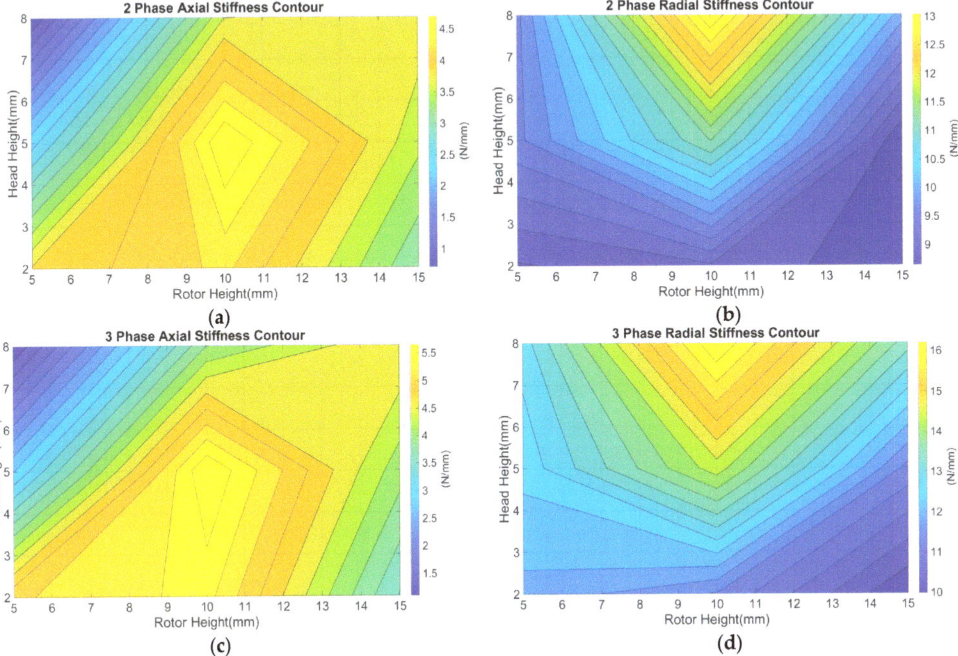

Figure 9. The relationship between rotor height and head height. (**a**) The two-phase axial stiffness contour, (**b**) the two-phase radial stiffness contour, (**c**) the three-phase axial stiffness contour, and (**d**) the three-phase radial stiffness contour. Note that, because the current force was monotonic for these geometric combinations, it was omitted here.

Figure 9 demonstrates the interdependent relationship between the two geometric features. We found that the axial stiffness is maximized when the dimensions of the rotor height and head height are 10 mm and 5 mm, respectively. This indicates that a rotor-height-to-head-height ratio of approximately 2:1 may be desirable for maximizing the axial stiffness. Conversely, we did not observe a similar trend for the radial stiffness because its visible maximum is when the dimensions of the rotor height and head height are 10 mm and 8 mm, respectively. These trends suggest that the combination of the rotor height and head height for maximum radial stiffnesses lay outside of the dimensions simulated in this study. To speculate, the combination of the two geometric features which results in the maximum radial stiffness is likely 10 mm and 10 mm. This would then suggest that a 1:1 ratio maximizes the radial stiffness.

4. Conclusions

In this study, two-phase and three-phase bearingless permanent magnet slice motor configurations were compared. Initially, two nominal designs were manufactured, and their performance characteristics were empirically tested. The three characteristics of the axial stiffness, radial stiffness, and current force were used as the metrics by which to evaluate these nominal designs. FEA models were created based upon the nominal devices and were validated by the empirical results. Consequently, simulations were used to vary seven different geometric features of the model to determine the sensitivity of the two-phase and three-phase configurations. Our findings showed that the nominal three-phase had a higher passive axial stiffness, but that was accompanied by a higher radial stiffness, which resulted in a modestly (6.4%) lower axial-to-radial-stiffness ratio. This is a meager advantage for the two-phase design. Furthermore, we showed that the nominal two-phase design has a higher (45.4%) current force, but, when considering the force generated as a function of power, the nominal three-phase design was slightly superior (2.7%). Bear in mind the three-phase nominal design would require more power to counteract the radial forces due to its higher radial stiffness.

To determine which configuration to use for a blood pump, designers would have to consider aspects such as the size, air gap, and prevalence of off-the-shelf controller electronics for each configuration. As mentioned previously, size is a major consideration when designing an implantable blood pump as there is the desire for them to be compact. One of the limitations as to how small a BPMSM blood pump can be made relates to the fact that BPMSMs require space between the stator arms for parts such as coils and sensors. With that said, the three-phase configuration presented in this study has more stator arms than its two-phase counterpart. While having more stator arms does not inherently impact the performance of a three-phase device, it does limit the extent to which it can be scaled down relative to a comparable two-phase device. This is simply because a three-phase device would have less room between the arms while maintaining adequate performance characteristics. It is, then, in this regard that the two-phase configuration presented here would have the advantage for usage in an implantable blood pump.

For the air gap, it was shown that the performance characteristics of both configurations are highly sensitive to even minor changes in the dimension. The three-phase configuration was shown to have a higher sensitivity to air gap changes, but this stems from its larger surface area compared to the two-phase design [28]. Furthermore, it is important to note that the air gap itself does not directly affect this surface area. Consequently, neither the two-phase nor three-phase configuration would inherently be better equipped to handle a larger air gap than the other.

The last thing to consider is the availability of off-the-shelf components because these can save on both development and manufacturing costs. In general, traditional electric three-phase motors are much more established in the modern era than two-phase motors [32]. As a result of this, the research and development of electric motors and their peripherals are predominantly centered around the three-phase configuration. This results in a myriad of off-the-shelf robust motor drive controllers which can control the rotational portion of the device that only work with three-phase configurations [25]. This becomes the only major drawback of the two-phase configuration as there is the need to develop a bespoke two-phase motor drive controller. Overall, the two-phase configuration's benefits outweigh its drawbacks for the application of an implantable blood pump within the context of this study.

Regardless of configuration, both showed consistent trends in their sensitivity to dimensional changes in specific geometric features. We determined that changes in the air gap and rotor length lead to the most drastic deviations in the three performance characteristics. Conversely, the rotor outer radius provided a negligible effect on the performance characteristics. While the majority of changes in the geometric features contained tradeoffs, increasing the head protrusion only benefited the overall performance. Furthermore, we elucidated the interdependent nature of the geometric features of the head height and rotor height. This may prove extremely useful to future motor designers.

Supplementary Materials: The following are available online at https://www.mdpi.com/article/10.3390/act13050179/s1, Figure S1: (a) A cut side view of the rotor and stator where the ϕ denotes the tilt angle. The black dot denotes the axis of about which the rotor tilts. (**b**) A top view of the rotor and stator where the dotted line indicates the axis of tilt; Figure S2: The numerical results for the rotor torque as a function of tilt angle for the two-phase nominal design. The slope of Figure S2 is the tilting stiffness which in this case is 0.0339 (Nm/°). thus demonstrating the equation's ability to approximate the tilting stiffness based upon axial stiffness; Figure S3: A general schematic of the connections of the various components of the force testing rig. An arrow between blocks indicate signal direction as well as the fact that a physical connection and electrical connection exists between these two components. Solid lines between blocks indicate that only a physical connection exists between the two components. The dotted line demonstrates the magnetic coupling between the rotor and stator and thus no physical or electrical connection; Figure S4: Comparison of numerical and empirical data for axial stiffness, radial stiffness, and current force of the nominal design. (a) The two-phase nominal design and (b) the three-phase nominal design; Figure S5: The sensitivity of the following performance characteristics as a function of the geometric features, (a) axial stiffness, (b) radial stiffness, (c) current force, and (d) axial to radial stiffness; Table S1: The sensitivity values for each performance characteristic. Additionally the percent difference for each performance characteristic between the two-phase and three-phase configurations. * The value for this position is much smaller than the rest with it only being 0.00009 (N/mm^2).

Author Contributions: Conceptualization, J.E.M.L. and S.W.D.; methodology, J.E.M.L. and S.W.D.; software, J.E.M.L.; validation, J.E.M.L. and S.W.D.; formal analysis, J.E.M.L.; investigation, J.E.M.L.; resources, A.L.T. and S.W.D.; data curation, J.L., G.C.M., A.L.T. and S.W.D.; writing—original draft preparation, J.E.M.L.; writing—review and editing, J.L., G.C.M., A.L.T. and S.W.D.; visualization, J.L., G.C.M., A.L.T. and S.W.D.; supervision, A.L.T. and S.W.D.; project administration, A.L.T. and S.W.D.; funding acquisition, A.L.T. All authors have read and agreed to the published version of the manuscript.

Funding: The authors wish to acknowledge the financial support as provided by the National Heart, Lung, and Blood Institute of the National Institutes of Health, under Award Number: R01HL153536. The content is solely the responsibility of the authors and does not necessarily represent the official views of the National Institutes of Health. Partial support was also provided from Drexel University's "Research and Engineering for Pediatrics by Interdisciplinary Collaboration Leveraging Education and Partnerships for Pediatric Healthcare (R-EPIC LEAP for Pediatric Healthcare)" from the U.S. Department of Education's Graduate Assistance in Areas of National Need Program (Fellowship Award: G. Matlis).

Data Availability Statement: The data presented in this study are available upon request from the corresponding author.

Acknowledgments: We thank Liliana Ramos and Elysia Hentschel for their illustrations. We thank Andrew Roof for his code development. Finally, we thank Arthur Thompson Johnson for his technical help.

Conflicts of Interest: The authors declare no conflict of interest.

References

1. Cheng, S.; Olles, M.W.; Burger, A.F.; Day, S.W. Optimization of a hybrid magnetic bearing for a magnetically levitated blood pump via 3-D FEA. *Mechatronics* **2011**, *21*, 1163–1169. [CrossRef]
2. Cheng, S.; Olles, M.W.; Olsen, D.B.L.; Joyce, D.; Day, S.W. Miniaturization of a magnetically levitated axial flow blood pump. *Artif. Organs* **2010**, *34*, 807–815. [CrossRef]
3. Hirschhorn, M.D.; Lawley, J.E.M.; Roof, A.J.; Johnson, A.P.T.; Stoddard, W.A.; Stevens, R.M.; Rossano, J.; Arabia, F.; Tchantchaleishvili, V.; Massey, H.T.; et al. Next generation development of hybrid continuous flow pediatric total artificial heart technology: Design–build–test. *ASAIO J.* **2023**, *69*, 1090–1098. [CrossRef] [PubMed]
4. Hirschhorn, M.D.; Catucci, N.; Day, S.W.; Stevens, R.M.; Tchantchaleishvili, V.; Throckmorton, A.L. Channel impeller design for centrifugal blood pump in hybrid pediatric total artificial heart: Modeling, magnet integration, and hydraulic experiments. *Artif. Organs* **2023**, *47*, 680–694. [CrossRef] [PubMed]
5. Fox, C.S.; Palazzolo, T.; Hirschhorn, M.; Stevens, R.M.; Rossano, J.; Day, S.W.; Tchantchaleishvili, V.; Throckmorton, A.L. Development of the centrifugal blood pump for a hybrid continuous flow pediatric total artificial heart: Model, make, measure. *Front. Cardiovasc. Med.* **2022**, *9*, 886874. [CrossRef]
6. Fox, C.; Chopski, S.; Murad, N.; Allaire, P.; Mentzer, R.; Rossano, J.; Arabia, F.; Throckmorton, A. Hybrid continuous-flow total artificial heart. *Artif. Organs* **2018**, *42*, 500–509. [CrossRef]
7. Fox, C.; Sarkisyan, H.; Stevens, R.; Arabia, F.; Fischer, W.; Rossano, J.; Throckmorton, A. New versatile dual-support pediatric heart pump. *Artif. Organs* **2019**, *43*, 1055–1064. [CrossRef]
8. James, S.L. Global, regional, and national incidence, prevalence, and years lived with disability for 354 diseases and injuries for 195 countries and territories, 1990–2017: A systematic analysis for the global burden of disease study 2017. *Lancet* **2018**, *392*, 1789–1858. [CrossRef]
9. Savarese, G.; Becher, P.M.; Lund, L.H.; Seferovic, P.; Rosano, G.M.C.; Coats, A.J.S. Global burden of heart failure: A comprehensive and updated review of Epidemiology. *Cardiovasc. Res.* **2022**, *118*, 3272–3287. [CrossRef] [PubMed]
10. Das, B. Current State of Pediatric Heart Failure. *Children* **2018**, *5*, 88. [CrossRef]
11. Kanter, K.R. Management of single ventricle and Cavopulmonary Connections. In *Sabiston and Spencer's Surgery of the Chest*; Elsevier: Amsterdam, The Netherlands, 2016; pp. 2041–2054.
12. Bauersachs, J. Heart failure drug treatment: The fantastic four. *Eur. Heart J.* **2021**, *42*, 681–683. [CrossRef]
13. Bowen, R.E.; Graetz, T.J.; Emmert, D.A.; Avidan, M.S. Statistics of heart failure and mechanical circulatory support in 2020. *Ann. Transl. Med.* **2020**, *8*, 827. [CrossRef]
14. Palazzolo, T.; Hirschhorn, M.; Garven, E.; Day, S.; Stevens, R.M.; Rossano, J.; Tchantchaleishvili, V.; Throckmorton, A. Technology landscape of pediatric mechanical circulatory support devices: A systematic review 2010–2021. *Artif. Organs* **2022**, *46*, 1475–1490. [CrossRef]
15. Jorde, U.P.; Saeed, O.; Koehl, D.; Morris, A.A.; Wood, K.L.; Meyer, D.M.; Cantor, R.; Jacobs, J.P.; Kirklin, J.K.; Pagani, F.D.; et al. The Society of Thoracic Surgeons intermacs 2023 annual report: Focus on magnetically levitated devices. *Ann. Thorac. Surg.* **2024**, *117*, 33–44. [CrossRef]
16. Zhou, A.L.; Etchill, E.W.; Giuliano, K.A.; Shou, B.L.; Sharma, K.; Choi, C.W.; Kilic, A. Bridge to transplantation from mechanical circulatory support: A narrative review. *J. Thorac. Dis.* **2021**, *13*, 6911–6923. [CrossRef]
17. Bourque, K.; Gernes, D.B.; Loree, H.M., 2nd; Richardson, J.S.; Poirier, V.L.; Barletta, N.; Fleischli, A.; Foiera, G.; Gempp, T.M.; Schoeb, R.; et al. Heartmate III: Pump Design for a centrifugal LVAD with a magnetically levitated rotor. *ASAIO J.* **2001**, *47*, 401–405. [CrossRef]
18. Zhang, J.; Gellman, B.; Koert, A.; Dasse, K.A.; Gilbert, R.J.; Griffith, B.P.; Wu, Z.J. Computational and experimental evaluation of the fluid dynamics and hemocompatibility of the Centrimag Blood Pump. *Artif. Organs* **2006**, *30*, 168–177. [CrossRef]
19. Köhne, I. Haemolysis induced by mechanical circulatory support devices: Unsolved problems. *Perfusion* **2020**, *35*, 474–483. [CrossRef] [PubMed]
20. Puentener, P.; Schuck, M.; Kolar, J.W.; Steinert, D. Comparison of bearingless slice motor topologies for pump applications. In Proceedings of the 2019 IEEE International Electric Machines & Drives Conference (IEMDC), San Diego, CA, USA, 12–15 May 2019.
21. Zhang, S.; Luo, F. Direct control of radial displacement for bearingless permanent-magnet-type synchronous motors. *IEEE Trans. Ind. Electron.* **2009**, *56*, 542–552. [CrossRef]
22. Weinreb, B.S.; Noh, M.; Fyler, D.C.; Trumper, D.L. Design and implementation of a novel interior permanent magnet bearingless Slice Motor. *IEEE Trans. Ind. Appl.* **2021**, *57*, 6774–6782. [CrossRef] [PubMed]
23. Liu, X.; Qu, H.; Chen, Q.; Wang, C.; Liu, J.; Wang, Q. Electromagnetic characteristics analysis of a novel bearingless permanent magnet slice motor with Halbach Arrays for Artificial Heart Pump. In Proceedings of the 2021 Photonics & Electromagnetics Research Symposium (PIERS), Hangzhou, China, 21–25 November 2021.
24. Zhang, Y.; Hu, L.; Su, R.; Ruan, X. Design method of bearingless permanent magnet slice motor for maglev centrifugal pump based on performance metric cluster. *Actuators* **2021**, *10*, 153. [CrossRef]
25. Zurcher, F.; Nussbaumer, T.; Gruber, W.; Kolar, J.W. Comparison of 2- and 3-phase Bearingless Slice Motor Concepts. *J. Syst. Des. Dyn.* **2009**, *3*, 471–482. [CrossRef]
26. Bartholet, M.T.; Nussbaumer, T.; Silber, S.; Kolar, J.W. Comparative evaluation of Polyphase bearingless slice motors for fluid-handling applications. *IEEE Trans. Ind. Appl.* **2009**, *45*, 1821–1830. [CrossRef]

27. Ahn, H.-J.; Vo, N.V.; Jin, S. Three-phase coil configurations for the radial suspension of a bearingless Slice Motor. *Int. J. Precis. Eng. Manuf.* **2022**, *24*, 95–102. [CrossRef]
28. Chiba, A. Electromagnetics and mathematical model of magnetic bearings. In *Magnetic Bearings and Bearingless Drives*; Elsevier: Oxford, UK, 2005; pp. 20–24.
29. Schoeb, R.; Barletta, N. Magnetic bearing. principle and application of a bearingless slice motor. *JSME Int. J. Ser. C-Mech. Syst. Mach. Elem. Manuf.* **1997**, *40*, 593–598. [CrossRef]
30. Barletta, R.N.; Hahn, J. The bearingless centrifugal pump—A perfect example of a mechatronics system. *IFAC Proc. Vol.* **2000**, *33*, 443–448. [CrossRef]
31. Weinreb, B. A Novel Magnetically Levitated Interior Permanent Magnet Slice Motor. Master's Thesis, Massachusetts Institute of Technology, Cambridge, MA, USA, 2020; p. 193.
32. Fukao, T.; Chiba, A. *Introduction in Magnetic Bearings and Bearingless Drives*; Elsevier: Oxford, UK, 2005; pp. 4–7.

Article

Overvoltage Avoidance Control Strategy for Braking Process of Brushless DC Motor Drives with Small DC-Link Capacitance

Wei Chen [1], Jialong Wu [2], Xinmin Li [1,*] and Chen Li [3]

[1] School of Electrical Engineering, Tiangong University, Tianjin 300387, China
[2] School of Control Science and Engineering, Tiangong University, Tianjin 300387, China
[3] Advanced Electrical Equipment Innovation Center, Zhejiang University, Hangzhou 311107, China; lichen_hz@zju.edu.cn
* Correspondence: lixinmin@tju.edu.cn

Abstract: Single-phase input rectifier brushless DC motor drives with a small film capacitor have many advantages, such as high power density and high reliability. However, when the motor system operates in regenerative braking mode, the dc-link capacitor with reduced capacitance may suffer from overvoltage without adding additional hardware circuits. At the same time, the braking torque control of the motor will be affected by speed variations. In order to ensure smooth and reliable operation of the motor system, an anti-overvoltage braking torque control method is proposed in this article. The relationship among the dc-link capacitance, the dc-link capacitor voltage, and the speed during regenerative braking is analyzed quantitatively, and the speed at which the regenerative braking is switched to the plug braking is obtained, which in turn consumes the capacitor energy to avoid dc-link overvoltage. Additionally, based on the relationship between the controllability of the braking torque and the speed, a reference value of the braking current that matches the speed is designed. The proposed method makes use of the capacitor's energy storage during regenerative braking. Meanwhile, it mitigates the impact of motor speed on braking torque. Finally, the effectiveness of the proposed method is verified on a motor platform equipped with the dc-link film capacitor.

Keywords: brushless DC motor (BLDCM); braking mode; small capacitor; anti-overvoltage control

Citation: Chen, W.; Wu, J.; Li, X.; Li, C. Overvoltage Avoidance Control Strategy for Braking Process of Brushless DC Motor Drives with Small DC-Link Capacitance. *Actuators* **2024**, *13*, 185. https://doi.org/10.3390/act13050185

Academic Editor: Leonardo Acho Zuppa

Received: 8 April 2024
Revised: 9 May 2024
Accepted: 10 May 2024
Published: 13 May 2024

1. Introduction

Due to the advantages of small size, simple structure, and high power density, the brushless DC motor (BLDCM) has been widely used in industrial transmission, household appliances, and other fields [1–4]. Traditional single-phase diode rectifier input motor systems usually use large-capacity electrolytic capacitors in the dc-link circuit; however, the service life of electrolytic capacitors decreases significantly under extreme high temperature and high humidity environments, and about 60% of drive circuit failures are related to the electrolytic capacitor, which reduces the reliability of the motor drive system [5,6]. Compared with an electrolytic capacitor, the film capacitor allows a higher current ripple and has a longer lifetime, enhances reliability, and reduces the failure of the drive system [7]. Hence, the electrolytic capacitor in the conventional dc-link circuit can be replaced by the film capacitor [8–11].

For the BLDCM drive with a single-phase rectifier, the dc-link circuit using a small capacitor in series with the IGBT scheme is mentioned in references [12–14], and the topology circuit diagram of the brushless DC motor system with the small capacitor is shown in Figure 1. In reference [14], based on the topology circuit of Figure 1, in each rectification cycle, when the capacitor voltage rises to the maximum value of the ac supply voltage, the capacitor charging vectors are applied at the same time, which will further increases the capacitor voltage, so that the motor input energy is not affected even if the dc-link capacitance is reduced.

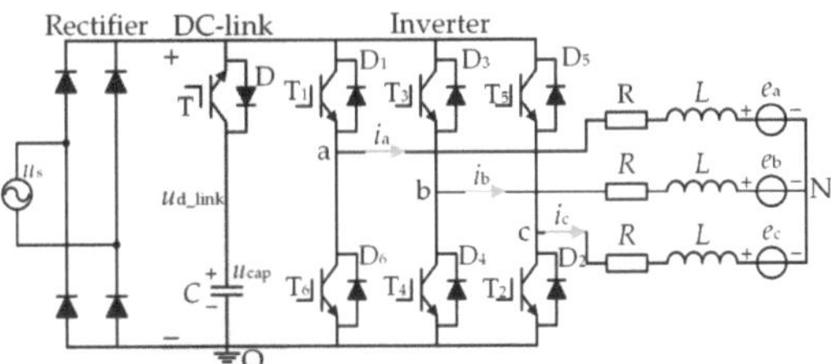

Figure 1. The equivalent circuit of BLDCM drives with small dc-link capacitance. The color-coded area in the figure is dc-link circuit.

However, when the single-phase rectifier BLDCM drives with reduced dc-link capacitance operate in regenerative braking mode, the dc-link capacitor is at risk of overvoltage. Because the dc-link capacitance in the drive system is reduced, and the diode rectifier is unable to realize the bidirectional flow of energy, the motor, during regenerative braking, can only feed energy back into the dc-link capacitor, and the capacitor voltage will rise rapidly. During regenerative braking, the copper loss of the motor can be harnessed to dissipate a portion of the motor's mechanical energy. This serves to prevent the capacitor voltage from rapidly pumping and mitigates the occurrence of dc-link overvoltage [9,10].

Therefore, the implementation of an anti-overvoltage braking control scheme using hardware devices already available on the control system is an avenue worth exploring, which will expand the range of applications for drive systems with reduced dc-link capacitance.

In comparison to the electric mode, the braking mode represents another crucial operational state of the motor. During braking mode, the motor conducting phase current is in opposition to the phase back electromotive force (back-EMF), thereby generating an electromagnetic torque in the opposite direction of motor rotation. When the BLDCM operates in the electrical braking mode, the available electrical braking modes include regenerative braking, dynamic braking, and plug braking. Regenerative braking is a process whereby the mechanical energy of the motor system is converted into electrical energy and subsequently fed back to the energy storage elements of the power supply (e.g., batteries or supercapacitors) via a hardware circuit [15–19]. Dynamic braking is a technique that involves connecting switches and braking resistors in series in the dc-link circuit. This is used to prevent the dc-link voltage from exceeding a threshold value set by the control system. When the voltage reaches this threshold, the switch is turned on, and the braking resistor absorbs the energy from the motor. This helps slow down the motor and prevent any damage from the excessive dc-link voltage. During plug braking mode, the motor can reverse the input voltage or introduce specific inverter control patterns to change the motor's conducting phase sequence, generating a braking torque in the opposite direction of the motor rotation and achieving a rapid drop in motor speed [20].

In the braking mode, the current flow circuit in the BLDCM drive is different from the normal electric mode, and the controllability of the braking torque is affected by the speed, so the control of the motor during braking is changed accordingly. Reference [20] proposes a braking torque control scheme in the whole speed range of the motor. Dynamic braking is used in the high-speed range of the motor to utilize the braking resistor to absorb the energy generated by the motor and to avoid the occurrence of dc-link overvoltage. Meanwhile, plug braking is used in the motor's low-speed range. The hybrid application of the two braking modes realizes the smooth control of the motor braking torque at the whole speed range. In reference [21], for a hybrid energy storage system with a supercapacitor, it is proposed that combining the optimal selection of switching vectors, thereby realizing the recovery of braking energy and the control of braking torque.

In order to ensure the smooth and reliable operation of a small capacitor BLDCM system under braking mode, an anti-overvoltage braking torque control method is proposed in this paper. The relationship among the dc-link capacitance, the dc-link capacitor voltage, and the speed during regenerative braking is analyzed quantitatively, and the speed at which the regenerative braking is switched to the plug braking is obtained, which in turn consumes the capacitor energy to mitigate the phenomenon of overvoltage. Meanwhile, the braking torque controllability in different braking modes was analyzed, and the braking torque controllable speed interval was obtained. According to the obtained controllable speed interval of braking torque, a reference value of braking torque matching the motor speed is designed. The method proposed in this paper, on the one hand, makes use of the capacitor's energy storage during regenerative braking and avoids the occurrence of capacitor overvoltage on the basis of the existing hardware of the system; on the other hand, it reduces the influence of the motor's speed on the braking torque and takes into account the controllability of the braking torque, which ensures that the braking process of the motor is smooth and rapid.

The rest of this article is organized as follows. Section 2 introduces the operation principle of the single-phase diode rectifier brushless DC motor drive with a small capacitance. Section 3 analyzes the relationship among the dc-link capacitance, the dc-link capacitor voltage, and the speed during regenerative braking. In Section 4, the corresponding relationship between the braking torque control performance and the speed is analyzed; thus, an anti-overvoltage braking torque control method is proposed. The experimental results to verify the proposed method are given in Section 5. Section 6 concludes this paper.

2. Single-Phase Diode Rectifier BLDCM Drives with Small Dc-Link Capacitance

The equivalent circuit diagram of the single-phase diode rectifier BLDCM drives with the reduced dc-link capacitance is illustrated in Figure 1. It primarily consists of a diode rectifier, dc-link circuit, inverter, and motor three-phase windings. The inverter is composed of six IGBTs, denoted as T_1–T_6. The voltage of the dc-link capacitor is denoted as u_{cap}, the instantaneous voltage of the ac power source is represented as u_s, and the dc-link voltage is denoted as u_{d_link}. The phase inductance, phase resistance, and neutral point of the three-phase windings of the BLDCM are L, R, and N, respectively. Figure 1 shows the direction of the current flow. According to the direction of the phase current, the three-phase windings can be redefined as positive conducting phase p, negative conducting phase n, and nonconducting phase o (p, n, $o \in$ {a, b, c}). The actual current direction of the positive conducting phase aligns with the specified positive direction. Conversely, the actual current direction of the negative conducting phase is opposite to the specified positive direction. The mathematical model of the BLDCM in the two conducting phases is expressed as follows:

$$\begin{cases} u_p = i_p R + L\frac{di_p}{dt} + e_p + u_N \\ u_n = i_n R + L\frac{di_n}{dt} + e_n + u_N \end{cases} \tag{1}$$

where e_p is back-EMF of the positive conducting phase, e_n is back-EMF of the negative conducting phase, i_p is the positive conducting phase current, i_n is the negative conducting phase current, u_p is the positive conducting phase terminal voltage, and u_n is the negative conducting phase terminal voltage.

For a single-phase diode rectifier BLDCM drive with small dc-link capacitance, the motor can achieve rated operation only if the input voltage complies with the rated voltage requirement. Taking the "$p+n-$" commutation period as an example, according to Equation (1), the phase current of the motor $I_N = i_p = -i_n$, and the back electromotive force amplitude $E = e_p = -e_n$. Therefore, under the rated operating conditions of the motor, it is imperative that the dc-link voltage (u_{d_link}) adhere to the following equation:

$$\begin{cases} U_{pnN} \geq 2E + 2I_N R \\ u_{d\_link} \geq U_{pnN} \end{cases} \tag{2}$$

where u_{d_link} is the dc-link voltage, U_{pnN} is the minimum value of dc-link voltage, E_N is the rated phase back-EMF amplitude, and I_N is the rated phase current of the motor.

To meet the requirements of Equation (2), the rectifier output voltage $|u_s|$ can be divided into three zones based on a single rectification period T_R. Although the capacitance decreases, the motor input voltage always meets the rated voltage requirements. As depicted in Figure 2, where the amplitude of the ac power supply voltage is U_{m}, the frequency of the ac power source is f, and the rectification period is T_{R}. The logic for dividing the specific rectification period zone is as follows: Zone 1: $|u_{\mathrm{s}}| \geq U_{pn\mathrm{N}}$ where $|u_{\mathrm{s}}|$ monotonically increases until U_{m}. Zone 2: $|u_{\mathrm{s}}| \geq U_{pn\mathrm{N}}$ where $|u_{\mathrm{s}}|$ monotonically decreases until 0. Zone 3: $|u_{\mathrm{s}}| \leq U_{pn\mathrm{N}}$.

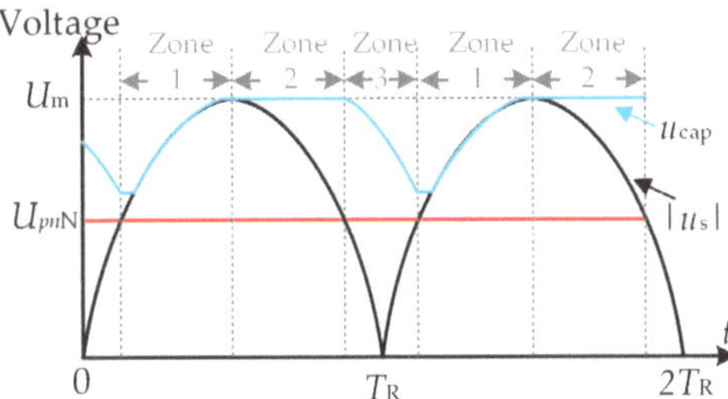

Figure 2. Regional division diagram of the output voltage of the rectifier.

The dc-link switch T is turned off in zone 1 and zone 2, while in zone 3, the switch T is turned on, and the capacitor alone supplies power to the motor. Subsequently, the capacitor voltage gradually declines until exiting zone 3. The dc-link voltage $u_{\mathrm{d\_link}}$ is given by the following equation:

$$u_{\mathrm{d\_link}} = \begin{cases} u_{\mathrm{cap}}, & \mathrm{T} = 1 \\ |u_{\mathrm{s}}|, & \mathrm{T} = 0 \end{cases} \tag{3}$$

where T = 1 and T = 0 indicate the on and off states of the dc-link switch, u_{cap} is capacitor voltage, and $|u_{\mathrm{s}}|$ is the rectifier output voltage.

For the BLDCM drive with reduced dc-link capacitance, there is only one zone 3 energy release process of the capacitor under the control of the dc-link switch transistor T. Therefore, by designing the appropriate capacitance, it is possible to ensure that the dc-link voltage $u_{\mathrm{d\_link}}$ is always greater than $U_{pn\mathrm{N}}$ in a rectification period. This guarantees a smooth and continuous motor current. Assuming the capacitor voltage is precisely $U_{pn\mathrm{N}}$ at the end of the capacitor discharge in zone 3, to simplify the analysis, the process of capacitor voltage drop is approximated to be linear. Therefore, the dc-link capacitance is designed to satisfy the following equation:

$$C\frac{\mathrm{d}u_{\mathrm{cap}}}{\mathrm{d}t} = I_{\mathrm{dc}} \;\rightarrow\; C \geq \frac{I_{\mathrm{N}}\arcsin\left(\frac{U_{pn\mathrm{N}}}{U_{\mathrm{m}}}\right)}{\pi f\left(U_{\mathrm{m}} - U_{pn\mathrm{N}}\right)} \tag{4}$$

where I_{dc} represents the average dc-link current, and U_{m} represents the amplitude of ac power supply voltage.

3. Analysis of Electrical Braking for BLDCM Drives with Small Dc-Link Capacitance

3.1. Introduction to the Braking Mode of BLDCM

When switching from electric to braking mode in BLDCM operation mode, regenerative braking or plug braking are two commonly used electrical braking modes for motors. Regenerative braking converts the mechanical energy of the motor into electrical energy, operating like a generator, while plug braking consumes energy from the power supply to accelerate the braking process. Therefore, the main difference between regenerative braking and plug braking is whether the braking consumes energy from the power supply. In braking mode, the phase back EMFs of the motor can be expressed as:

$$E = k_{\mathrm{e}}\omega_{\mathrm{m}} \tag{5}$$

where k_e and ω_m, respectively, represent the phase back-EMF coefficient and mechanical speed; E represents the amplitude of the phase back-EMF.

When the BLDCM operates in the braking mode, one electrical cycle is divided into six sectors, denoted by I to VI. The corresponding relationship between the ideal phase back-EMFs and the phase current waveform is shown in Figure 3. The conducting phase p and the conducting phase n take different values in different rotor sectors. For example, in sector I, p = b and n = a, which is the "b+a−" working state, while when the motor rotates to sector II, p = c and n = a, which is the "c+a−" working state. The conducting phase current in each sector is opposite to the direction of phase back-EMFs, resulting in a negative product. Therefore, the output braking torque of the motor is the electromagnetic torque T_e, which is opposite to the direction of motor rotation, which is different from the electric mode. As shown in Equation (6), where I_{ave} represents the average value of the phase current flowing through the conducting phase before commutation, the braking electromagnetic torque is expressed as:

$$T_e = \frac{e_a i_a + e_b i_b + e_c i_c}{\omega_m} = -\frac{2E I_{ave}}{\omega_m} \tag{6}$$

where T_e is the electromagnetic torque; i_k, ek (k = a, b, c), respectively, represent the phase current and the phase back-EMF.

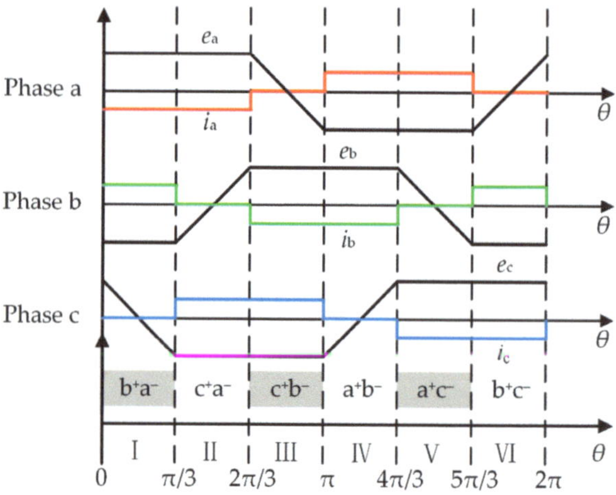

Figure 3. Correspondence between the phase back-EMFs and phase current in braking mode.

3.2. Relationship between Dc-Link Capacitor Voltage and Capacitance under Regenerative Braking

When the operating mode of the single-phase input rectifier BLDCM drives with small dc-link capacitance is shifted from the electric mode to the regenerative braking mode, a part of the mechanical energy W_m is converted into electrical energy W_C, and due to the cut-off effect of the rectifier diode, the electrical energy W_C will be stored in the dc-link capacitor, and the capacitor voltage u_{cap} will be increased rapidly. Because the dc-link voltage amplitude is closely related to the capacitance, the phenomenon of overvoltage is more likely to occur in the BLDCM drives with reduced dc-link capacitance. The other part of mechanical energy is converted into stator copper loss W_{Cu} and friction loss W_f between rotor and bearing. According to the BLDCM mechanical equation, the relationship between the regenerative braking electromagnetic torque T_{e1}, the dc-link capacitance C, the rotational inertia J, and the capacitor voltage during braking mode are mathematically quantitatively analyzed. The motor acceleration can be expressed as:

$$\frac{d\omega_m}{dt} = \frac{T_{e1} + T_L + B\omega_m}{J} \tag{7}$$

where J, T_L, and B denote the rotational inertia, load torque, and friction factor, respectively. In order to simplify the analysis, we ignore the effect of rotor bearing friction loss and load. The actual mechanical speed during regenerative braking mode is expressed as:

$$\omega_m = \omega - \frac{T_{e1}}{J}t \tag{8}$$

where ω is the initial mechanical speed of motor braking, ω_m is the actual mechanical speed, and t denotes the time taken for the speed to drop from ω_m to ω.

The motor mechanical energy W_m is satisfied:

$$W_m = \frac{1}{2}J(\omega^2 - \omega_m^2) = \frac{1}{2}C\left[U_C^2 - U_{C0}^2\right] + W_{Cu} \tag{9}$$

where U_C is the voltage of the dc-link capacitor during braking mode, and U_{C0} is the capacitor voltage during initial braking. The average regenerative braking torque can be expressed as $T_{e1} = 2I_1 k_e$, and as the braking process proceeds, the copper loss W_{Cu} can be derived as:

$$W_{Cu} = 2I_1^2 Rt = \frac{T_{e1}^2}{2k_e^2}Rt \tag{10}$$

where T_{e1} denotes the average value of regenerative braking torque; I_1 denotes the reference value of conducting phase current during regenerative braking.

According to Equation (6) to Equation (10), when the motor speed drops to the switch speed ω_c, the capacitor voltage is assumed to reach the maximum U_{CM} and the mathematical equation of the dc-link capacitance and the maximum value of capacitor voltage can be expressed as:

$$C = \frac{J(2k_e^2 T_{e1}\omega - T_{e1}^2 R)t - (T_{e1}k_e t)^2}{Jk_e^2(U_{CM}^2 - U_{C0}^2)} \tag{11}$$

where U_{CM} denotes the maximum value of capacitor voltage when the motor speed drops to the speed ω_c. U_{C0} denotes the initial value of the dc-link capacitor voltage when the motor speed is ω.

Assuming that the braking torque is constant during regenerative braking, we can conclude that the higher the initial mechanical speed ω, the larger capacitance that needs to be maintained so that the capacitor voltage does not exceed the threshold in order to be able to store more energy. In addition, if a larger dc-link capacitor is used, the amplitude of capacitor voltage rise will decrease.

4. Proposed Anti-Overvoltage Braking Torque Control Method for BLDCM Drives with Small Dc-Link Capacitance

This section systematically analyzes the electrical braking torque controllability of a BLDCM drive with small dc-link capacitance, providing a basis for the proposed anti-overvoltage braking torque control method. The equivalent circuit diagrams for the electrical braking modulation mode of the BLDCM drives in sectors I, III, and V are shown in Figures 4 and 5 below. Figure 4a,b shows the regenerative braking OFF_PWM modulation pattern, and Figure 5a,b shows the plug braking ON_PWM modulation pattern. S_{pH}, S_{pL}, S_{nH}, and S_{nL} indicate the switching state of positive conducting phase upper bridge arm IGBTs, positive conducting phase down bridge arm IGBTs, negative conducting phase upper bridge arm IGBTs, and negative conducting phase down bridge arm IGBTs. In the following analysis, the IGBT turn-on, turn-off, and duty cycle for one modulation period are denoted as "1", "0", and "D", respectively.

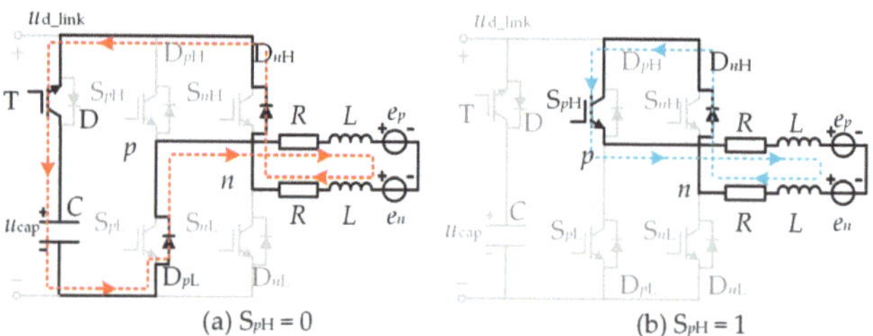

(a) $S_{pH} = 0$ (b) $S_{pH} = 1$

Figure 4. Equivalent circuits of regenerative braking OFF_PWM modulation pattern in an odd sector number (I, III, V). The direction of the arrow indicates the direction of current flow.

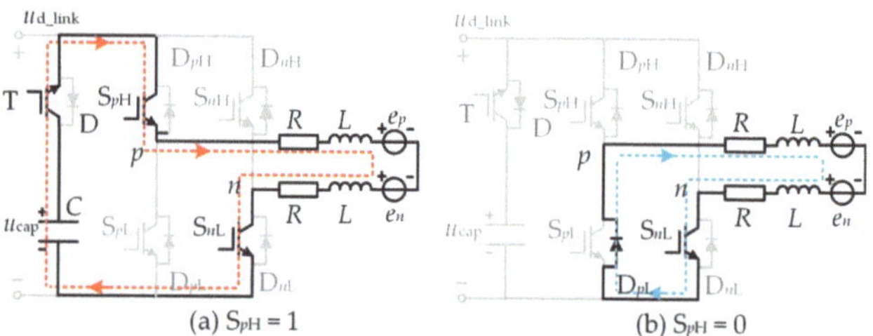

(a) $S_{pH} = 1$ (b) $S_{pH} = 0$

Figure 5. Equivalent circuits of plug braking ON_PWM modulation pattern in sectors (I, III, V). The direction of the arrow indicates the direction of current flow.

4.1. Controllability Analysis of Regenerative Braking Torque in Normal Conduction Period of the BLDCM

When the BLDCM drives with reduced dc-link capacitance adopt the regenerative braking OFF_PWM modulation pattern, the dc-link switch transistor T = 1. During this process, the capacitor can store a portion of the electrical energy converted from the mechanical energy. During an electrical cycle, when the rotor is located in an odd sector number (I, III, V), the switch transistor S_{pH}, where the positive conducting phase p is located, operates with a duty cycle D. At the same time, all other switch transistors of the inverter are turned off.

When $S_{pH} = 1$, under the excitation of the line back EMF e_{pn}, the positive conducting phase current i_p flows through S_{pH} and the antiparallel diode D_{nH}. Due to the presence of inductance in the phase winding, the winding inductance will store a portion of energy, forming a current path as shown in Figure 4b.

When $S_{pH} = 0$, under the action of line back EMF e_{pn} and inductor voltage, the positive conducting phase current i_p flows through the antiparallel diodes D_{pL} and D_{nH} and finally flows into the capacitor through the dc-link switch transistor T. At this time, the capacitor stores energy, forming another current flow path, as shown in Figure 4a. During the normal conduction period of the BLDCM, in order to simplify the analysis, we ignore the effect of the antiparallel diode freewheeling of the nonconducting phase, and the line voltage equation of the two-phase winding conduction can be expressed as:

$$u_{pn} = 2Ri_p + 2L\frac{di_p}{dt} + e_{pn} \tag{12}$$

where e_{pn} denotes the line back EMF.

When $S_{pH} = 1$, as shown in Figure 4b, in order to simplify the analysis and ignore the voltage drop effect of the diode, the line voltage $u_{pn} = 0$, combined with Equation (12), the rate of change of the positive conducting phase current i_p can be simplified as:

$$\frac{di_p}{dt} \approx \frac{k_e\omega_m - RI_0}{L} \tag{13}$$

where I_0 is the initial value of the positive conducting current i_p and $I_0 > 0$. By setting Equation (13) to 0, the mechanical speed ω_0 can be derived as:

$$\omega_0 = \frac{RI_0}{k_e} \tag{14}$$

When $S_{pH} = 1$, combining Equations (13) and (14), the mathematical relationship between the change rate of the positive conducting phase current and speed in regenerative braking mode can be expressed as:

$$\begin{cases} \dfrac{di_p}{dt} > 0, \omega_m > \omega_0 \\ \dfrac{di_p}{dt} < 0, \omega_m < \omega_0 \end{cases} \tag{15}$$

where ω_0 denotes the cut-off speed for the controlled rate of change of the positive conducting phase current when the conducting phase current reference is I_0, di_p/dt denotes the current change rate of the positive conducting phase during braking mode.

As illustrated in Figure 4a, when $S_{pH} = 0$, $u_{pn} = -u_{d\_link}$. Similarly, according to the derivation process of Equation (12) to Equation (13), the current change rate of the positive conducting phase during the capacitor charging process can be expressed as:

$$\frac{di_p}{dt} \approx \frac{2k_e\omega_m - 2RI_0 - u_{d\_link}}{2L} \tag{16}$$

Combining Equation (2) and Equation (16), $u_{d\_link} > 2(k_e\omega_{rate} + RI_N)$. So, when $S_{pH} = 0$, even if the motor starts braking from the rated speed ω_{rate}, $di_p/dt < 0$ satisfies any situation during the regenerative braking mode and is not affected by motor speed variations. Hence, in the capacitor charging and energy storage state illustrated in Figure 4a, the current flowing through the motor windings will gradually decrease, consequently reducing the output braking torque as well. When the rotor position is in an odd sector (I, III, V), if the motor speed satisfies the inequality $\omega_m > \omega_0$. During a modulation cycle, the change rate of conducting phase current can be controlled by chopping S_{pH}, achieving controllable braking torque. However, when the motor speed satisfies the inequality $\omega_m < \omega_0$, whether S_{pH} is turned on or turned off, the change rate of conducting phase current is always less than 0 within a modulation cycle. Therefore, the motor's braking torque output will continuously decrease. Similarly, when the rotor is located in an even sector (II, IV, VI), the negative conduction phase down bridge arm switch transistor S_{nL} chops with a duty cycle D. At the same time, all other switch transistors in the inverter are turned off. According to the symmetry principle, similar results can be deduced as an odd sector (II, IV, VI).

Therefore, the controllable speed range for braking torque during regenerative braking is $(\omega_0, \omega_{rate}]$. If the motor speed remains within the speed range $[0, \omega_0]$, the output braking torque will continuously decrease as the speed decreases due to the uncontrollable positive or negative rate of change in the conducting phase current.

4.2. Controllability Analysis of Plug Braking Torque in Normal Conducting Period of the BLDCM

When employing plug braking in the braking mode of the BLDCM drives with reduced dc-link capacitance, taking the ON_PWM modulation pattern as an example, the speed range of torque controllability for plug braking can be derived based on the torque controllability analysis method mentioned above for the regenerative braking OFF_PWM modulation pattern.

When using the plug braking ON_PWM modulation pattern, the switching logic of the dc-link switch transistor T is consistent with the control method in the electric mode described in Chapter 1. During one electrical cycle of the motor, when the rotor is in an odd sector number (I, III, V), the positive conducting phase switch transistor S_{pH} chops with a duty cycle D, while the negative conducting phase switch transistor S_{nL} remains on. When $S_{pH} = 1$, as shown in Figure 5a, the positive conducting phase current i_p flows through S_{pH} and S_{nL} under the combined action of line back EMF e_{pn} and dc-link voltage u_{d_link}. When

$S_{pH} = 0$, as shown in Figure 5b, due to the presence of inductance in the winding, under the action of line back EMF e_{pn}, the positive conducting phase current i_p flows through S_{pL} and the antiparallel diode D_{pL}.

To simplify the analysis and ignore the voltage drop effect of the diode, when $S_{pH} = 1$, $u_{pn} = u_{d\_link}$, combined with Equation (12), the change rate of the positive conducting phase current of the plug braking can be expressed as:

$$\frac{di_p}{dt} \approx \frac{u_{d\_link} + 2k_e\omega_m - 2RI_0}{2L} \tag{17}$$

When $S_{pH} = 1$, due to $u_{d\_link} > 2(k_e\omega_{rate} + RI_N)$, $di_p/dt > 0$ still holds even if the motor starts braking from the rated speed ω_{rate}, independent of the change in motor speed. Consequently, the conducting phase current will rise, causing a corresponding increase in the plug braking torque.

When $S_{pH} = 0$, as shown in Figure 5b, $u_{pn} = 0$. Analogous analysis shows that the current change rate of the conducting phase is consistent with that shown in Equation (15). Therefore, when S_{pH} is turned off, the increase or decrease of the motor conducting phase current is related to the speed, which affects the controllability of the motor braking torque. According to the symmetry principle, using a similar analysis method as above, when the rotor is located in even sector numbers (II, IV, VI), the same analysis results as above can be obtained. It can be seen that the controllable speed range of the plug braking torque is $[0, \omega_0]$, and within this range, the deviation between the absolute value of the output braking torque and the reference of the braking torque is small.

4.3. The Proposed Overvoltage Braking Torque Control Method for BLDCM Drives

According to the above analysis, the BLDCM drives with reduced dc-link capacitance adopt the regenerative braking mode. Due to the limitation of capacitance, the dc-link capacitor will experience overvoltage problems at a certain speed point ω_c during the regenerative braking process of the motor. To mitigate the phenomenon of overvoltage, the regenerative braking mode can be ended in advance, and the system can be switched to a plug braking mode to consume the energy of the capacitor and reduce the capacitor voltage. However, it should be noted that after switching to the plug braking, the performance of the plug braking torque control is related to the motor speed. Therefore, after switching to the plug braking, the reference of the braking torque needs to be redetermined.

When the motor performs regenerative braking with the braking torque T_{e1}, in a motor system with constant inertia and fixed capacitance C, after the capacitor is charged to the maximum voltage (U_{CM}), the motor speed is ω_c at this moment, and then the system is switched to the plug braking, and the capacitor voltage will drop rapidly, which mitigate the phenomenon of overvoltage. Combining Equations (6)–(10), the rotational speed ω_c at the end of regenerative braking satisfies the following equation.

$$\omega_c^2 - \frac{T_{e1}R}{k_e^2}\omega_c + \left[\frac{T_{e1}R\omega}{k_e^2} - \omega^2 + \frac{C(U_{CM}^2 - U_{C0}^2)}{J}\right] = 0 \tag{18}$$

where ω_c denotes the motor speed when the capacitor voltage rises to its maximum value U_{CM} during regenerative braking.

By solving Equation (18), we can see that since $\omega_c > 0$, ω_c can be deduced as:

$$\omega_c = \frac{RI_1}{k_e} + \sqrt{\frac{R^2I_1^2}{k_e^2} + \frac{J(k_e\omega^2 - 2I_1R\omega) - Ck_e(U_{CM}^2 - U_{C0}^2)}{Jk_e}} \tag{19}$$

According to Equation (19), when the capacitor voltage reaches U_{CM}, the rotational speed $\omega_c > (RI_1/k_e)$. Since the speed range of the controllable plug braking torque is $[0, RI_1/k_e]$, and $\omega_c > (RI_1/k_e)$, that is, the rotational speed ω_c at the end of the regenerative braking exceeds the speed range of the controllable plug braking torque. If the braking torque reference remains at T_{e1}, then the plug braking torque will not be controllable. Thus, to maintain torque control over the plug braking, it becomes necessary to increase the torque reference of the plug braking, which entails increasing the current reference, denoted as I_{ref}. The requirement of $I_{ref} > (\omega_c k_e / R)$ needs to be met, thus expanding the speed range of the motor under the premise of controllable plug braking torque to ensure smoothness during the motor braking process.

5. Experimental Results and Analysis

In order to verify the feasibility of the proposed method above, a washing machine with a washing weight of 1 kg is built as an experimental platform, as shown in Figure 6. The relevant parameters of the BLDCM system are calculated, and the parameters are listed in Table 1. The schematic diagram and flowchart of the proposed anti-overvoltage braking torque control method are shown in Figure 7. As shown in Figure 7a, i denotes the non-commutated phase current, which is used as the feedback value for the PI current loop control. The proportional coefficient P of the PI controller is 7.0, and the integral coefficient I is 0.09.

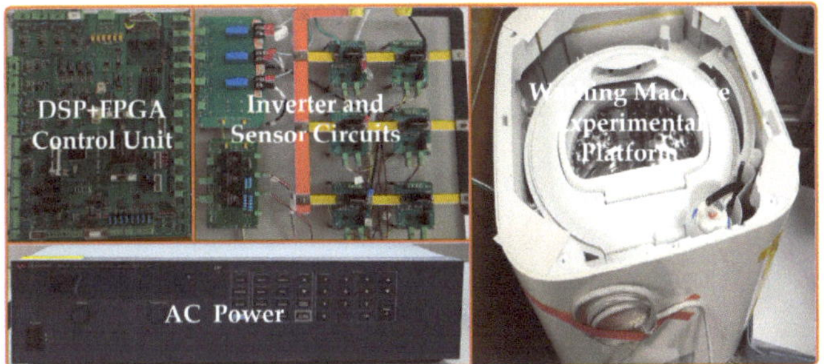

Figure 6. Experimental platform view.

Table 1. The BLDCM system parameters.

Parameter	Symbol	Valve
Rated current	I_N	0.26 A
Phase resistance	R	72 Ω
Phase inductance	L	120 mH
Back EMF coefficient	k_e	0.6685 V/(rad/s)
Rated speed	n_N	700 r/min
Poles pairs	P	4
Inertia	J	0.010762 kg·m^2
Sampling frequency	f_1	10 kHz
Mains supply voltage	U_s	220 Vrms
Mains supply frequency	f	50 Hz

The control method mainly consists of the main parts such as braking mode selection, zone identification, braking current reference selection, dc-link switch controller, and PI current controller. In the experiments, the proposed method in this paper uses DSP+FPGA as the control unit. According to the parameters listed in Table 1, a film capacitor of 70 μF is used in the experiment, the braking mode switching speed ω_c = 66 rad/s (n_c = 630 rpm), the rated operating voltage of the capacitor is 450 V, the frequency of the ac power supply is 50 Hz, and the peak value of the power supply voltage U_m is 311 V. According to Equation (2), we can set U_{pnN} = 252 V, which can meet the requirements of the motor in the rated operating conditions.

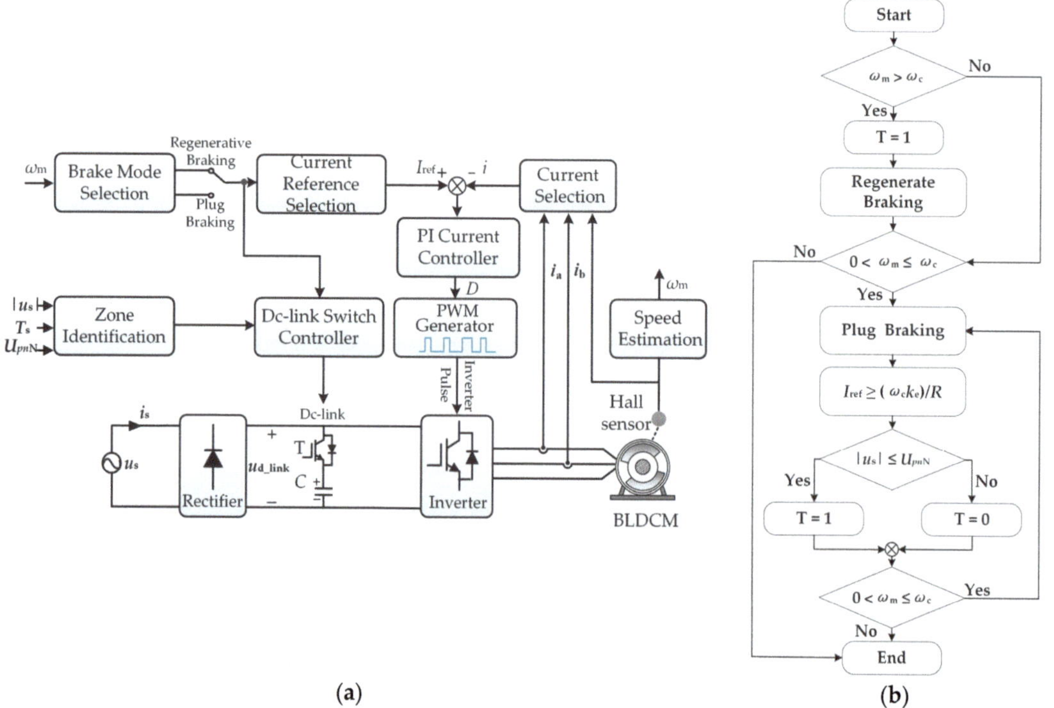

Figure 7. Control block diagram and flowchart of the proposed control method. (**a**) The control block diagram, and (**b**) the flowchart.

In order to demonstrate the overvoltage problem caused by BLDCM drives with small dc-link capacitance in regenerative braking, only the regenerative braking approach is applied to the constructed washing machine experimental platform.

The washing machine was left empty, with no objects placed on it. When the speed is decelerated from 700 rpm to 0 rpm, the reference for the braking torque is −0.35 N·m. This indicates that the absolute value of the current loop reference is 0.26 A. Figure 7 illustrates the overall control process. The experimental procedure involves first maintaining the washing machine motor speed at 700 rpm using the BLDCM electric control strategy with reduced dc-link capacitance, as described in Section 1. Next, the operating mode of the washing machine should be switched to braking mode. The regenerative braking OFF_PWM modulation pattern is used in the braking control process. When the dc-link switch T is turned on during the regenerative braking process, the dc-link voltage u_{d_link} is equal to u_{cap}. Figure 8 shows the experimental results, with waveforms representing speed, dc-link voltage, three-phase current, and electromagnetic torque from top to bottom.

Figure 8a shows that the washing machine braking process takes approximately 4.2 s from start to finish. The initial braking torque is smooth and controllable, and the capacitor voltage increases rapidly. The regenerative braking process causes the capacitor voltage to reach its maximum value. However, during the second half of the braking process, the amplitude of the motor's three-phase current gradually decreases while the motor's braking torque gradually diverges from the reference.

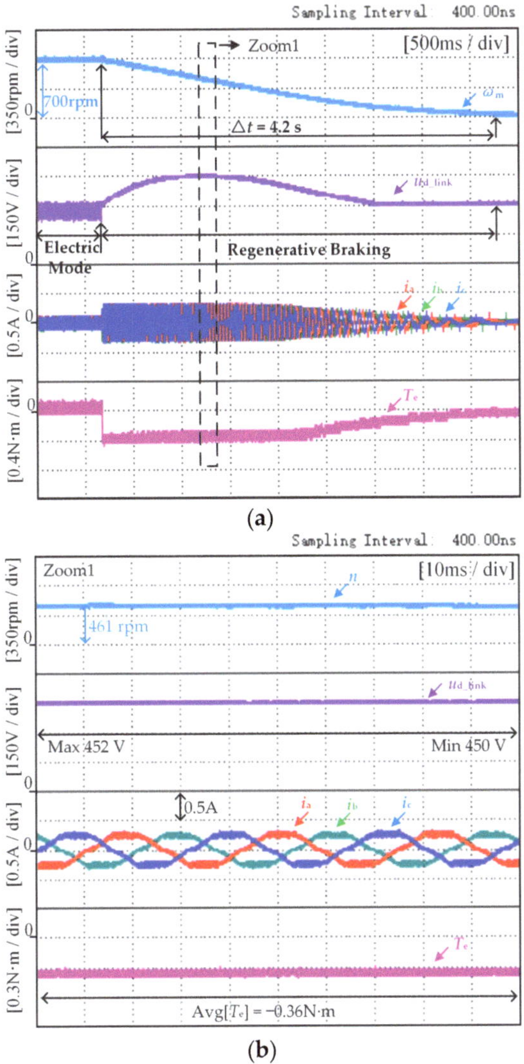

Figure 8. The experimental result with the torque reference is −0.35 N·m, and the speed decreases from 700 rpm to 0 rpm. (**a**) Overall view, and (**b**) the view of Zoom 1.

In Figure 8b, the motor is still in the braking torque controllable region, but the maximum value of the capacitor voltage exceeds 450 V. If the rated operating voltage of the capacitor is less than 450 V, the capacitor may be degraded or even damaged, which will affect the reliability of the motor system. Therefore, to prevent capacitor overvoltage without requiring an additional mechanical braking device for the washing machine, this paper proposes an anti-overvoltage braking torque control method. Experimental waveform diagrams for this method are shown in Figure 9. By setting the switching speed n_c = 630 rpm and ensuring that there is no load weighing 1 kg inside the washing machine, the dc-link voltage and torque variation characteristics of the motor system can be reflected in the experimental waveform in Figure 9.

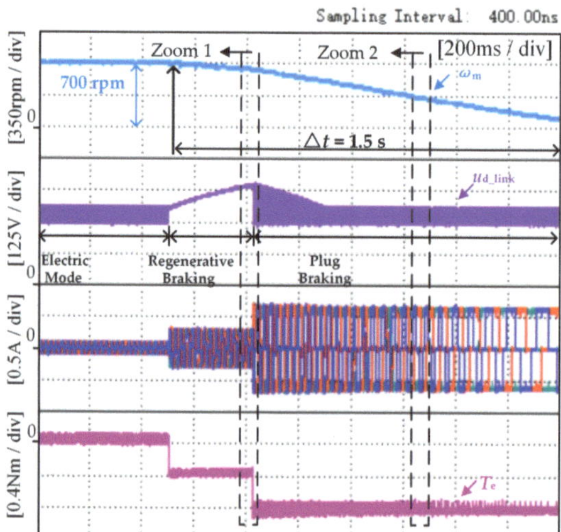

Figure 9. The overall figure of the proposed method when the speed drops from 700 rpm to 90 rpm under no-load conditions.

Figure 9 shows the experimental results of the proposed method of the washing machine under no-load conditions, in which the waveform graphs from top to bottom are the speed, three-phase current, dc-link voltage, and electromagnetic torque, respectively. During regenerative braking mode, the brake torque reference of the washing machine is −0.35 N·m. With the progress of regenerative braking, when the washing machine control system recognizes the speed ω_c and the braking mode changes to plug braking, the reference of plug braking torque is −0.83 N·m, and the modulation pattern is ON_PWM. As shown in Figure 9, three operating modes exist in the system: electric mode, regenerative braking, and plug braking. At first, the electric mode adopts a control method for the BLDCM drives with reduced capacitance, after which the control mode of the washing machine changes to regenerative braking. At the same time, once the system recognizes the rotational speed as ω_c, it will switch to the plug braking mode to consume the stored energy of the capacitor during the regenerative braking process and avoid overvoltage of the capacitor.

As can be seen from Figure 9, the washing machine takes approximately 1.5 s to decelerate from 700 rpm to 90 rpm during braking. This is a smooth and rapid braking process compared to the braking process shown in Figure 8, which only uses the regenerative braking mode. As seen in Figure 10a, the average value of regenerative braking torque is −0.36 N·m. The proposed method sets the switching speed of regenerative braking and plug braking at $n_c = 630$ rpm, at which time the maximum value of the capacitor voltage is 398 V, and then the braking mode is switched to plug braking. According to the previous analysis, the absolute value of the current loop reference converted from the regenerative braking torque reference of −0.35 N·m is not suitable for plug braking at this moment. Therefore, the absolute value of the plug braking current loop reference must be no less than 0.62 A; that is, the plug braking torque reference should be set to −0.83 N·m. As illustrated in Figure 10a, the average value of plug braking torque is −0.86 N·m. As the braking process continues, as depicted in Figure 10b, the average value of braking torque is −0.83 N·m. Hence, as the plug braking progresses, the final braking torque will tend to the set reference.

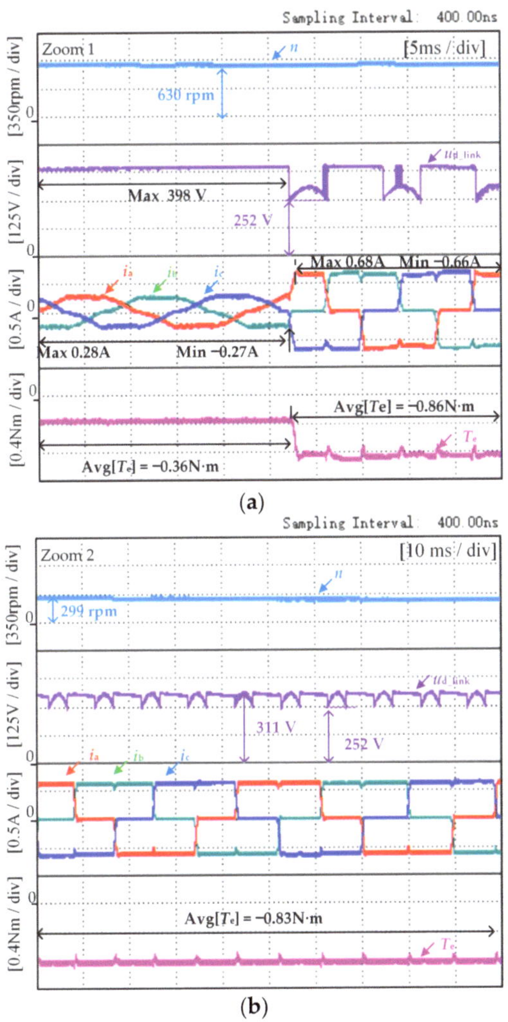

Figure 10. The experimental results of the proposed method when the speed drops from 700 rpm to 90 rpm under no-load conditions. (**a**) The view of Zoom 1, and (**b**) the view of Zoom 2.

In order to simulate the actual operating conditions of the washing machine, a 1 kg load object was added inside the washing machine, and the other experimental conditions were consistent with those mentioned in Figure 9. The results of the experiment are presented in Figure 11. Figure 11 presents the overall diagram of the system, which includes three operating modes. The washing machine takes approximately 2.4 s to brake when it decelerates from 700 rpm to 90 rpm, and the braking process remains stable. In Figure 12a, the maximum value of the capacitor voltage at the braking mode switching speed $n_c = 630$ rpm is 434 V, which is still within the rated range of the capacitor voltage. As seen in Figure 12b, the average value of braking torque reaches -0.83 N·m as the braking process proceeds, so the proposed method realizes prevention of capacitor overvoltage while taking into account the smoothness of braking torque.

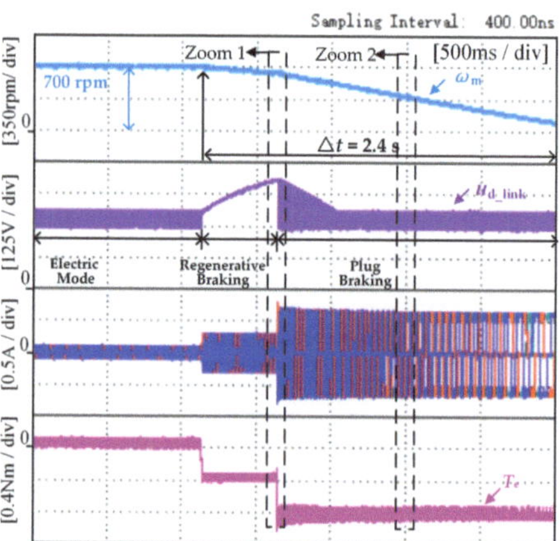

Figure 11. The experimental figure of the proposed method when the speed drops from 700 rpm to 90 rpm under a load condition of 1 kg.

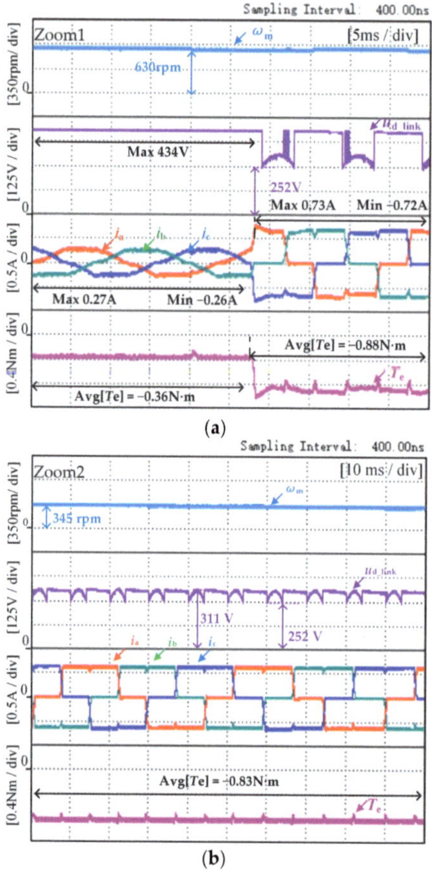

Figure 12. The experimental results of the proposed method when the speed drops from 700 rpm to 90 rpm under a load condition of 1 kg. (**a**) The view of Zoom 1, and (**b**) the view of Zoom 2.

6. Conclusions

This paper focuses on the dc-link overvoltage phenomenon in a single-phase input diode rectifier BLDCM drive with small dc-link capacitance during regenerative braking. The proposed method is designed to avoid capacitor overvoltage by detecting the motor speed at which the capacitor is overvoltage and then utilizing plug braking to reduce the energy of the capacitor and, at the same time, mitigate the impact of motor speed on braking torque controllability. The proposed anti-overvoltage braking torque control method is based on the existing hardware resources of the system, without adding additional mechanical braking devices or hardware circuits. Finally, the effectiveness of the proposed method is verified by an experimental platform of a washing machine equipped with a film capacitor.

Author Contributions: Data curation, J.W.; investigation, X.L. and W.C.; methodology, J.W. and C.L.; project administration, W.C.; resources, X.L.; validation, C.L. and J.W.; writing—original draft, J.W.; writing—review and editing, X.L. All authors have read and agreed to the published version of the manuscript.

Funding: This research was supported by the Joint Fund Key Program of the National Natural Science Foundation of China under Grant U23A20643 and in part by the National Natural Science Foundation of China under Grant 52077155.

Data Availability Statement: Data is contained within the article.

Conflicts of Interest: The authors declare no conflicts of interest.

References

1. Hemati, N.; Leu, M. A Complete Model Characterization of Brushless DC Motors. *IEEE Trans. Ind. Appl.* **1992**, *28*, 172–180. [CrossRef]
2. Yuan, T.; Chang, J.; Zhang, Y. Research on the Current Control Strategy of a Brushless DC Motor Utilizing Infinite Mixed Sensitivity Norm. *Electronics* **2023**, *12*, 4525. [CrossRef]
3. Zhou, Q.; Shu, J.; Cai, X.; Liu, Q.; Du, G. Improved PWM-OFF-PWM to Reduce Commutation Torque Ripple of Brushless DC Motor under Braking Conditions. *IEEE Access* **2020**, *8*, 204020–204030. [CrossRef]
4. Tan, B.; Hua, Z.; Zhang, L.; Fang, C. A New Approach of Minimizing Commutation Torque Ripple for BLDCM. *Energies* **2017**, *10*, 1735. [CrossRef]
5. Son, Y.; Ha, J. Direct Power Control of a Three-Phase Inverter for Grid Input Current Shaping of a Single-Phase Diode Rectifier with a Small DC-Link Capacitor. *IEEE Trans. Power Electron.* **2015**, *30*, 3794–3803. [CrossRef]
6. Zhao, N.; Wang, G.; Li, B.; Zhang, R.; Xu, D. Beat Phenomenon Suppression for Reduced DC-link Capacitance IPMSM Drives with Fluctuated Load Torque. *IEEE Trans. Ind. Electron.* **2019**, *66*, 8334–8344. [CrossRef]
7. Son, Y.; Ha, J. Discontinuous Grid Current Control of Motor Drive System with Single-Phase Diode Rectifier and Small DC-Link Capacitor. *IEEE Trans. Power Electron.* **2017**, *32*, 1324–1334. [CrossRef]
8. Din, D.; Xie, H.; Li, B.; Wang, G.; Gao, R.; Ren, Z.; Yue, W.; Xu, D. Harmonic Suppression Based on Rectified Current Regulation with DC-Link Voltage Decoupling for Electrolytic Capacitorless PMSM Drives. *IEEE Trans. Power Electron.* **2024**, *39*, 2213–2225. [CrossRef]
9. Din, D.; Wang, G.; Zhao, N.; Zhang, G.; Xu, D. An Antiovervoltage Control Scheme for Electrolytic Capacitorless IPMSM drives Based on Stator Surrent Vector Orientation. *IEEE Trans. Ind. Electron.* **2020**, *67*, 3517–3527. [CrossRef]
10. Din, D.; Zang, G.; Wang, G.; Xu, D. Dual Antiovervoltage Control Scheme for Electrolytic Capacitorless IPMSM Drives with Coefficient Autoregulation. *IEEE Trans. Power Electron.* **2020**, *35*, 2895–2907. [CrossRef]
11. Zang, C.; Xu, L.; Zhu, X.; Du, Y.; Quan, L. Elimination of DC-Link Voltage Ripple in PMSM Drives With a DC-Split-Capacitor Converter. *IEEE Trans. Power Electron.* **2021**, *36*, 8141–8154. [CrossRef]
12. Ransara, H.S.; Madawala, U.K. A Technique for Torque Ripple Compensation of a Low Cost BLDC Motor Drive. In Proceedings of the 2013 IEEE International Conference on Industrial Technology (ICIT), Cape Town, South Africa, 25–28 February 2013. [CrossRef]
13. Ransara, H.S.; Madawala, U.K. A Torque Ripple Compensation Technique for a Low-cost Brushless DC Motor Drive. *IEEE Trans. Ind. Electron.* **2015**, *62*, 6171–6182. [CrossRef]
14. Li, X.; Yuan, H.; Yan, Y.; Chen, W.; Shi, T.; Xia, C. A Novel Voltage-Boosting Modulation Strategy to Reduce DC-Link Capacitance for Brushless DC Motor Drives. *IEEE Trans. Power Electron.* **2022**, *37*, 15397–15410. [CrossRef]
15. Saha, B.; Singh, B. An Adaptive Delay Compensated Position Sensorless PMBLDC Motor Drive with Regenerative Braking for LEV Application. *IEEE Trans. Energy Convers.* **2023**, *38*, 1793–1802. [CrossRef]

16. Xu, Y.; Zhang, Z. Regenerated Energy Absorption Methods for More Electric Aircraft Starter/Generator System. *IEEE Trans. Power Electron.* **2023**, *38*, 7525–7534. [CrossRef]
17. Joice, C.; Paranjothi, S.; Kumar, V. Digital Control Strategy for Four Quadrant Operation of Three Phase BLDC Motor with Load Variations. *IEEE Trans. Ind. Inform.* **2013**, *9*, 974–982. [CrossRef]
18. Nian, X.; Peng, F.; Zhang, H. Regenerative Braking System of Electric Vehicle Driven by Brushless DC Motor. *IEEE Trans. Ind. Electron.* **2014**, *61*, 5798–5808. [CrossRef]
19. Naseri, F.; Farjah, E.; Ghanbari, T. An Efficient Regenerative Braking System Based on Battery/Supercapacitor for Electric, Hybrid, and Plug-In Hybrid Electric Vehicles With BLDC Motor. *IEEE Trans. Veh. Technol.* **2017**, *66*, 3724–3748. [CrossRef]
20. Zhou, X.; Fang, J. Precise Braking Torque Control for Attitude Control Flywheel with Small Inductance Brushless DC Motor. *IEEE Trans. Power Electron.* **2013**, *28*, 5380–5390. [CrossRef]
21. Cao, Y.; Shi, T.; Yan, Y.; Li, X.; Xia, C. Braking Torque Control Strategy for Brushless DC Motor with a Noninductive Hybrid Energy Storage Topology. *IEEE Trans. Power Electron.* **2020**, *35*, 8417–8428. [CrossRef]

 *actuators*

Article

Analysis of Parameter Matching on the Steady-State Characteristics of Permanent Magnet-Assisted Synchronous Reluctance Motors under Vector Control

Yu-Hua Lan [1,2], Wen-Jie Wan [1] and Jin Wang [3,*]

1   College of Electrical Engineering, Zhejiang University, Hangzhou 310027, China; 12110088@zju.edu.cn (Y.-H.L.); wwj258@zju.edu.cn (W.-J.W.)
2   Jiang Chao Motor Technology Co., Ltd., Huzhou 313014, China
3   National Engineering Research Center for REPM Electrical Machines, Shenyang University of Technology Shenyang, Shenyang 110870, China
*   Correspondence: wjcmf11@126.com

Abstract: In this paper, the impact of parameter matching on the steady-state performance of permanent magnet-assisted synchronous reluctance motors (PMaSynRM) under vector control is analyzed and discussed. First, based on the mathematical model of motors under the maximum torque per ampere (MTPA) control strategy, an analysis is conducted concerning two main parameters, i.e., the matching relationship between the back electromotive force (back-EMF) and the saliency ratio. The impact of these two parameters on the operational status of the motor is investigated. Then, the motor's voltage operating conditions are examined, and the operating curve under minimum voltage is derived. Furthermore, in the overvoltage region under the MTPA control strategy, the operation of the motor under the maximum torque per voltage (MTPV) control strategy is explored. This analysis illuminated the patterns of influence exerted by the back-EMF and the saliency ratio on the motor's voltage operating condition. Between these two control strategies, there remains scope for the motor to operate at its limits. An enhanced understanding of the effects of the back-EMF and saliency ratio within this range on motor performance was achieved, resulting in the optimal matching curve for the back-EMF and saliency ratio. Finally, a 45 kW PMaSynRM was designed, prototyped, and tested to validate the correctness of the design techniques, with the motor achieving IE5 efficiency.

Keywords: permanent magnet-assisted synchronous reluctance motor; vector control; saliency ratio; IE5

Citation: Lan, Y.-H.; Wan, W.-J.; Wang, J. Analysis of Parameter Matching on the Steady-State Characteristics of Permanent Magnet-Assisted Synchronous Reluctance Motors under Vector Control. *Actuators* **2024**, *13*, 198. https://doi.org/10.3390/act13060198

Academic Editor: Dong Jiang

Received: 28 April 2024
Revised: 18 May 2024
Accepted: 21 May 2024
Published: 22 May 2024

1. Introduction

The permanent magnet synchronous motor (PMSM) is known for its high efficiency and high power density [1,2], with its drawback lies in its high cost. Therefore, there is increasing attention on high-efficiency motor technologies with fewer or no rare-earth materials [3]. The permanent magnet-assisted synchronous reluctance motor (PMaSynRM) embeds neodymium-iron-boron (NdFeB) or ferrite magnets in its rotor magnetic barriers [4], with the main portion of the output torque being reluctance torque [5,6]. The added small number of permanent magnets primarily serves to improve the motor power factor [7,8]. Its good performance and relatively low cost [9,10] make it one of the potential alternatives comparable to PMSM. The rotor structure of PMaSynRM is complex [11,12], and the parameter matching is rich and diverse [13], which has prompted extensive research by many scholars.

Researchers from Zhejiang University conducted comparative studies on SynRM, IPMSM, and PMaSynRM motors [14]. The constant losses among these three types of motors are similar. The interior PMSM has the lowest copper loss, while SynRM has the highest; PMaSynRM's copper loss is intermediate, mainly due to differences in power factor and current. A study on the design of the saliency ratio of PMaSynRM indicated that the motor's saliency ratio should be designed above 2.73. Through optimization, the designed 5 kW motor can achieve an efficiency of 90.46% and a power factor of 0.73 [15]. Another research focused on a hybrid NdFeB and ferrite synchronous reluctance motor structure, optimizing a 28 kW prototype that achieved an efficiency of 93.42% at rated conditions, with a cost reduction of about 42 dollar compared to permanent magnet motors, although its peak torque is reduced [16]. Further studies investigated a hybrid permanent magnet assisted rotor structure, utilizing a multislot stator and a three-layer barrier rotor, with the first layer using NdFeB and the second and third layers using ferrite [17]. Its torque output and current are comparable to those of PMaSynRMs, with a high-efficiency zone close to that of PMSMs. However, the increase in types of motor materials leads to higher procurement management and production costs. Another research introduced a top-level optimization design concept for PMaSynRMs, where a 7.5 kW prototype achieved a power factor of 0.88 and an efficiency of 92.9%, surpassing IE4 efficiency levels [18]. A multi-degree-of-freedom parametric method was proposed, which allows for control over the barrier shape by altering coordinates on the barrier edge curve [19]. More degrees of freedom in control result in more diverse barrier shapes, aiding in finer optimization of motor performance, thus improving torque by 4.5% and enhancing the power factor. Italian researchers studied the reinforcement ribs of magnetic barriers [20]. Through the optimized design of these ribs, while maintaining the strength of the laminations, motor performance improved, increasing the saliency ratio by 31.3% and enhancing torque, efficiency, and power factor. In [21], a rotor structure incorporating V-shaped permanent magnets is analyzed, aligning the peak phase angles of permanent magnet torque and reluctance torque, thereby enhancing motor torque and reducing NdFeB usage by 4%.

The main research methods involve qualitatively pointing out first that placing permanent magnets in the barrier structure can enhance the motor's power facto and torque output, based on classical motor equations and vector diagrams. Further analysis considers how structural dimensions like barrier layers, magnet placement, split ratio, hybrid magnet structures, and axial hybrid rotor structures impact motor performance. The findings suggest diverse performance outcomes and complex conclusions, restricted to specific structures without forming clear principles for parameter matching design.

The paper is organized as follows. First, the relationship between back electromotive force (back-EMF) and saliency ratio under the maximum torque per ampere (MTPA) control strategy is derived and analyzed in Section 2, based on the mathematical model of the motor under vector control. Then, the impact of parameter matching under the maximum torque per voltage (MTPV) control strategy on motor operating states is investigated in Section 3. The optimal matching curve of back EMF and saliency ratio under vector control for PMaSynRM is discussed in Section 4, proposing principles for parameter matching that achieve high performance and cost-effectiveness. In Section 5, a 45 kW PMaSynRM is developed, with tests showing that it could achieve IE5 energy efficiency while effectively reducing costs. Finally, some conclusions are given in Section 6.

2. Typical Operational State of PMaSynRMs under MTPA Control Strategy

SynRM without permanent magnet excitation lacks a permanent magnet flux linkage. Define i_d, i_q, L_d, and L_q as the current and inductances in d and q-axis, accordingly. e_0 is the back-EMF. The torque curve in the $i_d - i_q$ coordinate plane draws a hyperbolic shape, symmetrically distributed along the plane's diagonal as the axis of symmetry, as shown in Figure 1a. The intersection of the current circle and torque curve is always at the current phase angle $\beta = 135°$, regardless of changes in L_d and L_q. The angle of current remains constant, even when operating at maximum torque per ampere. However, L_d and L_q affect the magnitude of voltage and power angle. Larger inductances result in higher voltages, and greater differences in inductance result in smaller power angles, with a limit approaching 90°. SynRMs exhibit a low power factor under maximum torque per ampere conditions accompanied by significant reactive current.

For PMaSynRM with permanent magnet excitation and permanent magnet flux linkage and back-EMF, the torque curve on the $i_d - i_q$ coordinate shifts rightward as depicted in Figure 1b. The intersection point of the current circle and torque curve no longer remains fixed at $\beta = 135°$ but also shifts rightward. The power factor increases, reactive current decreases, and operating current is reduced. The specific intersection point depends on the matching of e_0, L_d, and L_q. As the back-EMF increases and the saliency ratio decreases, the power factor of the motor operating at MTPA shifts from low to high. Figure 1c shows the torque curve and current circle trajectory for a power factor of 1.

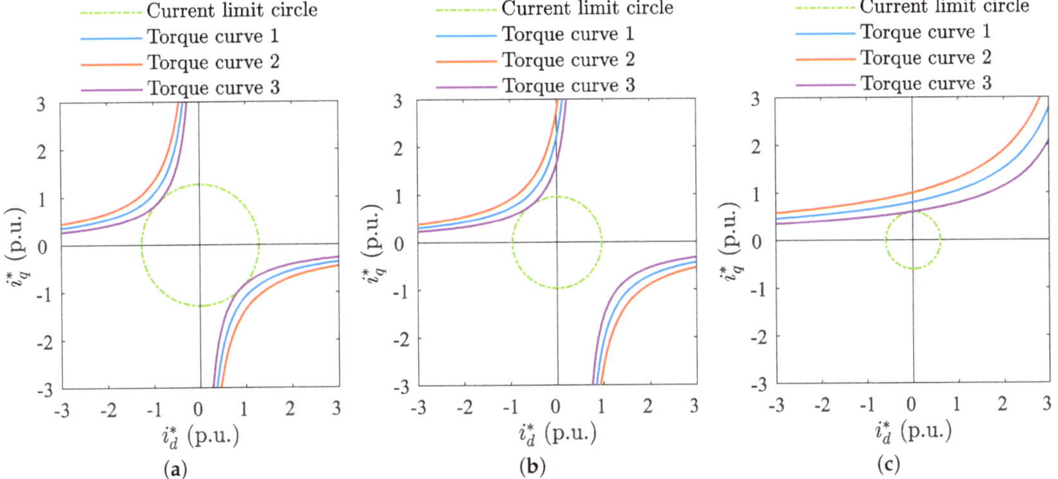

Figure 1. Torque trajectories of three types of motors: (**a**) torque trajectory of SynRM; (**b**) torque trajectory of PMaSynRM; (**c**) torque trajectory of PMaSynRM when power factor = 1.

The analysis indicates that to operate the motor at MTPA while reducing reactive current, it is necessary to increase the back-EMF and reduce the saliency ratio. Thus, during the MTPA state, the purpose of pursuing a high power factor is contradictory to enhancing saliency ratio and reluctance torque and reducing the use of permanent magnets. Therefore, aiming for an extreme power factor of 1 is not advisable. However, there should be an optimal operational state under the MTPA control strategy.

3. Analysis of Operational State of PMaSynRM under MTPA Control Strategy

3.1. Mathematical Model of Motors under MTPA Control

Based on the voltage, flux linkage, and torque equations of PMaSynRMs, the steady-state motor equations can be derived:

$$
\begin{cases}
u_d = -\omega L_q i_q \\
u_q = \omega L_d i_d + e_0 \\
i_d = \dfrac{-e_0/\omega + \sqrt{(e_0/\omega)^2 + 4(L_d - L_q)^2 i_q^2}}{2(L_d - L_q)} \\
T_{em} = p[\dfrac{e_0}{\omega} i_q + (L_d - L_q)i_d i_q] \\
i_s^2 = i_d^2 + i_q^2 \\
u_s^2 = u_d^2 + u_q^2
\end{cases}
\tag{1}
$$

where i_s and u_s are the stator current voltage. i_d, i_q, u_d, u_q, and L_d, L_q are the current, voltage, and inductances in the d and q-axis, accordingly. e_0 is the back-EMF. Once the motor's pole number p, target torque T_{em}, and speed ω are set, these parameters are considered constants. Meanwhile, a six-variable quadratic equation for variables u_d, u_q, u_s, i_d, i_q is formed in (1). The magnitude and phase of u_s and i_s are correspondingly related.

Let

$$B = -e_0/\omega + \sqrt{(e_0/\omega)^2 + 4(L_d - L_q)^2 i_q^2} \tag{2}$$

Then,

$$i_d = \frac{B}{2(L_d - L_q)} \tag{3}$$

yields the q-axis voltage

$$u_q = \frac{\omega L_d B}{2(L_d - L_q)} + e_0 \tag{4}$$

Combined with the d-axis voltage to obtain the stator voltage

$$u_s^2 = (\omega L_q i_q)^2 + \left[\frac{\omega L_d B}{2(L_d - L_q)} + e_0\right]^2 \tag{5}$$

further leads to

$$u_s^2 = (\omega L_q i_q)^2 + \frac{(\omega L_d B)^2}{4(L_d - L_q)^2} + \frac{\omega L_d B e_0}{(L_d - L_q)} + e_0^2 \tag{6}$$

The torque equation, when substituted with the d-axis current, gives

$$T_{em} = p[\frac{e_0}{\omega} i_q + (L_d - L_q)\frac{B}{2(L_d - L_q)} i_q] \tag{7}$$

and further leads to

$$T_{em} = p i_q (\frac{e_0}{\omega} + \frac{B}{2}) \tag{8}$$

which yields the q-axis current

$$i_q = \frac{T_{em}}{p(\frac{e_0}{\omega} + \frac{B}{2})} \tag{9}$$

connected with (2) to obtain

$$B - \sqrt{(e_0/\omega)^2 + 4(L_d - L_q)^2 \left[\frac{T_{em}}{p(\frac{e_0}{\omega} + \frac{B}{2})}\right]^2} + e_0/\omega = 0 \tag{10}$$

The equation has a solution for B, depending on L_d, L_q, e_0, T_{em}, and ω. Once the motor's design target is set, T_{em} and ω can be considered constants. At this time, the solution for B depends on L_d, L_q, and e_0. When different parameter matchings are determined, B can be seen as a constant.

The q-axis current can also be expressed as

$$i_q^2 = i_s^2 - \frac{B^2}{4(L_d - L_q)^2} \tag{11}$$

Substituting into (6) results in

$$u_s^2 = \omega^2 L_q^2 i_s^2 + \frac{\omega^2 B^2 (L_d^2 + L_q^2)}{4(L_d - L_q)^2} + \frac{\omega L_d B e_0}{(L_d - L_q)} + e_0^2 \tag{12}$$

From (12), the corresponding relationship between u_s and i_s is closely related to e_0, L_d, and L_q (which can be represented by two more intuitive parameters E_0/U and saliency ratio ρ), changing as the parameter matching changes.

3.2. Impact of Back-EMF and Saliency Ratio on Motor Operating Characteristics under MTPA Control Strategy

Results derived from mathematical models include theoretically all possible operating states. However, due to limitations in manufacturing processes or design constraints, not all states can be realized in actual prototypes. To make the analysis more applicable to real situations, this paper is based on a 45 kW, four-pole motor operating at a speed of 1500 rpm. The following is the analysis of operating states under different parameter matchings.

The variation in current phase angle β under different back-EMF and saliency ratio matches is shown in Figure 2a.

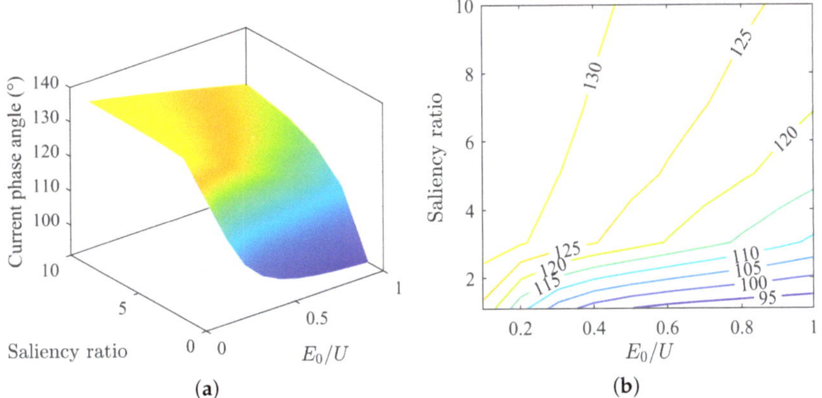

Figure 2. Variation of current phase angle under MTPA: (**a**) surface plot of current phase angle; (**b**) contour map of current phase angle.

From the current phase angle surface plot, the maximum and minimum values of the current phase angle are generally located at the two ends of the diagonal, decreasing gradually from a point of high saliency ratio and low back-EMF to a point of low saliency ratio and high back-EMF. This decrease is not linear but nonlinear, with the magnitude of decrease accelerating over time. The nonlinear decrease in current is not uniform across the entire surface. As shown in Figure 2b, the overall trend is that the current angle decreases with a decrease in saliency ratio and increases in back-EMF. When the saliency ratio is high, the increase in back-EMF has a smaller reduction in current angle. And when the saliency ratio is low, the reduction is more significant.

The variation in power angle θ under different back-EMF and saliency ratio matches is shown in Figure 3a. From the surface plot, the maximum power angle is near the point of minimum back-EMF and saliency ratio, and the minimum power angle is near the point of maximum back-EMF and minimum saliency ratio. The surface is relatively flat, and the variation in power angle is more uniform compared to the current angle.

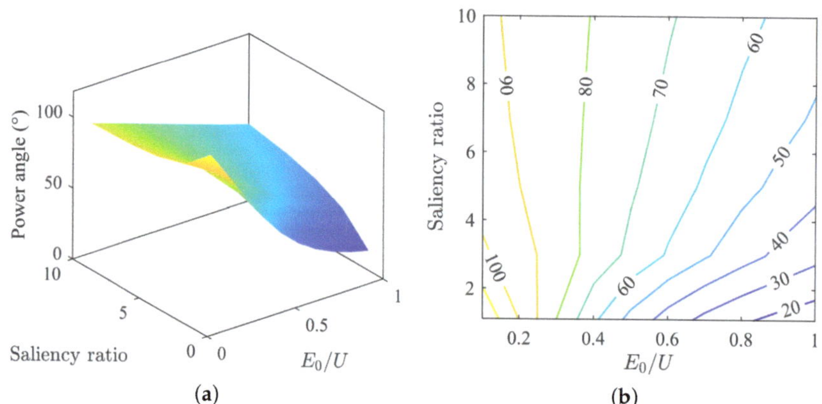

Figure 3. Variation of power angle under MTPA: (**a**) surface plot of power angle. (**b**) contour map of power angle.

As shown in Figure 3b, the power angle decreases with an increase in back-EMF and increases with an increase in saliency ratio. At higher saliency ratios, the decrease in power angle with increasing back-EMF is relatively linear, and the range of decrease is smaller. At lower saliency ratios, the decrease is more nonlinear and the range of decrease is larger.

The power factor and power factor angle corresponding to each other vary under different parameter matchings, as shown in Figure 4a. From the surface plot, as back-EMF increases, the power factor also increases. As shown in Figure 4b, when E_0/U is less than 0.5, the power factor increases with an increase in saliency ratio. When E_0/U is greater than 0.6, the power factor decreases with an increase in saliency ratio. Between E_0/U of 0.5 and 0.6, the power factor does not change significantly. From the perspective of power factor changes, under the MTPA control strategy, an increase in back-EMF leads to an increase in power factor, but an increase in saliency ratio does not necessarily lead to an increase in power factor. It also depends on the match with back-EMF.

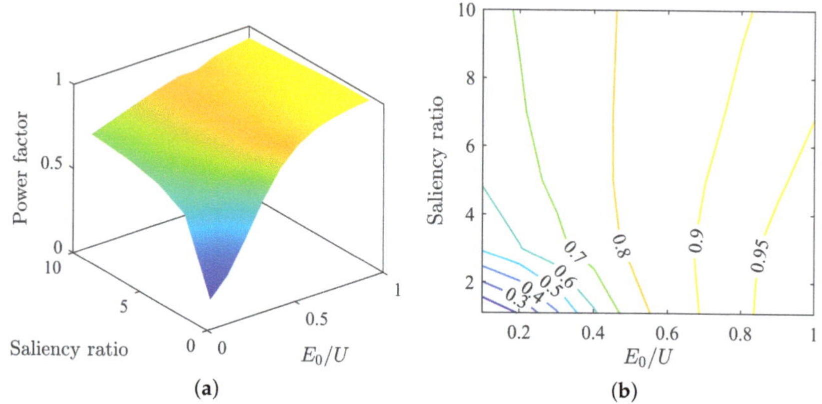

Figure 4. Variation of power factor under MTPA: (**a**) surface plot of power factor; (**b**) contour map of power factor.

The variation in d-axis current under different parameter matchings is shown in Figure 5. The plot indicates that the d-axis current increases with an increase in back-EMF. The trend in d-axis current is not consistent with an increase in saliency ratio. When E_0/U is greater than 0.2, the d-axis current first increases then decreases with an increase in saliency ratio, with a smaller fluctuation range. When E_0/U is less than 0.2, the d-axis current decreases with an increase in saliency ratio, with a larger fluctuation range.

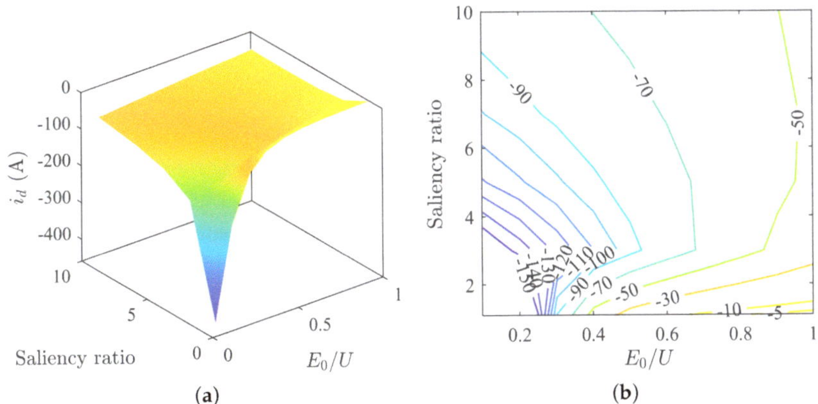

Figure 5. Variation in d-axis current under MTPA: (**a**) surface plot of d-axis current; (**b**) contour map of d-axis current.

The variation in q-axis current under different parameter matchings is shown in Figure 6. The q-axis current decreases consistently with an increase in back-EMF and saliency ratio. When the saliency ratio is high, the range of change in the q-axis current with changing back-EMF is relatively small. Conversely, when the saliency ratio is low, the range of change in the q-axis current with changing back-EMF is relatively large.

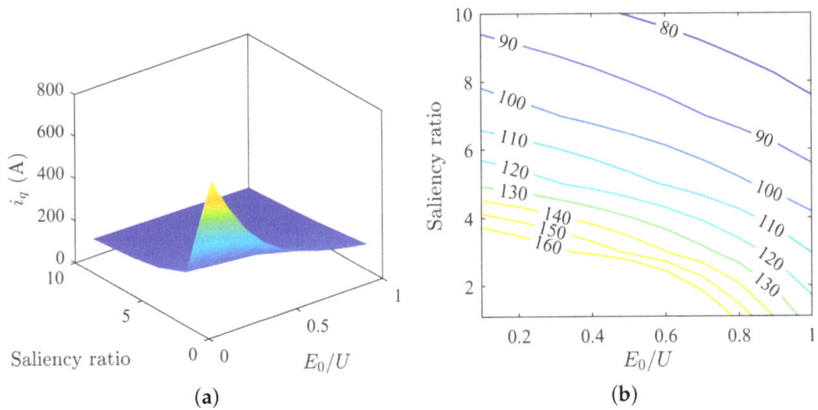

Figure 6. Variation in q-axis current under MTPA: (**a**) surface plot of q-axis current; (**b**) contour map of q-axis current.

The variation in stator current under different parameter matchings is shown in Figure 7. It can be observed that the maximum and minimum values of the stator current are generally located at the two ends of the diagonal, and the overall trend is a gradual decrease from the point of low saliency ratio and low back-EMF to the point of high saliency ratio and high back-EMF. The surface plot reveals steep and relatively flat slopes, indicating that the change in stator current is nonlinear. The smaller the saliency ratio and back-EMF, the greater the change. The larger the saliency ratio and back-EMF, the smaller the change. From Figure 7, it is evident that increasing the back EMF can reduce the stator current, and increasing the saliency ratio can also decrease the stator current.

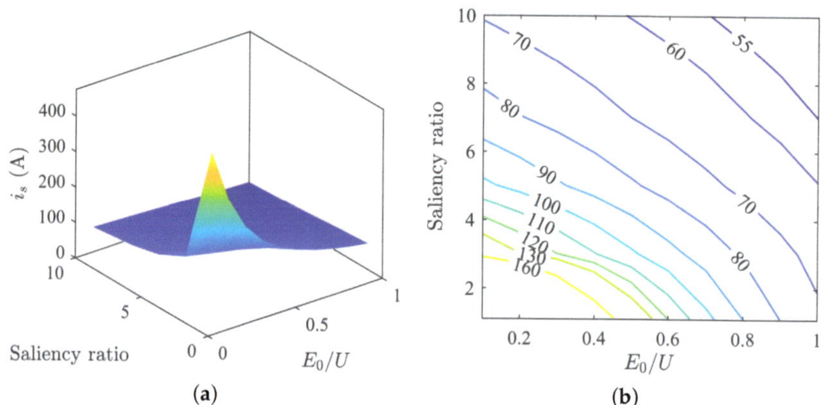

Figure 7. Changes in stator current under MTPA: (**a**) surface plot of stator current; (**b**) contour map of stator current.

The d-axis voltage varies with parameter matching as shown in Figure 8. Overall, the d-axis voltage increases as the back-EMF decreases. The change in d-axis voltage with saliency ratio is not consistent. When E_0/U is less than 0.2, the d-axis voltage first decreases then increases with increasing saliency ratio. And when E_0/U is greater than 0.2, the d-axis current increases with increasing saliency ratio.

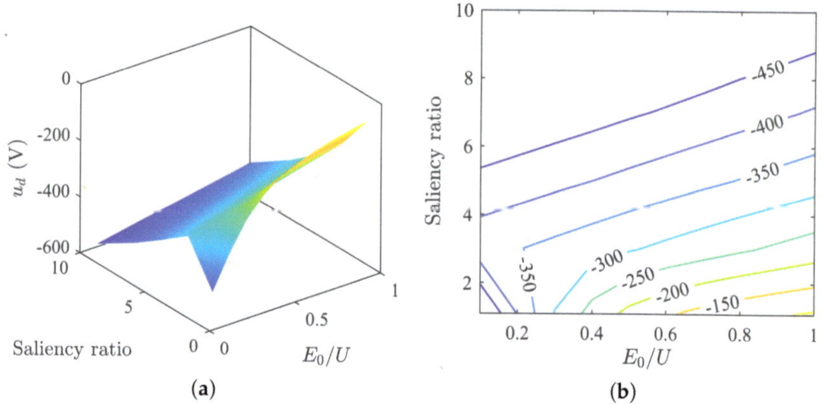

Figure 8. Changes in d-axis voltage under MTPA: (**a**) surface plot of d-axis voltage; (**b**) contour map of d-axis voltage.

The q-axis voltage changes with parameter matching as depicted in Figure 9. It is apparent that as the back-EMF decreases, the q-axis current gradually reduces until it reaches the zero crossing point, after which the q-axis current reverses and increases. As the saliency ratio increases, the overall trend of the q-axis voltage initially decreases and then increases. This variation is relatively mild when the back-EMF is high and more fluctuant when it is low.

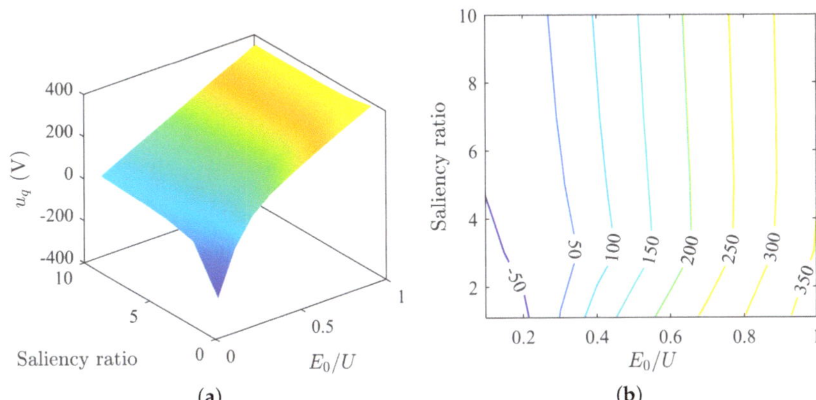

Figure 9. Variation of q-axis voltage under MTPA: (**a**) surface plot of q-axis voltage; (**b**) contour map of q-axis voltage.

Under different parameter matching conditions, the change in stator voltage is shown in Figure 10. From the surface map, it is evident that the overall stator voltage presents a state where the side with a higher saliency ratio is higher and the side with a lower saliency ratio is lower. The corners of the surface are higher at the sides with small and large back-EMF, and lower in the middle value of back-EMF. The entire surface resembles the shape of a "slide", with a curve across the surface where each point is a local minimum. The contour map shows that as the saliency ratio increases, the stator voltage increases, and as the back EMF increases, the stator voltage first decreases and then increases.

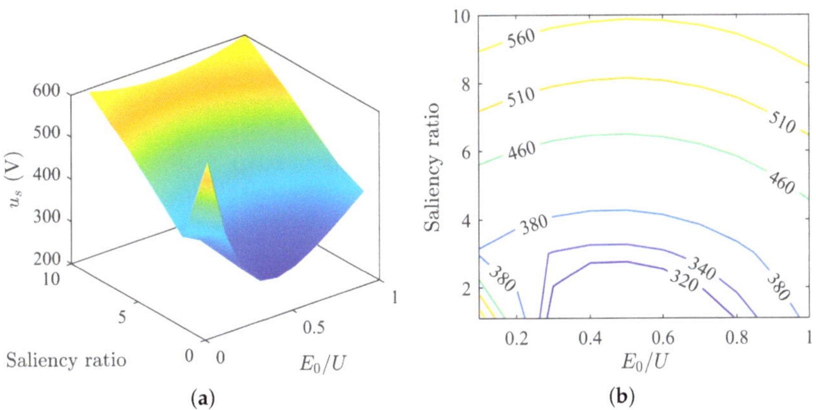

Figure 10. Variation of stator voltage under MTPA: (**a**) surface plot of stator voltage; (**b**) contour map of stator voltage.

From Figure 10b, it can be seen that the contour lines are higher in the middle and lower on both sides, with the highest points of each contour line generally located at $E_0/U = 0.5$, indicating the presence of a minimum voltage curve when the back-EMF is half of the rated voltage.

The relationship between current and voltage under different back EMF and saliency ratio matches is shown in Figure 11.

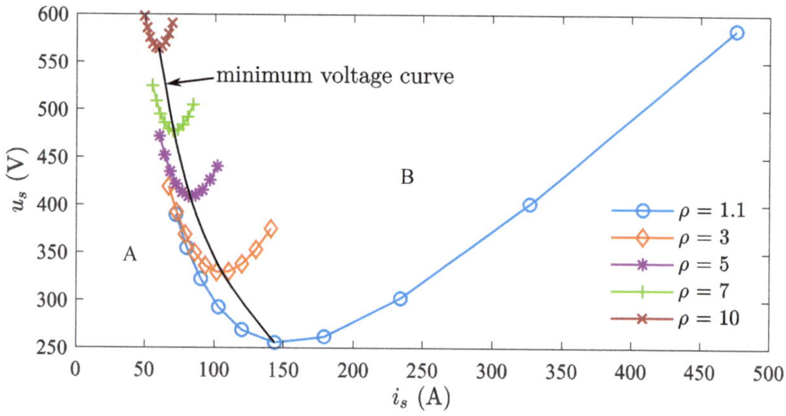

Figure 11. MTPA voltage–current curves under different saliency ratios.

It can be seen that each curve has a minimum, and connecting these minimum points forms the minimum voltage curve. With the minimum voltage curve as a boundary, the plane can be divided into two areas, A and B. In area A, the current is smaller, while in area B, the current is larger. The voltage in both areas gradually increases from the minimum voltage towards both sides. Looking at different saliency ratios, when the saliency ratio is small, the voltage is generally in a lower range. And when it is large, the voltage is generally in a higher range. When the saliency ratio is small, the current is generally larger. And when it is large, the current is generally smaller. As the saliency ratio changes, the overall distribution trend of current and voltage is not consistent, and the best parameter match should be either low voltage and low current, or when the voltage does not exceed the limit voltage and the current is small. The state in area B is not ideal, being either low-voltage and high-current or high-voltage and high-current. The state in area A is either low-current and low-voltage or low-current and high-voltage, with overlapping voltage curves at different saliency ratios, indicating an optimal state.

As shown in Figure 12, the stator voltage increases with the saliency ratio and shows a linear change. This indicates that using MTPA control strategy, there exists a minimum voltage curve under different back-EMF and saliency ratio matches, and as the saliency ratio increases, the minimum voltage linearly increases.

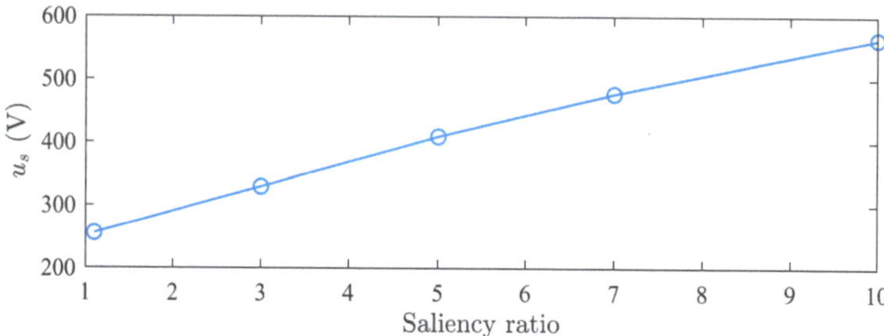

Figure 12. Minimum voltage curve under MTPA.

3.3. Analysis of Parameter Matching under MTPA Control Strategy

The selection of optimal parameter matching under MTPA current control strategy is crucial. Through the above analysis, the motor voltage and current states are obtained, and based on their distribution patterns, the optimal matching of back-EMF and saliency ratio can be determined. The voltage and current contour lines on the entire plane are divided into four regions, with the horizontal dividing line selected as the curve with the lowest voltage and the vertical dividing line selected as the line tangent to the 380 V voltage contour, as shown in Figure 13. The plane is divided into four regions, labeled as ① ② ③ ④.

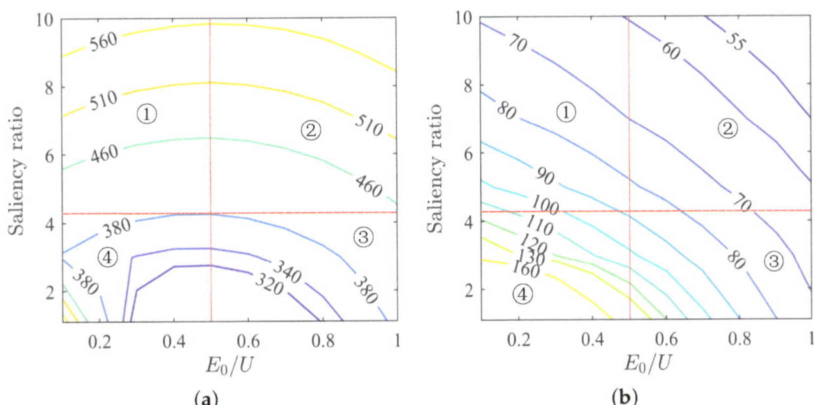

Figure 13. Voltage and current contours under MTPA: (**a**) voltage contour under MTPA; (**b**) current contour under MTPA.

Region ① has relatively low back-EMF and a high saliency ratio, with high voltage and low current.

A lower back-EMF requires fewer permanent magnets, resulting in lower motor costs. However, a higher saliency ratio imposes relatively higher structural and manufacturing requirements on the motor. Despite the lower current, the high voltage in this region may easily exceed the voltage limit.

Region ② exhibits higher back-EMF and saliency ratio with the lowest current and higher voltage. While the high back-EMF requires more permanent magnets, leading to higher motor costs, the high saliency ratio and associated manufacturing complexities pose challenges. Although this region features the lowest current, its advantages are offset by high costs, manufacturing complexities, and the risk of exceeding voltage limits.

Region ③ demonstrates higher back-EMF with smaller saliency ratio, relatively smaller current, and lower voltage. While this region offers better motor performance with lower manufacturing complexities, the increased usage of permanent magnets leads to higher costs.

Region ④ shows lower back-EMF and smaller saliency ratio with the highest current and lower voltage. Despite the lower usage of permanent magnets and lower manufacturing complexities, the high operating current in this region may not meet the motor's performance requirements. Moreover, considering the effects of saturation, achieving the required torque might be challenging.

Considering the above analysis, Region ④ exhibits an excessively high operating current, failing to meet the motor's performance requirements. Region ②, with its high costs, manufacturing complexities, and high voltage, is challenging to overcome. Therefore, neither Region ② nor Region ④ are suitable parameter-matching regions. Region ③, with its lower current, voltage meeting the voltage limit requirement, and lower manufacturing complexities, appears to be a better parameter-matching region from a performance perspective. However, the increased usage of permanent magnets and higher costs present challenges, making it not the optimal state pursued for permanent magnet-assisted synchronous reluctance motors.

Looking at Region ①, with fewer permanent magnets, lower costs, and smaller current, it represents the targeted area for PMaSynRMs. If we can manufacture rotor laminations with higher saliency ratios within our existing manufacturing capabilities, the only remaining issue would be the high voltage. There are two potential solutions:

Firstly, increasing the limit voltage of the inverter to enable the motor to operate normally. However, this method is only feasible for customization and may not be applicable to general industrial motors. Secondly, stopping operating the motor under the MTPA state to meet the voltage limit requirements, which would result in increased operating current. Figure 13b indicates that this region has relatively smaller currents, providing some room for current increase. If this method proves feasible, it could facilitate the application of general industrial motors.

4. Optimal Parameter Matching of PMaSynRM under Vector Control Strategy

4.1. Analysis of Operating States under MTPV Control Strategy

Under MTPA control, Region ① exhibits excessively high operating voltage. Analyzing the voltage and current states under MTPV control, as shown in Figures 14–16. The higher the saliency ratio, the lower the voltage at the same current, making it less prone to overvoltage during motor operation. Overall, within Region ①, under the MTPV control strategy, the motor operates at low voltage and high current. Although it is not an ideal operating state, it indicates the existence of non-overvoltage operating states within Region ①.

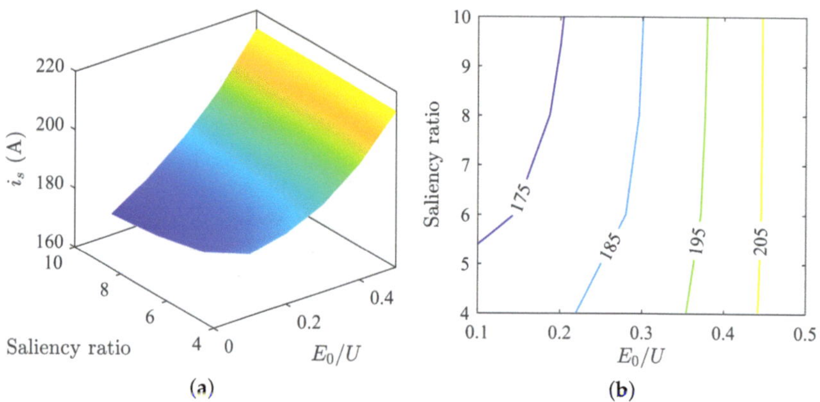

Figure 14. Variation in stator current under MTPV: (**a**) surface plot of stator current; (**b**) contour plot of stator current.

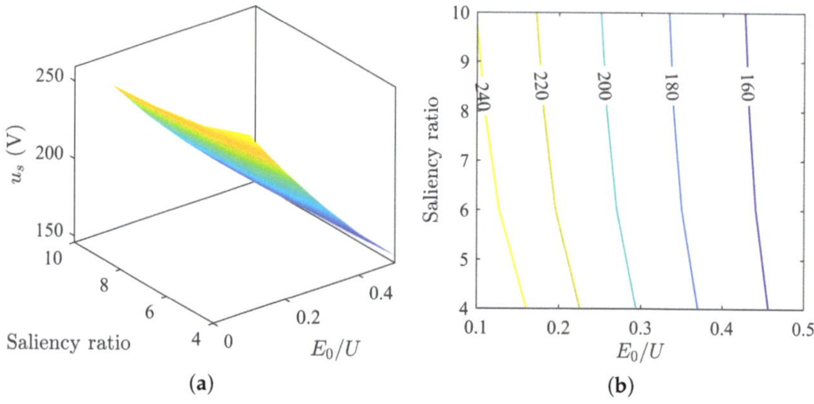

Figure 15. Variation in stator voltage under MTPV: (**a**) surface plot of stator voltage; (**b**) contour plot of stator voltage.

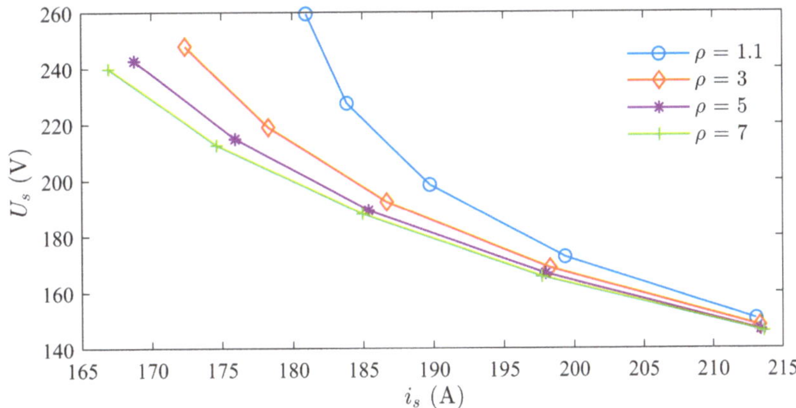

Figure 16. MTPV voltage–current curve.

4.2. Optimal Parameter-Matching Curve under Vector Control

It is explained that there exist operating states between the MTPV control strategy and the MTPA control strategy that satisfy the limit voltage, as shown in Figure 17. From the figure, it can be observed that there is a significant space between the MTPA voltage ellipse and the MTPV voltage ellipse, including many voltage states, among which the voltage limit exists. The intersection of the voltage ellipse and the torque curve represents the current at that voltage state.

It is indicated that under parameter matching in Region ①, operation at the voltage limit state is feasible. Furthermore, whether parameter matching in Region ① is desirable depends on whether the current level under operation at the voltage limit state can reach the current level in Region ③.

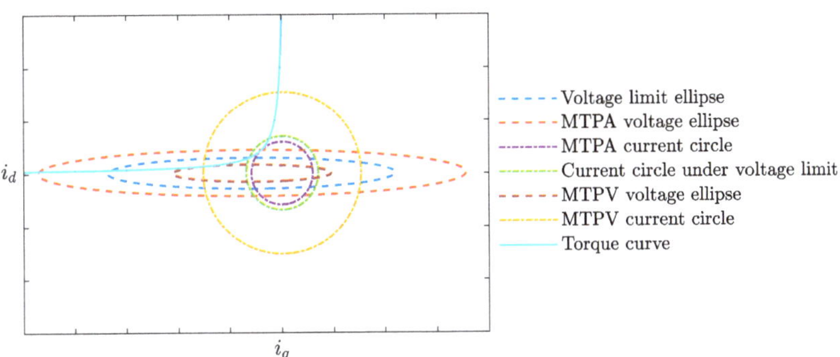

Figure 17. Motor operating state diagram.

Following is the further analysis of the limit voltage operating state in Region ①. Firstly, the optimal operating state in Region ③ is selected as the reference target, with $E_0/U = 0.9$ and $\rho = 2.55$ on the 380 V contour considered the optimal operating state in Region ③. By calculating the motor characteristics under different parameter matchings in Region ① at 380 V voltage state, points where the current reaches the optimal operating state in Region ③ are found, and the specific operating states of these points are as follows.

Overall, the power angle is less than 90°, and the voltage vector is in the second quadrant. The trend of the power angle varies inconsistently with the changes in saliency ratio and back-EMF. The power angle decreases as the back-EMF increases and increases with the increase in saliency ratio (Figure 18).

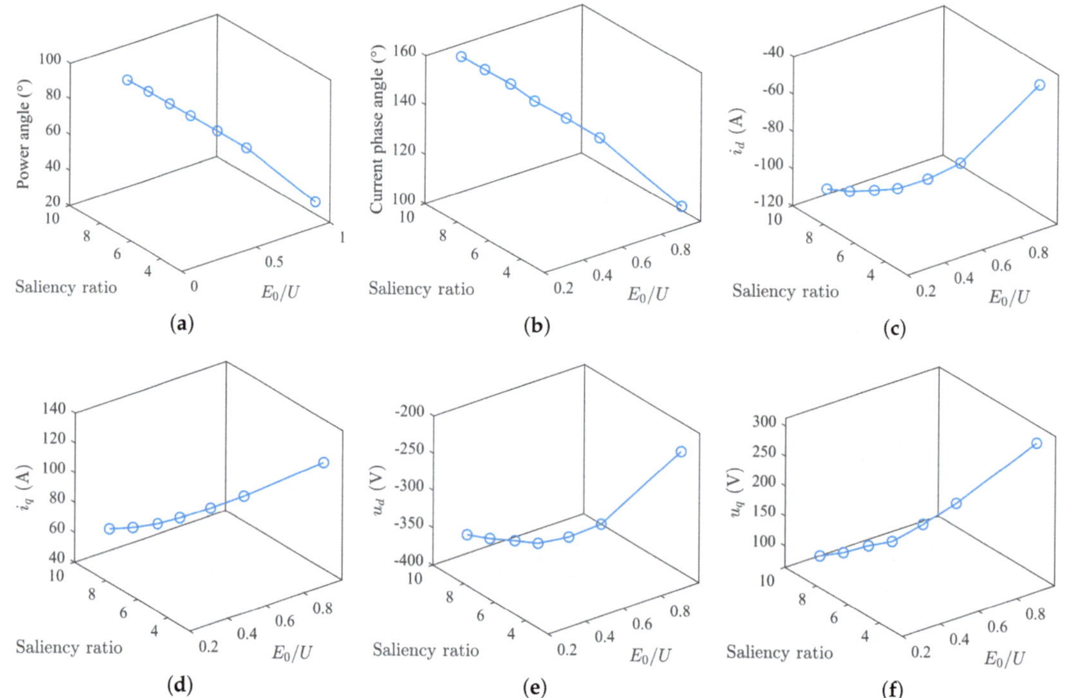

Figure 18. Variation of MTPV stator voltage: (**a**) power angle curve; (**b**) current angle curve; (**c**) direct axis current curve; (**d**) quadrature axis current curve; (**e**) direct axis voltage curve; (**f**) quadrature axis voltage curve.

The current phase angle ranges mostly below 160°, and the current vector is in the second quadrant. When the saliency ratio is large and the back-EMF is small, the current phase angle exceeds 135°, deviating from the trajectory of maximum torque ratio current, and the current increases. This is to avoid the motor voltage exceeding the voltage limit, caused by changes in the magnitude and angle of the current vector. The current phase angle decreases as the back-EMF increases and increases with the increase in saliency ratio. It can be foreseen that with the continued increase in the saliency ratio, the current phase angle will gradually increase until it approaches 180°. Therefore, as the saliency ratio increases, the magnitude of the increase in current phase angle is not consistent. When the saliency ratio is small, the magnitude of the increase in current phase angle is larger, while when the saliency ratio is large, the magnitude of the increase in current angle is smaller. The list of currents for the aforementioned points is presented in Table 1. It can be observed that the currents at these points are generally comparable to the point with $E_0/U = 0.9$ and $\rho = 2.55$.

Table 1. Comparison of stator current under different parameter matching.

Parameter	Stator Current/A
$E_0/U = 0.9, \rho = 2.55$	75.07
$E_0/U = 0.68, \rho = 5$	74.75
$E_0/U = 0.59, \rho = 6$	74.98
$E_0/U = 0.51, \rho = 7$	75.16
$E_0/U = 0.47, \rho = 8$	74.72
$E_0/U = 0.42, \rho = 9$	74.76
$E_0/U = 0.38, \rho = 10$	74.57

On the plane of back-EMF and saliency ratio, points with the same operating characteristics mentioned above are connected into curves, as shown in Figure 19.

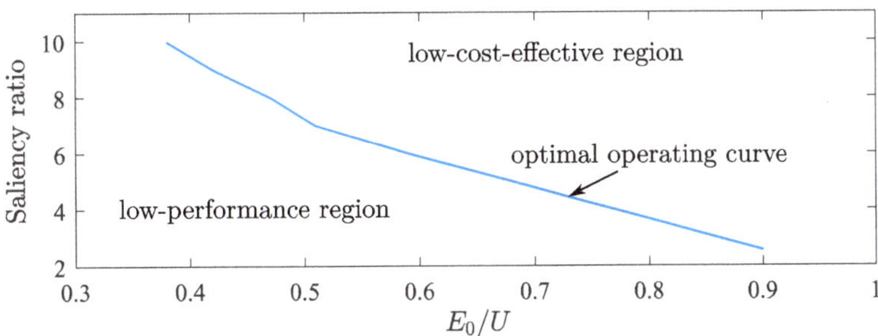

Figure 19. Optimal parameter matching curve.

From the figure, the upper right part represents the low-cost-effective region, while the lower left part represents the low-performance region, separated by a curve. Points on the curve represent the optimal parameter matching state. The low-performance region fails to meet the requirements of high-quality motors and is not a desirable parameter matching state. The low-cost-effective region requires more material and process costs and is also not a desirable parameter-matching state. Meanwhile, as the low-cost-effective region gradually shifts to the right, the performance will gradually decrease after reaching the theoretical optimum, making it even more uneconomical to enter this region. On the optimal operating curve, towards the lower right, the saliency ratio is smaller, and the back-EMF is higher, approaching traditional PMSM. As the saliency ratio increases towards the upper left, the back-EMF gradually decreases. When the saliency ratio is 7, the normalized value of back-EMF is around 0.5, entering Region ①, significantly reducing the cost of permanent magnet materials. With further increases in the saliency ratio, the amount of permanent magnet material is further reduced. The application of ferrite can further reduce costs, but its magnetic energy product is small, and its remanence is low, making it more suitable when the saliency ratio is above 9.

5. Results

Based on the analysis above of a 45 kW PMaSynRM, the motor's specifications are listed in Table 2, with its lamination structure shown in Figure 20. The no-load air-gap flux density waveform is shown in Figure 21a, along with the d-q axis inductance curves in Figure 21b.

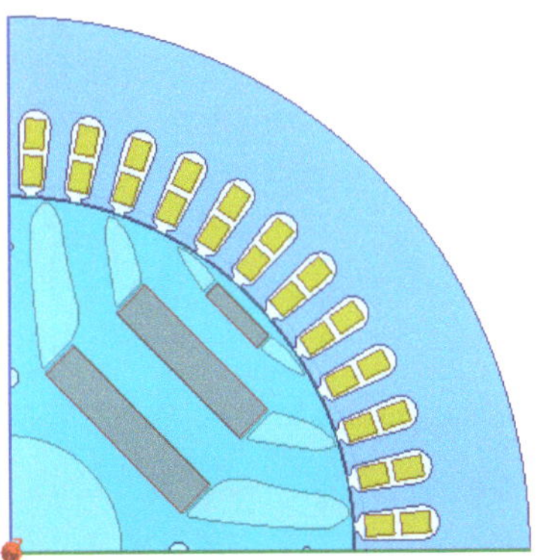

Figure 20. Sketch of the designed PMaSynRM.

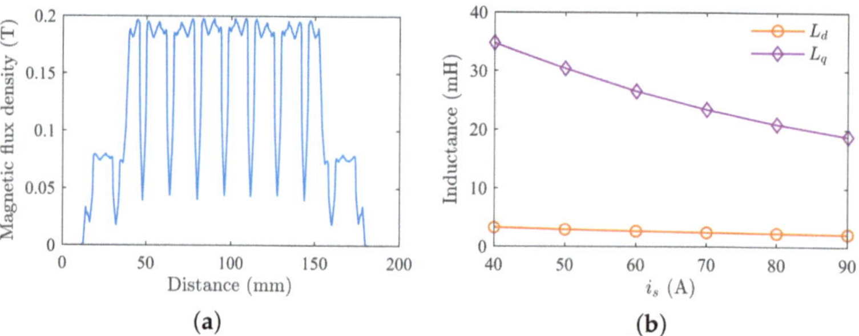

Figure 21. Characteristics of the designed prototype: (**a**) no-load air-gap flux density waveform by finite element simulation; (**b**) motor inductance curves.

Table 2. Main parameters of the prototype.

Parameter	Value
Power	45 kW
Speed	1500 rpm
Number of poles	4
Stator inner/outer diameter	368 mm/245 mm
Number of stator slots	48

The flux density waveform has a stepped shape, with a peak close to 0.2 T, which is significantly smaller compared to that of a PMSM. The permanent magnet torque is very small, with the reluctance torque being the main driving torque of the motor. As the current increases, both the d- and q-axis inductances decrease, with the quadrature axis inductance showing a large variation and significantly affected by core saturation. From the graph, it is evident that the motor's saliency ratio can reach above 9 under full load conditions.

An analysis of the torque composition at rated conditions of the motor is conducted. The permanent magnet torque accounts for 17% of the total torque, while the reluctance torque constitutes 83%, primarily providing the driving torque.

Following the design and manufacturing, a prototype is manufactured as shown in Figure 22.

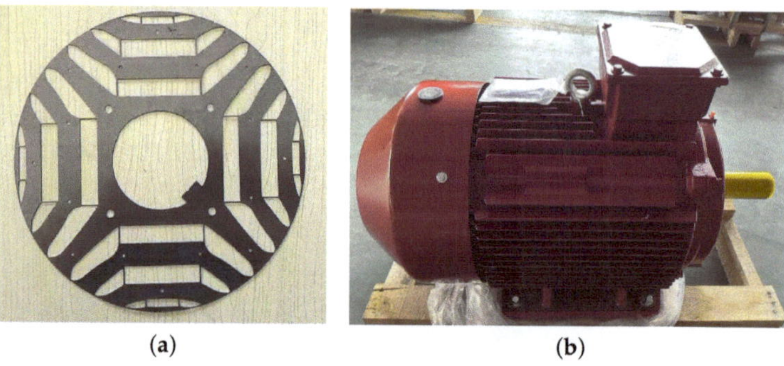

Figure 22. Manufactured prototype: (**a**) rotor core; (**b**) prototype assembly.

Experimental setup is depicted in Figure 23. The test motor is connected to a torque sensor and a load machine for testing its performance under various speeds and loads.

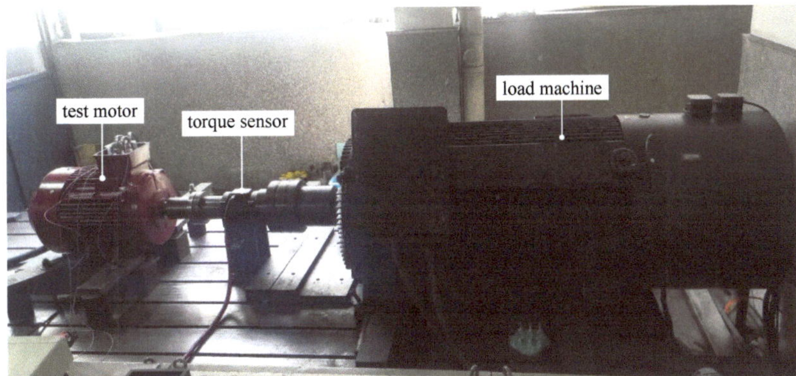

Figure 23. Experimental setup.

Load tests of the prototype is conducted, measuring the prototype's efficiency and power factor curves at different speed as shown in Figures 24–26. The motor's efficiency reaches 96.5%, achieving an IE5 energy efficiency level.

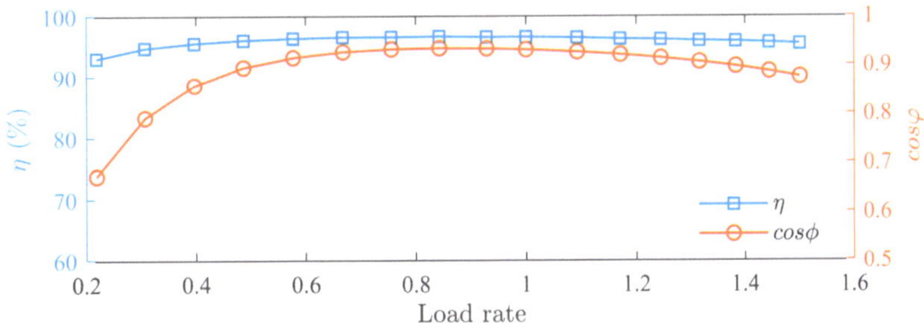

Figure 24. Efficiency and power factor curves at 1500 rpm.

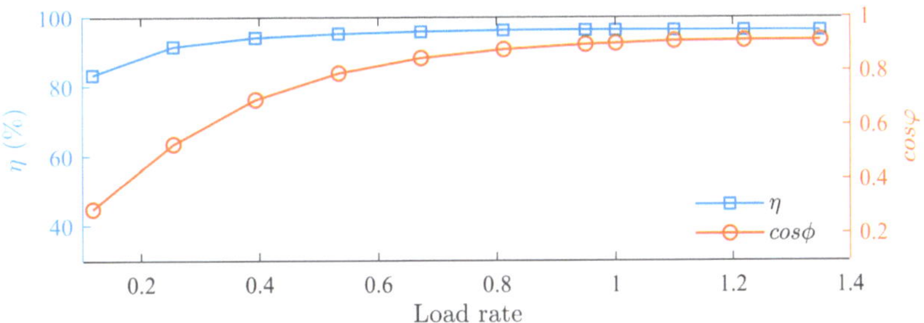

Figure 25. Efficiency and power factor curves at 1200 rpm.

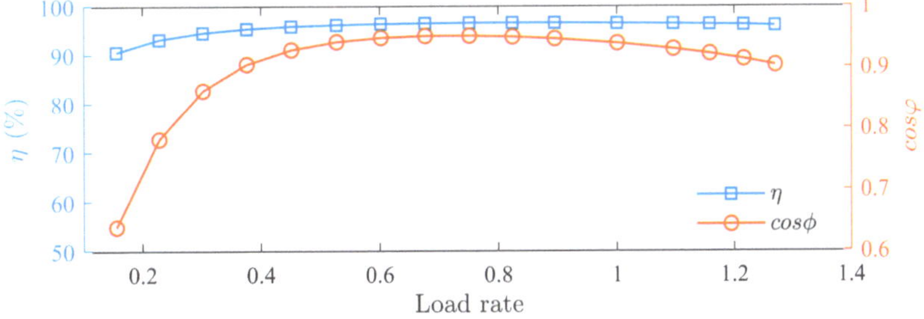

Figure 26. Efficiency and power factor curves at 1800 rpm.

6. Conclusions

This paper analyzes the impact of parameter matching on the steady-state characteristics of a PMaSynRM under vector control, summarizing its features and patterns. Based on this, a 45 kW prototype was designed, manufactured, and tested, achieving a high energy efficiency level, while also achieving the goal of not using rare-earth materials, thus reducing the cost of the motor. Specific conclusions are as follows:

Starting from the mathematical model of the motor under the MTPA control strategy, it was derived and analyzed that there is a corresponding relationship between the amplitude and phase of voltage and current, closely related to the parameters E_0/U and saliency ratio ρ. It was found that under different parameter matchings of MTPA control strategy, there exists a lowest voltage curve, with the corresponding back-EMF near 0.5 times the rated voltage. Operating voltages will be higher on either side of this. A higher saliency ratio with lower back-EMF results in a higher operating voltage.

The correspondence between voltage and current under MTPV control strategy was analyzed. It was found that under MTPV control, the higher the saliency ratio, the lower the operating voltage at the same current, staying below the limit voltage. This proves that there is a margin between MTPA and MTPV controls, allowing for a high saliency ratio and low back-EMF parameter matchings to operate without over-voltage. On this basis, the optimal parameter match curve for vector control was obtained. A 45 kW prototype is then designed, manufactured, and tested, achieving IE5 efficiency, and effectively reducing the cost of the motor.

Author Contributions: Conceptualization, Y.-H.L. and J.W.; methodology, Y.-H.L. and J.W.; software, Y.-H.L. and J.W.; validation, Y.-H.L. and J.W.; formal analysis, Y.-H.L.; investigation, Y.-H.L.; resources, Y.-H.L.; data curation, Y.-H.L.; writing—original draft preparation, Y.-H.L.; writing—review and editing, W.-J.W.; visualization, Y.-H.L. and W.-J.W.; supervision, J.W. All authors have read and agreed to the published version of the manuscript.

Funding: This research received no external funding.

Data Availability Statement: The data presented in this study are available on request from the corresponding author. The data are not publicly available due to privacy.

Conflicts of Interest: Author Yu-Hua Lan was employed by Jiang Chao Motor Technology Co., Ltd. The remaining authors declare that the research was conducted in the absence of any commercial or financial relationships that could be construed as a potential conflict of interest.

References

1. Zhang, J.; Zhu, D.; Jian, W.; Hu, W.; Peng, G.; Chen, Y.; Wang, Z. Fractional Order Complementary Non-Singular Terminal Sliding Mode Control of PMSM Based on Neural Network. *Int. J. Automot. Technol.* **2024**, *25*, 213–224. [CrossRef]
2. Zhang, J.; Chen, Y.; Gao, Y.; Wang, Z.; Peng, G. Cascade ADRC Speed Control Base on FCS-MPC for Permanent Magnet Synchronous Motor. *J. Circuits Syst. Comput.* **2021**, *30*, 2150202. [CrossRef]
3. Sun, Q.; Lyu, G.; Liu, X.; Niu, F.; Gan, C. Virtual Current Compensation-Based Quasi-Sinusoidal-Wave Excitation Scheme for Switched Reluctance Motor Drives. *IEEE Trans. Ind. Electron.* **2023**, 1–11. [CrossRef]
4. Lu, Y.; Li, J.; Xu, H.; Yang, K.; Xiong, F.; Qu, R.; Sun, J. Comparative study on vibration behaviors of permanent magnet assisted synchronous reluctance machines with different rotor topologies. *IEEE Trans. Ind. Appl.* **2021**, *57*, 1420–1428. [CrossRef]
5. Xu, M.; Liu, G.; Chen, Q.; Ji, J.; Zhao, W. Design and optimization of a fault tolerant modular permanent magnet assisted synchronous reluctance motor with torque ripple minimization. *IEEE Trans. Ind. Electron.* **2020**, *68*, 8519–8530. [CrossRef]
6. Guo, H.; He, X.; Xu, J.; Tian, W.; Sun, G.; Ju, L.; Li, D. Design of an aviation dual-three-phase high-power high-speed permanent magnet assisted synchronous reluctance starter-generator with antishort-circuit ability. *IEEE Trans. Power Electron.* **2022**, *37*, 12619–12635. [CrossRef]
7. Gallicchio, G.; Di Nardo, M.; Palmieri, M.; Marfoli, A.; Degano, M.; Gerada, C.; Cupertino, F. High speed permanent magnet assisted synchronous reluctance machines—Part I: A general design approach. *IEEE Trans. Energy Convers.* **2022**, *37*, 2556–2566. [CrossRef]
8. Di Nardo, M.; Gallicchio, G.; Palmieri, M.; Marfoli, A.; Degano, M.; Gerada, C.; Cupertino, F. High Speed Permanent Magnet Assisted Synchronous Reluctance Machines—Part II: Performance Boundaries. *IEEE Trans. Energy Convers.* **2022**, *37*, 2567–2577. [CrossRef]

9. Michalski, T.D.; Romeral, J.L.; Mino-Aguilar, G. Open-Phase Faulty Five-Phase Permanent Magnet Assisted Synchronous Reluctance Motor Model With Stepped Skew. *IEEE Trans. Magn.* **2021**, *57*, 1–7. [CrossRef]
10. Chai, W.; Yang, H.M.; Xing, F.; Kwon, B.I. Analysis and Design of a PM-Assisted Wound Rotor Synchronous Machine With Reluctance Torque Enhancement. *IEEE Trans. Ind. Electron.* **2021**, *68*, 2887–2897. [CrossRef]
11. Maroufian, S.S.; Pillay, P. Design and Analysis of a Novel PM-Assisted Synchronous Reluctance Machine Topology With AlNiCo Magnets. *IEEE Trans. Ind. Appl.* **2019**, *55*, 4733–4742. [CrossRef]
12. Huynh, T.A.; Hsieh, M.F.; Shih, K.J.; Kuo, H.F. An Investigation Into the Effect of PM Arrangements on PMa-SynRM Performance. *IEEE Trans. Ind. Appl.* **2018**, *54*, 5856–5868. [CrossRef]
13. Baka, S.; Sashidhar, S.; Fernandes, B.G. Design of an Energy Efficient Line-Start Two-Pole Ferrite Assisted Synchronous Reluctance Motor for Water Pumps. *IEEE Trans. Energy Convers.* **2021**, *36*, 961–970. [CrossRef]
14. Shen, J.X.; Cai, S.; Shao, H.; Hao, H. Evaluation of low-cost high-performance synchronous motors for ventilation application. In Proceedings of the 2015 International Conference on Sustainable Mobility Applications, Renewables and Technology (SMART), Kuwait City, Kuwait, 23–25 November 2015; pp. 1–6. [CrossRef]
15. Wu, W.; Zhu, X.; Quan, L.; Du, Y.; Xiang, Z.; Zhu, X. Design and Analysis of a Hybrid Permanent Magnet Assisted Synchronous Reluctance Motor Considering Magnetic Saliency and PM Usage. *IEEE Trans. Appl. Supercond.* **2018**, *28*, 5200306. [CrossRef]
16. Dong-liang, L.; LIA-qiang.; Wei, L.; Jian, W. Rotor optimization of permanent magnet-assisted synchronous reluctance motor for electric vehicles. *Electr. Mach. Control* **2022**, *26*, 119–129. [CrossRef]
17. Yingqian, L.; Yi, S.; Yunchong, W.; Jianxin, S. A Hybrid PM-Assisted SynRM with Ferrite and Rare-Earth Magnets. *Trans. China Electrotech. Soc.* **2022**, *37*, 1145–1157. [CrossRef]
18. Yi, S.; Shun, C.; Yingqian, L.; Yunchong, W.; Jianxin, S. Top-Level Design Pattern of PM-Assisted Synchronous Reluctance Machines. *Trans. China Electrotech. Soc.* **2022**, *37*, 2306–2318. [CrossRef]
19. Korman, O.; Nardo, M.D.; Degano, M.; Gerada, C. A Novel Flux Barrier Parametrization for Synchronous Reluctance Machines. *IEEE Trans. Energy Convers.* **2022**, *37*, 675–684. [CrossRef]
20. Credo, A.; Fabri, G.; Villani, M.; Popescu, M. Adopting the Topology Optimization in the Design of High-Speed Synchronous Reluctance Motors for Electric Vehicles. *IEEE Trans. Ind. Appl.* **2020**, *56*, 5429–5438. [CrossRef]
21. Mohammadi, A.; Mirimani, S.M. Design of a Novel PM-Assisted Synchronous Reluctance Motor Topology Using V-Shape Permanent Magnets for Improvement of Torque Characteristic. *IEEE Trans. Energy Convers.* **2022**, *37*, 424–432. [CrossRef]

 *actuators*

Article

A Reliable and Efficient I-f Startup Method of Sensorless Ultra-High-Speed SPMSM for Fuel Cell Air Compressors

Jilei Xing [1], Yao Xu [2,*], Junzhi Zhang [1], Yongshen Li [2] and Xiongwei Jiang [2]

[1] School of Vehicle and Mobility, Tsinghua University, Beijing 100084, China; xingjilei699@mail.tsinghua.edu.cn (J.X.); powertrain@tsinghua.edu.cn (J.Z.)
[2] School of Mechanical Engineering, Beijing Institute of Technology, Beijing 100081, China; liyongshen@bit.edu.cn (Y.L.); jiangxiongwei@bit.edu.cn (X.J.)
* Correspondence: xuyao@bit.edu.cn; Tel.: +86-181-1517-9369

Abstract: Extended back electromotive force (EEMF)-based position sensorless field-oriented control (FOC) is widely utilized for ultra-high-speed surface-mounted permanent magnet synchronous motors (UHS-SPMSMs) driven fuel cell air compressors in medium-high speed applications. Unfortunately, the estimated position is imprecise due to too small EEMF under low speed operation. Hence, current-to-frequency (I-f) control is more suitable for startup. Conventional I-f methods rarely achieve the tradeoff between startup acceleration and load capacity, and the transition to sensorless FOC is mostly realized in the constant-speed stage, which is unacceptable for UHS-SPMSM considering the critical requirement of startup time. In this article, a new closed-loop I-f control approach is proposed to achieve fast and efficient startup. The frequency of reference current vector is corrected automatically based on the active power and the real-time motor torque, which contributes to damping effect for startup reliability. Moreover, an amplitude compensator of reference current vector is designed based on the reactive power, ensuring the maximum torque per ampere operation and higher efficiency. Furthermore, the speed PI controller is enhanced by variable bandwidth design for smoother sensorless transition. These theoretical advantages are validated through experiments with a 550 V, 35 kW UHS-SPMSM. The experimental results demonstrated the enhanced startup performance compared with conventional I-f control.

Keywords: ultra-high-speed surface-mounted permanent magnet synchronous motor (UHS-SPMSM); current-frequency (I-f) control; sensorless control; startup strategy

Citation: Xing, J.; Xu, Y.; Zhang, J.; Li, Y.; Jiang, X. A Reliable and Efficient I-f Startup Method of Sensorless Ultra-High-Speed SPMSM for Fuel Cell Air Compressors. *Actuators* **2024**, *13*, 203. https://doi.org/10.3390/act13060203

Academic Editor: Leonardo Acho Zuppa

Received: 25 April 2024
Revised: 24 May 2024
Accepted: 27 May 2024
Published: 29 May 2024

1. Introduction

Due to high efficiency, high power density, and stable rotor structural strength [1,2], ultra-high-speed surface-mounted permanent magnet synchronous motors (UHS-SPMSMs) are being widely used to drive fuel cell air compressors [3–5], which in turn provide pressurized air to fuel cell stacks to ensure the power output demand and energy conversion efficiency of hydrogen fuel cell vehicles. UHS-SPMSMs and centrifugal compressors are integrated as the overall compressor system, whose rational speed is typically capable of 100 kr/min and above. Therefore, the mechanical rotor position sensor is replaced by position sensorless control technology in consideration of installation space and reliability, which also reduces maintenance cost.

For medium–high speed applications of UHS-SPMSM, the most commonly used position sensorless methods are based on back electromotive force (EMF) estimation, including the use of sliding mode observer [6,7], extended Kalman filter [8], extended state observer [9], etc. Unfortunately, the EMF is too small to be estimated accurately during low speed operation. For zero–low speed applications of UHS-SPMSM, the typical high frequency signal injection (HFSI) methods [10,11] will no longer be applicable considering the lack of motor saliency due to the equal dq-axes inductances, and the high-frequency noise and torque fluctuations of HFSI are undesirable. The rotor of UHS-SPMSM is carried by a self-acting gas bearing, in which the fluid film is formed by viscous shear forces resulting from the relative motion of journal and bushing under rotation [12]. Therefore, UHS-SPMSM needs to be started rapidly from standstill to a certain speed to ensure the normal performance of gas bearing, and UHS-SPMSM is generally designed with a minimum operating speed (i.e., idle speed) ranging from 20,000 r/min to 30,000 r/min. Under these requirements, the UHS-SPMSM do not operate at ultra-low speed except during startup. Preferably, an uncomplicated yet effective startup strategy, characterized by high dynamic performance, is more advantageous than HFSI methods for expediting the UHS-SPMSM to a speed conducive to accurate EMF estimation.

Voltage-to-frequency (V/f) control has been employed for the startup and speed regulation of UHS-SPMSM owing to its straightforward configuration and reduced number of control loops, including the open-loop optimal V/f control with design consideration to the stator resistance [13], and the closed-loop V/f control with correction of the voltage amplitude and phase [14]. Owing to uncontrollable current and torque ripple, open-loop V/f control is gradually replaced by closed-loop V/f control. In [15], correction loops are designed based on d-axis current error with integrated maximum torque per ampere (MTPA) block, which guarantees good dynamic performance especially at a very low speed. Additionally, a closed-loop V/f control is realized in the voltage vector plane [16]. However, the algorithm complexity increases similar to sensorless field-oriented control (FOC) due to load angle calculation in [15] and rotor position and speed estimation in [16], respectively. In summary, the V/f control has insufficient with-load startup ability, and the control accuracy and efficiency are limited in the medium–high speed range compared with FOC. Therefore, the V/f control is not recommended for the UHS-SPMSM used in hydrogen fuel cell vehicles.

In contrast, another simple and effective startup strategy known as the current-to-frequency (I-f) control is more widely used for the startup of UHS-SPMSM because of its advantages of strong robustness with controllable current loop. The UHS-SPMSM does not always start successfully under open-loop I-f control [17] due to large speed oscillation during the startup stage [18]. The inherent speed oscillation can be suppressed by correcting the frequency of the reference current vector based on active power fluctuation [19]. Meanwhile, with a summary of the change law of the fluctuation amplitude and frequency of torque angle with the motor parameters and the reference current vector [20], the method proposed in [19] can be adaptively designed. Additionally, any signal reflecting speed or torque perturbation can be used to obtain the frequency correction value [21], such as the power factor angle, the calculated torque current, the estimated electromagnetic torque [22], etc. In combination with adaptive back-stepping control [23], fully unknown parameters of SPMSM are able to be estimated accurately, ensuring the effectiveness of I-f control on variation of load and motor parameters.

The reference torque of UHS-SPMSM with conventional I-f startup method far exceeds actual load torque, leading to low motor efficiency. Although adjusting the amplitude of the reference current vector based on reactive power [24,25] and based on power factor [26] can actualize $i_d = 0$ control to enhance efficiency; this is only realized after the motor speed remains constant. The implementation mode is not applicable for UHS-SPMSM, which needs to complete the transition of the I-f control to EMF-based sensorless control during acceleration process to ensure the dynamic performance and high efficiency of the startup. In [27], variable reference current vector position is designed in the current compensation loop to avoid large transients. However, additional estimated rotor position is needed from EMF-based method. Moreover, the frequency ramp slope of the reference current vector can be dynamically adjusted during motor acceleration using an angle controller [24] or a torque controller [28].The load torque of UHS-SPMSM increases approximately quadratically in terms of the motor speed. In order to prevent the demand torque from exceeding the maximum output torque, the frequency ramp slope needs to be gradually reduced as a result of the dynamic adjustments, which significantly deteriorates the dynamic performance of UHS-SPMSM startup.

The overall requirement of the UHS-SPMSM startup process is high dynamic response with high efficiency. Considering that the frequency ramp slope cannot be set too high for avoiding startup failure, the transition speed from I-f control to sensorless FOC should be set as small as possible. Meanwhile, a smooth transition with a reduced convergence time and speed fluctuations needs to be addressed. The method proposed in [29] of adjusting the initial value of the integrator of the PI controller of the speed loop and current loop can reduce the speed fluctuations. The design of variable decreasing step size of reference current can contribute positively to reducing the transition time [17]. To improve the stability of the transition process, a feedback regulator is designed [30] to automatically adjust the reference current based on the position difference. However, merely with consideration of the constant speed operation, the above I-f control strategies have neglected the transition process during the motor acceleration, where the load torque increases apace in UHS-SPMSM application. The prevailing transition strategies, including the reduction of reference current [31], an adaptive transition algorithm [32], weighted processing method [33], and a first-order lag compensator of position [34], have hitherto placed insufficient emphasis on evaluating the performance of the I-f control preceding the transition.

The novelty in this article, as an extension of the conventional I-f control, is to propose a closed-loop I-f startup strategy with transition enhancement applicable to UHS-SPMSM to achieve a fast, efficient, and stable startup process, which is summarized as:

(1)   The speed convergence performance can be effectively improved by correcting the frequency of reference current vector adaptively based on not only the instantaneous active power but also the real-time motor torque, which reduces speed fluctuations distinctly.

(2)   The amplitude of reference current vector is compensated dynamically during speedup stage instead of during a constant speed stage, which is the basic requirement and distinctiveness for the UHS-SPMSM. Therefore, the reference startup torque can be reduced with improved efficiency under the premise of guaranteeing the startup rapidity.

(3)   The transition process to EMF-based sensorless FOC can be enhanced by designing a bandwidth-variable regulating scheme of speed loop PI controller, which achieves low transitional speed fluctuation. Moreover, the designed scheme can ensure high control stability over rated motor speed, compared with fixed-gain speed loop PI controller.

(4)   To the best of our knowledge, this is the first time that an innovative closed-loop I-f startup strategy is used for UHS-SPMSM startup, especially for driving fuel cell air compressors.

The rest of this article is organized as follows. Section 2 describes the mathematical model of UHS-SPMSM under I-f control. The stability performance of the conventional I-f startup strategy is analyzed in Section 3. Section 4 elaborates on the proposed closed-loop I-f startup strategy with transition enhancement. In Section 5, the effectiveness of the proposed method is verified by bench experiments with a 35 kW (at 95 kr/min) UHS-SPMSM. Finally, the paper is concluded in Section 6.

2. Mathematical UHS-SPMSM Model with I-f Control

Due to unknown actual rotor position information during I-f control, the actual d-q reference coordinate frame is also undetermined. Therefore, in order to analyze the characteristics of the controllable current vector, a γ-δ reference coordinate frame is introduced, in which the δ-axis is aligned with the current vector I, as shown in Figure 1. The angles from α-axis to δ-axis and to q-axis are referred to as θ_i and θ_e, respectively. The angle error between θ_i and θ_e is denoted as $\theta_{err} = \theta_e - \theta_i$. The frequencies ω_i and ω_e denote angular velocities of current vector and rotor, respectively. φ is the angle between the voltage vector

U and the current vector. Without loss of generality, the UHS-SPMSM voltage mathematical model [24] in the γ-δ reference coordinate frame can be expressed as:

$$\begin{cases} u_\gamma = R_s i_\gamma + L_d \frac{d}{dt} i_\gamma - \omega_i L_q i_\delta - \omega_e \varphi_f \sin \theta_{err} \\ u_\delta = R_s i_\delta + L_q \frac{d}{dt} i_\delta + \omega_i L_d i_\gamma + \omega_e \varphi_f \cos \theta_{err} \end{cases} \tag{1}$$

where u_γ, u_δ, i_γ, and i_δ denote the voltages and currents in the γ-δ reference coordinate frame, respectively. R_s, φ_f, L_d, and L_q denote stator resistance, rotor permanent magnet flux linkage, and d- and q-axis inductances, respectively. The condition $L_d = L_q$ holds throughout this article for the studied UHS-SPMSM. ω_e denotes the motor electrical frequency.

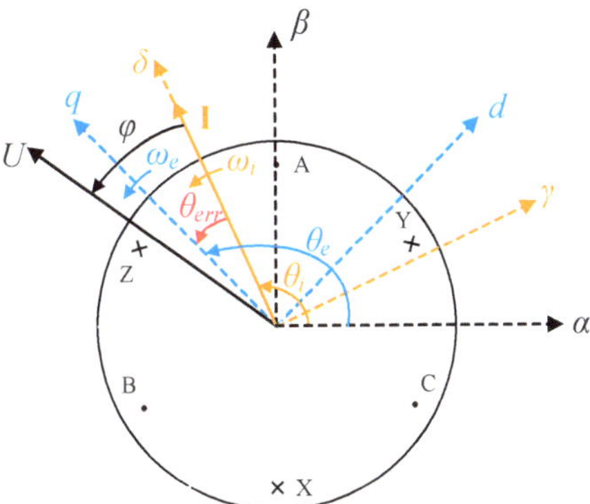

Figure 1. Vector diagram including d-q and γ-δ frames for UHS-SPMSM.

Considering that $i_\gamma = 0$, (1) can be rewritten as:

$$\begin{cases} u_\gamma = -\omega_i L_q i_\delta - \omega_e \varphi_f \sin \theta_{err} \\ u_\delta = R_s i_\delta + L_q \frac{d}{dt} i_\delta + \omega_e \varphi_f \cos \theta_{err} \end{cases} \tag{2}$$

The frequency error between ω_i and ω_e can be obtained from the derivative of angle error.

$$\Delta\omega = \omega_e - \omega_i = \frac{d}{dt}\theta_{err} = \frac{d}{dt}(\theta_e - \theta_i) \tag{3}$$

In addition, the UHS-SPMSM electromagnetic torque [24] can be expressed by the current vector amplitude I_m and angle θ_{err} as follows with $L_d = L_q$.

$$\begin{aligned} T_e &= 1.5 n_p i_q [\varphi_f + (L_d - L_q) i_d] \\ &= 1.5 n_p \varphi_f I_m \cos \theta_{err} \end{aligned} \tag{4}$$

where n_p denotes the number of pole pairs.

Therefore, the motion equation [24] of UHS-SPMSM is expressed as:

$$J\frac{d\omega_m}{dt} = T_e - T_L - B\omega_m \tag{5}$$

where $\omega_m = \omega_e/n_p$ is the rotor mechanical frequency. J is the motor inertia, B is the damping coefficient, and T_L denotes the load torque.

3. Conventional I-f Startup Method

I-f control establishes the relationship between reference current and reference frequency under the current closed-loop control so that the output torque is load-capable. Conventionally, the current vector amplitude is set at a large constant value, while the current vector frequency ramps up with time at a fixed ramp slope, as shown in Figure 2. The integration of the frequency yields the rotor position information needed by park and anti-park transformation, i.e.,

$$\theta_i = \int \omega_i dt. \tag{6}$$

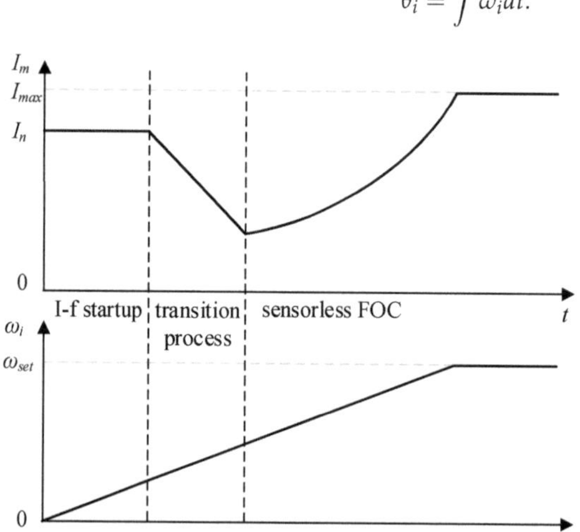

Figure 2. Current vector amplitude and frequency of conventional I-f control.

Since the electrical response is much faster than the mechanical effect, the actual current vector can be considered to be the same as the reference current vector, where the δ-axis reference current is equal to the reference current vector amplitude (e.g., the rated value I_n), while the γ-axis reference current is set to 0, i.e.,

$$i_\gamma = 0, \ i_\delta = I_n. \tag{7}$$

When the UHS-SPMSM exceeds the transition speed during the ramp up stage, according to the self-stabilization mechanism [31], the θ_{err} decreases with the reduction of the reference current amplitude. Therefore, the transition from I-f control to sensorless FOC can be successfully accomplished when the θ_{err} is smaller than a certain threshold value (e.g., 0.05 rad). It is worth noting that the estimated rotor position $\hat{\theta}_e$ by extended EMF (EEMF)-based method proposed in [35] can be used to replace the actual one during the transition process:

$$\theta_{err} = \theta_e - \theta_i \approx \hat{\theta}_e - \theta_i. \tag{8}$$

Generally, after successful transition, the UHS-SPMSM operates in double closed-loop control of speed and current. However, the inherent properties of speed oscillation and prone asynchronous operation of conventional I-f control cause incidental startup failures.

3.1. Inherent Speed Oscillation

Small signal model is aptly utilized to analyze the performance of UHS-SPMSM under I-f control. Therefore, the torque Equation (4) is linearized to represent the dynamic characteristics at a certain static operating point, i.e.,

$$\Delta T_e = K_I \Delta I - K_\theta \Delta \theta_{err} \tag{9}$$

with:

$$\begin{cases} K_I = \frac{\partial T_e}{\partial I_m} = 1.5 n_p \varphi_f \cos \theta_{err} \\ K_\theta = -\frac{\partial T_e}{\partial \theta_{err}} = 1.5 n_p \varphi_f I_m \sin \theta_{err}. \end{cases} \tag{10}$$

Further based on (3) and (5), small-signal models for angular error and motion equation can be obtained:

$$\Delta\theta_{err} = \frac{1}{s}(\Delta\omega_e - \Delta\omega_i),\tag{11}$$

$$\Delta\omega_e = \frac{n_p}{Js + B}(\Delta T_e - \Delta T_L).\tag{12}$$

The small signal model is schematically shown in Figure 3 with combination of (9)–(12).

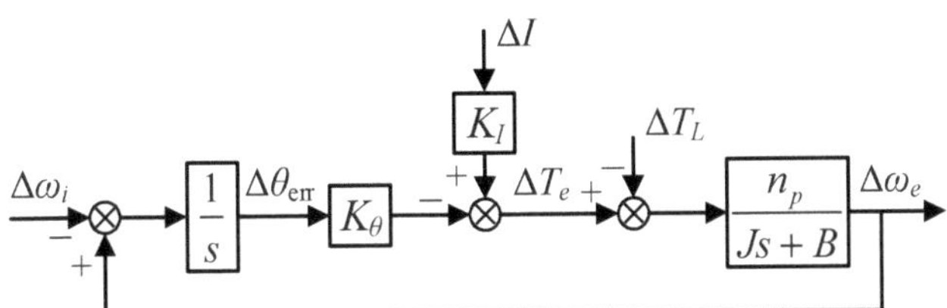

Figure 3. Small signal model of UHS-SPMSM under conventional I-f control.

With $\Delta\omega_i = 0$ and $\Delta I = 0$, the closed-loop transfer function from ΔT_L to $\Delta\omega_e$ is obtained as follows:

$$G(s) = \frac{\Delta\omega_e}{\Delta T_L} = \frac{-n_p s}{Js^2 + Bs + n_p K_\theta}.\tag{13}$$

The characteristic equation of the system under I-f control is obtained as:

$$Js^2 + Bs + n_p K_\theta = s^2 + 2\zeta\omega_n s + \omega_n^2 = 0\tag{14}$$

with:

$$\zeta = \frac{B}{2\sqrt{Jn_p K_\theta}}, \quad \omega_n = \sqrt{\frac{n_p K_\theta}{J}}\tag{15}$$

where ζ and ω_n denote the damping ratio and natural frequency.

ζ is very small with consideration to the relatively small B of UHS-SPMSM, which causes slow system response and large speed oscillation. At a certain static operating point, the load torque variation can be approximated considered as a step change, so the unit step response of the UHS-SPMSM according to (13) is derived as:

$$\begin{aligned}c(t) &= \mathcal{L}^{-1}(G(s)\cdot\tfrac{1}{s}) = \mathcal{L}^{-1}\left(\frac{-n_p s}{Js^2+Bs+n_p K_\theta}\cdot\tfrac{1}{s}\right)\\ &= -\frac{J\omega_n e^{-\zeta\omega_n t}}{K_\theta\sqrt{1-\zeta^2}}\sin\omega_d t = -A\sin\omega_d t\end{aligned}\tag{16}$$

with $\omega_d = \omega_n\sqrt{1-\zeta^2}$.

It is evident that $\Delta\omega_e$ and ΔT_L exhibit opposite changing trends, and the increase of ΔT_L leads to the inherent speed oscillation with oscillation frequency ω_d and oscillation amplitude A. Excessive speed oscillation during initial startup interval and transition process tends to cause startup failure and transition failure, respectively.

3.2. Prone Asynchronous Operation

The UHS-SPMSM is typically characterized by a steep ramp slope (e.g., $1000\,\pi\,\mathrm{rad/s^2}$) to facilitate rapid startup, and by considerable load torque considering the approximate quadratic correlation between motor load torque and speed. According (5), the demand torque T_d comprises the load torque T_L, the acceleration torque T_a, and the friction torque T_f.

$$\begin{aligned}T_d &= J\frac{d\omega_m}{dt} + T_L + B\omega_m\\ &= T_a + T_L + T_f\end{aligned}\tag{17}$$

Once T_d exceeds the maximum electromagnetic torque T_{max} under high-ramp-slope and heavy-load conditions, the I-f control may be prone to failure, leading to losing synchronization issues. Referring to (4), the variation between T_e and θ_{err} can be roughly illustrated in Figure 4. Depending on whether θ_{err} exceeds 0, the operating region can be categorized into synchronous region I and asynchronous region II.

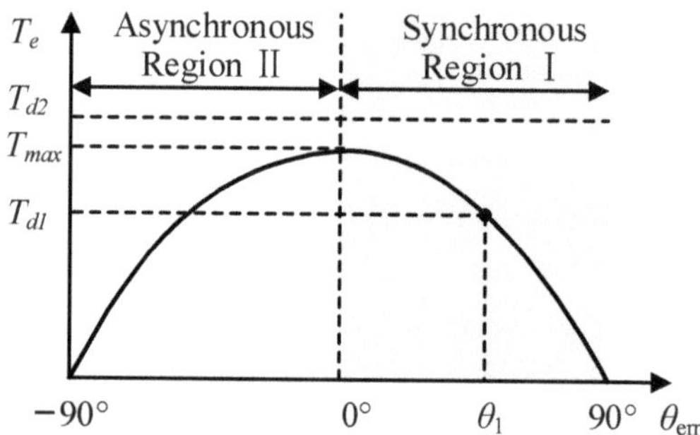

Figure 4. The variation between T_e and θ_{err}.

According to (4), when the electromagnetic torque T_e is balanced with the T_d, θ_{err} maintains constant, and the rotor synchronously follows the current vector. Upon a sudden increase in T_L, the rotor undergoes deceleration, leading to a reduction in θ_{err}. This prompts T_e and T_L to establish a new balance. Maintaining a constant current vector amplitude, as T_L gradually rises, θ_{err} continues to decrease, reaching a minimum value at $\theta_{err} = 0$, signifying synchronization with the δ-axis for the reference current vector. The above characteristics manifest in synchronous region I. Conversely, in asynchronous region II, θ_{err} and T_e have the same trend. Once T_e loses its capacity to match T_d, the motor loses synchronization. In summary, the permissible range of θ_{err} is defined as:

$$0° < \theta_{err} < 90°. \tag{18}$$

4. Proposed Closed-Loop I–f Startup Method with Transition Enhancement

4.1. Frequency Correction Design

According to active power equation, the motor frequency satisfies $\omega_e = P_e / T_e$, where P_e represents the active power. Thus, the derivative of motor frequency is obtained as:

$$\frac{d}{dt}\omega_e = \frac{d}{dt}\left(\frac{P_e}{T_e}\right) \approx \frac{HPF(P_e)}{T_e} \tag{19}$$

with $P_e = 1.5(u_\alpha^* i_\alpha + u_\beta^* i_\beta)$, where u_α^*, u_β^*, i_α, i_β are the voltage commands and currents in the α-β reference coordinate frame.

The prerequisite for the establishment of (19) is that T_e keeps constant under $\theta_{err} = 0$ during ramp up stage. The high pass filter (HPF) is utilized as a differentiator approximately under low frequency. Therefore, the following correction frequency $\Delta\omega_{i1}$ is designed proportional to the derivative of motor frequency in [24]:

$$\Delta\omega_{i1} = -k_1 \frac{d}{dt}\omega_e = -k_1 \frac{HPF(P_e)}{T_{e0}} \tag{20}$$

where $k_1 > 0$ denotes the active power based correction gain, and T_{e0} denotes the initial reference motor torque at startup, which is set as the constant rated value.

However, the reference motor torque changes significantly in the whole startup process due to amplitude compensation (see Section 4.2 for details) and sensorless transition. Therefore, the above frequency correction method is underperforming. The additional correction frequency $\Delta\omega_{i2}$ proportional to the electromagnetic torque variation is designed in this paper to enhance the startup performance:

$$\Delta\omega_{i2} = k_2 \frac{d}{dt} T_{e2} = k_2 HPF(T_{e1}) \tag{21}$$

where $k_2 > 0$ denotes the electromagnetic torque based correction gain, and T_{e1} is the real-time reference motor torque.

Therefore, the whole correction frequency can be expressed as:

$$\omega_i - \omega_{i0} = \Delta\omega_{i1} + \Delta\omega_{i2}. \tag{22}$$

where ω_{i0} is current vector frequency without frequency correction. The whole correction frequency is activated at zero motor speed. The corresponding small-signal equation of (22) can be obtained:

$$\Delta\omega_i = \Delta\omega_{i0} - k_1 \frac{d}{dt} \Delta\omega_e + k_2 \frac{d}{dt} \Delta T_e. \tag{23}$$

Substituting (23) into (11), the angle error θ_{err} with frequency correction can be obtained as:

$$\theta_{err} = \frac{1}{s}[(1 + k_1 s)\Delta\omega_e - \Delta\omega_{i0} - k_2 s \Delta T_e]. \tag{24}$$

With combination of (9), (10), (12), and (24), the closed-loop transfer function from ΔT_L to $\Delta\omega_e$ is renewed with frequency correction as follows:

$$G'(s) = \frac{\Delta\omega_e}{\Delta T_L} = \frac{(k_2 K_\theta - 1)n_p s}{(1 - k_2 K_\theta)Js^2 + B_e s + n_p K_\theta} \tag{25}$$

with $B_e = (1 - k_2 K_\theta)B + n_p k_1 K_\theta \approx n_p k_1 K_\theta (B \ll n_p k_1 K_\theta)$.

Therefore, the new system damping ratio ζ_e can be obtained as follows:

$$\zeta_e = \frac{n_p k_1 K_\theta}{2\sqrt{(1 - k_2 K_\theta)Jn_p K_\theta}}. \tag{26}$$

Compared with system damping ratio ζ_{e0} with only correction frequency $\Delta\omega_{i1}$ as follows [24]:

$$\zeta_{e0} = \frac{n_p k_1 K_\theta}{2\sqrt{Jn_p K_\theta}}, \tag{27}$$

ζ_e is larger with the increase of k_2 in a certain range. Thus, faster convergence with less oscillation can be guaranteed with the increase of system damping ratio.

4.2. Amplitude Compensation Design

Conventional I-f control necessitates setting the reference current amplitude I_m (proportional to reference motor torque) to a relatively large value to ensure that θ_{err} remains well within the stability domain boundaries. This is performed to provide the motor with an ample stabilization margin when encountering disturbances. However, such an approach significantly increases motor losses.

According to (4), T_e generated under the same I_m increases with the rise of $\cos\theta_{err}$. By maintaining $\cos\theta_{err}$ at its maximum value throughout the motor startup process, i.e., $\cos\theta_{err} = 1$ with $\theta_{err} = 0$, the UHS-SPMSM under I-f control can be effectively operated in the MTPA state. This approach reduces I_m, enhancing the motor's overall operating efficiency.

The information containing θ_{err} can be obtained from (2):

$$\omega_e \varphi_f \sin\theta_{err} = -u_\gamma - \omega_i L_q i_\delta \tag{28}$$

where u_γ can be derived from reactive power as follows [36]:

$$u_\gamma = -U\sin\varphi = \frac{1.5(u_\beta^* i_\alpha - u_\alpha^* i_\beta)}{I_m}. \tag{29}$$

It is worth stating that the used reactive power is calculated from voltages and currents in the α-β reference coordinate frame instead of γ-δ ones, which reduces the influence of inaccurate position information on the calculation result. Hence, a PI controller can be devised to regulate I_m with compensation of ΔI_m. For the ease of transition to sensorless FOC, in this paper, the regulation of reference motor torque is utilized with compensation of $\Delta T_{e,ref}$, as illustrated in Figure 5, where $\omega_e \varphi_f \sin\theta_{err}$ serves as the feedback value with a reference set at 0.

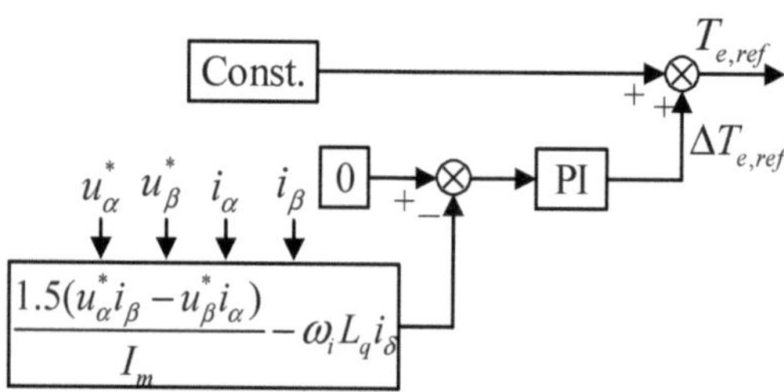

Figure 5. Amplitude compensation design.

Notably, u_α^* and u_β^* can be obtained from the α-β axes voltage commands for SVPWM. i_α and i_β can be obtained from the sampled three-phase currents through Clarke coordinate transformation, and thus I_m is also obtained. Additionally, i_δ can be obtained through Park coordinate transformation based on corrected rotor position information (integral of corrected frequency). L_q is known from the motor manufacturer. In view of the above known signals, $\omega_e \varphi_f \sin\theta_{err}$ can be calculated directly avoiding additional observation. Therefore, the amplitude-compensated PI controller is activated at zero speed. Utilizing the amplitude-compensated PI controller, $\omega_e \varphi_f \sin\theta_{err}$ can converge to 0, signifying the convergence of θ_{err} to 0. Referencing (10), it becomes apparent that $K_\theta = 0$ when $\theta_{err} = 0$. By substituting $K_\theta = 0$ to (25), the proposed I-f control with frequency correction can be likened to a first-order delay system. Therefore, adjustments in the control variables (current and frequency) influence the motor's electromagnetic torque and speed response over time, leading to a dampened response and diminishing speed oscillations effectively.

4.3. Sensorless Transition Enhancement

After transition from I-f startup strategy to EEMF-based sensorless FOC, the estimated rotor position and speed will be utilized to coordinate transformation and as feedback value of speed PI controller, respectively. Taking the first-order low-pass filtering effect for high-frequency noise suppression in digital sampling system into consideration, the closed-loop transfer function of the speed loop can be expressed as follows [37,38]:

$$\begin{aligned} G_s^c(s) &= \frac{\omega_m(s)}{\omega_m^*(s)} = \frac{K_T K_i}{Js^2 + (K_T K_p + B)s + K_T K_i} \\ &= \frac{\omega_{ns}^2}{s^2 + 2\zeta_s \omega_{ns} s + \omega_{ns}^2} \end{aligned} \tag{30}$$

with:

$$\zeta_s = \frac{K_T K_p + B}{2\sqrt{J K_T K_i}}, \quad \omega_{ns} = \sqrt{\frac{K_T K_i}{J}} \tag{31}$$

where K_p and K_i are the proportional and integral gains of speed PI controller; $K_T = 1.5 n_p \varphi_f$ denotes the torque constant; ζ_s and ω_{ns} denote the damping ratio and natural frequency of the speed loop control system.

The amplitude-frequency characteristic of (30) at the closed-loop bandwidth is:

$$M(\omega_b) = \frac{\omega_{ns}^2}{\sqrt{(\omega_{ns}^2 - \omega_b^2)^2 + (2\zeta_s\omega_{ns}\omega_b)}} = \frac{\sqrt{2}}{2}. \tag{32}$$

Therefore, speed loop bandwidth ω_b is derived with ζ_s and ω_{ns}:

$$\omega_b = \omega_{ns}\sqrt{1 - 2\zeta_s^2 + \sqrt{2 - 4\zeta_s^2 + 4\zeta_s^4}}. \tag{33}$$

Without variation of motor parameters, ω_b is only dependent on the control gains of speed PI controller. Generally, UHS-SPMSM operates over a wide frequency range with fast response requirement, a large ω_b is often designed more reasonably with matched constant K_p and K_i. When the UHS-SPMSM operates at low speed, particularly during the sensorless transition stage, lager speed overshoot and increased sensitivity to high-frequency noise result from an excessive ω_b, heightening the risk of transition failure. Therefore, in this paper, a variable bandwidth speed PI controller is designed based on variable PI control gains, which are designed as:

$$\begin{cases} K_p = -2.1176 \times 10^{-9}\omega_m + 2.4776 \times 10^{-4} \\ K_i = 2.2353 \times 10^{-8}\omega_m + 0.0016. \end{cases} \tag{34}$$

Considering the variation trend of K_p and K_i with ω_m ramping up, it is evident that ζ_s decreases and ω_{ns} varies oppositely. From (33), with ω_{ns} remaining constant, ω_b monotonically increases with the decrease in ζ_s. In summary, the design of (34) prompts the increasing trend of ω_b during motor speedup in the whole speed range. This mitigates the risk of transition failure from large speed oscillation after enabling speed PI controller. Hereto, the overall schematic diagram of the proposed I-f startup method with transition to FOC is illustrated as Figure 6.

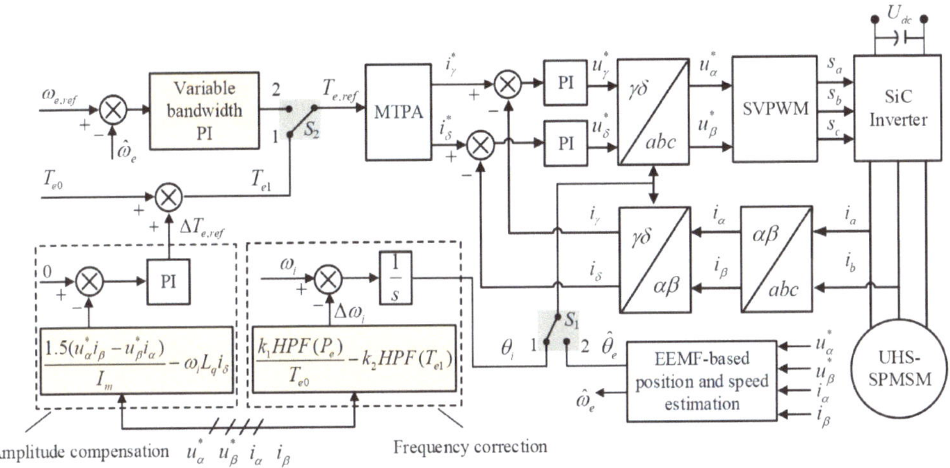

Figure 6. The overall schematic diagram of the proposed I-f startup method, including transition to FOC. The two switches (S_1 and S_2) are connected to terminal 1 for the proposed I-f control and 2 for sensorless FOC (where MTPA denotes maximum torque per ampere and SVPWM denotes space vector pulse width modulation, and EEMF-based position and speed estimation can be found in our previous research [35]).

5. Experimental Results

In order to validate the proposed startup method, the experiments are conducted on the test platform as shown in Figure 7, mainly including a test UHS-SPMSM, a homemade inverter, and some other ancillary equipment for control of cooling system, measuring system, and power supply system. The parameters of this UHS-SPMSM drive system are given in Table 1. For the homemade inverter, *Wolfspeed* CAB011M12FM3 Half-Bridge SiC power modules are designed as power switches and microprocessor TMS320F28335 DSP (provided by *Texas Instruments*) is designed as the control board. The switching frequency is 40 kHz and the sampling frequency is 20 kHz.

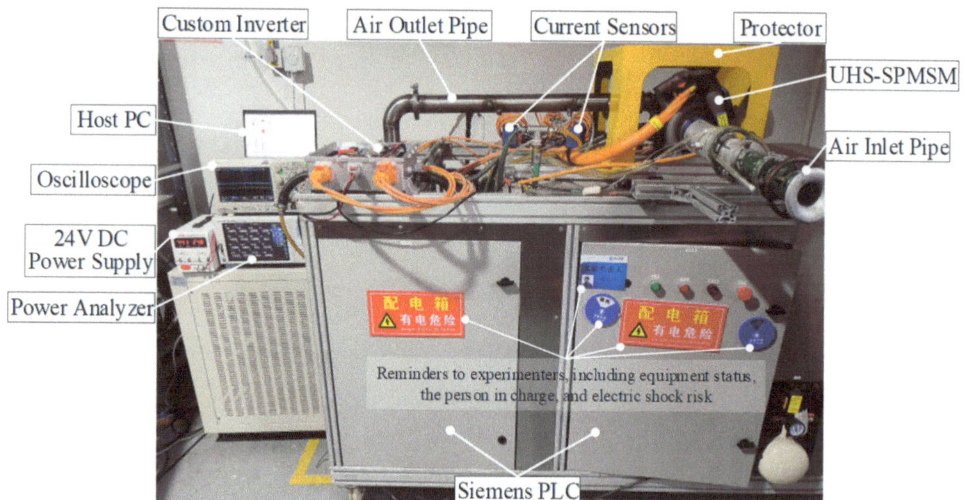

Figure 7. Overall experimental system test bench.

Table 1. Parameters of the UHS-SPMSM drive system.

Symbol	Parameter	Value
P_N	Rated power	35 kW
$\omega_{e,rated}$	Rated speed	9245 rad/s (90 kr/min)
$\omega_{e,max}$	Maximum speed	9948 rad/s (95 kr/min)
$\omega_{e,min}$	Minimum speed	3142 rad/s (30 kr/min)
ω_N	Fundamental frequency	1.583 kHz
n_p	Pole pairs	1
R_s	Stator resistance	0.0085 Ω
L_d	d-axis inductance	66.46 μH
L_q	q-axis inductance	66.46 μH
φ_f	Flux linkage	0.02387 Wb
J	Motor inertia	0.0005672 kg·m^2
U_{DC}	DC bus voltage	550 V
f_{sw}	Switching frequency	40 kHz
f_c	Control frequency	20 kHz
T_s	Sampling period	0.00005 s

The main control parameters used are: the PI controllers for regulating $\gamma\delta$-axes currents are $K_{p,cur} = 0.7$ and $K_{i,cur} = 0.002$, and for regulating motor speed (without variable bandwidth design) are $K_{p,spd} = 0.0000095$ and $K_{i,spd} = 0.0095$; for frequency correction design, the correction gains are designed as $k_1 = 3$ and $k_2 = 35$; for amplitude compensation design, the PI control parameters are $K_{p,amp} = 0.000006$ and $K_{i,amp} = 0.005$. All the results presented in this section are analyzed via CAN communication. That is, $\gamma\delta$-axes currents and motor speed are logged, parsed, and plotted via *Busmaster*, *CANoe*, and *Origin*, respectively. After transition to FOC, EEMF-based sensorless control method proposed in our previous research [35] works. While the estimated position may be distorted at low speeds, the estimated speed remains precise, effectively representing the actual motor speed without reliance on mechanical speed measurement.

In fact, without a position sensor installation for the UHS-SPMSM, the actual *dq*-axes currents cannot be obtained in the experiments due to lack of real rotor position information. Therefore, the $\gamma\delta$-axes currents are used in this section for effectiveness validation of the proposed method. On one hand, before the sensorless transition, the $\gamma\delta$-axes currents are derived from the position θ_i in the γ-δ reference coordinate frame. On the other hand, after the sensorless transition, the $\gamma\delta$-axes currents are derived from the position $\hat{\theta}_e$ in the *d-q* reference coordinate frame (assuming that $\theta_{err} \approx 0$, γ-δ reference coordinate frame is aligned with *d-q* one).

5.1. Startup Performance Comparison with Conventional I–f Startup Strategy

To verify the superior startup performance of the proposed method compared with the conventional open-loop I-f control, the UHS-SPMSM is started to 7000 r/min from a standstill.

Figure 8 shows the startup performance of the conventional I-f control. The δ-axis reference current i_δ^* for the I-f control is generally chosen to be a large value for a reliable startup with large output torque, and $i_\delta^* = 70$A (80% rated current) is used here. In Figure 8a, the actual δ-axis current i_δ converges to the initial set value of reference current under the whole startup process, and oscillates with a maximum amplitude of 9.8 A in steady state. In addition, even the actual γ-axis current converges to 0 A under current loop PI control, the fluctuations are noticeable with an average amplitude over 10 A. This implies that an obvious position error exists between the γ-δ and *d-q* reference coordinate frame without amplitude compensation of reference current vector, which is not desired. Furthermore, under conventional I-f control, motor speed fluctuations occur both in dynamic process and steady-state operation, and the root mean square errors (RMSEs) are 296 r/min and 348 r/min, respectively.

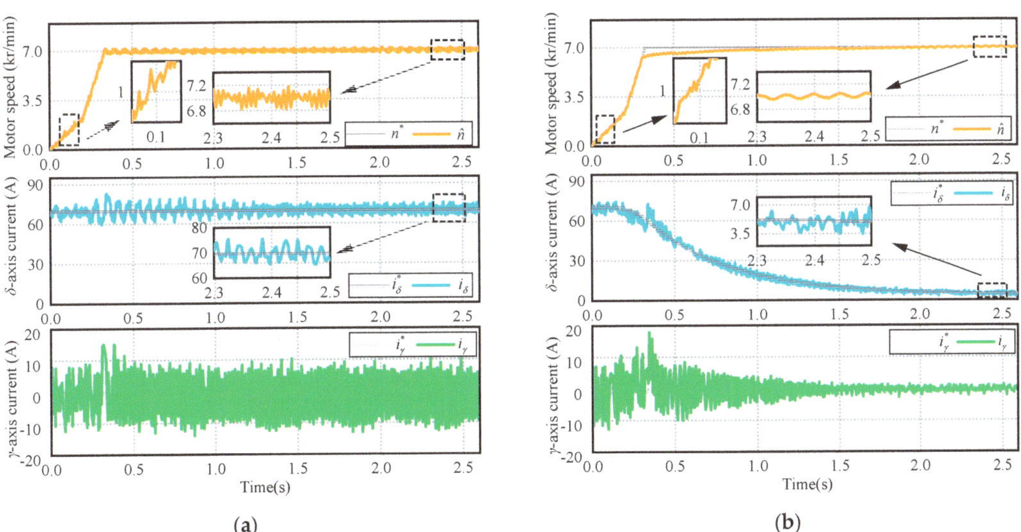

Figure 8. Startup performance from standstill to 7000 r/min. (**a**) Conventional I-f control. (**b**) Proposed closed-loop I-f control.

Figure 8b shows the startup performance of the proposed I-f control. The initial δ-axis reference current i_δ^* is set the same as the conventional I-f control; however, i_δ^* decreases gradually with motor speedup. Only 5.1 A δ-axis current can maintain motor's steady-state operation; compared with a conventional I-f control, it decreased 92.7%. This is due to the designed amplitude compensation of reference current. Although the effect of amplitude compensation is not so obvious at zero speed, the preliminary calculation of reactive power after motor startup is also helpful for the subsequent adjustment of amplitude compensator with motor speedup. Notably, the steady-state current oscillation of i_δ is significantly decreased with maximum amplitude of only 3.2 A, which decreased 67.3% compared with the conventional I-f control. Under proposed I-f control, stable $i_\gamma = 0$ operation is achieved easily, which leads to higher motor efficiency avoiding too large current amplitude fluctuations. Except slightly poorer dynamic speed tracking performance, the proposed method also has smaller dynamic and steady-state speed fluctuation of just 130 r/min and 78 r/min, respectively.

It is worth noting that to ensure fast startup, in practical engineering applications, the I-f control is only used to drive the UHS-SPMSM from standstill to a certain speed, and then transitions to the closed-loop FOC during the acceleration process. In other words, the I-f control does not enable under steady-state operating conditions. The transition process will be demonstrated in the Section 5.2. Therefore, the proposed I-f control is unaffected from the slightly poor dynamic convergence performance in practical applications.

5.2. Transition Performance Comparison from I–f Startup to Sensorless FOC

To ensure the fast startup ability of the UHS-SPMSM, the transition from I-f control to sensorless FOC is usually accomplished during the acceleration process, rather than in the steady-state condition. The transition speed should be set as low as possible under the premise of position estimation accuracy of the sensorless control. Generally, the UHS-SPMSM is only operated above the idle speed (30,000 r/min for the test motor) in the actual working condition for extending the life of air bearing. Therefore, Figure 9 shows the transition performance comparison, where the UHS-SPMSM is started from a standstill to the idle speed, and the transition speed is set to 12,000 r/min.

Figure 9a shows the transition performance of conventional I-f control, where the speed PI controller is set fixed-gain. During the conventional I-f control stage, the actual δ-axis current converges poorly to the constant reference δ-axis current. The $\gamma\delta$-axes current tracking error increases with the gradual rise of motor speed, which indicates that the used rotor position gained from I-f control exists, growing the position error as the motor speed increases. This leads to considerable current amplitude and the increasing possibility of motor shutdown due to the overcurrent protection trigger. In addition, significant speed oscillation is evident. This is due to the poor damping as analyzed in Section 3. At the transition instant of $t = 0.46$ s, the reference δ-axis current has an obvious pulse due to position mismatch between the I-f control and sensorless control, which may lead to transition failure. After successful transition to the sensorless FOC, the actual motor speed has a large overshoot in the process of tracking the reference speed. The dynamic overshoot gradually increases, and the steady-state overshoot value reaches 3984 r/min.

Figure 9b shows the transition performance of the proposed closed-loop I-f control, where the speed PI controller is also set fixed-gain. During the proposed I-f control stage, the δ-axis current gradually decreases with motor speedup. This is attributed to the amplitude compensation design. In other words, the UHS-SPMSM can achieve the same motor speed under the proposed I-f control at a smaller current, which is conductive to improve the motor efficiency. Meanwhile the $\gamma\delta$-axes current fluctuation and speed oscillation are significantly reduced compared to the conventional I-f control, which confirms the damping effect of the frequency correction. The UHS-SPMSM can instantly switch to the sensorless FOC without any reference δ-axis current pulses, which indicates more precision in used rotor position and contributes to smoother transition.

During initial sensorless FOC stage, the δ-axis current showed a certain extent of fluctuation, which brought the simultaneous speed fluctuation. Compared with the conventional I-f control, the average δ-axis current from the transition instant to the steady state is significantly reduced, which also reflects the improvement of the motor efficiency, and its steady-state speed overshoot value is very small, only 2204 r/min. It is worth noting that the fixed-gain speed PI controller, after transition, featuring in a large bandwidth of effectiveness in the whole speed range, holds redundant bandwidth at the transition speed of 12,000 r/min. This causes the speed and current fluctuation.

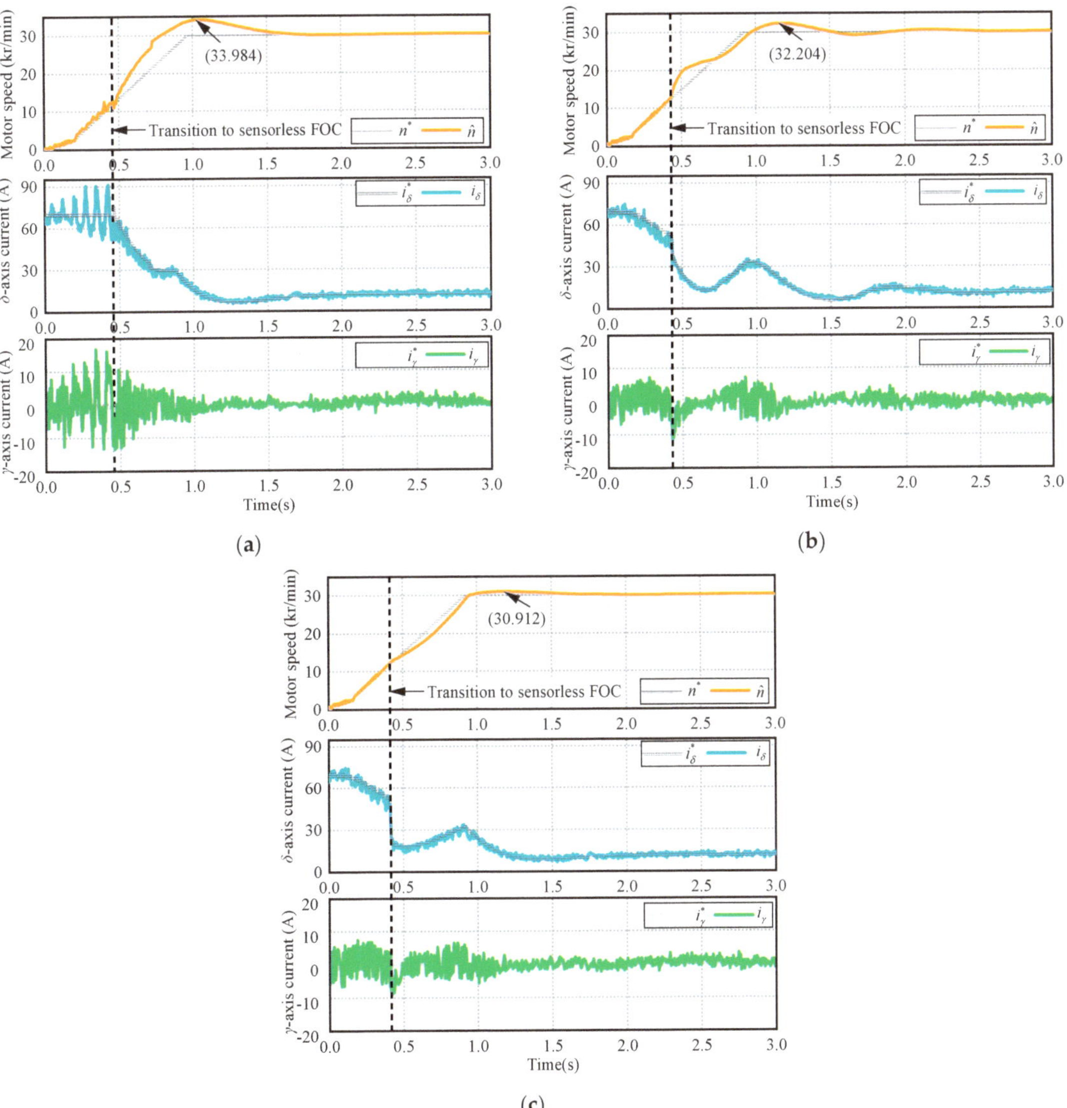

Figure 9. Startup performance from standstill to 30,000 r/min with transition from I–f control to sensorless FOC at 12,000 r/min. (**a**) Conventional I-f control with fixed-gain speed PI controller. (**b**) Proposed closed-loop I-f control with fixed-gain speed PI controller. (**c**) Proposed closed-loop I-f control with variable bandwidth speed PI controller.

Figure 9c shows the transition performance of the proposed closed-loop I-f control with variable bandwidth speed PI controller. Clearly, the δ-axis current and speed fluctuation are significantly reduced after transition, and the steady-state speed overshoot is reduced to 912 r/min. Sensorless transition enhancement obtains a smooth transition compared with fixed-bandwidth speed PI controller. Therefore, transition failure can be avoided for high startup reliability.

6. Conclusions

In this article, a reliable and efficient I-f control method with transition enhancement is introduced for UHS-SPMSMs driving fuel cell air compressors. The startup reliability and transition smoothness are visibly enhanced. The main conclusions are summarized as follows:

(1) During the speed up stage, the proposed frequency correction of reference current vector considering the active power and motor torque simultaneously, which reduces the speed oscillation significantly, while decreasing the possibility of startup failure.

(2) The proposed amplitude compensator of reference current vector obviously reduces the current amplitude under the same startup condition, which ensures MTPA operation. This characteristic improves the motor efficiency significantly and prerequisites smoother transition from I-f startup to sensorless FOC due to minimizing the current vector angle.

(3) After the transition to sensorless FOC with an enabling speed PI controller, the proposed variable bandwidth scheme ensures reduction of current amplitude and speed fluctuation, avoiding transition failure.

Generally, the contributions investigated in this paper can be extendedly applied to UHS-SPMSM driving gas turbines, electric turbochargers, flywheels for energy storage, and turbo-electric distributed propulsion system for aircraft, rather than merely driving fuel cell air compressors.

Author Contributions: Conceptualization, J.X.; Data curation, Y.X.; Formal analysis, Y.L.; Funding acquisition, J.Z.; Investigation, J.X.; Methodology, Y.X.; Project administration, J.X.; Resources, J.Z.; Software, Y.X.; Supervision, J.Z.; Validation, Y.X.; Visualization, X.J.; Writing—original draft, Y.X.; Writing—review and editing, J.X. and Y.X. All authors have read and agreed to the published version of the manuscript.

Funding: This research was funded by the National Key R&D Program of China sponsored by Ministry of Science and Technology of the People's Republic of China (MoST), grant number 2022YFB25031013.

Data Availability Statement: The data that support the findings of this study are available on request from the corresponding author. The data are not publicly available due to privacy restrictions.

Conflicts of Interest: The authors declare no conflicts of interest.

References

1. Zhang, J.; Peng, F.; Huang, Y.; Yao, Y.; Zhu, Z. Online Inductance Identification Using PWM Current Ripple for Position Sensorless Drive of High-Speed Surface-Mounted Permanent Magnet Synchronous Machines. *IEEE Trans. Ind. Electron.* **2022**, *69*, 12426–12436. [CrossRef]

2. Zhang, Q.; Yu, R.; Li, C.; Chen, Y.-H.; Gu, J. Servo Robust Control of Uncertain Mechanical Systems: Application in a Compressor/PMSM System. *Actuators* **2022**, *11*, 42. [CrossRef]

3. Looser, A.; Kolar, J.W. An Active Magnetic Damper Concept for Stabilization of Gas Bearings in High-Speed Permanent-Magnet Machines. *IEEE Trans. Ind. Electron.* **2014**, *61*, 3089–3098. [CrossRef]

4. Kim, J.H.; Kim, D.M.; Jung, Y.H.; Lim, M.S. Design of Ultra-High-Speed Motor for FCEV Air Compressor Considering Mechanical Properties of Rotor Materials. *IEEE Trans. Energy Convers.* **2021**, *36*, 2850–2860. [CrossRef]

5. Xu, Y.; Lin, C.; Xing, J.; Li, X. Extended State Observer-Based Position Sensorless Control for Automotive Ultra-high-Speed PMSM. In *Proceedings of the China SAE Congress 2022: Selected Papers*; Springer: Singapore, 2023; pp. 787–799.

6. Chi, W.-C.; Cheng, M.-Y. Implementation of a sliding-mode-based position sensorless drive for high-speed micro permanent-magnet synchronous motors. *ISA Trans.* **2014**, *53*, 444–453. [CrossRef]

7.   Zhang, Q.; Yang, D.; Kong, Q.; Sun, X.; Hu, Z. A super-high-speed PM motor drive for centrifugal air compressor used in fuel cell unmanned aerial vehicle. *IET Electr. Power Appl.* **2023**, *17*, 1459–1468. [CrossRef]
8.   Niedermayr, P.; Alberti, L.; Bolognani, S.; Abl, R. Implementation and Experimental Validation of Ultrahigh-Speed PMSM Sensorless Control by Means of Extended Kalman Filter. *IEEE J. Emerging Sel. Top. Power Electron.* **2022**, *10*, 3337–3344. [CrossRef]
9.   Hu, M.; Yu, W.; Lei, J.; Wu, Z.; Hua, W.; Hu, Y. Sensorless Control of a High-Speed PMSM with Rapid Acceleration for Air Compressors using a High-order Extended State Observer. In Proceedings of the 2021 IEEE Energy Conversion Congress and Exposition (ECCE), Virtual Conference, 10–14 October 2021; pp. 4781–4787.
10.   Du, B.; Wu, S.; Han, S.; Cui, S. Application of Linear Active Disturbance Rejection Controller for Sensorless Control of Internal Permanent-Magnet Synchronous Motor. *IEEE Trans. Ind. Electron.* **2016**, *63*, 3019–3027. [CrossRef]
11.   Griffo, A.; Drury, D.; Sawata, T.; Mellor, P.H. Sensorless starting of a wound-field synchronous starter/generator for aerospace applications. *IEEE Trans. Ind. Electron.* **2012**, *59*, 3579–3587. [CrossRef]
12.   Looser, A.; Tüysüz, A.; Zwyssig, C.; Kolar, J.W. Active Magnetic Damper for Ultrahigh-Speed Permanent-Magnet Machines with Gas Bearings. *IEEE Trans. Ind. Electron.* **2017**, *64*, 2982–2991. [CrossRef]
13.   Zhao, L.; Ham, C.H.; Han, Q.; Wu, T.X.; Zheng, L.; Sundaram, K.B.; Kapat, J.; Chow, L. Design of optimal digital controller for stable super-high-speed permanent-magnet synchronous motor. *IEE Proc. Electr. Power Appl.* **2006**, *153*, 213–218. [CrossRef]
14.   Ancuti, R.; Boldea, I.; Andreescu, G.-D. Sensorless V/f control of high-speed surface permanent magnet synchronous motor drives with two novel stabilising loops for fast dynamics and robustness. *IET Electr. Power Appl.* **2010**, *4*, 149–157. [CrossRef]
15.   Tang, Z.; Li, X.; Dusmez, S.; Akin, B. A New V/f-Based Sensorless MTPA Control for IPMSM Drives. *IEEE Trans. Power Electron.* **2016**, *31*, 4400–4415. [CrossRef]
16.   Datta, S.; Chandra, A.; Chowdhuri, S. High performance sensor-less V/f control of surface PMSM in voltage vector plane with ZVV injection and SMO-based position estimation method. *Electr. Eng.* **2022**, *104*, 657–666. [CrossRef]
17.   Xu, Y.; Lin, C.; Xing, J.; Zeng, Q.; Sun, J. I-f Starting Rapid and Smooth Transition Method of Full-Speed Sensorless Control for Low Current Harmonic Ultra-high-speed PMSM. In Proceedings of the 2022 IEEE Applied Power Electronics Conference and Exposition (APEC), Houston, TX, USA, 20–24 March 2022; pp. 1820–1826.
18.   Xing, J.; Qin, Z.; Lin, C.; Jiang, X. Research on Startup Process for Sensorless Control of PMSMs Based on I-F Method Combined with an Adaptive Compensator. *IEEE Access* **2020**, *8*, 70812–70821. [CrossRef]
19.   Haichao, F.; Boyang, S.; Lizhen, G. A closed-loop I/f vector control for permanent magnet synchronous motor. In Proceedings of the 2017 9th International Conference on Modelling, Identification and Control (ICMIC), Kunming, China, 10–12 July 2017; pp. 965–969.
20.   Xu, G.; Zhao, F.; Liu, T. I–f closed-loop starting strategy of high-speed PMSM based on current vector adaptive regulation. *IET Power Electron.* **2023**, *16*, 2724–2738. [CrossRef]
21.   Liu, J.; Nondahl, T.A.; Schmidt, P.B.; Royak, S.; Rowan, T.M. Generalized Stability Control for Open-Loop Operation of Motor Drives. *IEEE Trans. Ind. Appl.* **2017**, *53*, 2517–2525. [CrossRef]
22.   Liu, J.; Nondahl, T.A.; Dai, J.; Royak, S.; Schmidt, P.B. A Seamless Transition Scheme of Position Sensorless Control in Industrial Permanent Magnet Motor Drives with Output Filter and Transformer for Oil Pump Applications. *IEEE Trans. Ind. Appl.* **2020**, *56*, 2180–2189. [CrossRef]
23.   Yu, Y.; Chang, D.; Zheng, X.; Mi, Z.; Li, X.; Sun, C. A Stator Current Oriented Closed-Loop I--f Control of Sensorless SPMSM with Fully Unknown Parameters for Reverse Rotation Prevention. *IEEE Trans. Power Electron.* **2018**, *33*, 8607–8622. [CrossRef]
24.   Song, Z.; Yao, W.; Lee, K.; Li, W. An Efficient and Robust I-f Control of Sensorless IPMSM with Large Startup Torque Based on Current Vector Angle Controller. *IEEE Trans. Power Electron.* **2022**, *37*, 15308–15321. [CrossRef]
25.   Chen, D.; Lu, K.; Wang, D. An I-f Startup Method for Back-EMF based Sensorless FOC of PMSMs with Improved Stability During the Transition. In Proceedings of the 2020 International Symposium on Industrial Electronics and Applications (INDEL), Banja Luka, Bosnia and Herzegovina, 4–6 November 2020; pp. 1–6.
26.   Consoli, A.; Scelba, G.; Scarcella, G.; Cacciato, M. An Effective Energy-Saving Scalar Control for Industrial IPMSM Drives. *IEEE Trans. Ind. Electron.* **2013**, *60*, 3658–3669. [CrossRef]
27.   Chen, D.; Lu, K.; Wang, D.; Hinkkanen, M. I-F Control with Zero D-Axis Current Operation for Surface-Mounted Permanent Magnet Synchronous Machine Drives. *IEEE Trans. Power Electron.* **2023**, *38*, 7504–7513. [CrossRef]
28.   Nair, S.V.; Hatua, K.; Prasad, N.V.P.R.D.; Reddy, D.K. A Quick I-f Starting of PMSM Drive with Pole Slipping Prevention and Reduced Speed Oscillations. *IEEE Trans. Ind. Electron.* **2021**, *68*, 6650–6661. [CrossRef]
29.   Na, H.; Cho, K.-Y.; Kim, H.-W.; Kim, S.-H. Stable sensorless transition algorithm for PM synchronous motors under load. *J. Power Electron.* **2024**, *24*, 608–617. [CrossRef]
30.   Ning, B.; Zhao, Y.; Cheng, S. An Improved Sensorless Hybrid Control Method of Permanent Magnet Synchronous Motor Based on I/F Startup. *Sensors* **2023**, *23*, 635. [CrossRef] [PubMed]
31.   Wang, Z.; Lu, K.; Blaabjerg, F. A Simple Startup Strategy Based on Current Regulation for Back-EMF-Based Sensorless Control of PMSM. *IEEE Trans. Power Electron.* **2012**, *27*, 3817–3825. [CrossRef]
32.   Tang, Q.; Chen, D.; He, X. Integration of Improved Flux Linkage Observer and I–f Starting Method for Wide-Speed-Range Sensorless SPMSM Drives. *IEEE Trans. Power Electron.* **2020**, *35*, 8374–8383. [CrossRef]
33.   Liu, L.; Jin, D.; Si, J.; Liang, D. A novel nonsingular fast terminal sliding mode observer combining I-F method for wide-speed sensorless control of PMSM drives. *IET Power Electron.* **2023**, *16*, 843–855. [CrossRef]

34. Fatu, M.; Teodorescu, R.; Boldea, I.; Andreescu, G.D.; Blaabjerg, F. I-F starting method with smooth transition to EMF based motion-sensorless vector control of PM synchronous motor/generator. In Proceedings of the 2008 IEEE Power Electronics Specialists Conference, Rhodes, Greece, 15–19 June 2008; pp. 1481–1487.
35. Xu, Y.; Lin, C.; Xing, J. Transient Response Characteristics Improvement of Permanent Magnet Synchronous Motor Based on Enhanced Linear Active Disturbance Rejection Sensorless Control. *IEEE Trans. Power Electron.* **2023**, *38*, 4378–4390. [CrossRef]
36. Chen, D.; Lu, K.; Wang, D. An I-f Startup Method with Compensation Loops for PMSM with Smooth Transition. *IEEJ J. Ind. Appl.* **2020**, *9*, 263–270. [CrossRef]
37. Du, C.Y.; Yu, G.R. Optimal PI Control of a Permanent Magnet Synchronous Motor Using Particle Swarm Optimization. In Proceedings of the Second International Conference on Innovative Computing, Informatio and Control (ICICIC 2007), Kumamoto, Japan, 5–7 September 2007; p. 255.
38. Choi, J.W.; Lee, S.C. Antiwindup Strategy for PI-Type Speed Controller. *IEEE Trans. Ind. Electron.* **2009**, *56*, 2039–2046. [CrossRef]

Article

Robustness Improved Method for Deadbeat Predictive Current Control of PMLSM with Segmented Stators

Shijie Gu [†], Peng Leng [†], Qiang Chen, Yuxin Jin, Jie Li and Peichang Yu *

College of Intelligence Science and Technology, National University of Defense Technology, Changsha 410073, China; gusj2001@nudt.edu.cn (S.G.); lengpeng17@nudt.edu.cn (P.L.); chenqiang08@nudt.edu.cn (Q.C.); jinyuxin18@nudt.edu.cn (Y.J.); jieli@nudt.edu.cn (J.L.)
* Correspondence: yupeichang@nudt.edu.cn
[†] These authors contributed to the work equally and should be regarded as co-first authors.

Abstract: Permanent magnet linear synchronous motors (PMLSMs) with stator segmented structures are widely used in the design of high-power propulsion systems. However, due to the inherent delay and segmented structure of the systems, there are parameter disturbances in the inductance and flux linkage of the motors. This makes the deadbeat predictive current control (DPCC) algorithm for a current loop less robust in the control system, leading to a decrease in control performance. Compensation methods such as compensation by observer and online estimation of parameters, are problematic to apply in practice due to the difficulty of parameter adjustment and the high complexity of the algorithm. In this paper, a robustness-improved incremental DPCC (RII-DPCC) method—which uses incremental DPCC (I-DPCC) to eliminate flux linkage parameters—is proposed. The stability of the current loop was evaluated through zero-pole analysis of the discrete transfer function. Current feedforward was introduced to improve the stability of I-DPCC. The inductance stability range of I-DPCC was increased from 0.8–1.25 times to 0–2 times the actual value, and the theoretical stability range was increased more than 4 times, effectively improving the robustness of the predictive model to flux linkage and inductance parameters. Finally, the effectiveness of the proposed method was verified through numerical simulation and experiment.

Keywords: electromagnetic propulsion; PMLSM; incremental model; DPCC; parameter robustness

Citation: Gu, S.; Leng, P.; Chen, Q.; Jin, Y.; Li, J.; Yu, P. Robustness Improved Method for Deadbeat Predictive Current Control of PMLSM with Segmented Stators. *Actuators* **2024**, *13*, 300. https://doi.org/ 10.3390/act13080300

Academic Editor: Dong Jiang

Received: 11 July 2024
Revised: 31 July 2024
Accepted: 5 August 2024
Published: 6 August 2024

1. Introduction

PMLSMs have a high thrust weight ratio, high precision and fast dynamic response [1], and have advantages over induction motors in the design of high-power electromagnetic propulsion systems [2]. In order to solve the problems of high electromagnetic loss caused by excessive phase resistance and inductance under long stroke, the high-capacity requirements of energy storage inverter systems and the low overall operation efficiency caused by large instantaneous driving power, the structure of segmented stators is generally adopted. By dividing the stator winding into several power units, and switching the power supply between sections of the inverter system using the section switch, a better balance of the design cost, operation loss and work efficiency of the electromagnetic propulsion system is achieved [3].

To ensure the dynamic response of the current loop at high speeds, scholars have introduced DPCC in addition to the conventional PI current control. This method makes it easy to adjust parameters and to track the reference current in two control cycles, which realizes the improvement of control bandwidth and dynamic performance. However, conventional DPCC (C-DPCC) is more sensitive to motor parameter disturbances and system delay. When there are mismatches and delays, steady-state errors and current harmonic interference may occur in the current loop, and it may even become unstable in severe cases [4]. Due to the structure of the segmented stator, the stator inductance and mover flux linkage parameters of PMLSM are inevitably mismatched when the rotor crosses the boundary between adjacent segments [5]. In order to reduce the influence of motor parameter mismatch on predictive control performance, it is necessary to improve the C-DPCC algorithm.

Scholars have conducted a great deal of research—generally in three areas—in connection with the robustness of predictive control to parameter mismatch. The first area of research regards parameter mismatch as a disturbance, realizes disturbance compensation through observer, and reduces the dependence on the exact parameters of the model. The disadvantage of this research is that the observer parameters are difficult to adjust, and the complexity of the control system is thus greatly increased. Many scholars regard the parameter mismatch, external disturbances and other factors as aggregate perturbations of the system, and achieve better anti-perturbation performance by designing a sliding mode observer [6,7], an internal mode perturbation observer [8,9], a Luenberger perturbation observer [10,11], etc., for perturbation compensation. The second area of research focuses on online identification of motor parameters and real-time correction of model parameters, such as parameter identification by the parameter adaptive observer [12,13], recursive least-squares-based method [14,15] and inductance extraction method based on a sliding mode perturbation observer [16], etc. However, due to the under rank problem of PMSM [17,18], only part of the parameters can be identified. The third area of research seeks to improve the DPCC algorithm itself [19–21], typically by establishing an incremental model, which is able to eliminate the flux linkage parameter to achieve better parameter robustness, but which leads to a smaller stabilization range of the inductance parameter. In [19], a method was proposed to replace the feedback current in the voltage equation with a feedforward current, which broadened the stability domain, but did not achieve full stability. Other research [20] improved the dynamic performance and robustness under heavy load conditions by extending the single step prediction of the traditional DPCC algorithm to multi-step prediction of transient processes. However, multi-step calculations increase the computational complexity. A current error feedforward method combining an inductance correction algorithm was proposed in [21] to improve the robustness of the IPMSM current loop. However, too many coefficients require adjusting, resulting in high complexity in practice.

In order to solve the problems of inherent delay and segmented structure leading to a disturbance of the inductance and flux linkage parameters and degradation of the current loop performance, an I-DPCC algorithm considering delay compensation was proposed on the basis of the C-DPCC algorithm, which eliminates flux linkage in the predictive model. A stable range of inductance for the I-DPCC method was achieved by theoretical analysis of the current loop transfer function. The transfer function of the current loop and the stable range of its inductance parameters considering the feedforward weight factor were analyzed, based on the current feedforward method. The inductance stability range of I-DPCC increased from 0.8–1.25 times to 0–2 times the actual value, and the theoretical stability range increased more than 4 times, effectively improving the robustness of the predictive model while avoiding the complexity of designing observer and identification algorithms. The current integral compensation method was adopted to address the steady-state error caused by parameter disturbances. Finally, the feasibility of the method was verified through simulation and experiments.

2. PMLSM Modeling for Segmented Designs

The stator voltage equation for PMLSM in the dq coordinate system is as follows [22]:

$$\begin{cases} u_d = Ri_d + L_d \frac{di_d}{dt} - \omega_e L_q i_q \\ u_q = Ri_q + L_q \frac{di_q}{dt} + \omega_e \left(L_d i_d + \psi_f \right) \end{cases}, \tag{1}$$

where u_d, u_q are the dq-axis components of stator voltage; i_d, i_q are the dq-axis components of stator current; L_d, L_q are the dq-axis components of stator inductance; R is the stator resistance; ω_e is the electric angular velocity; ψ_f is the flux linkage of the permanent magnet. For the surface-mounted PMLSM used in this paper, it was taken that $L_d = L_q = L$.

To implement the predictive current control algorithm in a digital system, the above stator voltage equation was discretized using the forward Eulerian method. The discretized stator voltage predictive model for PMLSM is described as follows:

$$\begin{cases} u_{\mathrm{d}}(k) = \left(R - \frac{L}{T}\right)i_{\mathrm{d}}(k) + \frac{L}{T}i_{\mathrm{d}}(k+1) - \omega_{\mathrm{e}}(k)Li_{\mathrm{q}}(k) \\ u_{\mathrm{q}}(k) = \left(R - \frac{L}{T}\right)i_{\mathrm{q}}(k) + \frac{L}{T}i_{\mathrm{q}}(k+1) + \omega_{\mathrm{e}}(k)[Li_{\mathrm{d}}(k) + \psi_{\mathrm{f}}] \end{cases} \tag{2}$$

where T denotes the control period and k denotes the control moment.

The electromagnetic propulsion system targeted in this article adopts a long stator air-core PMLSM structure, and its stator adopts a segmented stator design, as shown in Figure 1.

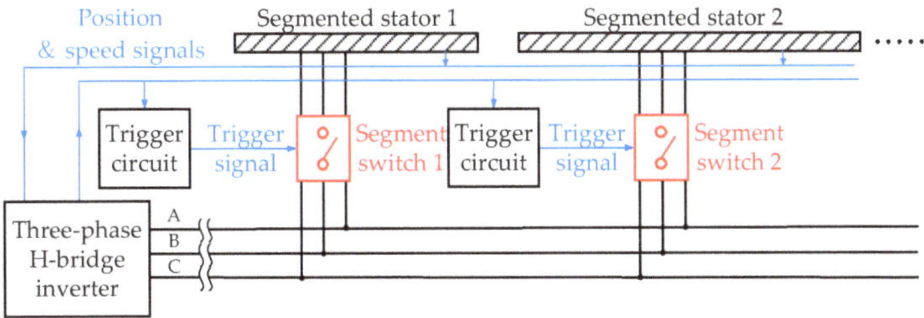

Figure 1. Segmented stator structure for an air-core PMLSM.

Considering the stator segmented design and operating conditions, the actual values of PMLSM model parameters $\left\{R, L, \psi_f\right\}$ are inevitably disturbed during operation, which affects the accuracy of the predictive model and causes a decrease in the control performance of the DPCC algorithm [3]. The subscripts $_0$ in this article represent the nominal values of the motor parameters used in the predictive model. If not, they represent the actual values.

3. Incremental Models for Flux Linkage Perturbation and Delay Optimization

3.1. DPCC with Delay Compensation

The principle of DPCC is that the predicted current at the next moment can track the present reference current, i.e., $i_d(k+1) = i_d^{ref}(k)$ and $i_q(k+1) = i_q^{ref}(k)$, thus the predicted voltage value at kth moment can be calculated via the predictive model. Through Equation (2), the predictive equation for stator current can be written as:

$$\begin{cases} i_d(k+1) = \left(1 - \frac{RT}{L}\right)i_d(k) + \frac{T}{L}u_d(k) + \omega_e(k)Ti_q(k) \\ i_q(k+1) = \left(1 - \frac{RT}{L}\right)i_q(k) + \frac{T}{L}u_q(k) - \omega_e(k)\left[Ti_d(k) + \frac{T}{L}\psi_f\right] \end{cases} \tag{3}$$

In controllers such as DSP, a one-step delay compensation of DPCC is required due to the delay caused by the PWM modulation link, sampling link, etc. The reference voltage at the $k+1$th moment should be calculated at the kth moment, i.e., $u_d(k+1) = u_d^*(k+1)$ and $u_q(k+1) = u_q^*(k+1)$. The stator current at the $k+2$th moment is regarded as the reference current at the kth moment, so that the output current tracks the reference current in two control cycles, i.e., $i_{\mathrm{d}}(k+2) = i_{\mathrm{d}}^{ref}(k)$ and $i_q(k+2) = i_q^{ref}(k)$. A timing diagram of DPCC considering delay compensation is shown in Figure 2.

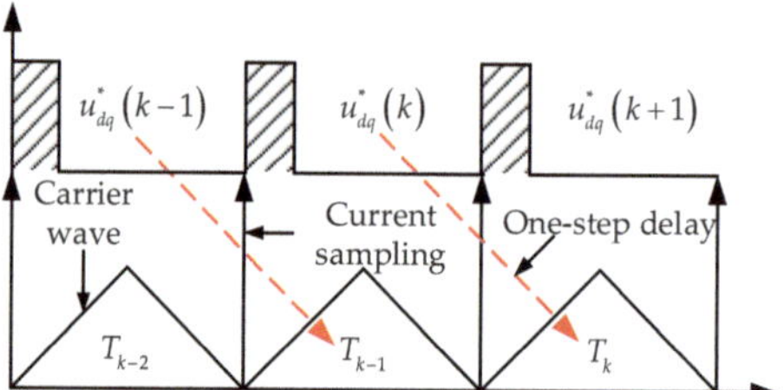

Figure 2. Timing diagram of DPCC with delay compensation.

It is assumed that the electrical angular velocity remains constant during the two adjacent control periods due to the short control period. Therefore, from Equations (2) and (3), the predictive equation of C-DPCC after one-step delay compensation is obtained as:

$$\begin{cases} i_d^p(k+1) = \left(1 - \frac{R_0 T}{L_0}\right) i_d(k) + \frac{T}{L_0} u_d^*(k) + \omega_e T_0 i_q(k) \\ i_q^p(k+1) = \left(1 - \frac{R_0 T}{L_0}\right) i_q(k) + \frac{T}{L_0} u_q^*(k) - \omega_e \left[T i_d(k) + \frac{T}{L_0} \psi_{f0} \right] \end{cases} \tag{4}$$

$$\begin{cases} u_d^*(k+1) = \left(R_0 - \frac{L_0}{T}\right) i_d^p(k+1) + \frac{L_0}{T} i_d^{ref}(k) - \omega_e L_0 i_q^p(k+1) \\ u_q^*(k+1) = \left(R_0 - \frac{L_0}{T}\right) i_q^p(k+1) + \frac{L_0}{T} i_q^{ref}(k) + \omega_e \left[L_0 i_d^p(k+1) + \psi_{f0} \right] \end{cases} \tag{5}$$

where nominal value of the predictive model is replaced by true value; $i_d^p(k+1)$ and $i_q^p(k+1)$ denotes the one-shot predictive value of the dq-axis currents; $u_d^*(k)$, $u_q^*(k)$ and $u_d^*(k+1)$, $u_q^*(k+1)$ denote the output values of dq-axis voltage at the previous and current moments respectively; $i_d^{ref}(k)$ and $i_q^{ref}(k)$ denotes the reference value of dq-axis currents.

3.2. IDPCC without Flux Linkage Parameters

It can be seen from C-DPCC predictive equations that the accuracy of three motor parameters affect the control performance. Parametric robustness of the system can be enhanced by constructing an incremental model that eliminates the mover flux linkage term.

By differentiating between adjacent moments, the current predictive equation, which eliminates the flux linkage term and considers one-step delay compensation, can be obtained from Equation (3).

$$\begin{cases} i_d(k+1) = \left(2 - \frac{RT}{L}\right) i_d(k) - \left(1 - \frac{RT}{L}\right) i_d(k-1) + \\ \qquad T\omega_e \left[i_q(k) - i_q(k-1) \right] + \frac{T}{L} \left[u_d(k) - u_d(k-1) \right] \\ i_q(k+1) = \left(2 - \frac{RT}{L}\right) i_q(k) - \left(1 - \frac{RT}{L}\right) i_q(k-1) - \\ \qquad T\omega_e \left[i_d(k) - i_d(k-1) \right] + \frac{T}{L} \left[u_q(k) - u_q(k-1) \right] \end{cases} \tag{6}$$

The voltage predictive equation with the elimination of the flux linkage term is similarly obtained from Equation (2) as:

$$\begin{cases} u_d(k+1) = u_d(k) + R\left[i_d(k+1) - i_d(k) \right] - \omega_e L \left[i_q(k+1) - i_q(k) \right] + \\ \qquad \frac{L}{T} \left[i_d(k+2) - 2 i_d(k+1) + i_d(k) \right] \\ u_q(k+1) = u_q(k) + R\left[i_q(k+1) - i_q(k) \right] + \omega_e L \left[i_d(k+1) - i_d(k) \right] + \\ \qquad \frac{L}{T} \left[i_q(k+2) - 2 i_q(k+1) + i_q(k) \right] \end{cases} \tag{7}$$

Similar to C-DPCC above, an I-DPCC algorithm with one-step delay compensation can be obtained by Equations (6) and (7). It is noted that I-DPCC does not necessitate flux linkage parameters, thereby eliminating any potential impact of the mismatch of flux linkage parameters on the control performance.

4. Parametric Robustness Improvement of Inductance

4.1. Inductance Stability Analysis of I-DPCC

The impact of parameter inconsistency on the steady-state error of the predictive current model has been examined in [23,24], demonstrating that the mismatch in resistor characteristics in the I-DPCC algorithm exerts a minimal influence on the equilibrium error and stability of the current loop, while the mismatch of the inductance and flux linkage has a more pronounced effect on the current control. The disturbances caused by the flux linkage parameters were eliminated by the incremental model in the preceding section; accordingly, this section focuses on the impact of inductance parameter mismatch on current loop stability.

When PMLSM operates in steady state at constant thrust, it can be assumed that the dq-axis currents reach stabilization, i.e., $i_d(k-1) = i_d(k)$ and $i_q(k-1) = i_q(k)$. Furthermore, considering that the sampling time T is small enough for the cross-coupling terms in the predictive equations to be neglected [25], the dq-axis is completely decoupled in this case, and Equations (6) and (7) can be combined into the following set of simplified equations:

$$i_{dq}^p(k+1) = \left(2 - \frac{RT}{L}\right)i_{dq}(k) - \left(1 - \frac{RT}{L}\right)i_{dq}(k-1) + \frac{T}{L}\left[u_{dq}(k) - u_{dq}(k-1)\right], \quad (8)$$

$$\frac{T}{L}\left[u_{dq}(k+1) - u_{dq}(k)\right] = i_{dq}(k+2) - 2i_{dq}^p(k+1) + i_{dq}(k) + \frac{RT}{L}\left[i_{dq}^p(k+1) - i_{dq}(k)\right]. \quad (9)$$

where Equation (8) represents the incremental delay compensation current predictive equation and Equation (9) represents the incremental voltage predictive equation. After substituting the nominal values, the dq-axis decoupled predictive equation can be obtained as:

$$\hat{i}_{dq}^p(k+1) = \left(2 - \frac{R_0 T}{L_0}\right)i_{dq}(k) - \left(1 - \frac{R_0 T}{L_0}\right)i_{dq}(k-1) + \frac{T}{L_0}\left[u_{dq}(k) - u_{dq}(k-1)\right], \quad (10)$$

$$\frac{T}{L_0}\left[u_{dq}^*(k+1) - u_{dq}^*(k)\right] = i_{dq}^{ref}(k) - 2\hat{i}_{dq}^p(k+1) + i_{dq}(k) + \frac{R_0 T}{L_0}\left[\hat{i}_{dq}^p(k+1) - i_{dq}(k)\right]. \quad (11)$$

Assuming that the stator voltage is equal to the output voltage, i.e., $u_{dq}^* = u_{dq}$, since R_0 has a large order of magnitude difference compared to $1/L_0$ and the sampling time T is small, the term containing $R_0 T / L_0$ can be ignored and Z-transformed respectively, and the discrete transfer function from the output voltage to the stator current be obtained from Equations (8) and (9) as:

$$G_{ui}(z) = \frac{i_{dq}(z)}{u_{dq}(z)} = \frac{T/L}{z-1}. \quad (12)$$

Further through Equations (10) and (11), the discrete transfer function from the error of the reference current to the output voltage can be obtained as follows:

$$G_{eu}(z) = \frac{u_{dq}(z)}{i_{dq}^{ref}(z) - i_{dq}(z)} = \frac{L_0/T \cdot z(z-1)}{z^3 + (2l-3)z + 2 - 2l}. \quad (13)$$

A closed-loop discrete transfer function from the reference current to real stator current can be derived into Equation (14), as can an incremental transfer function block diagram as shown in Figure 3.

$$G(z) = \frac{i_{dq}(z)}{i_{dq}^{ref}(z)} = \frac{l \cdot z}{z^3 - 3(1-l)z + 2(1-l)}.$$

(14)

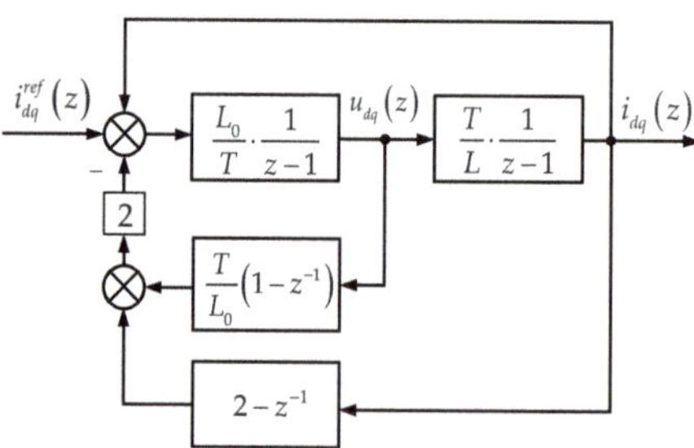

Figure 3. Closed-loop control block diagram of I-DPCC.

In order to stabilize the current loop, poles of the discrete transfer function should lie within the unit circle. Assuming that the inductance mismatch ratio is $l = L_0/L$, which indicates the ratio of the nominal value of the inductance to the true value, it can be used to measure the stability domain of the inductance parameter. Afterwards, the bilinear transformation through the Routh stability criterion can be calculated for the range of values of l under the premise of closed-loop transfer function stabilization:

$$0.8 < l < 1.25 \,.$$

(15)

Figure 4 shows the distribution of the closed-loop zero poles with an inductance mismatch ratio of I-DPCC algorithm under steady state operation. It can be seen that the incremental predictive model requires high accuracy for the inductance value, and that I-DPCC will fail to converge when it exceeds this limit. In contrast, the C-DPCC algorithm has a stable range for inductance of $0 < l < 2$ [21]. It can be concluded that the I-DPCC reduces the error brought by the flux linkage parameters, but puts higher requirements on the accuracy of the inductance parameters.

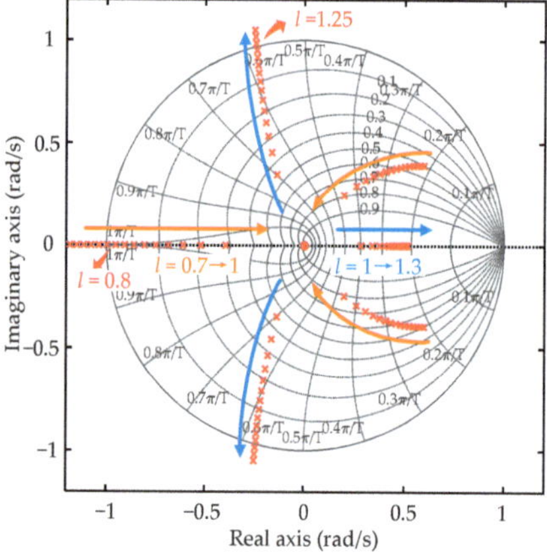

Figure 4. Closed-loop zero-pole plot of I-DPCC.

4.2. Principle of RII-DPCC

As may be seen in the analysis in the previous section, I-DPCC imposes high accuracy requirements on the inductance parameters—the error range is limited to about 20%, which greatly reduces the usability of this algorithm. To seek the widening of the stabilization interval of the inductance parameters, this section realizes the widening of the stabilization range of the inductance by introducing a robustness-improved I-DPCC algorithm combining feedforward and feedback currents, and by designing the weighting factors a and b, where $a + b = 1$. The improved closed-loop control block diagram is shown in Figure 5, and the improved voltage predictive equation is as follows:

$$\begin{cases} u_d^*(k+1) = u_d^*(k) + R_0\left[i_d^r(k+1) - i_d(k)\right] + \\ \quad \frac{L_0}{T}\left[i_d^{ref}(k) - 2i_d^r(k+1) + i_d(k)\right] - \omega_e L_0\left[i_q^r(k+1) - i_q(k)\right] \\ u_q^*(k+1) = u_q^*(k) + R_0\left[i_q^r(k+1) - i_q(k)\right] + \\ \quad \frac{L_0}{T}\left[i_q^{ref}(k) - 2i_q^r(k+1) + i_q(k)\right] + \omega_e L_0\left[i_d^r(k+1) - i_d(k)\right] \\ i_{dq}^r(k+1) = a\cdot i_{dq}^{p}(k+1) + b\cdot i_{dq}^{ref}(k-1) \end{cases} , \qquad (16)$$

where a one-step predictive current term is calculated as above, but is replaced by a modified delay compensation current term that considers a combination of feedforward and feedback when substituting into the voltage predictive equation.

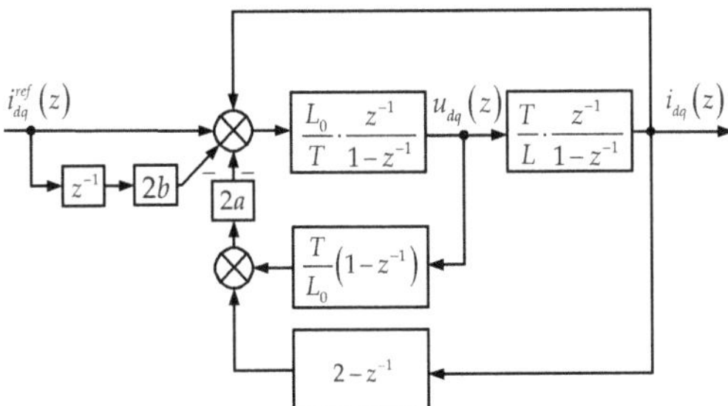

Figure 5. Closed-loop control block diagram of RII-DPCC.

4.3. Stability Analysis of RII-DPCC

The current loop transfer function using a RII-DPCC is analyzed using a similar approach as in Section 4.1. In order to obtain a more concise expression, the derivation is uniformly calculated using a. The improved system transfer function can be obtained as:

$$G(z) = \frac{l\cdot(z - 2 + 2a)}{z^3 + (2a - 2)z^2 + (1 - 4a)(1 - l)z + 2a(1 - l)}. \qquad (17)$$

It can be seen that the introduction of feedforward compensation changes the zero point at the origin of the closed-loop transfer function into an adjustable zero point with a feedforward weight factor, also changing the position of the poles. According to the Routh criterion, the range of the weighting factor a for the stabilization premise of the closed-loop transfer function is calculated as $0.5 \le a < 1$. This indicates that for RII-DPCC, the feedforward substitution cannot account for more than half of the feedback amount, otherwise the necessary condition for system stabilization will no longer be satisfied.

From the stability criterion, it is further calculated that, with the change of the weight factor a, the sufficient and necessary condition for the value of the inductance mismatch ratio l for the stabilization of the current loop transfer function is shown in Equation (19). The trend of the change is shown in Figure 6.

$$\frac{8a-4}{6a-1} < l < \frac{1+4a^2}{4a^2}.$$

(18)

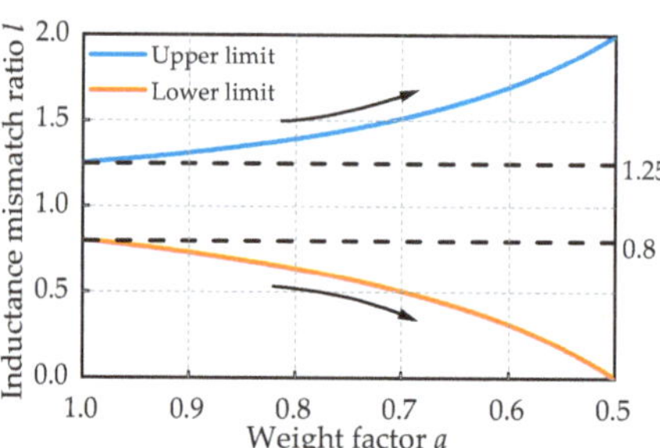

Figure 6. Trend of inductance mismatch ratio l with weight factor a.

Substituting the conditions for the maximum case of the theoretical stability domain into Equation (17), with the introduction of feedforward, a set of pairwise eliminable zero poles at this point in the transfer function can be reduced to:

$$G(z) = \frac{l(z-1)}{(z^2+l-1)(z-1)}.$$

(19)

The $(z-1)$ term exists in the numerator and denominator of Equation (19). The zero-pole distribution of the improved closed-loop transfer function is shown in Figure 7 and the inductance stability domain is extended to $0 < l < 2$, which is consistent with the theoretical calculation of Equation (19). However, it is worth noting that it is difficult to achieve the accurate zero-pole cancellation of the transfer function in practical engineering applications, which may lead to oscillations in the output of the system. This extreme situation should be avoided as much as possible. For this reason $a = 0.55$ is considered as the upper limit of the feed-forward weights for the stability verification in the later simulation.

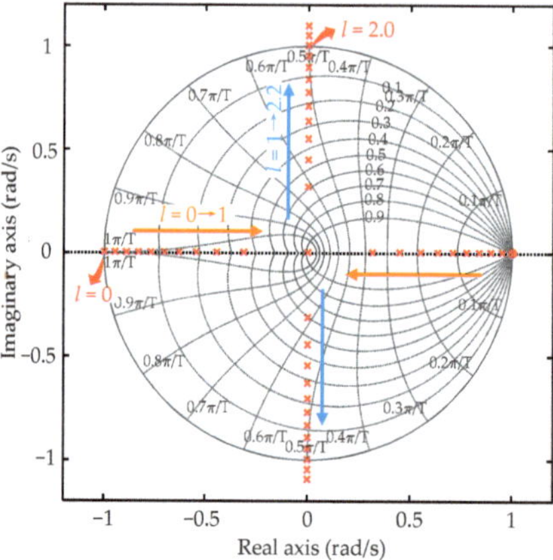

Figure 7. Closed-loop zero-pole plot of RII-DPCC when $a = 0.5$.

4.4. Steady-State Error Analysis

Substituting Equations (8) and (16) into Equations (9) and (11) respectively, and then calculating the difference, the steady state error is set to be $err = i_{dq}^*(k) - i_{dq}(k+2)$, and the steady state error at the $k+2$th moment can be obtained as:

$$err = 2(1-a)\left[i_{dq}(k-1) + i_{dq}^*(k-1) - 2i_{dq}(k)\right] + \left(\frac{T}{L_0} - \frac{T}{L}\right)\left[\Delta u_{dq}(k+1) + 2(1-a)\Delta u_{dq}(k)\right] \tag{20}$$

where Δu_{dq} denotes the amount of dq-axis voltage variation between two neighboring moments.

Ignoring the current and voltage ripples and other perturbations, when the PMLSM operates in the steady state, it can be assumed that the dq-axis currents are kept constant, and then the current term in the first half of *err* always converges to 0, which is not affected by weight *a*; while the voltage error term in the second half is itself affected by the inductance bias, which cannot be eliminated. In addition, the introduction of feedforward, will additionally introduce a voltage error term for the system which increases with the decrease in the weight *a* of the voltage error term, which in turn makes the steady state error larger on the basis of I-DPCC.

When $a = 0.5$, the steady state error reaches the theoretical maximum, which is twice as much as when feedforward compensation is not introduced. For the electromagnetic propulsion system, the importance of the current loop stability is obviously greater than the steady state error under the limit operating conditions. The steady state error can be eliminated by way of integral static differential compensation, which is not specifically derived in this paper.

5. Simulation and Experiment Results

To validate the RII-DPCC algorithm proposed in this paper, numerical simulations are carried out as follows: the propulsion system adopts an air-core PMLSM structure, the system block diagram is shown in Figure 8 and the nominal parameters are shown in Table 1. The inverter adopts an H-bridge cascade topology with a chopping frequency of 1 kHz and a control frequency of 4 kHz to verify the performance of the current control algorithm under constant load.

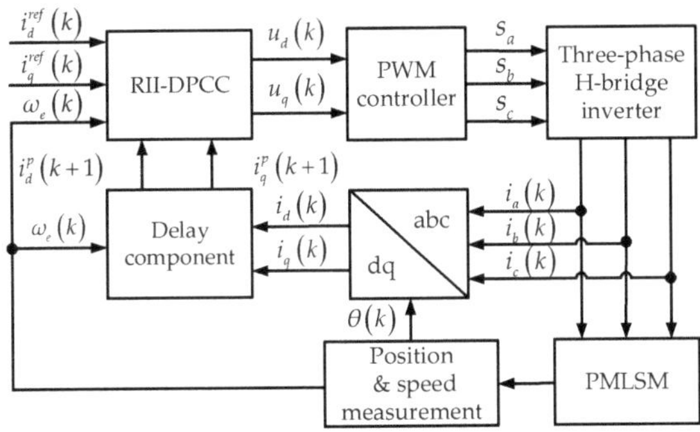

Figure 8. Block diagram of an air-core PMLSM electromagnetic propulsion system using RII-DPCC.

Table 1. Nominal parameters of the PMLSM simulation model.

Symbol	Parameters	Values
U	DC bus voltage	720 V
R_0	Nominal stator resistance	93.1 mΩ
L_0	Nominal stator inductance	55.6 mH
ψ_{f0}	Nominal mover flux linkage	1.065 Wb
m	Mover mass	215 kg
τ	Polar distance	0.54 m
p	Polar pairs	3

The experimental part does not use segmented motors, and due to the use of hollow core windings, the end effect is small, so it can be assumed that the actual inductance of the model hardly changes at all. To verify the range of l in the experimental conclusion, we changed the L_0 of the prediction model, which also validates the effectiveness of the method.

5.1. Effect of Delay Compensation on Control Performance

The I-DPCC algorithm is sensitive to delay. When there is no parameter mismatch, the control performance with or without delay compensation is verified by the joint simulation platform. The propulsion system is set to start with a constant q-axis current of 2000 A at 0.1 s, and reduced to 0 at 0.5 s. As shown in Figure 9, when the delay compensation is not considered, the current has a large overshoot at the moment of response. After delay compensation, the overshoot amplitude of the current is reduced by more than 80%, the response is fast and tracking is stable.

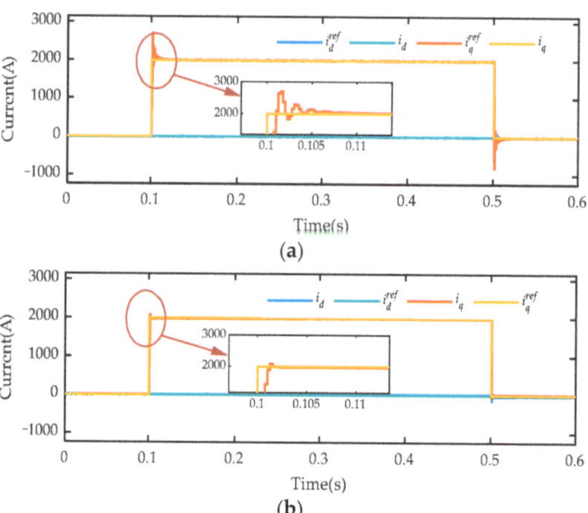

Figure 9. Comparison of dq-axis current of I-DPCC with and without delay compensation. (**a**) Not adding delay compensation. (**b**) Adding delay compensation.

5.2. Comparison of Flux Linkage Parameter Robustness

The electromagnetic propulsion system is set to start a 215 kg constant load with a constant q-axis current of 2000 A, and the current is increased by 500 A at 0.2 s and 0.4 s, ultimately reaching 3000 A, to verify the stability and dynamic performance of the current loop. To ensure system safety, the current loop is limited to 3300 A.

As is shown in Figure 10, due to the need for ψ_f in the C-DPCC algorithm, steady-state errors are inevitable when there is a mismatch in ψ_f. However, the RII-DPCC algorithm can avoid this problem.

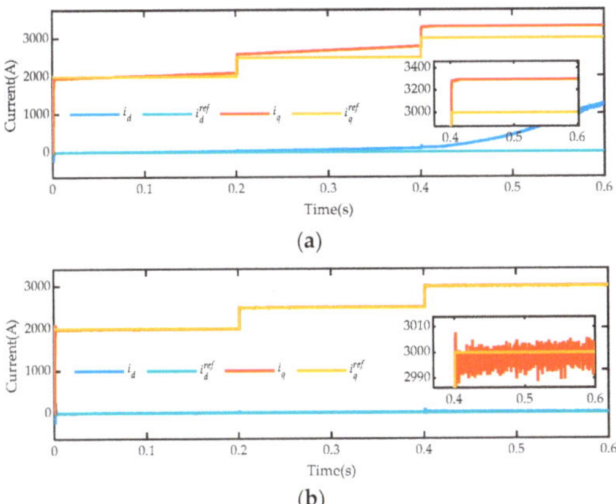

Figure 10. Comparison of the dq-axis currents at $\psi_{f0} = 2\psi_f$. (**a**) C-DPCC. (**b**) RII-DPCC.

5.3. Comparison of Inductance Parameter Robustness

The operating conditions in this section are consistent with those in the previous section. The nominal inductance parameters of the model are changed separately to verify the improvement in inductance stability of the I-DPCC algorithm before and after improvement. As shown in Figure 11, the dq-axis currents are obtained when the inductance mismatch reaches the actual stable upper and lower limits of the unimproved I-DPCC algorithm.

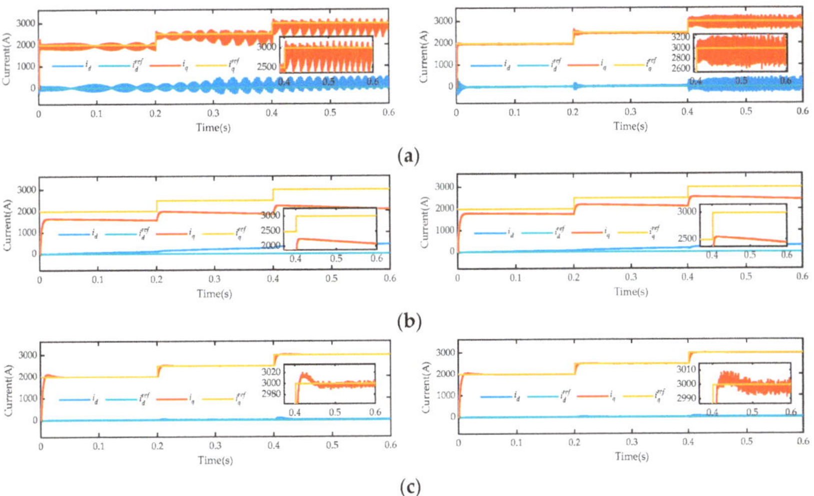

Figure 11. Comparison of the dq-axis current at $l = 0.65$ and 1.3. (**a**) I-DPCC with no improvement. (**b**) RII-DPCC when $a = 0.55$. (**c**) RII-DPCC with static error compensation.

Figure 11b shows the simulation results of the I-DPCC inductance mismatch calculated in Section 4.1, where the inductance stability domain is expanded only through expected current feedforward and the current loop reconverges. To avoid system oscillation caused by zero pole cancellation of critical stability, a feedforward weight factor is taken. From the simulation results, it can be observed that the addition of feedforward broadens the stability domain of the inductance and again stabilizes the current loop. However, the steady-state error caused by inductance mismatch still exists, which is consistent with the calculation in Section 4.4.

Figure 11c shows the RII-DPCC with integrated static error compensation added on the basis of expected current feedforward. From the simulation results, the current loop achieves zero static error current output while improving the robustness of inductance parameters.

The inductance robustness of RII-DPCC under extreme operating conditions is further verified. As shown in Figure 12, the simulation results of RII-DPCC with inductance mismatch ratios of 0.05 and 2 respectively show that the current loop still achieves stable and fast tracking without static errors. The improved incremental DPCC method improves the inductance stability domain through expected current feedforward, and compensates for steady-state errors through feedback voltage integration. The simulation results show that the stability domain of the current loop inductance has been significantly improved when compared with the results using the traditional incremental method.

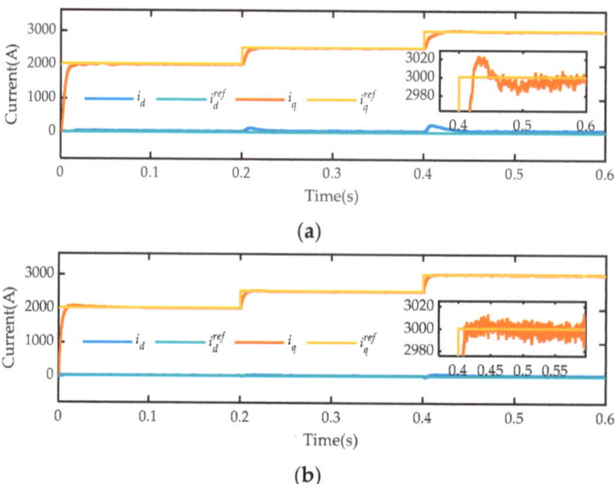

Figure 12. Inductance robustness of RII-DPCC under extreme operating conditions. (**a**) $l = 0.05$. (**b**) $l = 2$.

5.4. Experimental Results of the Proposed Method

To further verify the proposed method, a scaled-down electromagnetic propulsion platform was constructed, as shown in Figure 13. This platform included an air-core stator winding, permanent magnetic mover, grating positioning system, DC power supply powered by lithium batteries, DC/AC inverter, control board and PC. The main control chip in the control board adopts TMS320F28335 produced by Texas Instruments (Dallas, TX, United States). The current and speed measurement sensors are integrated on the control board and communicate with the PC through UART protocol.

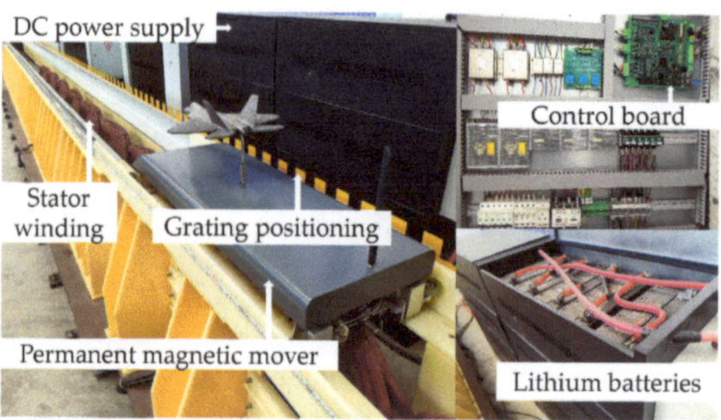

Figure 13. Scaled-down experimental platform of PMLSM electromagnetic propulsion system.

This platform maintains consistency with the simulated system in terms of the design of the inverter, stator winding, and rotor structures, with only specific circuit parameters differing. The specific parameters of the platform are shown in Table 2. The control frequency of the system is consistent with the chopping frequency of the inverter, both at 3 kHz.

Since the influence of flux linkage on the stability of predictive control has been dealt with in other studies, this section will not repeat it, and only the optimization of inductance stability will be experimentally verified. The operating conditions in the experiments are all 50 A current constant thrust, and the running time is 0.6 s.

Table 2. The parameters of the scaled-down electromagnetic propulsion platform.

Symbol	Parameters	Values
U	DC bus voltage	550 V
R_0	Nominal stator resistance	12.64 mΩ
L_0	Nominal stator inductance	22.2 mH
ψ_{f0}	Nominal mover flux linkage	0.1717 Wb
m	Mover mass	50 kg
τ	Polar distance	0.27 m
p	Polar pairs	2

Figure 14 shows the experimental comparison between I-DPCC and RII-DPCC under constant current. It is observed that the experimental results are consistent with the previous simulation, and the inductance stability domain of I-DPCC is relatively narrow. The current loop exhibits instability when l is less than 0.8 and greater than 1.25. In comparison, the current loop using RII-DPCC can achieve stable tracking without static error in both cases.

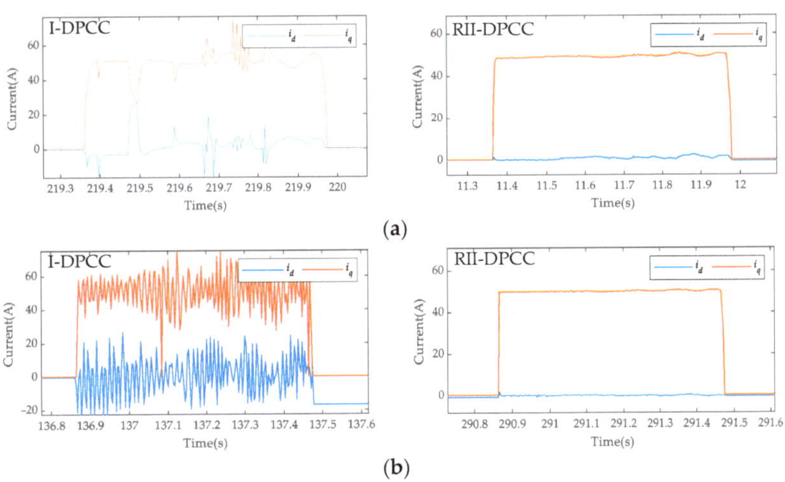

Figure 14. Experimental comparison between I-DPCC and RII-DPCC under constant 50 A current condition. (**a**) $l = 0.7$. (**b**) $l = 1.3$.

As is shown in Figure 15, the current loop stability under several extreme inductance mismatches on the electromagnetic propulsion platform has been verified. Theoretical analysis and simulation show that the inductance stability domain of RII-DPCC is $0 < l < 2$, thus the experimental results verify this conclusion. In two extreme cases, the current loop did not become unstable, but both showed some performance degradation. For example, when $l = 0.05$, the response speed of the current loop was slow and there was a significant ripple current; when $l = 2$, there was an overshoot peak in the current at the startup moment. When the nominal inductance in the model was changed even further (when $l = 2.1$), there was a divergence phenomenon in the current loop, which is consistent with the previous analysis.

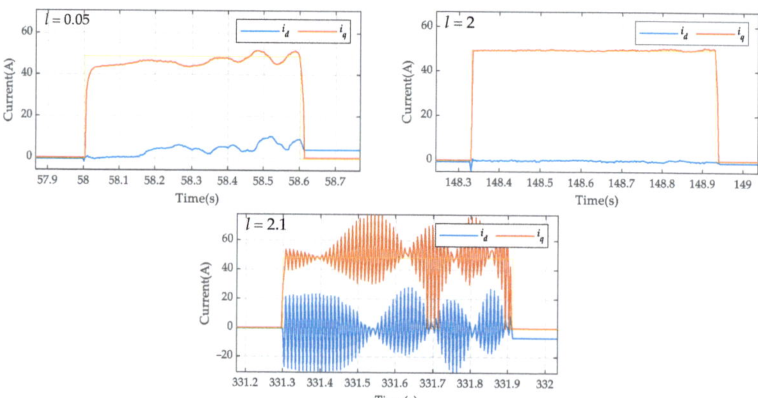

Figure 15. Experimental results of RII-DPCC under extreme operating conditions.

6. Conclusions

This article focused on the robustness of predictive control algorithms to L and ψ_f, which is caused by the inherent delay and structure of high-power electromagnetic propulsion systems. Firstly, the impact of delay on predictive algorithms was analyzed. Simulation results showed that the addition of delay compensation can effectively reduce the overshoot of I-DPCC. Under the model proposed in this paper, the overshoot value was reduced by more than 80%. Secondly, the robustness of I-DPCC to ψ_f was verified, and, when compared to DPCC that requires ψ_f parameters, steady-state errors caused by mismatch could be eliminated. Finally, regarding inductance stability, this article theoretically analyzed the stability range of I-DPCC from the perspective of transfer function, and proposed a feedforward compensation method to improve the stability of I-DPCC. Through transfer function calculation, this method can increase the inductance stability range of I-DPCC from 0.8–1.25 times the actual value to 0–2 times the actual value, and the theoretical stability range more than 4 times. Stable domain measurement was expanded and steady-state error was reduced combined with the integral compensation.

The simulated and experimental results demonstrate the effectiveness of the improved method, enhancing the robustness of the predictive method to L and ψ_f, and improving the stability and steady-state error of the system current loop.

Author Contributions: Conceptualization, S.G. and P.L.; methodology, S.G. and P.Y.; software, S.G. and P.L.; validation, P.Y. and P.L.; formal analysis, Q.C.; investigation, Y.J.; resources, P.Y; writing—original draft preparation, S.G.; writing—review and editing, S.G. and P.L.; visualization, Y.J.; supervision, Q.C.; project administration, P.Y.; funding acquisition, J.L. All authors have read and agreed to the published version of the manuscript.

Funding: This work is supported by National Key R&D Program of China [grant number 2016YFB1200601], and the Major Project of Advanced Manufacturing and Automation of Changsha Science and Technology Bureau under grant number kq1804037.

Data Availability Statement: Data are contained within the article.

Conflicts of Interest: The authors declare no conflicts of interest.

References

1. Tavana, N.R.; Shoulaie, A.; Dinavahi, V. Analytical Modeling and Design Optimization of Linear Synchronous Motor with Stair-Step-Shaped Magnetic Poles for Electromagnetic Launch Applications. *IEEE Trans. Plasma Sci.* **2012**, *40*, 519–527. [CrossRef]
2. Tan, Y.; Zhou, D.; Li, J.; Chen, Q.; Jia, Z.; Leng, P.; Liu, M. Dynamic Simulation Analysis of Null-Flux Coil Superconducting Electrodynamic Suspension System. *IEEE Trans. Intell. Veh.* **2024**, *9*, 1005–1016. [CrossRef]
3. Wang, M.; Kang, K.; Zhang, C.; Li, L. A Driver and Control Method for Primary Stator Discontinuous Segmented-PMLSM. *Symmetry* **2021**, *13*, 2216. [CrossRef]

4. Zhang, Z.; Xu, Q.; Wang, Y. A New Sliding-Mode Observer-Based Deadbeat Predictive Current Control Method for Permanent Magnet Motor Drive. *Machines* **2024**, *12*, 297. [CrossRef]
5. Zhang, T.; Mei, X. Research on Detent Force Characteristics of Winding Segmented Permanent Magnet Linear Synchronous Motor Based on Analytical Model. *Symmetry* **2022**, *14*, 1049. [CrossRef]
6. Wang, H.; Zhang, G.; Liu, X. Sensorless Control of Surface-Mount Permanent-Magnet Synchronous Motors Based on an Adaptive Super-Twisting Sliding Mode Observer. *Mathematics* **2024**, *12*, 2029. [CrossRef]
7. Zhang, X.; Zhao, Z. Model Predictive Control for PMSM Drives with Variable Dead-Zone Time. *IEEE Trans. Power Electron.* **2021**, *36*, 10514–10525. [CrossRef]
8. Zhang, R.; Yin, Z.; Du, N.; Liu, J.; Tong, X. Robust Adaptive Current Control of a 1.2-MW Direct-Drive PMSM for Traction Drives Based on Internal Model Control with Disturbance Observer. *IEEE Trans. Transp. Electrif.* **2021**, *7*, 1466–1481. [CrossRef]
9. Mohamed, Y.A.-R.I.; El-Saadany, E.F. A Current Control Scheme with an Adaptive Internal Model for Torque Ripple Minimization and Robust Current Regulation in PMSM Drive Systems. *IEEE Trans. Energy Convers.* **2008**, *23*, 92–100. [CrossRef]
10. Zhang, P.; Shi, Z.; Yu, B.; Qi, H. Research on a Sensorless ADRC Vector Control Method for a Permanent Magnet Synchronous Motor Based on the Luenberger Observer. *Processes* **2024**, *12*, 906. [CrossRef]
11. Andersson, A.; Thiringer, T. Motion Sensorless IPMSM Control Using Linear Moving Horizon Estimation with Luenberger Observer State Feedback. *IEEE Trans. Transp. Electrif.* **2018**, *4*, 464–473. [CrossRef]
12. Zhang, L.; Tao, R.; Bai, J.; Zeng, D. An Improved Sliding Mode Model Reference Adaptive System Observer for PMSM Applications. *Expert Syst. Appl.* **2024**, *250*, 123907. [CrossRef]
13. Wu, X.; Chen, H.; Liu, B.; Wu, T.; Wang, C.; Huang, S. Improved Deadbeat Predictive Current Control of PMSM Based on a Resistance Adaptive Position Observer. *IEEE Trans. Transp. Electrif.* **2024**. [CrossRef]
14. Naikawadi, K.M.; Patil, S.M.; Kalantri, K.; Dhanvijay, M.R. Comparative Analysis of Features of Online Numerical Methods Used for Parameter Estimation of PMSM. *IJPEDS* **2022**, *13*, 2172. [CrossRef]
15. Yuan, T.; Chang, J.; Zhang, Y. Parameter Identification of Permanent Magnet Synchronous Motor with Dynamic Forgetting Factor Based on H∞ Filtering Algorithm. *Actuators* **2023**, *12*, 453. [CrossRef]
16. Ahn, H.; Kim, S.; Park, J.; Chung, Y.; Hu, M.; You, K. Adaptive Quick Sliding Mode Reaching Law and Disturbance Observer for Robust PMSM Control Systems. *Actuators* **2024**, *13*, 136. [CrossRef]
17. Wang, L.; Zhang, S.; Zhang, C.; Zhou, Y. An Improved Deadbeat Predictive Current Control Based on Parameter Identification for PMSM. *IEEE Trans. Transp. Electrif.* **2024**, *10*, 2740–2753. [CrossRef]
18. Rafaq, M.S.; Jung, J.-W. A Comprehensive Review of State-of-the-Art Parameter Estimation Techniques for Permanent Magnet Synchronous Motors in Wide Speed Range. *IEEE Trans. Ind. Inf.* **2020**, *16*, 4747–4758. [CrossRef]
19. Zhang, X.; Huang, X. An Incremental Deadbeat Predictive Current Control Method for PMSM with Low Sensitivity to Parameter Variation. In Proceedings of the 2020 International Conference on Electrical Machines (ICEM), Gothenburg, Sweden, 23–26 August 2020; IEEE: New York, NY, USA, 2020; pp. 1060–1066.
20. Liu, Z.; Huang, X.; Hu, Q.; Li, Z.; Jiang, Z.; Yu, Y.; Chen, Z. A Modified Deadbeat Predictive Current Control for Improving Dynamic Performance of PMSM. *IEEE Trans. Power Electron.* **2022**, *37*, 14173–14185. [CrossRef]
21. Huang, X.; Hu, Q.; Liu, Z.; Li, W.; Yang, G.; Li, Z. A Robust Deadbeat Predictive Current Control Method for IPMSM. *IEEE Trans. Transp. Electrif.* **2023**. [CrossRef]
22. Blouh, F.; Bezza, M.; El Myasse, I. Modeling and Control of an Aerogenerator Utilizing DFIG and Direct Matrix Converter Technology. In Proceedings of the 2024 4th International Conference on Innovative Research in Applied Science, Engineering and Technology (IRASET), Fez, Morocco, 16–17 May 2024; IEEE: New York, NY, USA, 2024; pp. 1–6.
23. Zhao, M.; Zhang, S.; Li, X.; Zhang, C.; Zhou, Y. Parameter Robust Deadbeat Predictive Current Control for Open-Winding Surface Permanent Magnet Synchronous Motor Drives. *IEEE J. Emerg. Sel. Top. Power Electron.* **2023**, *11*, 3117–3126. [CrossRef]
24. Zhang, X.; Zhang, L.; Zhang, Y. Model Predictive Current Control for PMSM Drives with Parameter Robustness Improvement. *IEEE Trans. Power Electron.* **2019**, *34*, 1645–1657. [CrossRef]
25. Wang, W.; Xi, X. Current Control Method for PMSM with High Dynamic Performance. In Proceedings of the 2013 International Electric Machines & Drives Conference, Chicago, IL, USA, 12–15 May 2013; IEEE: New York, NY, USA, 2013; pp. 1249–1254.

Article

Addressing EMI and EMF Challenges in EV Wireless Charging with the Alternating Voltage Phase Coil

Zeeshan Shafiq [1], Tong Li [1], Jinglin Xia [2], Siqi Li [3,*], Xi Yang [4,*] and Yu Zhao [4]

[1] Department of Transportation Engineering, Kunming University of Science and Technology, Kunming 650500, China; shafiq.xeeshan@gmail.com (Z.S.); litongkm410@stu.kust.edu.cn (T.L.)

[2] Facility of Mechanical and Electrical Engineering, Kunming University of Science and Technology, Kunming 650500, China; xiajl22@kust.edu.cn

[3] Department of Electrical Engineering, Kunming University of Science and Technology, Kunming 650500, China

[4] Yunnan Energy Research Institute Co., Ltd., Kunming 650500, China; zhaoy32933@cnyeig.com

* Correspondence: lisiqi@kust.edu.cn (S.L.); yangxi@cnyeig.com (X.Y.)

Abstract: Wireless charging technologies are widely used in electric vehicles (EVs) due to their advantages of convenience and safety. Conventional wireless charging systems often use planar circular or square spiral windings, which tend to produce strong electric fields (E-fields), leading to electromagnetic interference (EMI) and potential health risks. These standard coil configurations, while efficient in energy transfer, often fail to address the critical balance between E-field emission reduction and power transfer effectiveness. This study presents an "Alternating Voltage Phase Coil" (AVPC), an innovative coil design that can address these limitations. The AVPC retains the standard dimensions of traditional square coils (400 mm in length and width, with a 2.5 mm wire diameter and 22 turns), but introduces a novel current flow pattern called Sequential Inversion Winding (SIW). This configuration of the winding significantly reduces E-field emissions by altering the sequence of current through its loops. Rigorous simulations and experimental evaluations have demonstrated the AVPC's ability to lower E-field emissions by effectively up to 85% while maintaining charging power. Meeting stringent regulatory standards, this advancement in the proposed coil design method provides a way for WPT systems to meet stringent regulatory standards requirements while maintaining transmission capability.

Keywords: electric vehicles (EVs); electromagnetic field exposure; wireless power transfer; Alternating Voltage Phase Coil (AVPC); E-field mitigation in charging systems; coil designs

Citation: Shafiq, Z.; Li, T.; Xia, J.; Li, S.; Yang, X.; Zhao, Y. Addressing EMI and EMF Challenges in EV Wireless Charging with the Alternating Voltage Phase Coil. *Actuators* **2024**, *13*, 324. https://doi.org/10.3390/act13090324

Academic Editor: Dong Jiang

Received: 4 July 2024
Revised: 17 August 2024
Accepted: 23 August 2024
Published: 26 August 2024

1. Introduction

Wireless charging technology, characterized by its non-contact method of energy transfer, has revolutionized the convenience and safety of powering devices across various applications [1]. This technology facilitates the transfer of power without physical connections [2], finding use in diverse fields such as biomedical devices for remote healthcare monitoring [3], logistic optimization through unmanned warehousing [4], and advanced navigation systems for expansive marine exploration [5]. Originating over a century ago with basic inductive charging methods [6], wireless charging has progressively evolved into sophisticated forms like resonant inductive coupling and capacitive coupling [7,8], broadening its applicability. Despite these advancements, the technology faces significant challenges that hinder its wider adoption, particularly in high-power applications such as electric vehicles (EVs) [9]. These challenges include electromagnetic interference (EMI) and electromagnetic field (EMF) leakage, especially the electric field (E-field) component, which poses safety risks and reduces system efficiency [10,11]. Innovative solutions are needed to offer high performance and environmental compatibility [12], focusing on mitigating E-field emissions while maintaining efficient power transfer [13].

Previous research efforts in wireless charging coil design have focused on optimizing coil geometry [14], investigating novel materials, and developing advanced control strategies [15] to enhance coupling efficiency, reduce EMF leakage, and minimize EMI [16,17]. However, existing approaches, such as the distributed compensation topology [18], have limitations in coil design flexibility and overall system performance [19,20]. Most state-of-the-art wireless charging systems rely on passive shielding techniques [21], like metal plates near the coils or the implementation of alternative coil geometries, to mitigate EMF exposure risks [15,22]. While these methods effectively shield EMFs [23], they increase the charging units' size and complexity, introducing challenges such as enhanced parasitic capacitance [24], which can exacerbate conductive common mode noise issues and complicate compliance with stringent EMC standards [25,26].

In addition to shielding, optimized coil design, and active field cancelation techniques, the spread spectrum technique has emerged as another effective approach for reducing leakage fields and EMI in wireless charging systems. This technique, based on either random or periodic modulations, has been explored in recent studies [27]. The spread spectrum approach works by distributing the energy of the electromagnetic fields over a wider frequency range, thereby reducing the peak field strengths and minimizing potential interference with other electronic devices. While this technique has shown promise in mitigating leakage fields and EMI, it often requires more complex control systems and may introduce challenges in terms of power transfer efficiency and compatibility with existing wireless charging standards [28], as well as compliance with the ICNIRP guidelines and the European Directive limit for public human exposure to EMF at 85 kHz, set at 87 V/m [29]. Nonetheless, the spread spectrum technique represents a valuable addition to the arsenal of methods available for addressing the critical issues of electromagnetic compatibility and safety in wireless charging applications.

In response to these challenges, this paper introduces the Alternating Voltage Phase Coil (AVPC), a novel coil design optimized for EV wireless charging systems. The AVPC addresses key issues in traditional wireless charging systems by significantly reducing EMI and the E-field component of EMF leakage [22,30]. This study prioritizes the reduction in E-field emissions over leakage magnetic fields because E-fields are the primary contributor to potential health risks [31] and electromagnetic compatibility issues in wireless charging systems [32,33], in line with the ICNIRP guidelines and the European Directive limit for public human exposure to EMF [29]. By focusing on E-field mitigation, the AVPC aims to address the most pressing safety concerns while maintaining efficient power transfer through the intentional preservation of the H-field [24]. This unique approach to mitigating E-field emissions sets the AVPC apart from other state-of-the-art solutions. While existing techniques, such as the use of ferrite shields [23,34] or the implementation of metamaterials [35], have shown promise in reducing electromagnetic interference, they often come with trade-offs in terms of system complexity, cost, and power transfer efficiency [25]. In contrast, the AVPC's innovative design achieves significant E-field reduction through its unique coil geometry and current flow pattern, without compromising on power transfer efficiency or requiring additional complex components [15]. This sets the AVPC apart as a potentially game-changing solution for the wireless charging industry, offering a simpler, more efficient, and more effective approach to addressing the essential challenges of electromagnetic compatibility and safety in high-power wireless charging applications [26]. Its superiority lies in its innovative geometry, which minimizes electromagnetic emissions while maintaining high power transfer, similar to traditional coils [36]. Additionally, by its unique coil structure, the AVPC does not need the additional capacitors for maintaining a uniform magnetic field (H-field) distribution similar to traditional coil designs, as has been carried out in the distributed compensation topologies [37], which is essential for the reduction in E-field and optimal energy transmission. The AVPC's innovative Sequential Inversion Winding (SIW) technique creates an alternating high and low voltage arrangement across adjacent turns, ensuring effectively out-of-phase voltages that lead to opposing and canceling electric fields, significantly reducing overall E-field emissions. The development of the AVPC offers practical advantages for wireless charging systems and contributes to the fundamental understanding of electromagnetic field behavior in these systems.

This study provides new insights into the relationship between coil geometry, current flow patterns, and the resulting electromagnetic field characteristics through the design and analysis of the AVPC. The architecture of EV wireless charging systems, as illustrated in Figure 1, involves several essential components, including grid frequency rectifiers, inverters, coupling coils, and network compensation, which must function cohesively to achieve efficient and safe power transmission. The coupling coil plays a central role in the wireless transfer of energy, and has been the focus of substantial research and development efforts aimed at enhancing its performance and safety features.

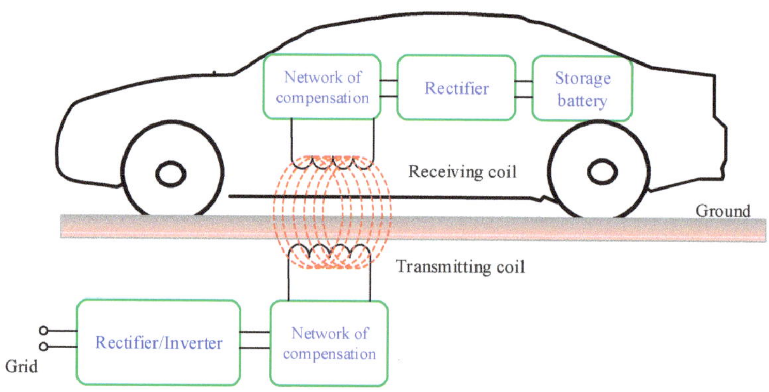

Figure 1. General wireless power transfer charging system for EVs.

To address the limitations of traditional wireless charging systems, particularly high EMI and E-field leakage, our study proposes the AVPC, which incorporates several innovative features. First, the AVPC's innovative coil configuration significantly reduces E-field emissions while maintaining the same power transfer as traditional coils, enhancing the system's operational reliability. Second, the AVPC's enhanced safety protocols primarily focus on minimizing EMI and E-field leakage, the main contributors to potential health risks associated with electromagnetic emissions. While the AVPC effectively reduces the E-field, it intentionally maintains the H-field (magnetic field) necessary for efficient power transfer, as the leakage magnetic field is not the primary concern in this study. The AVPC's ability to selectively reduce the E-field while maintaining the H-field is a key aspect of this research. Third, the AVPC demonstrates exceptional versatility and broad adaptability for various device applications, including EVs, mobile devices, and medical equipment, showcasing its potential beyond traditional applications.

Table 1 presents a detailed comparison of the AVPC and traditional coil designs, emphasizing the AVPC's significant improvements in E-field emission reduction [38], innovative current flow sequence, and broad adaptability to various devices while maintaining comparable energy transfer efficiency. Our experimental and simulation results demonstrate that the AVPC achieves an 85% reduction in E-field emissions compared to traditional coil designs while maintaining the same power transfer efficiency as traditional coils. These quantitative findings underscore the effectiveness of the AVPC in addressing the limitations of conventional wireless charging systems, positioning it as a superior solution for various wireless charging applications.

Table 1. Comparison of traditional coil and AVPC configuration.

Feature	Conventional Coil	AVPC
E-field emission reduction	Limited	Significantly improved
Energy transfer efficiency	High	Comparable to conventional coils
E-field neutralization method	Capacitors (complex)	Design configuration
Innovative current flow sequence	No	Yes
Sequential turn progression	Turns in sequence	Not applicable
Adaptability to various devices	Limited	Broad (EVs, mobiles, medical equipment)

The key objectives of this study are as follows:

1. Primary objective: Develop and validate the AVPC design, demonstrating its effectiveness in reducing EMI and E-field emissions while maintaining efficient power transfer, thus addressing the most pressing safety and performance issues in traditional wireless charging systems.
2. Experimental validation: Conduct comprehensive testing to measure the EMF emissions of the AVPC under various operating conditions and compare the results with those of traditional coil systems to quantify improvements in field management.
3. Safety and efficiency metrics: Assess the safety improvements brought by the AVPC, particularly through reduced EMF exposure, and analyze the system's efficiency in terms of power transfer and energy loss.
4. Design optimization: Explore new design variations of the AVPC based on the findings from the previous objectives, aimed at further enhancing its performance, safety, and compatibility with high-power wireless charging applications.

To experimentally validate the AVPC's enhanced capabilities, comprehensive testing will be conducted using advanced measurement techniques to assess the coil's efficiency and EMF emissions across various operational conditions. Additionally, simulations will be performed to investigate the impact of the AVPC's design on the system's performance and explore potential design optimizations.

The paper is structured as follows: Section 2, Proposed Coil Structure Compared to the Traditional Coil, presents the design methodology and developmental insights of the AVPC, contrasting it with traditional coil structures. Section 3, Simulation Analysis of the Coil, delves into comprehensive simulation studies to validate the performance of the AVPC. In Section 4, the paper transitions into the experimental validation, providing a detailed analysis of the experimental setup and discussing the results obtained from real-world testing scenarios. The paper culminates in Section 5, the conclusion, summarizing the key findings, drawing conclusions from both the simulated and experimental investigations, and offering perspectives for future research in EV wireless charging systems.

2. Proposed Coil Structure Compared to the Traditional Coil

2.1. Structural Design and Development of AVPC

This section elucidates the innovative design principles and structural characteristics of the AVPC, an innovative advancement in EV wireless charging technology. The AVPC's revolutionary design departs from conventional coil configurations by employing a unique turn sequencing pattern. In traditional coils, as shown in Figure 2a, the current flows sequentially through the turns in the order of L_1, L_2, L_3, L_4, L_5, and L_6. In contrast, the AVPC's electrical current follows a different path while maintaining the same direction of current flow. In the AVPC design depicted in Figure 2b, the current first flows through the odd-numbered turns L_1, L_3, and L_5 and then through the even-numbered turns L_2, L_4, and L_6. Similarly, in the AVPC design shown in Figure 2c, the current flows through L_1, L_3, and L_5, and then L_6, L_4, and L_2, maintaining the same direction of current flow throughout the coil. This innovative sequencing creates an alternating voltage pattern across the coil, with higher voltage differences between adjacent turns compared to traditional coils.

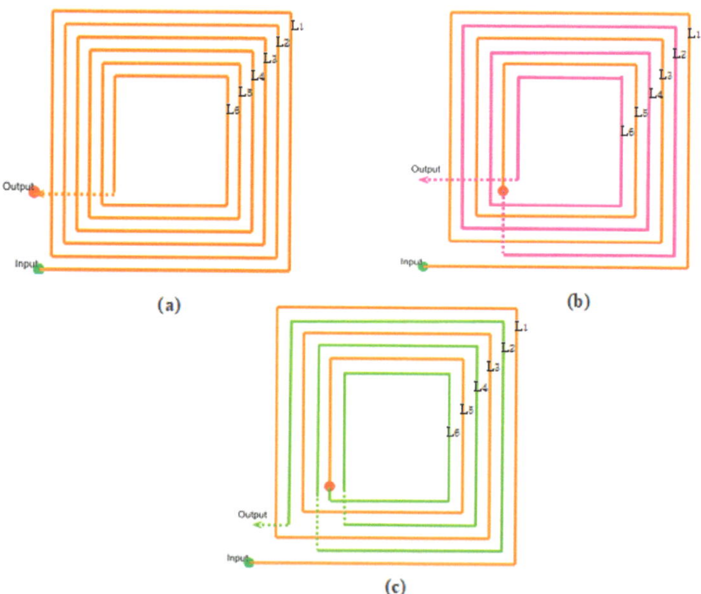

Figure 2. Structure of the coil prototype. (**a**) Traditional coil. (**b**) AVPC 1. (**c**) AVPC 2.

LTSpice simulations and mathematical analyses of the voltage distributions in the AVPC and traditional coil reveal a significant difference in their voltage patterns. In the AVPC, the voltage waveforms of adjacent turns exhibit a phase difference, when compared to the average voltage of the coil, which is used as the reference. This phase difference indicates substantial voltage differences between neighboring turns, resulting in alternating high and low voltage points along the coil. In contrast, the voltage waveforms of adjacent turns in the traditional coil are in-phase, suggesting a more uniform voltage difference distribution and smaller voltage differences between neighboring turns. The alternating voltage pattern created by the innovative sequencing in the AVPC design contributes to reduced field intensity in the vicinity of the coil, ultimately leading to a reduction in E-field emissions compared to traditional coil designs.

The design of the AVPC involves careful consideration of its geometry and parameters to optimize its performance. The key parameters that define the coil's geometry include the outer diameter D_{out}, inner diameter D_{in}, wire diameter d_w, and spacing between turns (s). These parameters are related to the number of turns (N) in the coil, as described by Equation (1):

$$N = \frac{(D_{Out} - D_{In})}{(2 \times d_w + 2 \times S)} \tag{1}$$

Equation (1) governs the relationship between the geometrical parameters and the number of turns in the coil. By carefully selecting these parameters, the geometry can be optimized while maintaining the same power transfer as traditional coils and achieving reduced E-field emissions, particularly in the case of AVPC.

The inductance of the coil (L) is calculated using Equation (2), where μ_0 is the permeability of free space ($4\pi \times 10^{-7}$ H/m), N is the number of turns, A is the cross-sectional area of the coil (m^2), and l is the length of the coil (m) [39]. The mutual inductance between adjacent turns (M) is given by Equation (3):

$$L = \frac{(\mu_0 \times N^2 \times A)}{l} \tag{2}$$

$$M = \frac{(\mu_0 \times N_1 \times N_2 \times A_m)}{L_m} \tag{3}$$

where N_1 and N_2 are the number of turns in adjacent coil segments, A_m is the effective cross-sectional area of the region where the magnetic fields of the adjacent segments interact, contributing to the mutual inductance effect, and L_m is the effective length of the mutual inductance path, representing the average distance between the adjacent coil segments contributing to the mutual inductance effect. The coupling coefficient between adjacent turns (k) is calculated using Equation (4), where M is the mutual inductance between adjacent turns and L_1 and L_2 are the self-inductances of the adjacent turns. The quality factor of the coil (Q) is given by Equation (5), where ω is the angular frequency $(2\pi \times f)$, L is the inductance of the coil, and R is the resistance of the coil. The skin depth of the wire (δ) is calculated using Equation (6), where ρ is the resistivity of the wire material $(\Omega \cdot m)$, f is the frequency of the alternating current (Hz), and μ is the magnetic permeability of the wire material (H/m).

$$K = \frac{M}{\sqrt{L_1 - L_2}} \tag{4}$$

$$Q = \frac{(\omega \times L)}{R} \tag{5}$$

$$\delta = \sqrt{\frac{\rho}{\pi \times f \times \mu}} \tag{6}$$

The AVPC's structural uniqueness lies in its strategic arrangement of the turn sequence, which harnesses the principle of electric field opposition. This design involves an innovative Sequential Inversion Winding (SIW) technique where the current flows through the turns in a specific sequence that creates an alternating voltage pattern across adjacent turns. Unlike traditional coil designs, where the voltage phases of adjacent turns are the same, leading to additive electric fields, the AVPC design arranges the turns such that the voltage potentials between adjacent turns are effectively out-of-phase. While direct measurements of the voltages (e.g., V_1 and V_2) at each turn might show them as in-phase due to the same AC source, the relative voltage differences between adjacent turns in the AVPC design create effective phase shifts. These phase shifts are determined by using the average voltage of the coil and extracting the voltages of individual nodes from this average voltage. By comparing the extracted node voltages, the phase shifts between adjacent turns can be clearly identified. Although each turn receives the same AC signal, the spatial arrangement and alternating voltage patterns cause adjacent turns to have voltages that effectively cancel each other out corresponding to the average voltage of the coil. This phase opposition results from the specific sequencing of the turns and the inherent geometric configuration of the coil. The cancelation effect can be described by the principle of superposition of electric fields. If E_1 and E_2 are the electric fields generated by adjacent turns with opposing voltages, the resultant electric field E_{Total} is given by the following:

$$E_{Total} = E_1 + (-E_2) \tag{7}$$

where E_1 is the electric field generated by one turn and $-E_2$ is the electric field generated by the adjacent turn with an opposite voltage polarity, effectively canceling out E_1. By optimizing the AVPC's design parameters, such as the number of turns, spacing between turns, and coil dimensions, this ingenious design feature enables the AVPC to significantly mitigate electromagnetic interference without relying on complex shielding or extensive modifications, ultimately simplifying the coil's architecture while enhancing its performance.

The AVPC's superior characteristics, as outlined in Table 1, position it as an environmentally friendly and user-safe solution, surpassing the limitations of conventional coil designs. The development of the AVPC represents a significant stride in addressing the pressing need for safer and more efficient wireless charging solutions. By reimagining the fundamental structure and voltage pattern of the coil, this innovative design paves the way for overcoming persistent challenges in the field, setting a new paradigm for future advancements in EV wireless charging technology.

2.2. Simulation Setup and Preliminary Results

The simulation parameters and outcomes, depicted in Figures 3 and 4, play a vital role in validating the operational advantages of the AVPC design. The coil parameters (inductance and coupling coefficients) presented in Figure 4 and Table 2 are derived from the ANSYS FEM simulation of the coil pad structure shown in Figure 3a. LTSpice simulations were conducted using these data for both the AVPC (Figure 3c) and the traditional coil (Figure 3b), focusing on a six-turn configuration to simplify the analysis and understand the reason behind the E-field reduction.

To maintain accuracy while simplifying the analysis, assumptions such as ideal component behavior were made in the SPICE simulations. These assumptions allow for a focused study on the fundamental behavior of the AVPC design compared to the traditional coil structure.

Table 2. Coil inductance and system parameters.

Coil	Inductance (uH)	Parameter	Value
L_1	0.120	Input AC current I	5 A
L_2	0.087469	Resonant frequency	85 KHz
L_3	0.060760	Shielding plate	74 mm × 74 mm × 1 mm
L_4	0.043119	Ferrite core 1	72 mm × 72 mm × 2 mm
L_5	0.030174	Coil dimension	70 mm × 70 mm × 3 mm
L_6	0.019673	Coil turn to turn spacing	1.5 mm

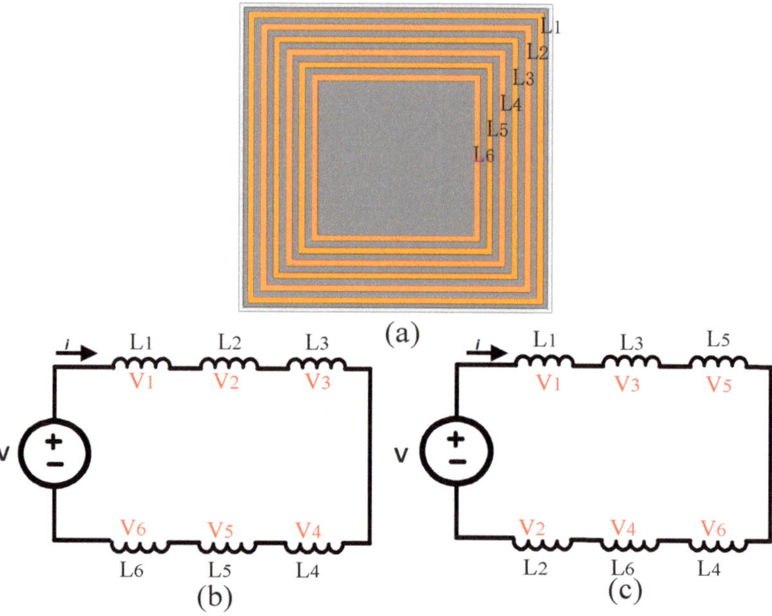

Figure 3. E-field mitigation in wireless charging systems. (**a**) Optimized coil design. (**b**) Spice circuit of traditional coil. (**c**) Spice circuit of the AVPC.

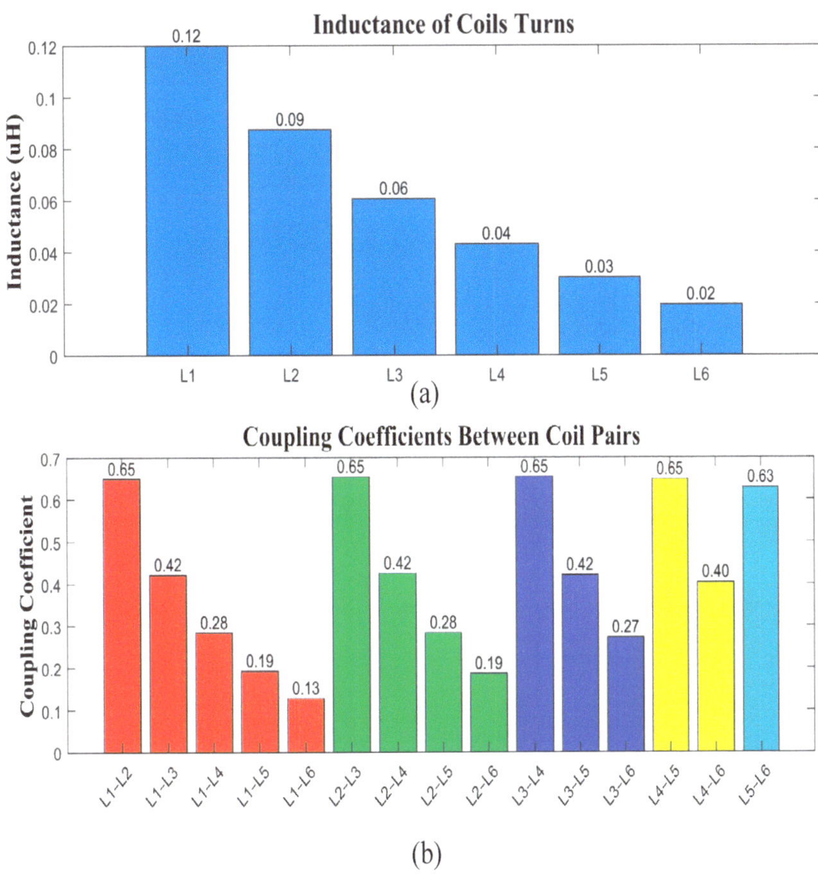

Figure 4. Coil Parameter. (**a**) Inductance. (**b**) Coupling coefficient.

As illustrated in Figure 3b,c, the traditional coil features a sequential arrangement, while the AVPC design employs an alternating coil arrangement to strategically reduce E-field emissions. The LTSpice simulations, measuring average voltage waveforms at different nodes, highlight the role of phase differences in the AVPC's E-field reduction capability. This comprehensive analysis of the voltage distribution and phase differences between the two designs demonstrates the AVPC's superior ability to suppress E-field emissions. The AVPC's unique configuration enhances E-field mitigation through phase differences and field cancelation, resulting in superior performance compared to the traditional coil while maintaining comparable power transfer.

2.3. Waveform Analysis and E-Field Mitigation through Spice Simulation

The voltage waveforms of the traditional and AVPC coil designs, obtained from SPICE simulations and presented in Figure 5a,b, are essential for understanding their electrical characteristics and inferring potential E-field emissions. The simplified equation $E = (V_{total}/2\pi\varepsilon_0 r)$ demonstrates how the total voltage across the coil correlates with the E-field strength. In the traditional coil design, the in-phase voltage waveforms across inductors L_1 to L_6, described by the relationship $V_i = J\omega L_i I_i$, enhance magnetic field generation, but also increase the potential for E-field emission due to the higher total voltage across the coil, which can be expressed as the sum of the voltages across each inductor.

$$V_{total} \; = \; V_1 + V_2 + \cdots + V_n \tag{8}$$

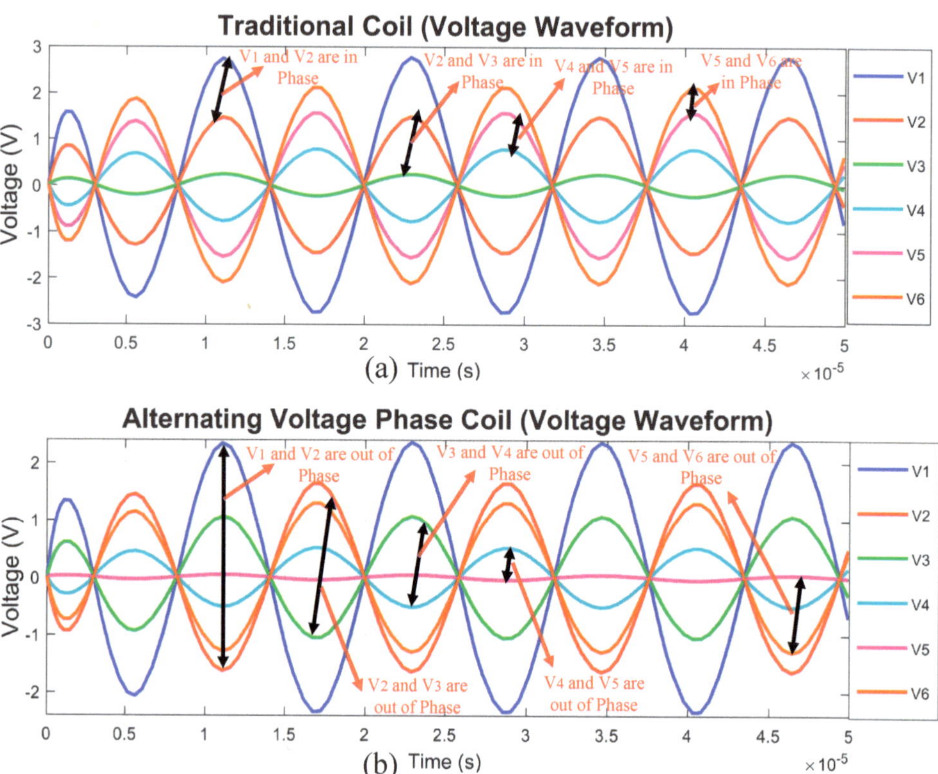

Figure 5. Coil wave forms. (**a**) Traditional coil. (**b**) AVPC.

In contrast, the AVPC design introduces a strategic phase shift to mitigate E-field emissions. The modified equation $V_i \; = \; J\omega L_i I_i - J\omega M_{ik} I_k$, where M_{ik} represents the mutual inductance between adjacent turns, leads to a phase shift between the voltages across adjacent turns, promoting phase opposition and reducing the total voltage and, consequently, the E-field emissions. The total voltage across the coil in the AVPC design can be expressed as the sum of the voltages across each inductor, considering the phase shift.

$$V_{total} \; = \; V_1 + V_2 e^{(j\theta_2)} + \cdots + V_n e^{(j\theta_n)} \tag{9}$$

The 180 degree phase shift is achieved through the specific arrangement and sequencing of the turns in the AVPC design, where the strategic placement of turns with respect to each other promotes the desired phase opposition. The formula used in the SPICE simulation is shown in Equation (12) and demonstrates how the voltages at different nodes are averaged, leading to an effective phase shift. This analysis reveals the phase difference in the voltage in the AVPC compared to the traditional coil, as shown in Figure 5a,b.

The average voltage across the turn is calculated as follows:

$$V_{avg} \; = \; \frac{1}{N}\sum_{i\,=\,1}^{N} V_i \tag{10}$$

The effective voltage at a specific turn, say V_i, is given by the following:

$$V_{effective} = V_1 - V_{avg} \tag{11}$$

The generalized form of the voltage at any turn V_{turn} is as follows:

$$V_{turn} = V_{turn} - \frac{1}{N}\sum_{i=1}^{N} V_{all\ turns} \tag{12}$$

Figure 5 illustrates the voltage waveforms for the traditional coil and the AVPC. In the traditional coil design (Figure 5a), the in-phase voltage waveforms V1 and V2 have positive peaks reaching approximately 2.8V, leading to higher overall voltage and potentially increased E-field emissions. Conversely, the AVPC design (Figure 5b) exhibits an out-of-phase relationship between V1 (positive peaks around 2.4 V) and V2 (negative peaks around −2.4 V), resulting in lower overall voltage and reduced E-field emissions. The E-field strength at a distance r from the coil can be approximated using Equation (13).

$$E = \frac{(V_{total})}{(2\pi\varepsilon_0 r)} \tag{13}$$

where V_{total} represents the total voltage across the coil and ε_0 is the permittivity of free space (8.85×10^{-12} F/m). This Equation provides a simplified estimation of the E-field strength based on the total voltage and the distance from the coil [40]. However, as presented in [41,42], a more detailed analysis of the E-field distribution is necessary to fully understand the electromagnetic behavior of these coil structures and accurately compare the E-field mitigation capabilities of the AVPC and traditional coil designs.

The LTSpice circuit model, based on the AVPC design, facilitates the desired phase cancelation through the innovative arrangement of coil turns, concentrating the magnetic field for efficient energy transfer while substantially diminishing E-field emissions. The essential difference between the two designs lies in the phase relationship between the voltages, with the AVPC demonstrating an out-of-phase relationship that enables E-field mitigation while maintaining energy transfer. This is achieved by ensuring that the current through the coil remains consistent, thereby sustaining the magnetic field necessary for energy transfer. Simulation results confirm the AVPC's enhanced ability to achieve a balance between energy transfer efficiency and E-field mitigation by leveraging phase shift cancelation to reduce E-field emissions while preserving core functionalities, setting a new standard in wireless charging technology that prioritizes both electromagnetic compatibility and safety.

3. Simulation Analysis of the Coil

3.1. Simulation Parameters and Setup

Finite Element Method (FEM) simulations, performed using a computational electromagnetics software package, were used to analyze the electromagnetic properties of the traditional coil and AVPC designs in EV wireless charging systems. Table 3 lists the key simulation parameters, including input current, resonant frequency, and dimensions of the shielding plate, ferrite core, and coil, which are selected to represent realistic operating conditions in EV wireless charging applications. While Table 2 focuses on the specific case of a six-turn coil to simplify the analysis and understand the E-field reduction mechanism, Table 3 provides a more comprehensive set of simulation parameters to evaluate the performance of the traditional coil and AVPC designs under various operating conditions. These simulation parameters were then used as a basis for the experimental setup to ensure consistency between the computational models and the physical prototype.

The FEM simulations discretize the domain using second-order tetrahedral elements and apply appropriate boundary conditions to represent the wireless charging pad, an essential component of the EV wireless charging system. The pad is modeled with realistic dimensions, materials, and operating conditions to accurately capture its electromagnetic behavior. The electromagnetic responses, such as magnetic field distribution, electric field intensity, and power transfer capability, are computed by solving Maxwell's equations iteratively. By accurately modeling the pad's electromagnetic properties, the FEM simulations provide valuable insights into the performance and safety aspects of the EV wireless charging system.

Table 3. System parameters of pad.

Parameters	Values
Input AC current	5 A
Resonant frequency	85 KHz
Shielding plate dimension	420 mm × 420 mm × 4 mm
Ferrite core dimension	404 mm × 404 mm × 5 mm
Coil dimension	400 mm × 400 mm × 2.5 mm
Number of turns	22 Turns
Coil inductance	336.4 uH

Key performance metrics, such as peak magnetic field strength, average electric field intensity, and power transfer capability, are extracted from the simulation results and compared between the traditional coil and AVPC designs. Visualizations of field distributions and current densities provide insights into the electromagnetic behavior of the coils. The FEM simulations offer a comprehensive assessment of the electromagnetic performance of AVPC designs compared to traditional coil setups, contributing to the understanding of their potential in enhancing the safety and capability of EV wireless charging systems.

3.2. Electromagnetic Field Distribution and the Impact of Conductive Materials

A comparative analysis of electromagnetic field distributions, focusing on electric field (E-field) emissions for both traditional and AVPC designs, demonstrates the AVPC's enhanced effectiveness in E-field mitigation. As demonstrated in Figure 6, the E-field intensity around the traditional coil is noticeably higher compared to the AVPC models, which exhibit a significant reduction in E-field intensity. This reduction is quantitatively supported by the data presented in Figure 7, where the AVPC designs consistently show lower E-field intensities at various measurement points.

The FEM analysis clearly illustrates the AVPC's ability to reduce E-field emissions, emphasizing a pivotal advancement in coil technology aimed at safer wireless charging solutions. This consistent reduction in E-field level underscores the AVPC's effectiveness in mitigating electromagnetic emissions across a wide spatial distribution.

Simulations were conducted with a foreign metal object (FMO) positioned 100 mm above the coils to mimic practical conditions in electric vehicle (EV) charging scenarios. Figure 8 illustrates the influence of the FMO on the E-field distribution for each coil design.

The presence of the FMO led to an increase in E-field intensity across all designs due to the metal's interaction with the electromagnetic fields. However, the AVPC configurations demonstrated remarkable resilience by suppressing the amplification of E-field levels. The AVPC 2 design demonstrates exceptional performance, maintaining significantly lower E-field intensities compared to the traditional coil at all measurement points, even in the presence of the FMO (Figure 7b).

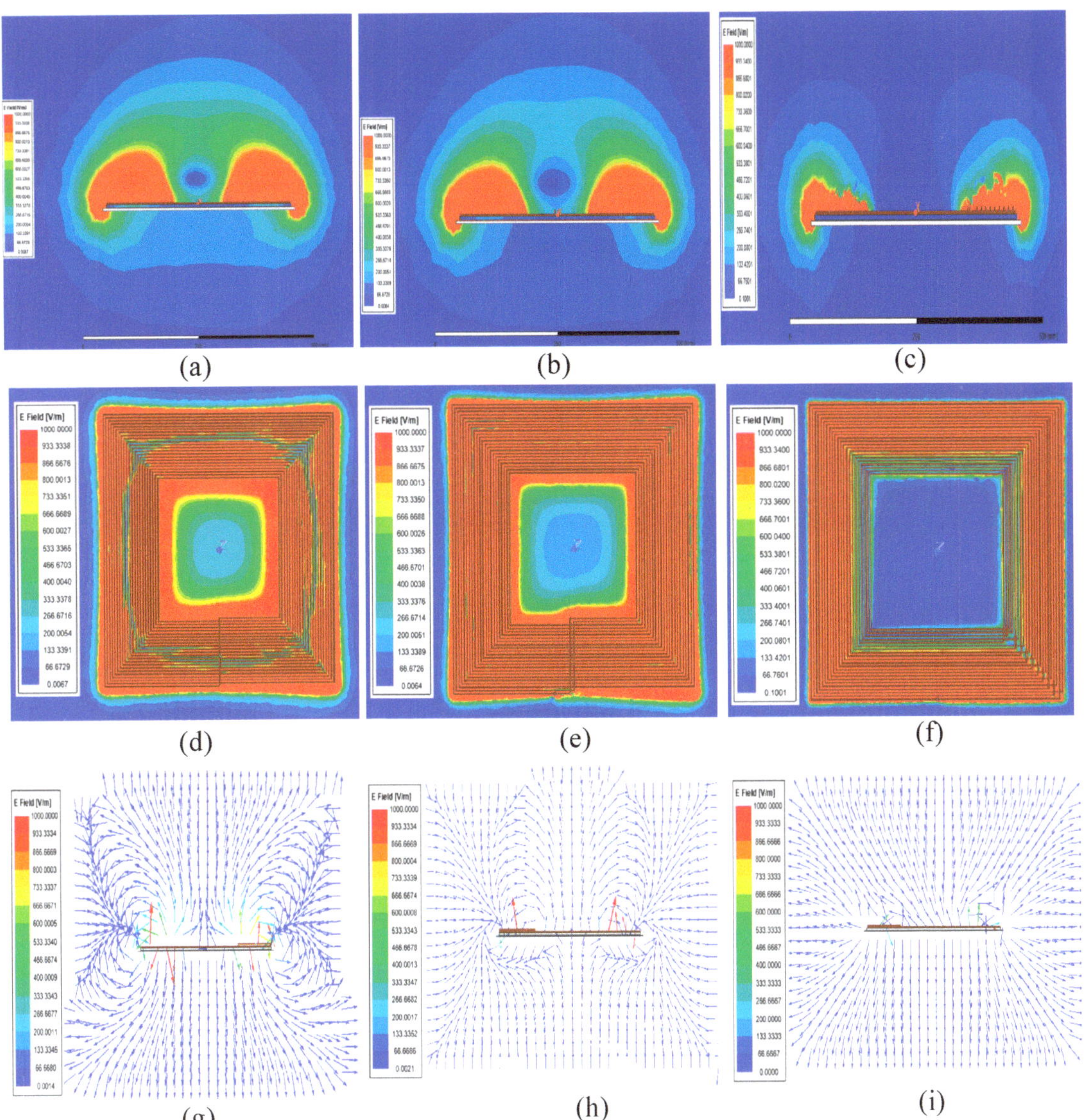

Figure 6. Comparative E-field distributions in coil designs without environmental effects. (**a**) Traditional coil side view. (**b**) AVPC 1 side view. (**c**) AVPC 2 side view. (**d**) Traditional coil top view. (**e**) AVPC 1 top view. (**f**) AVPC 2 top view. (**g**) Traditional coil side view vector distribution. (**h**) AVPC 1 side view vector distribution. (**i**) AVPC 2 side view vector distribution.

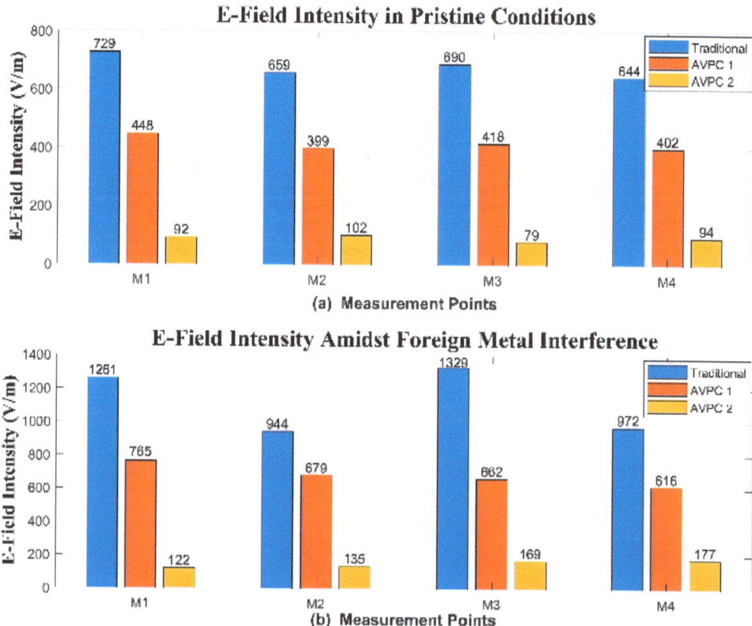

Figure 7. E−field intensity comparison at measurement points at 100 mm distance at Z axis. (**a**) E−Field Intensity in Pristine Conditions (**b**) E-Field Intensity Amidst Foreign Metal Interference.

3.3. Evaluating Electric Field Mitigation and Similar Magnetic Field Distribution in Coil Design

Figure 7 presents a detailed comparison of electric field (E-field) intensities measured at specific points, illustrating the effectiveness of the AVPC designs in managing electromagnetic fields under various conditions. The measurement points M1 (X = 0, Y = 120, Z = 100 mm), M2 (X = 0, Y = 160, Z = 100 mm), M3 (X = 0, Y = −120, Z = 100 mm), and M4 (X = 0, Y = −160, Z − 100 mm) are strategically located at different coordinates. Figure 7a provides baseline E-field measurements at a 100 mm distance along the Z-axis in an environment free from external influences, while Figure 7b reveals the E-field intensities in the presence of foreign metal objects, demonstrating the robust capability of the AVPC designs to mitigate E-field disturbances.

The AVPC designs, particularly AVPC 2, demonstrate a significant reduction in E-field intensities across all measurement points. Under pristine conditions (Figure 7a), the AVPC 2 design exhibits significantly lower E-field intensities compared to the traditional coil at each measurement point, showcasing its effectiveness in mitigating electromagnetic emissions across a wide spatial distribution. Moreover, despite the interference caused by conductive materials, the AVPC configurations, especially AVPC 2, consistently maintain their superior E-field mitigation performance, as evidenced by the results presented in Figure 7b. This reduction in E-field levels under both pristine and interfering conditions underscores the robustness of the AVPC designs in managing electromagnetic emissions.

The comparative analysis of the magnetic field distribution in the traditional coil and two AVPC designs, as depicted in Figure 9, reveals a similar magnetic field distribution within each coil design. The consistent color patterns indicate that the magnetic field intensity and distribution are comparable across all three designs, particularly in the central regions where the intensity is strongest. This similarity in patterns suggests that the AVPC designs maintain a magnetic field distribution that closely resembles that of the traditional coil.

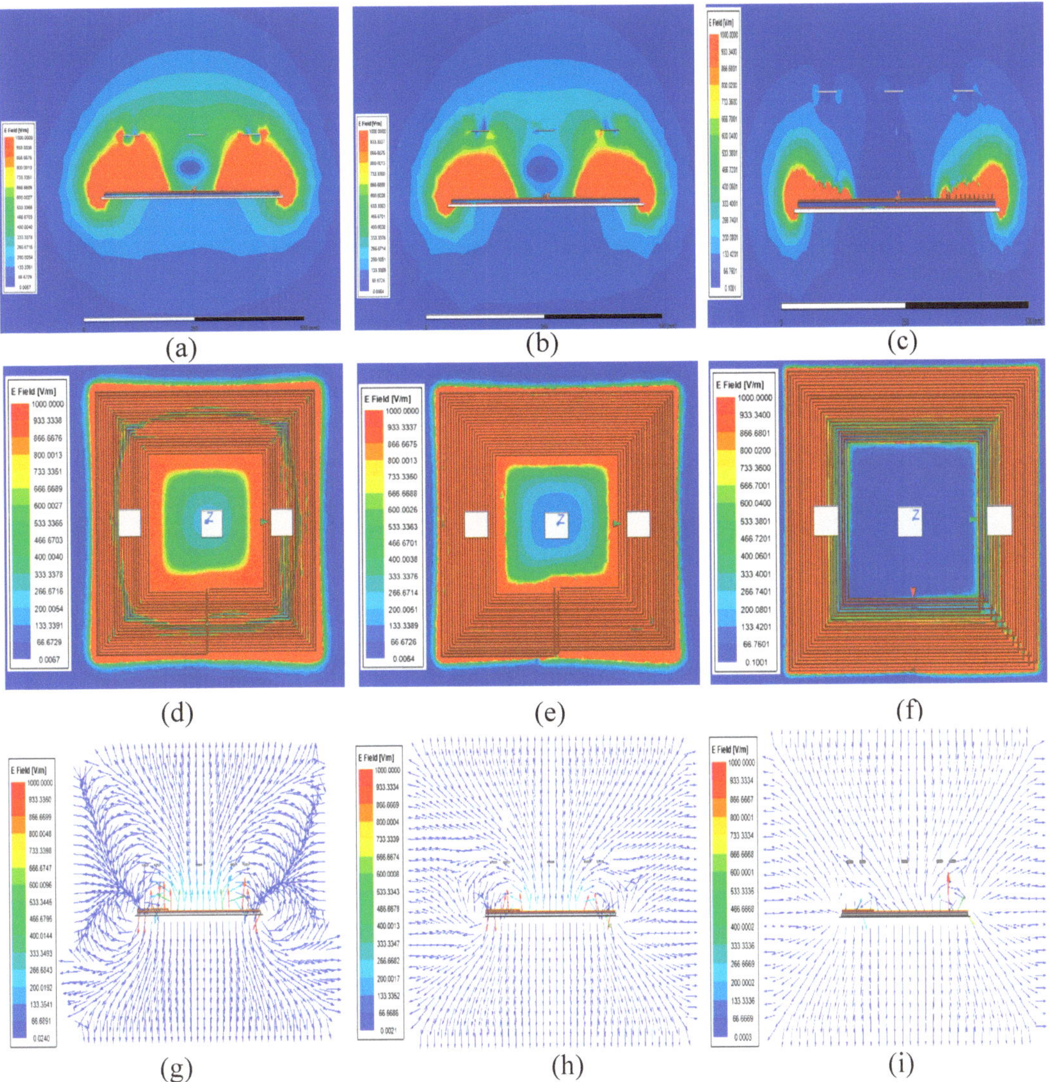

Figure 8. Comparison E-field distribution with a conductive aluminum plate at 100 mm distance. (**a**) Traditional coil side view. (**b**) AVPC 1 side view. (**c**) AVPC 2 side view. (**d**) Traditional coil top view. (**e**) AVPC 1 top view. (**f**) AVPC 2 top view. (**g**) Traditional coil side view vector distribution. (**h**) AVPC 1 side view vector distribution. (**i**) AVPC 2 side view vector distribution.

Observing similar magnetic field distributions across different coil configurations suggests that the AVPC designs can maintain a magnetic field comparable to the traditional coil. This analysis indicates that the enhancements made to reduce E-field emissions in AVPC designs do not significantly alter the magnetic field distribution, which is favorable for maintaining the effectiveness of power transfer while minimizing electromagnetic exposure through innovative coil design.

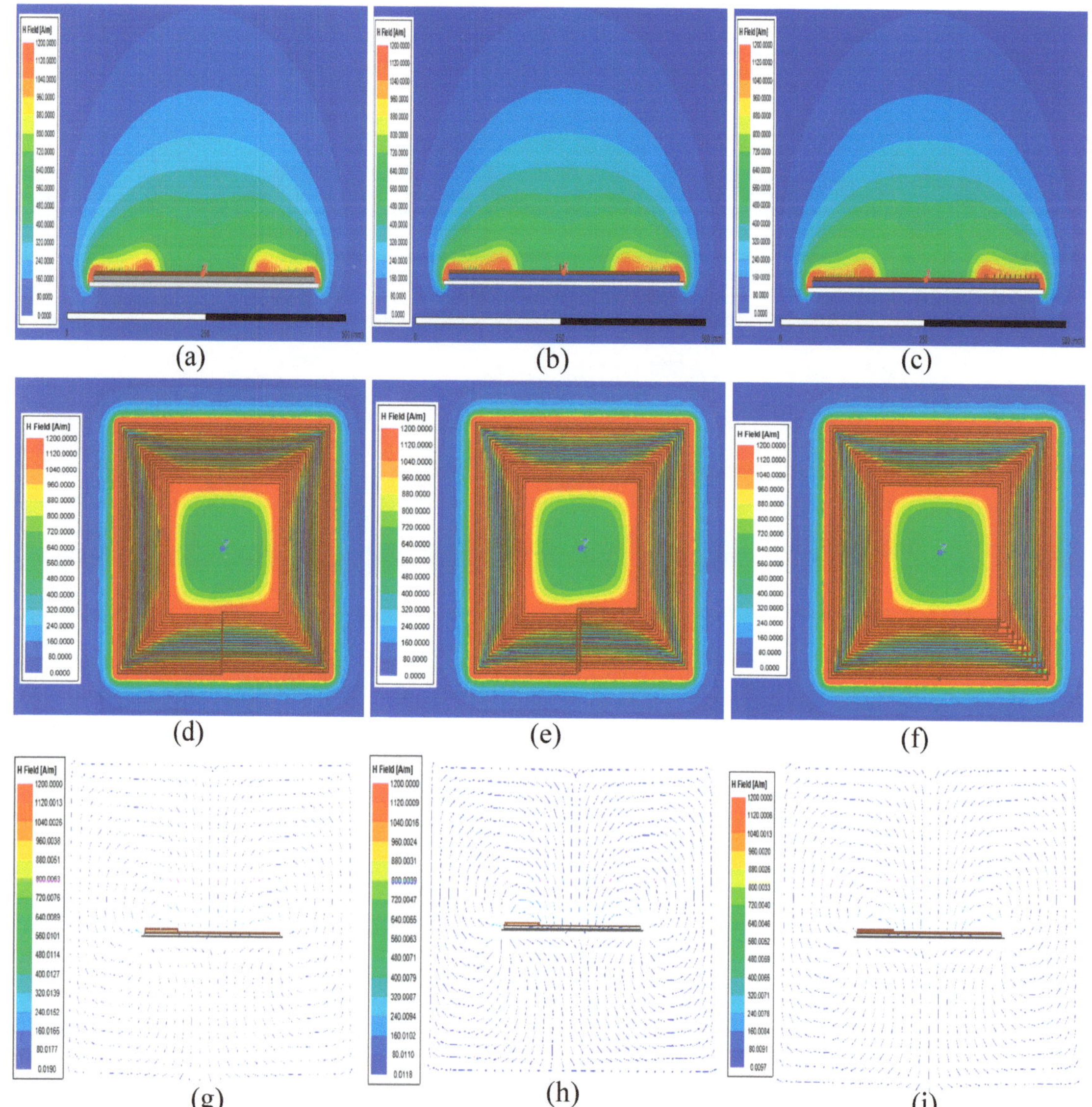

Figure 9. Comparison of the H-Field in the design coil paid. (**a**) Traditional coil side view. (**b**) AVPC 1 side view. (**c**) AVPC 2 side view. (**d**) Traditional coil top view. (**e**) AVPC 1 top view. (**f**) AVPC 2 top view. (**g**) Traditional coil side view vector distribution. (**h**) AVPC 1 side view vector distribution. (**i**) AVPC 2 side view vector distribution.

4. Experimental Setup of Coil Design and Result Verification

4.1. Methodical Assessment and Verification of Coil Design Performance

The practical performance of the innovative AVPC designs was thoroughly evaluated in a laboratory setting that closely mimicked real-world conditions to enhance safety and efficiency in electric vehicle wireless charging applications. Extensive testing simulated real-world EV charging conditions in a laboratory setting, comparing traditional coils against AVPC configurations, as shown in Figure 10. The experiments were conducted using three coils of identical size, number of turns, frequency, and input current for each design (traditional, AVPC 1, and AVPC 2). Each coil was tested to ensure the reliability and reproducibility of the results. The traditional and AVPC coils were energized using an AC source to mimic typical charging behaviors, with an NF-5035 spectrum sensor placed 100 mm above the coil to closely resemble the spacing in practical EV charging scenarios for accurate E-field data collection.

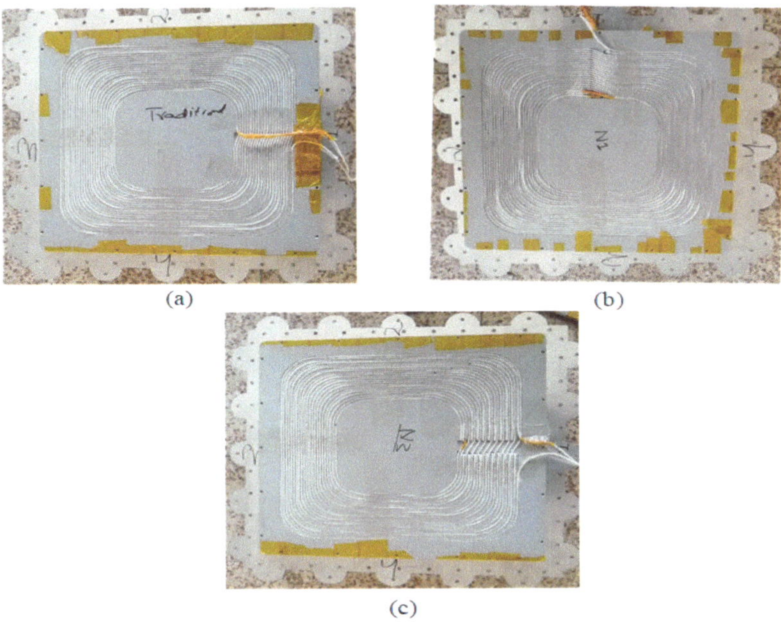

Figure 10. Comparative analysis of electromagnetic radiation distributions in innovative coil designs. (**a**) Traditional coil. (**b**) AVPC 1. (**c**) AVPC 2.

The NF-5035 spectrum sensor was selected for its proficiency in detailed electromagnetic detection, providing an accurate representation of EV charging scenarios. This enabled a focused analysis of the coils E-field emissions, specifically isolating the AVPC's inherent electromagnetic attributes. The collected data revealed that AVPC 1 and AVPC 2 designs exhibited a reduction in average E-field emissions as compared to the traditional coil design. Observations were systematically analyzed to identify the compliance and the distinct advantages of the AVPC design.

The experimental setup, depicted in Figure 11, maintained a consistent 100 mm distance between the coil and sensor to generate reliable and comparable data on electric field emissions from different coil designs. The E-field measurements were recorded at four different points around each coil, showing the variation but while still being comparable to the traditional coil, demonstrating the reduction in the E-field in the AVPC designs.

The AVPC employs an LCC compensation network to adjust the phase across the coil, facilitating optimal electromagnetic field interaction and E-field mitigation. While the LCC topology is used in experiments to validate the design, it is important to note that the inherent 180 degree phase shift primarily arises from the coil geometry itself, not the LCC network. Different topologies, such as series–series, series–parallel, parallel–series, and parallel–parallel, can be used, as the phase shift depends on the coil geometry rather than these topologies [43]. The phase shift, calculated using Equation (14), is essential for neutralizing fields emitted by adjacent turns, enhancing E-field mitigation.

$$\phi = \arctan\left(\frac{Imaginary(z)}{Real(z)}\right) \tag{14}$$

Ensuring that the compensation network resonates at the coil system's natural frequency, calculated using Equation (15), is essential for efficient energy transfer and preserving the geometrically induced phase shift.

$$f_0 = \frac{1}{2\pi\sqrt{LC}} \tag{15}$$

The deliberate incorporation of the LCC compensation network was essential in the experimental methodology, providing precise control over the electric field generated by the coil. While the experimental setup closely mimics real-world EV charging conditions, future research should focus on validating the AVPC's performance in actual EV charging scenarios to further strengthen this study's findings.

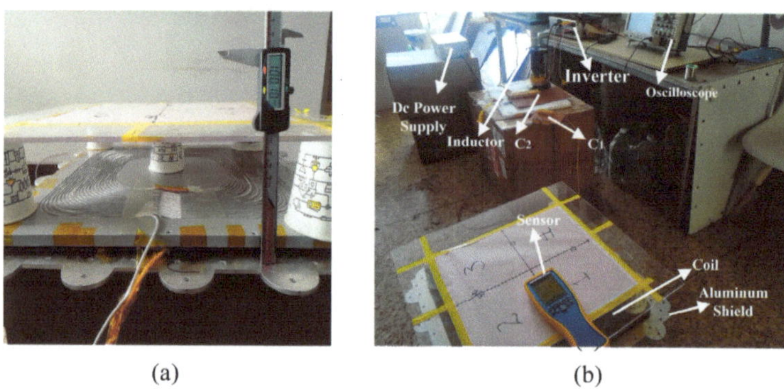

Figure 11. Shows experimental setup illustrating LCC compensation network and measurement apparatus. (**a**) Show distance between coil and the measuring points. (**b**) Experimental setup and equipment's.

4.2. Detailed Analysis of E-Field Emission Findings

The experimental study quantified the E-field emissions of both traditional and AVPC designs to compare their performance in reducing E-field radiation. The NF-5035 spectran sensor designed by aaronia AG, a German company in gewerbegebiet aaronia AG was used to measure E-field levels accurately, replicating typical EV charging scenarios. The measurement points were located at a height of 100 mm above the coil surface, with specific coordinates of (x = 0 mm, y = 120 mm, 160 mm, −120 mm, −160 mm) for each coil design. These locations were selected based on the high E-field intensities observed in the simulations (Figures 6 and 8), corresponding to regions where the E-field is expected to be the strongest. This allows for a comprehensive assessment of the AVPC designs' performance in mitigating E-field radiation. The simulations also guided the experimental setup by considering the impact of nearby conductive objects, such as an aluminum plate, on the E-field's distribution.

Figure 12 illustrates the specific measurement points on each coil design, providing a visual representation of the locations where E-field data were collected. The arrows in Figure 12a–c indicates the four strategic points around the traditional coil, AVPC 1, and AVPC 2, respectively, where the E-field measurements were recorded.

The experiments were conducted in a normal laboratory setting to maintain a realistic environment. However, variability in E-field readings was observed within specific ranges. This variability can be attributed to environmental factors, such as the presence of nearby metallic objects or electromagnetic interference from other devices, which are relevant to the European Directive limit for public human exposure to EMF at 85 kHz, set at 87 V/m. The presence of such factors in real-world EV charging scenarios could potentially impact the performance and safety of the charging system. To mitigate these effects, strategies such as shielding, adaptive control systems, site planning, and continuous monitoring can be implemented. These measures aim to ensure stable and safe charging conditions, even in the presence of environmental influences. Notably, the AVPC designs exhibited lower fluctuations compared to the traditional coil, suggesting better environmental adaptability.

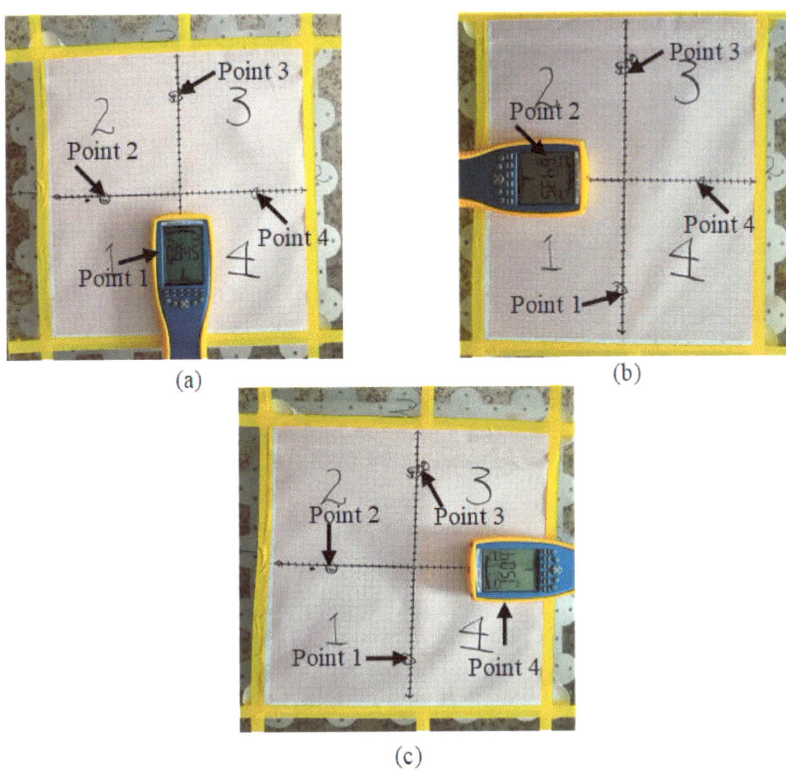

Figure 12. Measurement E-field emission comparisons across coil designs. (**a**) Traditional coil. (**b**) AVPC 1. (**c**) AVPC 2.

4.3. Quantitative Assessment of E-Field Mitigation in AVPC Designs

The simulation results shown in Figure 7 provided initial insights into the distribution, which were further supported by the experimental findings illustrated in Figure 13, demonstrating the strong correlation between the simulated and measured data. Figure 13 presents the corresponding E-field data collected from these measurement points, enabling a comparative analysis of the E-field intensity across the different coil designs. The graph clearly demonstrates the significantly lower E-field values observed across all measurement points for AVPC designs, particularly AVPC 2, compared to the traditional coil, highlighting the importance of minimizing E-field intensities in wireless charging solutions to reduce the overall cumulative exposure. This quantitative assessment highlights the effectiveness of the proposed AVPC designs in reducing E-field radiation. The experimental setup closely mimicked real-world EV charging conditions; however, measuring E-field values at specific points within a complete EV body structure can be challenging. Simulations are often used to obtain these measurements.

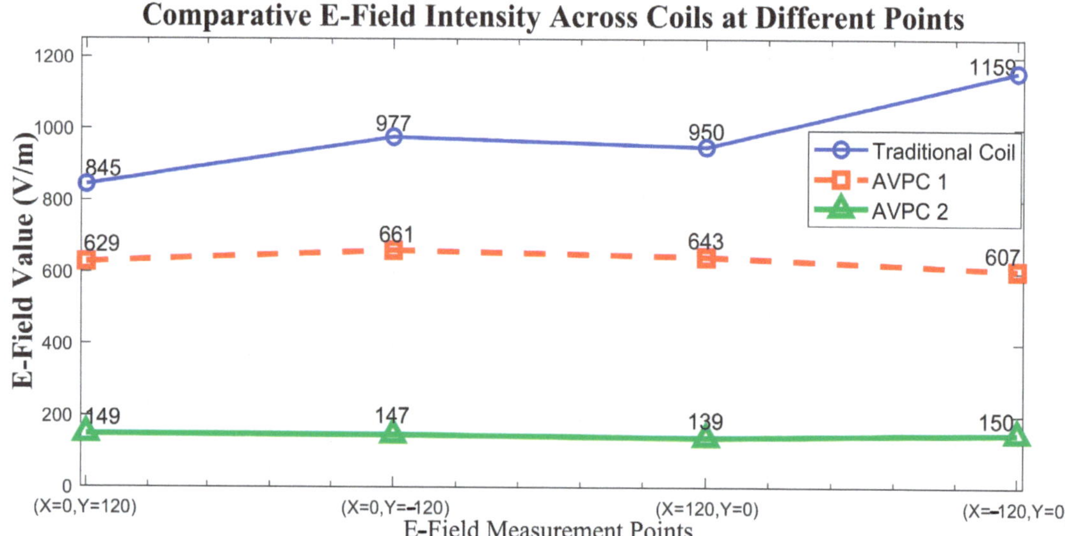

Figure 13. Comparative visualization of E−field intensity reduction across coil designs in EV charging applications.

To provide a comprehensive view, Figure 14 presents the average E-field intensities under different conditions for each coil configuration. The conditions evaluated include scenarios without an environmental effect (Without FMO), as well as those with an environmental effect introduced by the presence of foreign metal objects (With FMO). Practical experimental results were also considered, providing a triangulated approach to understanding E-field behavior in varying environmental contexts, including the presence of conductive materials like vehicle bodies. This understanding essential for the widespread adoption and public acceptance of wireless charging technology, particularly in the automotive industry.

In the absence of foreign metal objects, the traditional coil exhibits an E-field intensity of 680.5 V/m, serving as our reference point. Alternating voltage phase coils show a significant reduction in E-field intensity, registering at 416.8 V/m and 91.8 V/m, respectively. AVPC 1 achieved a notable reduction of 38.8%, while AVPC 2 demonstrated an impressive 86.5% reduction compared to the traditional design. Such marked reductions underscore the efficacy of the novel coil designs in mitigating E-field exposure, an important consideration for maintaining electromagnetic compatibility and ensuring user safety. When foreign metal objects are introduced, simulating a common environmental effect in electric vehicle (EV) wireless charging scenarios, we observe an overall increase in E-field intensity across all coil designs, highlighting the influence of conductive materials in proximity to the coils, an important variable in the design and deployment of wireless charging infrastructure. Despite this increase, the AVPCs continue to maintain lower E-field levels than the traditional coil, with AVPC 2 in particular showing resilience to the environmental change, increasing to 150.8 V/m but still remaining significantly lower than the traditional coil.

The experimental results validate the trends observed in simulated conditions, with AVPC1 and AVPC 2 exhibiting reduced E-field intensities of 635.3 V/m and 144.3 V/m, respectively. The close alignment between the experimental and simulated data affirms the reliability of our simulation model and reinforces the potential of AVPC designs in real-world applications. The findings underscore the importance of considering environmental factors, such as the presence of foreign metal objects and the distance between the transmitter coil and these objects in the design and implementation of wireless power transfer systems. These factors are particularly relevant in the context of the European Directive limit for public human exposure to EMF at 85 kHz, set at 87 V/m. By significantly reducing E-field emissions, AVPC offers a safer and more environmentally friendly alternative to traditional coil designs, successfully meeting our central research aim of enhancing the safety and compatibility of wireless charging systems and propelling advancements in EV charging technology. The experimental findings demonstrate the AVPC's potential for practical implementation in EV wireless charging systems. However, several challenges need to be addressed for widespread adoption, including the integration with existing EV charging infrastructure, the development of standardized testing protocols, and the assessment of long-term reliability and durability under various environmental conditions.

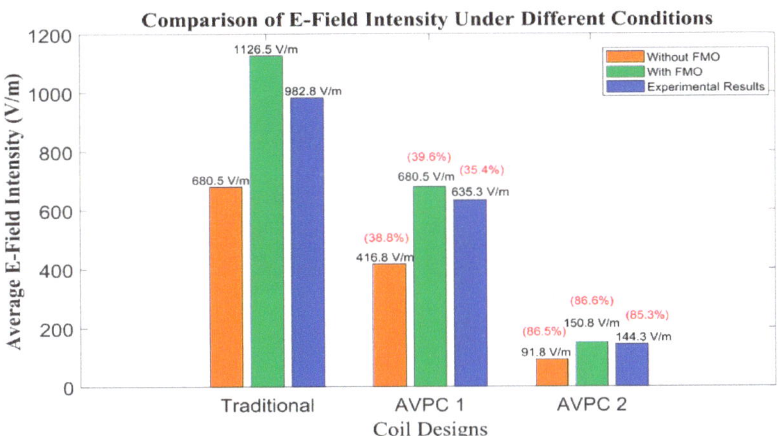

Figure 14. Average electric field intensities comparison.

5. Conclusions

In this paper, we presented the Alternating Voltage Phase Coil (AVPC), a novel coil design that effectively mitigates electromagnetic interference (EMI) in electric vehicle (EV) wireless charging systems. The AVPC's innovative geometry and current flow pattern, achieved through the Sequential Inversion Winding (SIW) configuration, significantly reduce E-field emissions by 85% while maintaining efficient power transfer comparable to traditional coils. The findings of this study clearly demonstrate the efficiency of the AVPC in reducing electric field intensity, an important factor for safer electromagnetic applications.

Comprehensive simulations and experimental validations establish that the strategic modifications in AVPC's turn geometry significantly mitigate E-field exposure, even in environments with foreign metal objects, enhancing electromagnetic compatibility. These results validate the AVPC as a promising approach for developing electromagnetic systems with stringent safety requirements.

The AVPC's adaptability to various devices underscores its broad applicability across multiple industries. This research contributes to the fundamental understanding of electromagnetic field behavior in wireless power transfer systems and opens up new avenues for optimization and innovation. Future work will focus on further enhancing the AVPC's performance, investigating alternative materials, and exploring its scalability for higher power applications.

The AVPC represents a significant advancement in EV wireless charging technology, offering a safer, more efficient, and eco-friendly solution. Future research should focus on validating the AVPC's performance in actual EV charging scenarios to further strengthen the study's findings and support the widespread adoption of this innovative technology. These findings provide a solid foundation for future research and development in designing more efficient and safer coil configurations, with the potential for widespread industrial application and integration. The AVPC is a highly important innovation that plays a vital part in ensuring conformity to exacting safety standards in the rapidly evolving field of electromagnetic technology. With its promising performance and potential for widespread adoption, the AVPC can contribute to the sustainable development of transportation and other sectors.

Author Contributions: Article outline and conceptualization, S.L.; methodology, experiment, and writing—original draft preparation, Z.S. and J.X.; investigation and writing—review and editing, T.L.; project administration and funding acquisition, S.L., Y.Z. and X.Y. All authors have read and agreed to the published version of the manuscript.

Funding: This work was supported in part by the National Natural Science Foundation of China under Grant: (No. 52067011) and in part by Yunnan Fundamental Research Project under Grant (No. 202301AT070429) and also supported by Yunnan Province Major Science and Technology Project (Project No.: 202402AF080001).

Data Availability Statement: Dataset available on reqeust from the authors.

Conflicts of Interest: Authors Xi Yang and Yu Zhao were employed by the company Yunnan Energy Research Institute Co., Ltd. The remaining authors declare that the research was conducted in the absence of any commercial or financial relationships that could be construed as a potential conflict of interest.

References

1. Jawad, A.M.; Nordin, R.; Gharghan, S.K.; Jawad, H.M.; Ismail, M. Opportunities and Challenges for Near-Field Wireless Power Transfer: A Review. *Energies* **2017**, *10*, 1022. [CrossRef]
2. Barman, S.D.; Reza, A.W.; Kumar, N.; Karim, M.E.; Munir, A.B. Wireless Powering by Magnetic Resonant Coupling: Recent Trends in Wireless Power Transfer System and Its Applications. *Renew. Sustain. Energy Rev.* **2015**, *51*, 1525–1552. [CrossRef]
3. Hui, S.Y.R.; Zhong, W.; Lee, C.K. A Critical Review of Recent Progress in Mid-Range Wireless Power Transfer. *IEEE Trans. Power Electron.* **2014**, *29*, 4500–4511. [CrossRef]
4. Covic, G.A.; Boys, J.T. Modern Trends in Inductive Power Transfer for Transportation Applications. *IEEE J. Emerg. Sel. Top. Power Electron.* **2013**, *1*, 28–41. [CrossRef]
5. Chopra, S.; Bauer, P. Driving Range Extension of EV With On-Road Contactless Power Transfer—A Case Study. *IEEE Trans. Ind. Electron.* **2013**, *60*, 329–338. [CrossRef]
6. Xia, J.; Yuan, X.; Lu, S.; Dai, W.; Li, T.; Li, J.; Li, S. A General Parameter Optimization Method for a Capacitive Power Transfer System with an Asymmetrical Structure. *Electronics* **2022**, *11*, 922. [CrossRef]
7. Dai, J.; Ludois, D.C. A Survey of Wireless Power Transfer and a Critical Comparison of Inductive and Capacitive Coupling for Small Gap Applications. *IEEE Trans. Power Electron.* **2015**, *30*, 6017–6029. [CrossRef]
8. Kim, J.; Kim, D.-H.; Park, Y.-J. Analysis of Capacitive Impedance Matching Networks for Simultaneous Wireless Power Transfer to Multiple Devices. *IEEE Trans. Ind. Electron.* **2015**, *62*, 2807–2813. [CrossRef]
9. Li, T.; Li, S.; Liu, Z.; Fang, Y.; Xiao, Z.; Shafiq, Z.; Lu, S. Enhancing V2G Applications: Analysis and Optimization of a CC/CV Bidirectional IPT System with Wide Range ZVS. *IEEE Trans. Transp. Electrific.* **2024**, 1. [CrossRef]
10. Cruciani, S.; Campi, T.; Maradei, F.; Feliziani, M. Wireless Charging in Electric Vehicles: EMI/EMC Risk Mitigation in Pacemakers by Active Coils. In Proceedings of the 2019 IEEE PELS Workshop on Emerging Technologies: Wireless Power Transfer (WoW), London, UK, 18–21 June 2019; IEEE: London, UK, 2019; pp. 173–176.
11. Lu, X.; Wang, P.; Niyato, D.; Kim, D.I.; Han, Z. Wireless Charging Technologies: Fundamentals, Standards, and Network Applications. *IEEE Commun. Surv. Tutor.* **2016**, *18*, 1413–1452. [CrossRef]
12. Kalwar, K.A.; Aamir, M.; Mekhilef, S. Inductively Coupled Power Transfer (ICPT) for Electric Vehicle Charging—A Review. *Renew. Sustain. Energy Rev.* **2015**, *47*, 462–475. [CrossRef]
13. Li, S.; Yu, X.; Yuan, Y.; Lu, S.; Li, T. A Novel High-Voltage Power Supply With MHz WPT Techniques: Achieving High-Efficiency, High-Isolation, and High-Power-Density. *IEEE Trans. Power Electron.* **2023**, *38*, 14794–14805. [CrossRef]
14. Yashima, Y.; Omori, H.; Morizane, T.; Kimura, N.; Nakaoka, M. Leakage Magnetic Field Reduction from Wireless Power Transfer System Embedding New Eddy Current-Based Shielding Method. In Proceedings of the 2015 International Conference on Electrical Drives and Power Electronics (EDPE), Tatranska Lomnica, Slovakia, 21–23 September 2015; IEEE: Tatranska Lomnica, Slovakia, 2015; pp. 241–245.
15. Deng, J.; Li, W.; Nguyen, T.D.; Li, S.; Mi, C.C. Compact and Efficient Bipolar Coupler for Wireless Power Chargers: Design and Analysis. *IEEE Trans. Power Electron.* **2015**, *30*, 6130–6140. [CrossRef]

16. Mei, Y.; Wu, J.; He, X. Common Mode Noise Analysis for Inductive Power Transfer System Based on Distributed Stray Capacitance Model. *IEEE Trans. Power Electron.* **2022**, *37*, 1132–1145. [CrossRef]
17. Estrada, J.; Sinha, S.; Regensburger, B.; Afridi, K.; Popovic, Z. Capacitive Wireless Powering for Electric Vehicles with Near-Field Phased Arrays. In Proceedings of the 2017 47th European Microwave Conference (EuMC), Nuremberg, Germany, 10–12 October 2017; IEEE: Nuremberg, Germany, 2017; pp. 196–199.
18. Mai, J.; Wang, Y.; Zeng, X.; Yao, Y.; Wu, K.; Xu, D. A Multi-Segment Compensation Method for Improving Power Density of Long-Distance IPT System. *IEEE Trans. Ind. Electron.* **2022**, *69*, 12795–12806. [CrossRef]
19. Shafiq, Z.; Xia, J.; Min, Q.; Li, S.; Lu, S. Study of the Induced Electric Field Effect on Inductive Power Transfer System. In Proceedings of the 2021 IEEE PELS Workshop on Emerging Technologies: Wireless Power Transfer (WoW), San Diego, CA, USA, 1–4 June 2021; IEEE: San Diego, CA, USA, 2021; pp. 1–5.
20. Kim, M.; Ahn, S.; Kim, H. Magnetic Design of a Three-Phase Wireless Power Transfer System for EMF Reduction. In Proceedings of the 2014 IEEE Wireless Power Transfer Conference, Jeju City, South Korea, 8–9 May 2014; IEEE: Jeju City, South Korea, 2014; pp. 17–20.
21. Liu, Z.; Li, T.; Li, S.; Mi, C.C. Advancements and Challenges in Wireless Power Transfer: A Comprehensive Review. *Nexus* **2024**, *1*, 100014. [CrossRef]
22. Choi, S.Y.; Gu, B.W.; Jeong, S.Y.; Rim, C.T. Advances in Wireless Power Transfer Systems for Roadway-Powered Electric Vehicles. *IEEE J. Emerg. Sel. Top. Power Electron.* **2015**, *3*, 18–36. [CrossRef]
23. Budhia, M.; Covic, G.A.; Boys, J.T. Design and Optimization of Circular Magnetic Structures for Lumped Inductive Power Transfer Systems. *IEEE Trans. Power Electron.* **2011**, *26*, 3096–3108. [CrossRef]
24. Kim, J.; Kim, J.; Kong, S.; Kim, H.; Suh, I.-S.; Suh, N.P.; Cho, D.-H.; Kim, J.; Ahn, S. Coil Design and Shielding Methods for a Magnetic Resonant Wireless Power Transfer System. *Proc. IEEE* **2013**, *101*, 1332–1342. [CrossRef]
25. Siqi, L.; Mi, C. Wireless Power Transfer for Electric Vehicle Applications. *IEEE J. Emerg. Sel. Top. Power Electron.* **2015**, *3*, 4–17. [CrossRef]
26. Campi, T.; Cruciani, S.; Maradei, F.; Feliziani, M. Near-Field Reduction in a Wireless Power Transfer System Using LCC Compensation. *IEEE Trans. Electromagn. Compat.* **2017**, *59*, 686–694. [CrossRef]
27. Stepins, D.; Zakis, J.; Padmanaban, P.; Deveshkumar Shah, D. Suppression of Radiated Emissions from Inductive-Resonant Wireless Power Transfer Systems by Using Spread-Spectrum Technique. *Electronics* **2022**, *11*, 730. [CrossRef]
28. Inoue, K.; Kusaka, K.; Itoh, J.-I. Reduction in Radiation Noise Level for Inductive Power Transfer Systems Using Spread Spectrum Techniques. *IEEE Trans. Power Electron.* **2018**, *33*, 3076–3085. [CrossRef]
29. Park, S. Evaluation of Electromagnetic Exposure During 85 kHz Wireless Power Transfer for Electric Vehicles. *IEEE Trans. Magn.* **2018**, *54*, 5100208. [CrossRef]
30. Duan, X.; Lan, J.; Kodera, S.; Kirchner, J.; Fischer, G.; Hirata, A. Wireless Power Transfer Systems With Composite Cores for Magnetic Field Shielding With Electric Vehicles. *IEEE Access* **2023**, *11*, 144887–144901. [CrossRef]
31. Li, T.; Yuan, Y.; Xiao, Z.; Fang, Y.; Xingpeng, Y.; Li, S. Large Space Wireless Power Transfer System That Meets Human Electromagnetic Safety Limits. In Proceedings of the 2023 IEEE Wireless Power Technology Conference and Expo (WPTCE), San Diego, CA, USA, 4–8 June 2023; IEEE: San Diego, CA, USA, 2023; pp. 1–6.
32. *Guidelines on Limiting Exposure to Non-Ionizing Radiation: A Reference Book Based on the Guidelines on Limiting Exposure to Non-Ionizing Radiation and Statements on Special Applications*; Matthes, R.; Bernhardt, J.H. (Eds.) ICNIRP; International Commission on Non-Ionizing Radiation Protection: Oberschleißheim, Germany, 1999; ISBN 978-3-9804789-6-0.
33. International Commission on Non-Ionizing Radiation Protection. Guidelines for limiting exposure to time-varying electric and magnetic fields (1 Hz to 100 kHz). *Health Phys.* **2010**, *99*, 818–836. [CrossRef]
34. Jo, M.; Sato, Y.; Kaneko, Y.; Abe, S. Methods for Reducing Leakage Electric Field of a Wireless Power Transfer System for Electric Vehicles. In Proceedings of the 2014 IEEE Energy Conversion Congress and Exposition (ECCE), Pittsburgh, PA, USA, 14–18 September 2014; IEEE: Pittsburgh, PA, USA, 2014; pp. 1762–1769.
35. Wang, B.; Yerazunis, W.; Teo, K.H. Wireless Power Transfer: Metamaterials and Array of Coupled Resonators. *Proc. IEEE* **2013**, *101*, 1359–1368. [CrossRef]
36. Sample, A.; Smith, J.R. Experimental Results with Two Wireless Power Transfer Systems. In Proceedings of the 2009 IEEE Radio and Wireless Symposium, San Diego, CA, USA, 18–22 January 2009; IEEE: San Diego, CA, USA, 2009; pp. 16–18.
37. Sample, A.P.; Meyer, D.A.; Smith, J.R. Analysis, Experimental Results, and Range Adaptation of Magnetically Coupled Resonators for Wireless Power Transfer. *IEEE Trans. Ind. Electron.* **2011**, *58*, 544–554. [CrossRef]
38. Shahjalal, M.; Shams, T.; Tasnim, M.N.; Ahmed, M.R.; Ahsan, M.; Haider, J. A Critical Review on Charging Technologies of Electric Vehicles. *Energies* **2022**, *15*, 8239. [CrossRef]
39. Hayt, W. Engineering Electromagnetics. In *Physical Constants*; McGraw-Hill Companies: New York, NY, USA, 1974.
40. Paul, C.R. *Introduction to Electromagnetic Compatibility*; John Wiley & Sons: Hoboken, NJ, USA, 2022.
41. Tumanski, S. Induction Coil Sensors—A Review. *Meas. Sci. Technol.* **2007**, *18*, R31–R46. [CrossRef]

42. Kurokawa, K. Power Waves and the Scattering Matrix. *IEEE Trans. Microw. Theory Techn.* **1965**, *13*, 194–202. [CrossRef]
43. Sirbu, I.-G.; Mandache, L. Comparative Analysis of Different Topologies for Wireless Power Transfer Systems. In Proceedings of the 2017 IEEE International Conference on Environment and Electrical Engineering and 2017 IEEE Industrial and Commercial Power Systems Europe (EEEIC/I&CPS Europe), Milan, Italy, 6–9 June 2017; IEEE: Milan, Italy, 2017; pp. 1–6.

Article

A Mechanical Fault Diagnosis Method for UCG-Type On-Load Tap Changers in Converter Transformers Based on Multi-Feature Fusion

Yanhui Shi [1], Yanjun Ruan [1], Liangchuang Li [1], Bo Zhang [1], Kaiwen Yuan [2], Zhao Luo [2], Yichao Huang [2], Mao Xia [2], Siqi Li [2] and Sizhao Lu [2,*]

[1] Guangzhou Bureau of EHV Power Transmission Company, China Southern Power Grid Co., Ltd., Guangzhou 510000, China; shiyanhui@ieee.org (Y.S.); ruanyanjun@ieee.org (Y.R.); liangchaungli@ieee.org (L.L.); bozhang1@ieee.org (B.Z.)
[2] Faculty of Electric Power Engineering, Kunming University of Science and Technology, Kunming 650500, China; 20222105010@stu.kust.edu.cn (K.Y.); 20170025@kust.edu.cn (Z.L.); 20232105012@stu.kust.edu.cn (Y.H.); xiamao@stu.kust.edu.cn (M.X.); lisiqi@kust.edu.cn (S.L.)
* Correspondence: lusz10@kust.edu.cn

Abstract: The On-Load Tap Changer (OLTC) is the only movable mechanical component in a converter transformer. To ensure the reliable operation of the OLTC and to promptly detect mechanical faults in OLTCs to prevent them from developing into electrical faults, this paper proposes a fault diagnosis method for OLTCs based on a combination of Particle Swarm Optimization (PSO) algorithm and Least Squares Support Vector Machine (LSSVM) with multi-feature fusion. Firstly, a multi-feature extraction method based on time/frequency domain statistics, synchrosqueezed wavelet transform, singular value decomposition, and multi-scale modal decomposition is proposed. Meanwhile, the random forest algorithm is used to screen features to eliminate the influence of redundant features on the accuracy of fault diagnosis. Secondly, the PSO algorithm is introduced to optimize the hyperparameters of LSSVM to obtain optimal parameters, thereby constructing an optimal LSSVM fault diagnosis model. Finally, different types of feature combinations are utilized for fault diagnosis, and the impact of these feature combinations on the fault diagnosis results is compared. Experimental results indicate that features of different types can complement each other, making the OLTC state information carried by multi-dimensional features more comprehensive, which helps to improve the accuracy of fault diagnosis. Compared with four traditional fault diagnosis methods, the proposed method performs better in fault diagnosis accuracy, achieving the highest accuracy of 98.58%, which can help to detect mechanical faults in the OLTC early and reduce the system's downtime.

Keywords: multi-feature extraction; on-load tap changers; least squares support vector machine; fault diagnosis; particle swarm optimization

Citation: Shi, Y.; Ruan, Y.; Li, L.; Zhang, B.; Yuan, K.; Luo, Z.; Huang, Y.; Xia, M.; Li, S.; Lu, S. A Mechanical Fault Diagnosis Method for UCG-Type On-Load Tap Changers in Converter Transformers Based on Multi-Feature Fusion. *Actuators* **2024**, *13*, 387. https://doi.org/10.3390/act13100387

Academic Editor: Giorgio Olmi

Received: 15 August 2024
Revised: 24 September 2024
Accepted: 28 September 2024
Published: 1 October 2024

1. Introduction

The On-Load Tap Changer (OLTC) is the only movable mechanical component in a converter transformer [1]. It simultaneously endures electrical and mechanical stress, making it one of the most vulnerable components in the converter transformer [2]. Relevant data indicate that OLTC failures account for more than 20% of the total transformer failures, with mechanical failures constituting over 95% of all OLTC failures [3,4]. Therefore, it is of great significance to carry out research on an OLTC mechanical fault diagnosis technology of converter transformer and identify the hidden dangers of OLTC faults as early as possible to maintain the stable operation of high-voltage DC transmission systems.

The mechanical vibration signals of OLTCs carry abundant information about the state of the OLTC. In 1996, Bengtsson first introduced vibration signal analysis techniques into the OLTC mechanical fault diagnosis [5]. The core concept of this method is to use vibration sensors to monitor the vibration signals generated during the operation of the OLTC in a non-intrusive way, thereby obtaining OLTC state information for OLTC operation evaluation and judgment.

Currently, most of studies on OLTC mechanical fault diagnosis focus on rotary-type OLTCs [1,6,7], while studies on swing-arm-type OLTCs remain relatively scarce [7]. OLTC mechanical fault diagnosis methods can be roughly divided into two categories: one is to realize fault diagnosis through manual feature extraction combined with classifiers; the other is to use deep learning models for automatic feature extraction and fault diagnosis.

The manual feature extraction methods for OLTCs include time-domain feature extraction, frequency-domain feature extraction, time-frequency feature extraction, and dynamic feature extraction. These approaches have been extensively investigated by many researchers. Kang et al. employed continuous wavelet transform to extract time-domain envelope lines and successfully identified typical faults such as contact aging, erosion, and looseness using "ridge distribution maps" as features [8,9]. Gao et al. combined empirical mode decomposition with energy entropy to successfully distinguish between normal states and contact erosion faults [10]. Zhang et al. designed a time-frequency matrix partitioning algorithm and extracted parameter features such as partition lines, kurtosis, and envelope spectrum entropy [11]. Qian et al. utilized variational mode decomposition to extract the energy features of OLTC vibrations [12]. Zhao et al. utilized phase space reconstruction techniques to map one-dimensional time series signals into high-dimensional space and defined the phase point space distribution coefficient as the feature [13]. However, the existing methods have certain limitations. For instance, wavelet transform relies on extensive experience and expertise [14], while empirical mode decomposition is prone to mode mixing and endpoint effects [15]. The effectiveness of variational mode decomposition is significantly influenced by the number of decomposition components and the penalty factor [16]. Meanwhile, phase space reconstruction methods are sensitive to noise and computationally intensive [17]. In OLTC fault diagnosis, commonly used classifiers include Support Vector Machine (SVM) [1], BP neural network [18], Long Short-Term Memory network (LSTM) [19], and Convolutional Neural Network (CNN) [20].

There are two main methods of applying deep learning-based automatic feature extraction in OLTC fault diagnosis. The first method directly employs the neural networks to process the OLTC vibration signals and uses the hierarchical structure of the network to automatically extract signal features to perform fault diagnosis [21]. The second method is to convert the OLTC vibration signals into images, and then extract key features from the image through deep learning methods such as CNNs to achieve fault diagnosis [22,23].

Deep learning models can automatically extract features from OLTC vibration signals, eliminating the need for complex preprocessing and feature engineering, and significantly simplifying the feature design process [24,25]. These models also exhibit strong generalization capabilities [26]. For instance, target transfer deep learning methods based on a distribution barycenter medium [27], along with label recovery and trajectory-designable deep transfer learning methods [28], can effectively achieve fault diagnosis between different devices of the same type. However, they often struggle to explain the specific relationships between features and faults [26]. Additionally, deep learning models typically require large amounts of labeled data for training, they are prone to overfitting or reduced diagnostic accuracy when training data are insufficient [29]. Moreover, due to their complex network structures, deep learning models consume significant computational resources and have long training times, making them unsuitable for real-time or resource-constrained applications [30].

Manual feature extraction relies on domain knowledge, with clear correspondence between features and actual physical phenomena, making the diagnostic results easier to interpret [31]. The extracted features are typically of lower dimensionality, consuming fewer computational resources, which is well-suited for real-time applications [32]. In cases where the data sample size is limited, the fault diagnosis methods based on the manual feature extraction often perform more consistently than deep learning models based on small datasets [33].

In practical operations, OLTC faults occur infrequently, making it a typical small-sample-size diagnostic problem. The Least Squares Support Vector Machine (LSSVM), with its high accuracy, fast training speed, simple parameter tuning, excellent small-sample handling capability, and outstanding nonlinear processing ability, has emerged as a powerful tool for addressing the difficulties in fault diagnosis [34]. However, in the LSSVM model, the penalty factor c and the kernel parameter g have a significant impact on the model performance, so these hyperparameters need to be optimized.

As an intelligent optimization algorithm, Particle Swarm Optimization (PSO) has the characteristics of few parameters, fast calculation speed, and strong global search ability [35]. Therefore, this study employed PSO to optimize the hyperparameters of LSSVM. To prevent PSO from falling into a local optimum, PSO was used to optimize the hyperparameters multiple times, so as to obtain the optimal hyperparameters of LSSVM.

This paper focuses on the UCG-type OLTC used in high-voltage DC converter transformers. The UCG-type OLTC employs a swing-arm switching mechanism, which differs from the rotary-type OLTC [6,7]. The vibration signals generated by these two types of OLTCs during the operation exhibit significant differences. Consequently, a multi-feature extraction method is proposed to address the issue of the low accuracy of fault diagnosis based on a single feature. Additionally, a PSO-LSSVM-based fault diagnosis model is developed to achieve accurate diagnosis of OLTCs' mechanical faults. Firstly, vibration signals under different OLTC fault conditions are collected through the experimental test platform, and time-domain and frequency-domain statistical features, singular-value features of time-frequency matrices, and multi-scale modal features are extracted. Secondly, multi-dimensional feature vectors are constructed, and the random forest algorithm is employed for feature selection to eliminate the influence of redundant features, thereby improving the accuracy and reliability of fault diagnosis. Finally, PSO is applied to optimize the penalty factor c and kernel parameter g of the LSSVM model, and a PSO-LSSVM fault diagnosis model is built.

The results demonstrate that the multi-feature extraction method can effectively capture the feature changes of OLTCs under different states, while the LSSVM model optimized by the PSO can accurately identify mechanical faults in OLTCs. This study provides an effective method for diagnosing mechanical faults in high-voltage DC converter transformer OLTCs, preventing early mechanical faults in OLTCs from evolving into electrical faults.

The contributions of this paper are listed as follows:

(1) This paper focuses on the mechanical fault diagnosis of swing-arm-type (UCG-type) OLTCs, which are employed in 800 kV high-voltage DC converter transformers.

(2) A multi-feature extraction method is proposed, which overcomes the shortcomings of insufficient information of a single feature, which leads to a low accuracy of fault diagnosis.

(3) The feature importance evaluation based on random forest algorithm is introduced to screen the features and eliminate redundant features, so as to find the most effective feature combination, which can further improve the accuracy of fault diagnosis.

(4) The parameters of LSSVM are optimized by the PSO algorithm, and a small-sample-size fault diagnosis model based on PSO-LSSVM is established, and then the model is applied to the fault diagnosis of a UCG-type OLTC; good fault diagnosis results are obtained.

2. Proposed Mechanical Fault Diagnosis Method

This study extracts multi-dimensional features of OLTC vibration signals from four perspectives: time-domain features, frequency-domain features, time-frequency matrix energy features, and multi-scale modal features. A PSO-optimized LSSVM model is then utilized for fault diagnosis. The specific process of the proposed method is illustrated in Figure 1, with the main steps as follows:

(1) Step I: An experimental test platform for the UCG-type converter transformer OLTC is constructed. Multiple vibration sensors are used to collect OLTC vibration signals under different states.

(2) Step II: Multi-dimensional feature extraction is performed on the collected OLTC vibration signals, creating a dataset of features. The random forest algorithm is used to screen the features to eliminate the influence of any redundant features.

(3) Step III: An OLTC fault diagnosis model based on PSO-optimized LSSVM is established. The detailed optimization process is discussed in Section 4.2. Firstly, the number of PSO optimization iterations is preset. After each iteration, an optimal combination of hyperparameters is obtained and it is used to conduct 20 fault diagnosis tests. Consequently, the average optimal accuracy rate is calculated. Then, the optimization LSSVM model is obtained by synthesizing the results of multiple optimizations and selecting the hyperparameters corresponding to the optimal accuracy. Finally, the optimal LSSVM model is employed for OLTC mechanical fault diagnosis, yielding the diagnostic results.

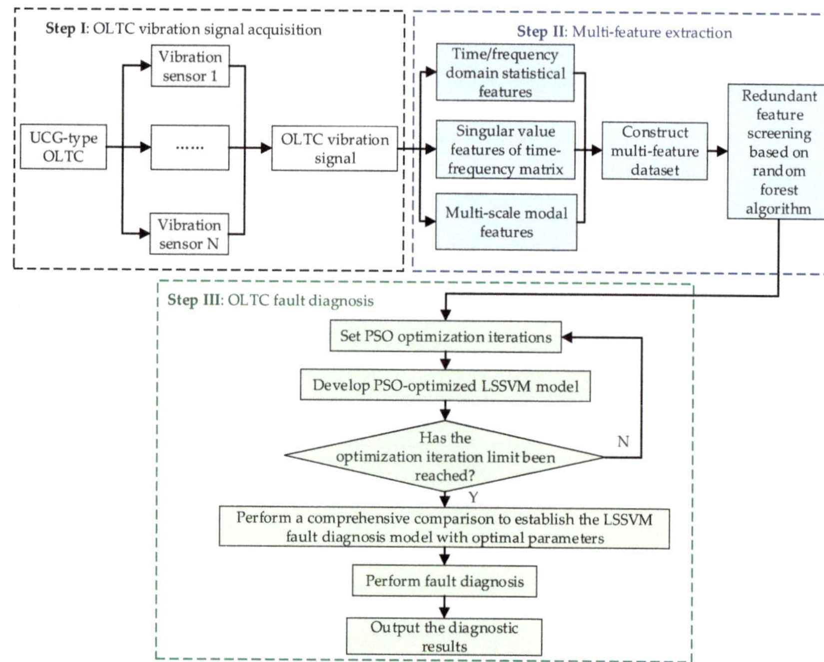

Figure 1. OLTC fault diagnosis process of the proposed method.

3. OLTC Vibration Signal Acquisition

In this study, the UCG-type OLTC manufactured by Hitachi Energy was used to build the experimental test platform. The tests were conducted under no-load conditions. The UCG-type OLTC consists of a motor drive unit, a diverter switch, and a selector switch, which is illustrated in Figure 2. The switching process of the UCG-type OLTC follows the sequence in which the selector switch first selects, and then the diverter switch operates.

The sensor adopts Beijing QUATRONIX ULT2001 piezoelectric vibration sensor (QUATRONIX Electronics Co., Ltd., Beijing, China) with a range of ± 50 g and a sensitivity of 100 mV/g, and the frequency response range is 0.5~15 kHz. A data acquisition card with a sampling frequency of up to 100 kHz is employed. The data acquisition process is illustrated in Figure 3.

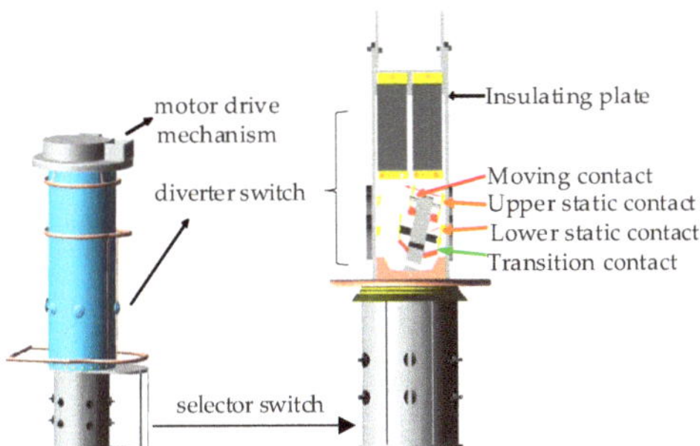

Figure 2. Internal structure of the UCG-type OLTC.

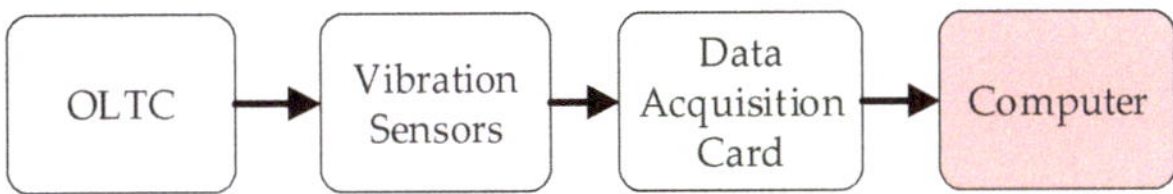

Figure 3. Data collection process.

The installation position of the vibration sensors is shown in Figure 4, in which four vibration sensors are installed on the top of the OLTC.

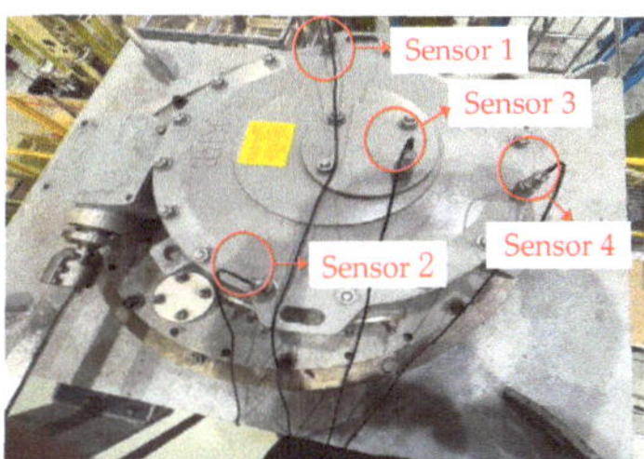

Figure 4. Installation position of vibration sensors. Figure 4 illustrates the installation positions of the four Beijing QUATRONIX ULT2001 vibration sensors. Vibration sensors 1, 2, and 4 are mounted on the edge of the OLTC top cover, while sensor 3 is positioned on the protrusion of the top cover.

Figure 5 illustrates the setting of four fault conditions. In Figure 5a, the fault of the upper stationary contact loosening is set by loosening the screws of the upper stationary contact. In Figure 5b, the fault of the lower stationary contact loosening is set by loosening the screws of the lower stationary contact. In Figure 5c, the fault of the insulation plate loosening is set by loosening the screws of the insulation plate. In Figure 5d, the fault of the moving contact loosening is set by cutting the spring that connects the moving contact. Data from five conditions were collected, including normal, upper stationary contact loosening, lower stationary contact loosening, insulation plate loosening, and moving contact loosening conditions.

The UCG-type OLTC vibration signals under different operating conditions are shown in Figure 6. By comparing Figure 6c,e,g,i with Figure 6a, it can be observed that when different types of loosening faults occur, the vibration signal intensity in the time domain changes, and the waveform shows significant differences. A comparison of Figure 6d,f,h with Figure 6a reveals that in the frequency domain, when the upper static contact, lower static contact, or insulation plate become loose, a significant number of low-frequency components appears in the vibration signal. When the moving contact becomes loose, both high-frequency and low-frequency components increase. Compared to normal conditions, the peak value of the vibration intensity decreases. Therefore, the mechanical fault diagnosis of the UCG-type OLTC can be achieved by extracting the frequency distribution characteristics of the UCG-type OLTC vibration signals.

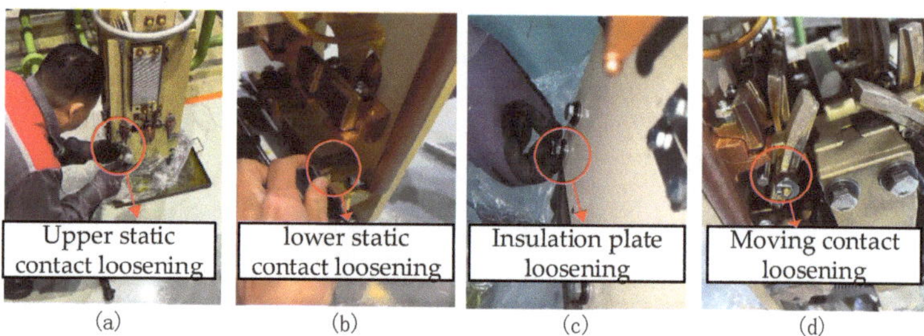

Figure 5. Different fault condition settings: (**a**) upper static contact loosening, (**b**) lower static contact loosening, (**c**) insulation plate loosening, (**d**) moving contact loosening. In (**a**), the red area shows how the upper static contact loosening is set up. In (**b**), the red area indicates the setup and position of the lower static contact loosening. In (**c**), the red area illustrates the setup and location of the insulation plate loosening. In (**d**), the red area shows the setup and location for the moving contact loosening.

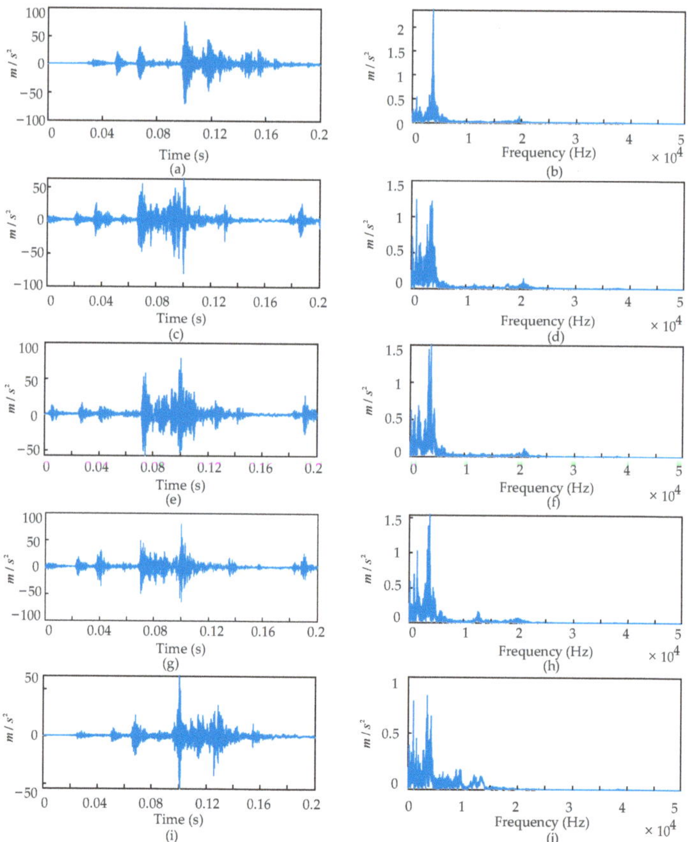

Figure 6. UCG-type OLTC vibration signals under different operating conditions. (**a**) Normal condition time-domain signal, (**b**) normal condition frequency-domain signal, (**c**) upper static contact loosening time-domain signal, (**d**) upper static contact loosening frequency-domain signal, (**e**) lower static contact loosening time-domain signal, (**f**) lower static contact loosening frequency-domain signal, (**g**) insulation plate loosening time-domain signal, (**h**) insulation plate loosening frequency-domain signal, (**i**) moving contact loosening time-domain signal, (**j**) moving contact loosening frequency-domain signal.

4. Multi-Feature Extraction Method for OLTC Vibration Signals

To obtain the feature variations of OLTC vibration signals under different operating conditions, this study extracted both time-domain and frequency-domain features of the OLTC vibration signals, focusing on the overall changes in the vibration signals. By employing synchrosqueezed wavelet transform (SWT), detailed features of the time-frequency energy distribution were extracted. The multi-scale decomposition of the signals was performed to focus on the local features. The complementary nature of different feature types enhances the accuracy of fault diagnosis.

4.1. Statistical Feature Extraction in the Time/Frequency Domain

OLTC vibration signals typically manifest in the form of impulse waves. When a fault occurs in the OLTC, the statistical features of the vibration signal in both the time and frequency domains will change due to the change of fault type and fault severity, which can be used as the basis for fault diagnosis. When a fault occurs in the OLTC, the amplitude and the distribution of vibrational energy change in the time-domain, while the signal frequency and energy distribution also change in the time-domain. Table 1 shows the calculation methods for time-domain features, while Table 2 illustrates the calculation methods for frequency-domain features. In these two tables, x_i represents the raw OLTC vibration signal, and N denotes the sampling length of the signal. X_i represents the frequency-domain signal of x_i, and f_i denotes the frequency values of each frequency-domain point.

Table 1. Time-domain statistical features [36].

Feature	Calculation Formula	Feature	Calculation Formula
Mean	$S_1 = \frac{1}{N} \sum_{i=1}^{N} \lvert x(n) \rvert$	Crest Factor	$S_5 = S_4/S_3$
Root Amplitude	$S_2 = \left(\frac{1}{N} \sum_{i=1}^{N} \sqrt{\lvert x(n) \rvert} \right)^2$	Form Factor	$S_6 = S_3/S_1$
Root Mean Square	$S_3 = \sqrt{\frac{1}{N} \sum_{i=1}^{N} x_i^2}$	Impulse Factor	$S_7 = S_4/S_1$
Peak-to-Peak Value	$S_4 = \max(x_i) - \min(x_i)$	Margin Indicator	$S_8 = S_4/S_2$

Table 2. Frequency-domain feature parameters [36].

Feature	Calculation Formula	Feature	Calculation Formula
Average Energy	$P_1 = \sum_{i=1}^{N} X_i / N$	Frequency Standard Deviation	$P_3 = \sqrt{\sum_{i=1}^{K} (f_i - P_2)^2 \times X_i / \sum_{i=1}^{N} X_i}$
Central Frequency	$P_2 = \sum_{i=1}^{N} (f_i \times X_i) / \sum_{i=1}^{N} X_i$	Dispersion Factor	$P_4 = \sum_{i=1}^{N} (X_i - P_2)^3 / \sum_{i=1}^{N} X_i \times (P_2)^{\frac{3}{2}}$

4.2. Singular Value Feature Extraction of Time-Frequency Matrix

4.2.1. Synchrosqueezed Wavelet Transform

In 2011, Daubechies et al. proposed a synchronous compression transform method based on continuous wavelet transform, which is SWT [37]. This method enhances frequency aggregation and energy concentration by reordering the frequency series, thus achieving more efficient compression transformation. The formula for SWT is obtained from reference [37]. First, the continuous wavelet transform of the signal *s(t)* is expressed as

$$W_s^\psi(a, b) = \frac{1}{\sqrt{a}} \int_{-\infty}^{+\infty} s(t) \psi\left(\frac{t - b}{a}\right) dt \tag{1}$$

where a is the scale factor, b is the translation factor, $\psi(t)$ is the wavelet basis function, and $\overline{\psi(t)}$ is the conjugate function of $\psi(t)$.

For any $W_s^\psi(a,b) \neq 0$ of (a, b), the instantaneous frequency $\omega_s(a,b)$ of the signal $s(t)$ is given by the following equation:

$$\omega_s(a,b) = -j\frac{\partial_b W_s^\psi(a,b)}{W_s^\psi(a,b)} \tag{2}$$

where $\partial_b W_s^\psi(a,b)$ is the partial derivative of the continuous wavelet transform $W_s^\psi(a,b)$ with respect to the translation factor b.

Based on the mapping relationship $(a,b) \rightarrow (\omega_s(a,b),b)$, in the time-frequency plane, the continuous wavelet transform coefficient $W_s^\psi(a,b)$ can be reordered using synchronous compression calculation as follows:

$$T_s(\omega,b) = \int_{-\infty}^{+\infty} W_s^\psi(a,b)a^{-3/2}\delta[\omega_s(a,b) - \omega]da \tag{3}$$

4.2.2. Singular Value Decomposition Feature Extraction

The singular values obtained by the Singular Value Decomposition (SVD) of the matrix contain important information of the matrix and are widely used for the feature extraction of signals [38]. For any given matrix M, there are orthogonal matrices U and V, and it can be decomposed as

$$M_{m \times n} = USV^T \tag{4}$$

$$S = diag(\sigma_1, \sigma_2, \cdots, \sigma_r, \cdots, 0) \tag{5}$$

where $\sigma_1, \sigma_2, \cdots, \sigma_r, \cdots, 0$ are the singular values of the matrix M.

The time-frequency matrix of the OLTC vibration signal is obtained by SWT. Each element of this matrix is modulated, and then the SVD is performed. Therefore, the obtained singular values reflect the features of the OLTC vibration signal, and the corresponding feature matrix is determined. Since the number of singular values is affected by the number of rows and columns of the amplitude matrix, which is usually larger, further processing is required to highlight the relative changes in the OLTC vibration signal in different states. This paper extracts the maximum and average singular values of the OLTC vibration signal in different states as features, which are expressed as

$$T_1 = \max(\sigma_1, \sigma_2, \cdots, \sigma_r) \tag{6}$$

$$T_2 = \frac{\sigma_1 + \sigma_2 + \cdots + \sigma_r}{r} \tag{7}$$

4.3. Multi-Scale Mode Feature Extraction

To enhance the sensitivity to subtle changes in OLTC states, the Improved Complete Ensemble Empirical Mode Decomposition with Adaptive Noise (ICEEMDAN) is used to decompose the OLTC vibration signal [39]. ICEEMDAN is an improved signal decomposition method, which effectively solves the modal aliasing problem in the EEMD algorithm by introducing Gaussian white noise and redefining the modal mean, thereby significantly improving the accuracy and stability of signal decomposition [39]. The specific process is given as follows, the formula for ICEEMDAN is obtained from reference [39].

(1) Step I: Add m sets of white noise $\omega^{(m)}$ to the original sequence x to obtain a new sequence, which is given by

$$X_1^{(m)} = x + B_1 E_1(\omega^{(m)}) \tag{8}$$

where B_k is the noise of the r-*th* sequence, and $E_k(\cdot)$ represents the k-*th* mode component of the EMD decomposition.

(2) Step II: Calculate the first residual signal r_1 and mode component IMF_1, which are given by

$$r_1 = \left\langle X_1^{(m)} - E_1(X_1^{(m)}) \right\rangle \tag{9}$$

$$IMF_1 = x - r_1 \tag{10}$$

where $M(\cdot)$ represents the local mean of the generated signal, and $\langle \cdot \rangle$ is the averaging operator.

(3) Step III: Continue to add white noise to r_1, and then find the mean value to obtain $X_2^{(m)} = x + B_1 E_1(\omega^{(m)})$, and calculate the modal component $IMF2$ of the second set of residual signal r_2, which are given by

$$r_2 = \left\langle X_2^{(m)} - E_2(X_2^{(m)}) \right\rangle \tag{11}$$

$$IMF_2 = r_1 - r_2 \tag{12}$$

(4) Step IV: Repeat *Step III* to obtain the k-*th* residual signal and the k-*th* mode component, which can be obtained by

$$r_k = \left\langle X_k^{(m)} - E_k(X_k^{(m)}) \right\rangle \tag{13}$$

$$IMF_k = r_{k-1} - r_k \tag{14}$$

(5) Step V: Return to *Step IV* until the residual signal can no longer be decomposed or meets the stop criterion, obtaining all mode components.

After ICEEMDAN decomposition, the original signal is decomposed into multiple IMF components, each of which has a different correlation with the original signal. Therefore, the Pearson correlation coefficient method is used to screen out the first k modes IMF_k with higher correlation to the original signal, reducing the interference of irrelevant components. Then, the frequency-domain mean value M_{IMF} and the central frequency G_{IMF} of each IMF are extracted, and they are combined to form the multi-scale mode features. The specific calculation formulas can be expressed as

$$M_{IMF} = \sum_{k=1}^{N} IMF(k) \bigg/ N \tag{15}$$

$$G_{IMF} = \sum_{k=1}^{N} IMF(k) \times f_k \bigg/ \sum_{k=1}^{N} IMF(k) \tag{16}$$

where f_k is the frequency corresponding to each component spectrum, and N is the sample size of the IMF sequence.

4.4. Feature Screening Based on Random Forest Algorithm

The extracted UCG-type OLTC vibration signal features have high dimensionality, and there may be redundant features. By utilizing the random forest algorithm for feature screening, the impact of redundant features on the accuracy of fault diagnosis can be effectively avoided [40]. The random forest algorithm can evaluate the importance of features by measuring their contribution in each decision tree and averaging them. This allows for the features to be ranked and the influence of each feature to be compared [41].

4.4.1. Construction Principles of the Random Forest Algorithm

The construction process of the random forest model is shown in Figure 7. Suppose the dataset contains H samples, and each sample has M features. The specific process is given as follows:

(1) Step I: Use the bootstrap method to randomly select $2H/3$ of the samples from the dataset to create a training set.
(2) Step II: Generate a decision tree for each training set. Randomly select M features (without repetition) as candidate features, and use these M features to determine the optimal feature that results in the best splitting effect.
(3) Step III: Repeat step I and step II until k decision trees are generated.
(4) Step IV: The trained random forest is used to make predictions on the testing set, and the final prediction result is determined by a voting mechanism.

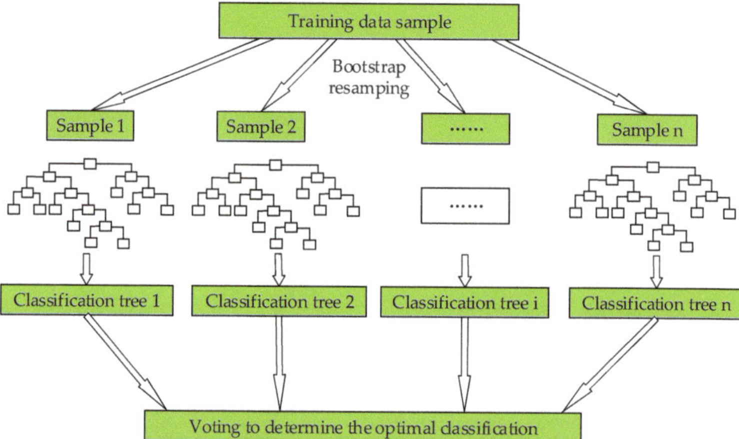

Figure 7. Construction principles of the random forest algorithm.

4.4.2. Feature Importance Evaluation

When the random forest algorithm is used to evaluate the feature importance, its core idea is to quantify the contribution of each feature to the model performance. In random forest algorithms, Out-Of-Bag (OOB) error is often used as an indicator to assess the contribution of each feature [41]. The Feature Importance Measure (FIM) reflects the contribution of each feature to the classification accuracy.

The $FIM_{km}^{(OOB)}$ of feature F_m in the k-th decision tree is calculated as follows [41]:

$$FIM_{km}^{(OOB)} = \frac{\sum_{p=1}^{n_0^k} I(Y_p, Y_p^k)}{n_0^k} - \frac{\sum_{p=1}^{n_0^k} I(Y_p, Y_{p,\pi m}^k)}{n_o^k} \tag{17}$$

where I(x,y) is an indicator function, which is defined as $I(x, y) = \begin{cases} 1, x = y \\ 0, x \neq y \end{cases}$. It is used to determine whether the classification result is correct. n_o^k is the number of OOB samples in the k-th decision tree, Y_p is the true label of the p-th sample, and Y_p^k is the predicted result of the p-th sample in the k-th decision tree without feature permutation. $Y_p^{k,m}$ is the predicted result of the p-th sample after feature F_m has been permuted in the k-th decision tree.

By comparing the classification accuracy of OOB samples before and after permuting feature F_m, the contribution of the feature to the classification accuracy can be measured. If the classification accuracy decreases significantly after the permutation, it indicates that this feature plays a crucial role in the random forest algorithm.

To obtain the overall importance score of feature F_m in the random forest, $FIM_{km}^{(OOB)}$ in k-th decision tree is averaged and divided by the standard deviation to reduce the uncertainty. It is calculated as follows [41]:

$$FIM_m^{(OOB)} = \frac{\sum_{k=1}^{K} FIM_{km}^{(OOB)}}{K\sigma} \tag{18}$$

where K is the total number of trees in the random forest, and σ is the standard deviation of the importance score $FIM_{km}^{(OOB)}$ for feature F_m.

4.4.3. Feature Screening Process Based on the Random Forest Algorithm

The steps for feature screening using the random forest algorithm are given as follows:

(1) Step I: Extract features from the UCG-type OLTC vibration signals.
(2) Step II: Construct a random forest model.
(3) Step III: Calculate the importance scores of the features.
(4) Step IV: Based on the feature importance scores, screen features step by step according to the ranking from high to low.

5. PSO-LSSVM-Based OLTC Fault Diagnosis

5.1. LSSVM

The main difference between LSSVM and the traditional Support Vector Machine (SVM) lies in the modification of the objective function and the constraint conditions [42]. In traditional SVM, the optimization problem is a convex quadratic programming problem that includes inequality constraints; whereas in LSSVM, the optimization changes to a linear programming problem, which greatly simplifies the calculation process.

The formula for LSSVM is obtained from reference [42]. The optimization problem of LSSVM can be expressed as

$$\begin{cases} \min_{\omega,b,e} J(\omega,e) = \frac{1}{2}\omega^T\omega + \frac{c}{2}\sum_{i=1}^{N} e_u^2 \\ y_i = \omega^T\phi(h_u) + b + e_u \forall u = 1,\cdots,N \end{cases} \tag{19}$$

where ω is the weight vector, b is the bias, e is the error term, and c is the regularization parameter.

The optimization process of LSSVM is given as follows:

(1) Step I: By introducing Lagrange multipliers α_i, the optimization problem can be transformed into solving a linear programming problem, thereby improving the computational efficiency. The constructed Lagrange function is given by

$$L(\omega,b,e,\alpha) = \frac{1}{2}\omega^T\omega + \frac{c}{2}\sum_{u=1}^{N} e_u^2 - \sum_{u=1}^{n} \alpha_u\left[y_u - \omega^T\phi(h_u) - b - e_u\right] \tag{20}$$

(2) Step II: Solve the Karush–Kuhn-Tucker (KKT) conditions. The KKT conditions can be obtained by taking derivatives of ω,b,e,α in Equation (20) and setting them to zero, which are given by

$$\begin{cases} \omega = \sum_{u}^{.n} \alpha_u\phi(h_u) \\ \sum_{i}^{.n} \alpha_u = 0 \\ \alpha_u = ce_u \\ \omega^T\phi(h_u) + b + e_u = y_u \end{cases} \tag{21}$$

(3) Step III: Solve the system of linear equations. By expressing the linear equations in matrix form, the optimal ω and b can be obtained using matrix operations.

5.2. PSO-Optimized LSSVM for OLTC Fault Diagnosis

The performance of LSSVM highly depends on the selection of hyperparameters, including the regularization parameter c and kernel parameter g. The predictive ability of the model will be affected if the parameters are not selected properly. In addition, manual tuning of these parameters is cumbersome and time-consuming, and it usually requires the use of methods such as grid search or random search. These methods are inefficient in the space of high-dimensional parameters. Moreover, the objective function of LSSVM is usually non-linear and non-convex, which is easy to fall into the local optimum, making it difficult for traditional optimization methods to find the global optimal solution.

This study employs PSO [35] to optimize the parameters of the LSSVM model. The fundamental principles of PSO-based optimization for LSSVM are given as follows:

(1) Step I: Initialize the parameters of PSO. The search range for parameter c is set to [0.1, 300], and the search range for parameter g is set to [0.1, 10]. The optimization parameter dimension D is equal to 2, the maximum number of iterations is 20, and the particle swarm size N is equal to 5. The initial maximum velocity of the particle is set to 10, and its minimum velocity is set to -10. The formula for PSO is obtained from reference [43]. The initialization processes of the particle positions Q_i and velocities v_i are given as follows:

$$\begin{cases} Q_i = rand(N,D)\bullet(L_{maxj} - L_{minj}) + L_{minj} \\ v_i = rand(N,D)\bullet(V_{max} - V_{min}) + V_{min} \end{cases} \tag{22}$$

where $rand(\bullet)$ represents a random vector, L_{minj} and L_{maxj} denote the lower and upper bounds of the problem, respectively. Q_i represents the current position of the i-th particle, and the velocity v_i indicates the direction and rate of position changes.

(2) Step II: Calculate the initial fitness value. The dataset is divided into a training set and testing set in an 8:2 ratio. The current particle Q_i is used to train the LSSVM model, and the classification accuracy of the training set is obtained. The fitness value of the particle is then calculated by using the classification error rate of the testing set, which is given by the following formula:

$$f(Q_i) = 1 - Accuracy \tag{23}$$

where Accuracy refers to the classification accuracy of the testing set.

(3) Step III: Update the individual optimal position $p_{best,i}$ of the particle. Let the current position of the particle be Q_i, and its current fitness value be $f(Q_i)$. If $f(Q_i) < f(P_{best,i})$, then update $P_{best,i} = Q_i$.

(4) Step IV: Update the global optimal position G_{best} of the particle. For each generation, find the particle with the smallest $f(P_{best,i})$. The current global best fitness value of the particle is $f(P_{best,i})$. If the individual optimal position of any particle $f(P_{best,i}) < f(G_{best})$, then the global optimal position is updated to $G_{best} = P_{best,i}$.

(5) Step V: Update the position and velocity of the particle. The velocity update equation for the particle is given as follows [43]:

$$v_i(t+1) = \omega \cdot v_i(t) + c_1 \cdot r_1 \cdot (p_{best} - Q_i(t)) + c_2 \cdot r_2 \cdot (p_{best} - Q_i(t)) \tag{24}$$

The position update formula for the particle is given as follows [43]:

$$Q_i(t+1) = Q_i(t) + v_i(t+1) \tag{25}$$

where r_1 and r_2 are random numbers in the range of [0, 1], c_1 and c_2 are learning factors whose value is set to 1.5, and ω is the inertia weight, which is set to 0.8.

(6) Step VI: Repeat steps (3) to (5). When the maximum number of iterations or other stopping criteria are met, the optimal combination of parameters c and g are obtained.

When the PSO algorithm is applied, PSO might fall into local optima during the single optimization process. This paper proposes a method to improve the process of PSO-optimized LSSVM. After each optimization, multiple fault diagnoses are performed using the optimized LSSVM to further determine whether the optimized parameters make the model perform the best. The optimal combination of c and g is obtained by combining the results of multiple optimizations, which can not only make full use of the global search ability of PSO, but also avoid PSO from falling into local optimization in a single optimization. Consequently, the optimal hyperparameters of LSSVM can be found by the improved optimization method. The flowchart of the improved optimization method is given in Figure 8, which includes four main steps as follows:

(1)   Step I: Setting the number of optimizations.
(2)   Step II: Optimizing the regularization parameter c and kernel parameter g of LSSVM using PSO to obtain the optimal parameter combination for each optimization.
(3)   Step III: Using the optimized LSSVM parameters to perform 20 training and testing iterations, calculating and recording the average accuracy of the test set.
(4)   Step IV: To avoid PSO falling into local optima, the parameter combination with the highest average accuracy of LSSVM is found by the multiple optimizations, and the optimal LSSVM model is obtained.

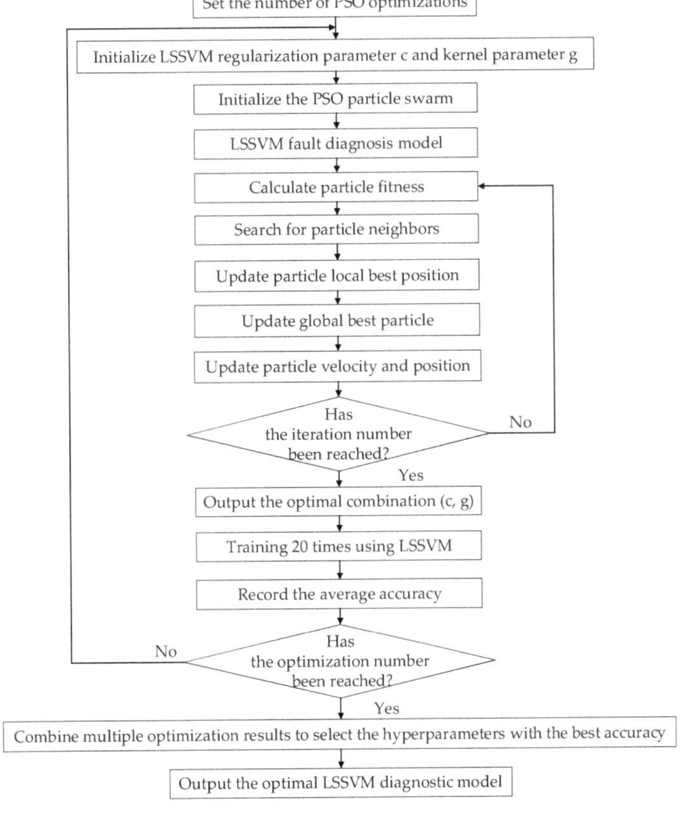

Figure 8. Flowchart of the improved PSO-Optimized LSSVM.

6. Experiment and Result Analysis

6.1. Pre-Processing of OLTC Vibration Signals

During normal operation, the OLTC performs a typical tap position change (from the tap position 20 to 19), the vibration signals collected by different measurement points are shown in Figure 9. It can be seen that the vibration amplitudes of different positions are quite different. Among them, the amplitudes of measurement points #1 and #3 are similar, and the amplitudes of measurement points #2 and #4 are similar. Since measuring point #3 is closest to the OLTC actuator and is more sensitive to the change of the OLTC's mechanical state, the vibration signal of measuring point #3 is mainly used in the subsequent analysis.

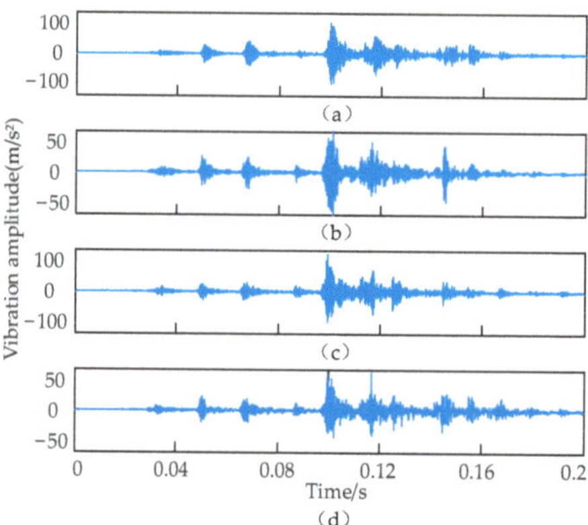

Figure 9. Measured OLTC vibration signals at different measuring points. (**a**) Point #1. (**b**) Point #2. (**c**) Point #3. (**d**) Point #4.

To eliminate the dimensional differences between different signals and reduce the data instability, a normalization process is adopted, which is given by

$$Z_i = \frac{x_i - \mu}{\sigma} \tag{26}$$

where x_i is the original signal, μ is the mean value of the original signal, σ is the standard deviation of the original signal, and Z_i is the result after normalization.

6.2. Comparative Analysis of Different Feature Combinations

To more intuitively compare the advantages of the multi-dimensional feature fusion method over single-dimensional features, this paper combines different types of features and calculates the average accuracy of 20 fault diagnosis iterations under different combinations. The results are tabulated in Table 3. From Table 3, it can be seen that the performance of single features is quite limited. Different types of features contain various types of fault information and can complement each other. By using multi-feature fusion for fault diagnosis, the accuracy of fault diagnosis can be significantly improved.

Table 3. Comparison of the average accuracy of the training set and the test set composed of different feature combinations.

Feature Name	Average Training Accuracy	Average Testing Accuracy
Time/Frequency Domain	$94.98 \pm 3.21\%$	$90.11 \pm 5.19\%$
SWT-SVD	$75.53 \pm 3.58\%$	$71.72 \pm 6.72\%$
Multi-Scale Modal Features	$97.85 \pm 2.25\%$	$90.69 \pm 4.71\%$
Time/Frequency Domain + SWT-SVD	$98.71 \pm 0.65\%$	$92.35 \pm 3.84\%$
Time/Frequency Domain + Multi-Scale Modal Features	$99.22 \pm 0.28\%$	$94.01 \pm 3.25\%$
SWT-SVD + Multi-Scale Modal Features	$99.18 \pm 0.35\%$	$95.23 \pm 3.57\%$
Time/Frequency Domain + SWT-SVD+ Multi-Scale Modal Features	$100.00 \pm 0.00\%$	$97.98 \pm 2.14\%$

6.3. Feature Analysis Based on Random Forest Algorithm

Using the multi-feature extraction method presented in Section 4, the time-domain features $S_1 \sim S_8$, frequency-domain features $P_1 \sim P_4$, singular-value features of the time-frequency matrix $T_1 \sim T_2$, multi-scale modal frequency-domain energy features $M_1 \sim M_5$, and multi-scale modal center of gravity frequency $G_1 \sim G_5$ are obtained, and a total of 24-dimensional features are extracted. These features may contain some redundant features. Therefore, the random forest algorithm is used to screen out features with high sensitivity to faults, and redundant features are eliminated to improve the accuracy of fault diagnosis.

The dataset was divided into a training set and a testing set with a ratio of 8:2. The training set was used to train a random forest model. After training, feature importance scores were calculated for each feature. The importance scores of each feature calculated by the random forest algorithm are shown in Figure 10. The higher the score, the greater the role of the feature in the model. According to the order of the important score, the selected features were combined, and the LSSVM was used for fault diagnosis. The average accuracy rate of 20 repeated tests was calculated and is shown in Figure 11.

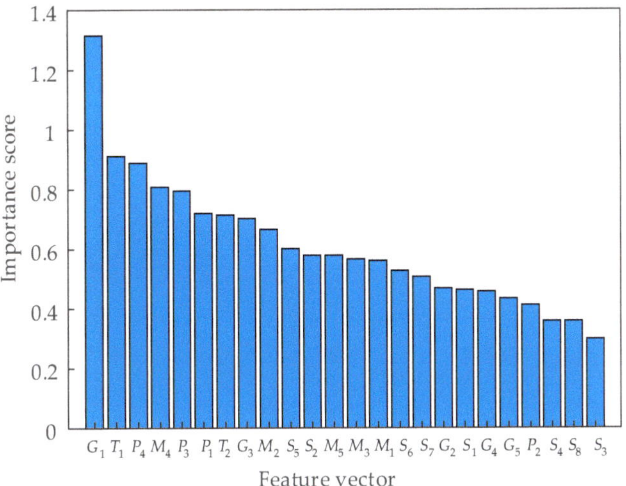

Figure 10. Importance scores for different features.

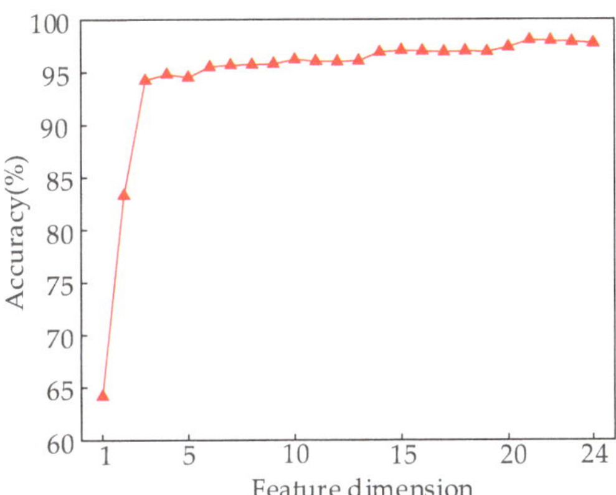

Figure 11. Average accuracy of different combinations of feature dimensions.

From Figure 10, it can be seen that the importance scores of each feature calculated by the random forest algorithm were all above 0.2. Using the first three features of the importance scores, G_1, T_1, and P_4, the average diagnostic accuracy reached 94.16%, which is shown in Figure 11. When using the first 21 dimensions of features, the fault diagnosis accuracy reached as high as 98.58%. However, when the feature dimensions increased from 21 to 24, the fault diagnosis accuracy decreased to 97.98%. Redundant features reduced the accuracy of fault diagnosis by 0.6%.

It can be concluded that the random forest algorithm can be used to screen the features, and the features with high sensitivity to faults can be found. The information between different types of features is complementary, which can provide more OLTC state information. Consequently, this approach can be used to construct a dataset with more comprehensive feature information and eliminate the influence of redundant features on the accuracy of fault diagnosis. Therefore, the first 21 features were used to construct the sample set in this study.

6.4. Experimental Results and Comparative Analysis

Based on the multi-feature dataset, with 50 samples for each OLTC operating condition, a random sampling method was used to divide the dataset into an 80% training set and a 20% testing set. Using the classification error rate of the testing set as the fitness function, the iterative curve of PSO during a single search process is shown in Figure 12. Figure 12 indicates that PSO rapidly converges to a relatively optimal solution within the first two iterations, leading to a sharp decrease in the classification error rate, which then reaches a steady-state value. This demonstrates that PSO can effectively enhance the classification performance of LSSVM.

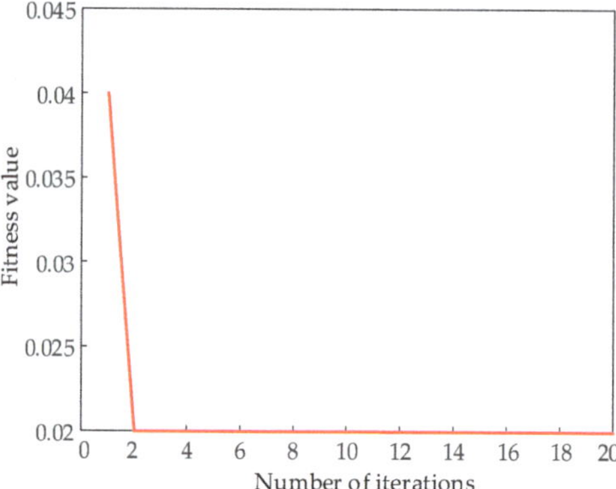

Figure 12. PSO iteration curve.

PSO was used to perform a 10-times parameter optimization, and the results are shown in Table 4. The combination of optimal parameters c and g was (182.83, 1.40), and the accuracy of fault diagnosis using this parameter reached 98.58%. Based on the PSO optimization parameters, the difference in accuracy between the optimal and worst results of the LSSVM model was 1.64%. It clearly shows that some PSO optimization parameters are only local optimal parameters. By comparing the optimization results multiple times, the local optimal parameters can be avoided. The optimization method proposed in this paper can quickly and directly find the optimal parameters of LSSVM so as to improve the performance of the LSSVM fault diagnosis model.

Table 4. Parameter optimization results based on PSO.

Parameter Combination of (c, g)	AverageAccuracy	Parameter Combination of (c, g)	Average Accuracy
(286.63, 7.02)	97.20 ± 2.08%	(154.98, 3.05)	97.62 ± 2.28%
(182.83, 1.40)	98.58 ± 1.84%	(166.75, 1.86)	98.32 ± 1.98%
(189.88, 9.44)	96.96 ± 3.17%	(142.74, 0.19)	96.94 ± 3.34%
(114.21, 5.82)	97.06 ± 2.34%	(289.86, 1.16)	98.54 ± 2.01%
(290.85, 5.76)	97.28 ± 2.21%	(293.28, 8.22)	97.10 ± 2.23%

To further analyze the diagnostic results for each fault category, the confusion matrix is employed to display the outcomes. The confusion matrix is defined as follows [25]:

$$C_{ij} = \sum_{k=1}^{m} \delta(y_{true,k} = i) \cdot \delta(y_{pred,k} = j) \tag{27}$$

where m represents the total number of samples, and $\delta(\cdot)$ is the indicator function, which returns 1 if the condition inside the parentheses is true; otherwise, it returns 0. If the true label of the k-th sample is i, then $\delta(y_{true,k} = i)$ returns 1; otherwise, it returns 0. If the predicted label of the k-th sample is j, then $\delta(y_{pred,k} = i)$ returns 1; otherwise, it returns 0. C_{ij} represents the number of samples whose true label is i and predicted label is j.

The results of fault diagnosis using the optimized LSSVM are shown in Figure 13. The identifying accuracy for the 4th type of fault (insulating board loosening) is 90%, while the identification accuracy for all other faults reaches 100%. This diagnostic result shows that the optimized LSSVM has excellent identification performance on most fault types and can accurately distinguish different fault types. This can improve system reliability and maintenance efficiency, reducing equipment downtime and maintenance costs.

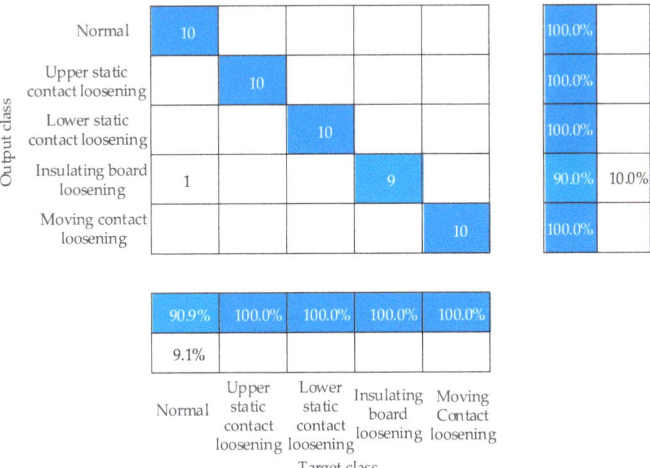

Figure 13. Fault diagnosis results using the optimized LSSVM.

To further compare the effectiveness of PSO-LSSVM in OLTC diagnosing faults with small sample sizes, four fault diagnosis models were constructed, including SVM, CNN, LSTM, and BP neural networks. These models were used to diagnose OLTC mechanical faults. The diagnosis results of these models are shown in Table 5. The method proposed in this paper achieved the highest accuracy in fault diagnosis. Compared with the four traditional OLTC fault diagnosis models of SVM [1], CNN [20], LSTM [18], and BP [18] neural network, the proposed method improved the accuracy of fault diagnosis by 18.58%, 2.48%, 4.58%, and 3.68%, respectively.

Table 5. Diagnosis results of different diagnostic models.

Diagnostic Model	Average Training Accuracy	Average Testing Accuracy
The proposed model	$100.00 \pm 0.00\%$	$98.58 \pm 1.84\%$
SVM [1]	$80.50 \pm 5.18\%$	$80.00 \pm 2.92\%$
CNN [20]	$99.58 \pm 0.05\%$	$96.10 \pm 3.12\%$
LSTM [19]	$98.75 \pm 0.85\%$	$94.00 \pm 4.58\%$
BP [18]	$98.15 \pm 0.91\%$	$94.90 \pm 3.98\%$

To evaluate the performance of the proposed method in the case of small sample sizes, the dataset was re-divided into training and testing sets in a ratio of 2:8. Consequently, the size of the training samples set was decreased from 40 to 10 and the number of testing samples set was increased from 10 to 40. The diagnostic accuracy results and the training times are tabulated in Table 6. From Table 6, it can be seen that the proposed method achieves high accuracy in fault diagnosis, with 100% accuracy on the training set and 95.5% on the testing set, which is only a 3.08% decrease compared to the previous results. In contrast, the SVM model's training set accuracy remained unchanged, but the testing set accuracy was decreased by 3.5%. The CNN, LSTM, and BP models all achieved 100% accuracy on the training set, but their testing set accuracies dropped by 6.1%, 6.1%, and 8.4%, respectively. The large discrepancy between the training set accuracies and testing set accuracies indicates that these deep learning models exhibited overfitting in the case of small sample sizes. Meanwhile, the proposed model achieved the fastest training time.

Table 6. Diagnosis results of different diagnostic models in the case of small sample sizes.

Diagnostic Model	Average Training Accuracy	Average Testing Accuracy	Average Training Time
The proposed model	$100.00 \pm 0.00\%$	$95.5 \pm 3.73\%$	0.315 s
SVM	$80.00 \pm 3.05\%$	$76.5 \pm 3.21\%$	0.327 s
CNN	$100.00 \pm 0.00\%$	$90.0 \pm 6.25\%$	2.619 s
LSTM	$100.00 \pm 0.00\%$	$88.0 \pm 11.97\%$	4.291 s
BP	$100.00 \pm 0.00\%$	$86.5 \pm 10.32\%$	0.789 s

7. Conclusions

The diagnosis method of the OLTC mechanical fault is investigated in this paper. Based on the vibration signals acquired from the UCG-type OLTC experimental test platform, a feature extraction and fault diagnosis method is established, and the conclusions are drawn as follows:

(1) Compared with single-dimensional features, multi-dimensional features contain more abundant fault information, and the multi-dimensional feature fusion method can significantly improve the accuracy of fault diagnosis. The accuracy of fault diagnosis can reach 97.98% by using a combination of time/frequency domain, SWT-SVD, and multi-scale modal features.

(2) The redundant features in multi-dimensional features affect the accuracy of the LSSVM fault diagnosis. The random forest algorithm was used to eliminate the influence of redundant features, and the accuracy of the LSSVM fault diagnosis was further improved by 0.6%.

(3) By introducing the PSO algorithm to optimize the hyperparameters of LSSVM, and comprehensively comparing multiple optimization results, the optimal hyperparameters of the LSSVM model were obtained, which effectively improved the diagnostic performance of the proposed model. Compared with the traditional SVM, CNN, LSTM, and BP neural network models, the method proposed in this paper achieved the highest accuracy of 98.58%, indicating that it can effectively identify OLTC faults.

Author Contributions: Conceptualization, Y.R. and S.L. (Sizhao Lu); methodology, Y.S. and Y.R.; software, Y.R.; validation, Z.L. and Y.R.; formal analysis, Y.R.; investigation, L.L.; resources, B.Z.; data curation K.Y. and Y.H.; writing—original draft preparation, Y.S.; writing—review and editing, Y.S. and S.L. (Sizhao Lu); visualization, Y.H. and M.X.; supervision, S.L. (Siqi Li). and Z.L.; project administration, S.L. (Sizhao Lu). All authors have read and agreed to the published version of the manuscript.

Funding: This research was funded by the Yunnan applied basic research program (Grant No. 202201AT070155); the project was funded by the Guangzhou Bureau of EHV Power Transmission Company Technology Project (Grant No. 0101002022030301SB00054).

Data Availability Statement: The data presented in this study are available on request from the corresponding authors as the data are part of an ongoing research study.

Conflicts of Interest: Authors Yan-Hui Shi, Yan-Jun Ruan, Liang-Chuang Li, Bo Zhang were employed by the company Guangzhou Bureau of EHV Power Transmission Company, China Southern Power Grid Co., Ltd. The remaining authors declare that the research was conducted in the absence of any commercial or financial relationships that could be construed as a potential conflict of interest.

References

1. Cichoń, A.; Włodarz, M. OLTC Fault detection Based on Acoustic Emission and Supported by Machine Learning. *Energies* **2024**, *17*, 220. [CrossRef]
2. Liu, J.; Wang, G.; Zhao, T.; Zhang, L. Fault Diagnosis of On-Load Tap-Changer Based on Variational Mode Decomposition and Relevance Vector Machine. *Energies* **2017**, *10*, 946. [CrossRef]
3. Bengtsson, C. Status and trends in transformer monitoring. *IEEE Trans. Power Deliv.* **1996**, *11*, 1379–1384. [CrossRef]
4. Simas, F.E.F.; de Almeida, L.A.L.; de C. Lima, A.C. Vibration monitoring of on-load tap changers using a genetic algorithm. In Proceedings of the IEEE Instrumentation & Measurement Technology Conference, Ottawa, ON, Canada, 16–19 May 2005; pp. 2288–2293.
5. Bengtsson, C.; Kols, H.; Martinsson, L.; Foata, M.; Leonard, F.; Rajotte, C.; Aubin, J. Acoustic diagnosis of on tap changers. In Proceedings of the CIGRE 1996 Session, Paris, France, 21–28 August 1996.
6. Yan, Y.; Ma, H.; Song, D.; Feng, Y.; Duan, D. OLTC Fault Diagnosis Method Based on Time Domain Analysis and Kernel Extreme Learning Machine. *J. Comput.* **2022**, *33*, 91–106. [CrossRef]
7. Duan, R.; Wang, F. Fault diagnosis of on-load tap-changer in converter transformer based on time–frequency vibration analysis. *IEEE Trans. Ind. Electron.* **2016**, *63*, 3815–3823. [CrossRef]
8. Rivas, E.; Burgos, J.C.; Garcia-Prada, J.C. Condition assessment of power OLTC by vibration analysis using wavelet transform. *IEEE Trans. Power Deliv.* **2009**, *24*, 687–694. [CrossRef]
9. Kang, P.; Birtwhistle, D. Condition monitoring of power transformer on-load tap-changers. Part I: Automatic condition diagnostics. *IEEE Proc.-Gener. Transm. Distrib.* **2001**, *148*, 301–306.
10. Kang, P.; Birtwhistle, D. Condition monitoring of power transformer on-load tap-changers. Part II: Detection of ageing from vibration signatures. *IEEE Proc.-Gener. Transm. Distrib.* **2001**, *148*, 307–311.
11. Gao, P.; Ma, H.; Zhang, H.; Chen, K.; Wang, C. Comparison of EMD entropy and wavelet entropy in vibration signals of OLTC. *Proc. CSU-EPSA* **2012**, *24*, 48–53. (In Chinese)
12. Zhang, Z.; Chen, W.; Tang, S.; Wang, Y.; Wan, F. State Feature Extraction and Anomaly Diagnosis of On-Load Tap-Changer Based on Complementary Ensemble Empirical Mode Decomposition and Local Outlier Factor. *Trans. China Electrotech. Soc.* **2019**, *34*, 4508–4518. (In Chinese)
13. Qian, G.; Wang, S. Fault Diagnosis of On-Load Tap-Changer Based on the Parameter-adaptive VMD and SA-ELM. In Proceedings of the 2020 IEEE International Conference on High Voltage Engineering and Application (ICHVE), Beijing, China, 6–10 September 2020; IEEE: New York, NY, USA, 2020; pp. 1–4.
14. Zhao, T.; Zhang, L.; Li, Q. Feature Analysis Methodology for Phase Portrait Structure of Mechanical Vibration Signals of On-load Tap Changers in Multidimensional Space. *High Volt. Eng.* **2007**, *33*, 102–105. (In Chinese)
15. Chen, S.; Zhu, H.Y. Wavelet transform for processing power quality disturbances. *EURASIP J. Adv. Signal Process.* **2007**, *2007*, 047695. [CrossRef]
16. Lian, J.; Liu, Z.; Wang, H.; Dong, X. Adaptive variational mode decomposition method for signal processing based on mode characteristic. *Mech. Syst. Signal Process.* **2018**, *107*, 53–77. [CrossRef]
17. Nazari, M.; Sakhaei, S.M. Successive variational mode decomposition. *Signal Process.* **2020**, *174*, 107610. [CrossRef]
18. Dong, J.; Valzania, L.; Maillard, A.; Pham, T.; Gigan, S.; Unser, M. Phase retrieval: From computational imaging to machine learning: A tutorial. *IEEE Signal Process. Mag.* **2023**, *40*, 45–57. [CrossRef]
19. Jiang, C. Fault recognition of on-load tap-changer based on improved BP neural network. In Proceedings of the 2017 3rd International Conference on Computational Systems and Communications (ICCSC 2017), Tokyo, Japan, 24–26 November 2017; pp. 24–28.

20. Cui, L.; Chen, Z.; Liu, D. On-load Tap Changer Fault Diagnosis Based on Improved Whale Algorithm Optimized LSTM Neural Network. In Proceedings of the 2024 IEEE 7th Advanced Information Technology, Electronic and Automation Control Conference (IAEAC), Chongqing, China, 15–17 March 2024; IEEE: New York, NY, USA, 2024; Volume 7, pp. 1085–1089.

21. Dong, Y.; Zhou, H.; Sun, Y.; Liu, Q.; Wang, Y. On-Load Tap-Changer Mechanical Fault Diagnosis Method Based on CEEMDAN Sample Entropy and Improved Ensemble Probabilistic Neural Network. In Proceedings of the 2021 IEEE 4th International Electrical and Energy Conference (CIEEC), Wuhan, China, 28–30 May 2021; IEEE: New York, NY, USA, 2021; pp. 1–6.

22. Liang, X.; Wang, Y.; Gu, H. A mechanical fault diagnosis model of on-load tap changer based on same-source heterogeneous data fusion. *IEEE Trans. Instrum. Meas.* **2021**, *71*, 1–9. [CrossRef]

23. Han, S.; Gao, F.; Wang, B.; Liu, Y.; Wang, K.; Wu, D.; Zhang, C. Audible Sound Identification of on Load Tap Changer Based on Mel Spectrum Filtering and CNN. *Power Syst. Technol.* **2021**, *45*, 3609–3617.

24. Yan, J.; Cheng, Y.; Wang, Q.; Liu, L.; Zhang, W.; Jin, B. Transformer and graph convolution-based unsupervised detection of machine anomalous sound under domain shifts. *IEEE Trans. Emerg. Top. Comput. Intell.* **2024**, *8*, 2827–2842. [CrossRef]

25. Cheng, Y.; Chen, B.; Zhang, W. Enhanced spectral coherence and its application to bearing fault diagnosis. *Measurement* **2022**, *188*, 110418. [CrossRef]

26. Liu, L.; Zhang, W.; Gu, F.; Song, D.; Tang, G.; Cheng, Y. An unsupervised transfer network with adaptive input and dynamic channel pruning for train axle bearing fault diagnosis. *Struct. Health Monit.* **2024**.

27. Mumuni, F.; Mumuni, A. Automated data processing and feature engineering for deep learning and big data applications: A survey. *J. Inf. Intell.* **2024**.

28. Yang, B.; Lei, Y.; Li, X.; Li, N. Targeted transfer learning through distribution barycenter medium for intelligent fault diagnosis of machines with data decentralization. *Expert Syst. Appl.* **2024**, *244*, 122997. [CrossRef]

29. Yang, B.; Lei, Y.; Li, X.; Li, N. Label recovery and trajectory designable network for transfer fault diagnosis of machines with incorrect annotation. *IEEE CAA J. Autom. Sin.* **2024**, *11*, 932–945. [CrossRef]

30. Castiglioni, I.; Rundo, L.; Codari, M.; Di Leo, G.; Salvatore, C.; Interlenghi, M.; Gallivanone, F.; Cozzi, A.; D'Amico, N.C.; Sardanelli, F. AI applications to medical images: From machine learning to deep learning. *Phys. Medica* **2021**, *83*, 9–24. [CrossRef] [PubMed]

31. Yu, C.; Kang, M.; Chen, Y.; Wu, J.; Zhao, X. Acoustic modeling based on deep learning for low-resource speech recognition: An overview. *IEEE Access* **2020**, *8*, 163829–163843. [CrossRef]

32. Zhang, C.; Mousavi, A.A.; Masri, S.F.; Gholipour, G.; Yan, K.; Li, X. Vibration feature extraction using signal processing techniques for structural health monitoring: A review. *Mech. Syst. Signal Process.* **2022**, *177*, 109175. [CrossRef]

33. Dargan, S.; Kumar, M.; Ayyagari, M.R.; Kumar, G. A survey of deep learning and its applications: A new paradigm to machine learning. *Arch. Comput. Methods Eng.* **2020**, *27*, 1071–1092. [CrossRef]

34. De Lange, M.; Aljundi, R.; Masana, M.; Parisot, S.; Jia, X.; Leonardis, A.; Slabaugh, G.; Tuytelaars, T. A continual learning survey: Defying forgetting in classification tasks. *IEEE Trans. Pattern Anal. Mach. Intell.* **2021**, *44*, 3366–3385.

35. Gao, S.; Zhou, C.; Zhang, Z.; Geng, J.; He, R.; Yin, Q.; Xing, C. Mechanical fault diagnosis of an on-load tap changer by applying cuckoo search algorithm-based fuzzy weighted least squares support vector machine. *Math. Probl. Eng.* **2020**, *2020*, 3432409. [CrossRef]

36. Marini, F.; Walczak, B. Particle swarm optimization (PSO). A tutorial. *Chemom. Intell. Lab. Syst.* **2015**, *149*, 153–165. [CrossRef]

37. Lu, L.; Shan, C.; Zhou, H.; Su, Z.; Sun, H.; Li, X.; Wang, Y. Research on Multi-modal feature extraction and covolutional neural network for deterioration identification of GIS equipment. In Proceedings of the 2023 IEEE 4th China International Youth Conference on Electrical Engineering (CIYCEE), Chengdu, China, 8–10 December 2023; IEEE: New York, NY, USA, 2023; pp. 1–7.

38. Daubechies, I.; Lu, J.; Wu, H.T. Synchrosqueezed wavelet transforms: An empirical mode decomposition-like tool. *Appl. Comput. Harmon. Anal.* **2011**, *30*, 243–261. [CrossRef]

39. Hao, J.; Yang, Y.; Zhou, Z.; Wu, S. Fetal electrocardiogram signal extraction based on fast independent component analysis and singular value decomposition. *Sensors* **2022**, *22*, 3705. [CrossRef]

40. Emeksiz, C.; Tan, M. Wind speed estimation using novelty hybrid adaptive estimation model based on decomposition and deep learning methods (ICEEMDAN-CNN). *Energy* **2022**, *249*, 123785. [CrossRef]

41. Rigatti, S.J. Random forest. *J. Insur. Med.* **2017**, *47*, 31–39. [CrossRef] [PubMed]

42. Lee, W.M. *Python Machine Learning*; John Wiley & Sons: Hoboken, NJ, USA, 2019.

43. Ismail, S.; Shabri, A.; Samsudin, R. A hybrid model of self-organizing maps (SOM) and least square support vector machine (LSSVM) for time-series forecasting. *Expert Syst. Appl.* **2011**, *38*, 10574–10578. [CrossRef]

Article

An Improved Adaptive Finite-Time Super-Twisting Sliding Mode Observer for the Sensorless Control of Permanent Magnet Synchronous Motors

Mingchen Luan, Jiuhong Ruan *, Yun Zhang, Haitao Yan and Long Wang

School of Rail Transportation, Shandong Jiaotong University, Jinan 250357, China;
22121003@stu.sdjtu.edu.cn (M.L.); zhangyun@sdjtu.edu.cn (Y.Z.); 23221027@stu.sdjtu.edu.cn (H.Y.);
23221021@stu.sdjtu.edu.cn (L.W.)
* Correspondence: ruanjh2011@163.com

Abstract: In order to improve the observation accuracy of rotor positions in the sensorless control of permanent magnet synchronous motors and to simplify the parameter adjustment process, this paper proposes an improved finite-time adaptive super-twisting sliding mode observer. First, a linear gain term is introduced into the conventional super-twisting sliding mode observer model as a way of improving the identification accuracy of the observer. Then, for the multi-parameter variable problem in the traditional observer model, a rotational speed variable function design is presented, which simplifies the multi-variables into a single adaptive variable. This reduces the complexity of the observer model while further improving the observation accuracy and stability of the improved observer algorithm (which is verified using Lyapunov's stability theory). A new back EMF filter and an adaptive phase-locked loop are then used to improve the model's speed tracking capability. Finally, through simulation and experimental tests, the improved algorithm's ability to quickly observe changes in rotor position and speed, as well as its fast convergence, small jitter and high accuracy characteristics, are verified.

Keywords: permanent magnet synchronous motor; position sensorless; super-twisting sliding mode observer; speed variable function

Citation: Luan, M.; Ruan, J.; Zhang, Y.; Yan, H.; Wang, L. An Improved Adaptive Finite-Time Super-Twisting Sliding Mode Observer for the Sensorless Control of Permanent Magnet Synchronous Motors. *Actuators* **2024**, *13*, 395. https://doi.org/10.3390/act13100395

Academic Editor: Ioan Ursu

Received: 26 August 2024
Revised: 26 September 2024
Accepted: 28 September 2024
Published: 3 October 2024

1. Introduction

In recent years, due to the characteristics of the permanent magnet synchronous motor (PMSM), such as high efficiency, high power density, and wide speed range, it has been widely used in the fields of robotics, electric vehicles, and electric airplanes, which have high requirements for the performance of motors. PMSM is a typical nonlinear multivariable coupled system, and the most commonly used control strategy is field oriented control (FOC), in which the rotor position and speed information needs to be obtained first [1].

Currently, the traditional way to obtain motor rotor information is to measure the motor rotor angle using mechanical sensors, such as encoders and resolvers, but this method causes the control system to rely on the accuracy of the sensors. Therefore, the PMSM position sensorless control method has attracted the attention of many researchers. Position sensorless control refers to the sampling of motor signals using highly stable voltage and current sensors, which in turn calculate the rotor information of the motor. This method abandons the traditional mechanical position sensors and therefore reduces the system's cost and installation difficulties and improves the control system's reliability [2,3]. Position sensor-less control methods can be categorized into two main groups, depending on the speed range in which the motor operates. One class is the medium and high-speed position sensorless control methods, including the sliding mode observer (SMO) [4–7], the model reference adaptive system (MRAS) [8] and the extended Kalman filter (EKF) [9], among others. The other category is the low-speed position sensorless control method, which mainly tracks the rotor position by means of a high-frequency signal injection [10,11]. The advantage the sliding mode observer algorithm has over other control algorithms is that it is simple and easy to implement. The observation results are also independent of the external perturbations of the motor, which means the system can adapt to the various operating conditions of the motor. This type of system also exhibits a high degree of stability. However, due to the structural characteristics of the sliding mode algorithm, it will lead to the discontinuity of the control target; this in turn leads to the shivering phenomenon, which affects the control accuracy of the system [12–14]. Therefore, international scholars have conducted a lot of research into the phenomenon of jitter vibration in the sliding mode system, usually using Sigmoid function or hyperbolic tangent function combined a with low-pass filter instead of the traditional sign function. It has been shown experimentally that the jitter amplitude of the observations will be significantly reduced by this method, but this method is prone to phase shift the system, which leads to a long response time for the observer [15,16]. The literature [17] proposes an adaptive iterative SMO, where the system is able to autonomously perform parameter identification during motor startup, improving the algorithmic porting of multiple motor parameters. A high order super-twisting sliding mode observer has also been used in the literature [18] to decouple the perturbation quantities and reduce the jitter phenomenon by increasing the order of the system. In the literature [19], a parallel super-twisting Algorithm Sliding-mode observer (STA-SMO) was used for the first time to estimate the load torque, and the parameter tuning process was simplified by designing the gain parameters. The use of traditional arctan function calculation in the process of rotor information estimation will lead to the introduction of high frequency noise in the system's estimation results and will even destabilize the system [20–22]. Nowadays, phase-locked loops are commonly used to extract rotor information, filtering out the system's high-frequency noise, while also being able to quickly track changes in rotor information [23,24]. However, the conventional phase-locked loop will also have an impact on the observation results, due to the limiting nature of the fixed bandwidth, which results in the motor speed switching process not being able to take into account both the estimation accuracy and the stability of the system [25,26].

In this paper, an improved sensorless control strategy is proposed to estimate the rotor information by combining the novel adaptive finite-time super-twisting sliding-mode observer, synchronized reference system filter, and the novel adaptive phase-locked loop outlined in this paper. The main contributions of this paper are as follows:

(1) An improved equivalent sliding mode model is proposed. By analyzing the traditional sliding mode model, the observation accuracy of the sliding mode observer is improved by adding a linear term and defining a perturbation term.

(2) An adaptive gain finite time super-twisting algorithm sliding mode observer (AGFSTA-SMO) is proposed. Compared with the traditional Linear Super-twisting Algorithm Sliding-mode Observer (LSTA-SMO), the observer algorithm proposed in this paper requires only one parameter to be designed, and the parameter is a rotational speed adaptive function with gain self-adjustment capability.

(3) A novel counter electromotive force optimization strategy is employed. Since the presence of high harmonics in the extended back EMF waveforms leads to a reduction in estimation accuracy from the phase-locked loop, a synchronous reference frame filter (SRFF) is designed in this paper to filter out the high harmonics in the extended back EMF. The addition of the back EMF optimization process will reduce the influence of high harmonics on the estimation results, decrease the jitter phenomenon, and improve the estimation accuracy compared to the current position of sensorless control strategy.

(4) A novel Adaptive quadrature phase-locked loop (AQPLL) is used to estimate the rotor information. In order to solve the problems of low estimation accuracy and poor stability of the traditional phase-locked loop during the switching of motor speed, an improved adaptive quadrature phase-locked loop is used in this paper. The inverse potential normalization method is first used to eliminate the effect of

speed variables on the phase-locked loop bandwidth, followed by the inclusion of a parameter adaptive tuning module based on the stochastic gradient descent method. Therefore, the new adaptive quadrature phase-locked loop is able to improve the speed tracking performance during speed switching and always maintain the stability of the motor.

2. PMSM Mathematical Model

The voltage equation of a surface mounted permanent magnet synchronous motor in a stationary α-β coordinate system can be expressed as:

$$\begin{cases} u_\alpha = Ri_\alpha + L\dfrac{di_\alpha}{dt} + E_\alpha \\ u_\beta = Ri_\beta + L\dfrac{di_\beta}{dt} + E_\beta \end{cases} \tag{1}$$

The expression for the back EMF is given by:

$$\begin{cases} E_\alpha = -\omega_e\psi_f \sin\theta_e \\ E_\beta = \omega_e\psi_f \cos\theta_e \end{cases} \tag{2}$$

Rewrite Equation (1) as a current expression:

$$\begin{cases} \dfrac{di_\alpha}{dt} = \dfrac{1}{L}(u_\alpha - Ri_\alpha - E_\alpha) \\ \dfrac{di_\beta}{dt} = \dfrac{1}{L}(u_\beta - Ri_\beta - E_\beta) \end{cases} \tag{3}$$

Equation (3) is discretized using the forward Euler method:

$$\begin{cases} \dfrac{i_\alpha(k+1)-i_\alpha(k)}{T_s} = \dfrac{1}{L}[u_\alpha(k) - Ri_\alpha(k) - E_\alpha(k)] \\ \dfrac{i_\beta(k+1)-i_\beta(k)}{T_s} = \dfrac{1}{L}[u_\beta(k) - Ri_\beta(k) - E_\beta(k)] \end{cases} \tag{4}$$

Equation (4) is further simplified to Equation (5):

$$\begin{cases} i_\alpha(k+1) = \left[1 - \dfrac{T_s R}{L}\right]i_\alpha(k) + \dfrac{T_s}{L}[u_\alpha(k) - E_\alpha(k)] \\ i_\beta(k+1) = \left[1 - \dfrac{T_s R}{L}\right]i_\beta(k) + \dfrac{T_s}{L}[u_\beta(k) - E_\beta(k)] \end{cases} \tag{5}$$

In Equation (5), u_α and u_β are the α-β axis components of the stator voltage; $L = L_d = L_q$ is the d-q axis component of the stator inductance. R is the stator resistance; ω_e is the electrical angular velocity; i_α, i_β are the α-β axis components of the stator current. E_α, E_β are the extended back EMFs in the α-β axis system; Ψ_f is the rotor flux linkage; θ_e is the electrical angle, and t is the sampling time. $i(k)$, $u(k)$ and $E(k)$ are the values of current, voltage, and extended back EMF, respectively, at time k. T_s denotes the sampling period of the discrete-time system.

In practice, the $i_d^* = 0$ A control method is usually applied to surface-mounted permanent magnet synchronous motor drive systems. Figure 1 shows the overall block diagram of the control system in this paper. For the sensorless control system of the permanent magnet synchronous motor, in order to achieve its high-precision control requirements, the primary goal is to obtain the rotor information of the motor [27] and then feedback the position error to the controller for system control.

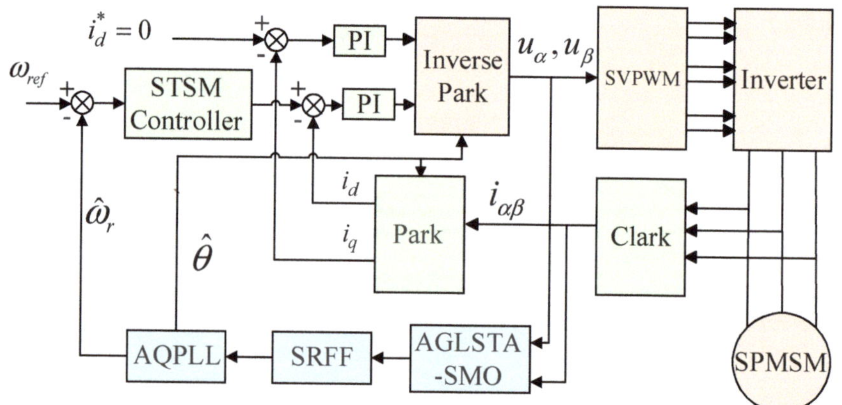

Figure 1. Block diagram of permanent magnet synchronous motor control system.

3. Adaptive Sliding Mode Algorithm

3.1. Traditional Super-Twisting Algorithms

The conventional super-twisting sliding mode observer model is shown in Equation (6):

$$\begin{cases} \dot{\hat{x}}_1 = -l_1|\overline{x}_1|^{1/2}sign(\overline{x}_1) + x_2 + \rho_1 \\ \dot{\hat{x}}_2 = -l_2sign(\overline{x}_1) + \rho_2 \end{cases} \tag{6}$$

where $\dot{\hat{x}}_i$ is the estimated value of the state variable.

It has been shown in the literature [28] that the system converges and remains stable in finite time when Equation (6) satisfies the conditions of Equation (7) and l_1, l_2 satisfy the conditions of Equation (8):

$$\begin{cases} |\rho_1| \leq \delta_1|x_1|^{1/2} \\ |\rho_2| \leq \delta_2 \end{cases} \tag{7}$$

where δ_1, δ_2 are constants greater than zero.

$$\begin{cases} l_1 > 2\delta_1 \\ l_2 > k_1 \dfrac{5\delta_1 l_1 + 6\delta_2 + 4\left(\delta_1 + \frac{\delta_2}{l_1}\right)^2}{2(l_1 - 2\delta_1)} \end{cases} \tag{8}$$

where k_1, k_2 are the observer gain parameters.

3.2. Gain Adaptive LSTA-SMO

Let $1/L = j$, $\boldsymbol{u_s} = [u_\alpha \ u_\beta]^T$, $\boldsymbol{i_s} = [i_\alpha \ i_\beta]^T$, and $\boldsymbol{E_s} = [E_\alpha \ E_\beta]^T$. Since $\boldsymbol{i_s}$ and $\boldsymbol{E_s}$ are state variables, Equation (3) can be rewritten as Equation (9):

$$\begin{cases} \dot{\boldsymbol{i}_s} = j(\boldsymbol{u_s} - R\boldsymbol{i_s} - \boldsymbol{E_s}) \\ \dot{\boldsymbol{E}_s} = g \end{cases} \tag{9}$$

where g denotes the derivative of the back EMF with respect to time.

Based on Equation (9) and the super-twisting algorithm model, the basic observer is designed as follows:

$$\begin{cases} \dot{\hat{\boldsymbol{i}}}_s = j\left[\boldsymbol{u_s} - R\hat{\boldsymbol{i}}_s - \hat{\boldsymbol{E}}_s - \kappa_1|\overline{\boldsymbol{i}}_s|^{1/2}sign(\overline{\boldsymbol{i}}_s)\right] \\ \dot{\hat{\boldsymbol{E}}}_s = \kappa_2 sign(\overline{\boldsymbol{i}}_s) \end{cases} \tag{10}$$

where $\hat{i}_s$ and $\hat{E}_s$ are the estimated values of is and Es, respectively, $\tilde{i}_s = \hat{i}_s - i_s, \kappa_1$ and κ_2 are the sliding mode gains.

According to the literature [28], it is known that the introduction of an additional linear gain term improves the recognition accuracy of the conventional super-twisting sliding mode algorithm. Therefore, an additional linear gain term is introduced in Equation (10) to obtain the LSTA–SMO expression (11):

$$
\begin{cases}
\dot{\hat{i}}_s = j\left[u_s - R\hat{i}_s - \hat{E}_s - k_1\left|\tilde{i}_s\right|^{1/2}sign(\tilde{i}_s) - k_2\tilde{i}_s\right] \\
\dot{\hat{E}}_s = k_3 sign(\tilde{i}_s) + k_4\tilde{i}_s
\end{cases}
\tag{11}
$$

where kI have made modifications, kI have made modifications, I have made modifications, kI have made modifications are the observer gains.

The traditional STA–SMO and LSTA–SMO gains need to be solved in advance for the boundaries of the back EMF derivatives, which makes the algorithm debugging process cumbersome because it is difficult to find suitable gains to match them with different system boundaries.

3.3. AGFSTA–SMO

Based on the above gain mismatch, which leads to difficulties in debugging the algorithm, AGFSTA–SMO is proposed with the expression of Equation (12):

$$
\begin{cases}
\dot{\hat{i}}_s = j\left[u_s - R\hat{i}_s - \hat{E}_s - \frac{1}{4}\lambda\left|\tilde{i}_s\right|^{1/2}sign(\tilde{i}_s) - \frac{1}{4}\lambda\tilde{i}_s\right] \\
\dot{\hat{E}}_s = \frac{1}{2}\lambda^2 sign(\tilde{i}_s) + \lambda^2\tilde{i}_s
\end{cases}
\tag{12}
$$

where λ is the time-varying gain function with respect to $\tilde{i}_s$. The conditions in Equation (13) need to be satisfied when designing the gain function.

$$
\begin{cases}
\lambda = |h| \\
\dot{h} = \frac{\sigma}{e^{-\tilde{i}_s}}sign(\tilde{i}_s)
\end{cases}
\tag{13}
$$

where h is a time-varying gain function with respect to $\tilde{i}_s$ and σ is a proportional integration factor that is a constant greater than zero.

In Equation (13), the initial value of λ needs to satisfy that $\lambda > 0$ is constant. Comparison of Equations (11) and (12) reveals that $k_1 = k_2 = 0.25\lambda$; $k_3 = 0.5\lambda^2$, and $k_4 = \lambda^2$.

Subtracting Equation (12) from Equation (9) gives the error Equation (14) for AGFSTA–SMO:

$$
\begin{cases}
\dot{\tilde{i}}_{\alpha\beta} = j\left[-\overline{E}_s - \frac{1}{4}\lambda\left|\tilde{i}_s\right|^{1/2}sign(\tilde{i}_s) - \frac{1}{4}\lambda\tilde{i}_s\right] \\
\dot{\overline{E}}_s = \frac{1}{2}\lambda^2(\tilde{i}_s)sign(\tilde{i}_s) + \lambda^2(\tilde{i}_s)\tilde{i}_s - g
\end{cases}
\tag{14}
$$

where $\overline{E}_s = \hat{E}_s - E_s$.

$\dot{\overline{E}}_s$ in Equation (14) is bounded. The parameter g in Equation (14) ranges from $|g| \leq \varepsilon_1 + \varepsilon_2|\tilde{i}_s|$, where ε_1 and ε_2 are positive numbers.

The conclusion can be drawn from Equation (14):

For any initial values $\tilde{i}_s$ and $\overline{E}_s$ of $\tilde{i}_s(0)$ and $\overline{E}_s(0), \tilde{i}_s$ and $\overline{E}_s$ converge in finite time if the designed time-varying gain function λ satisfies Equation (15).

$$
\lambda > MAX\left(\sqrt{\frac{32}{17}\varepsilon_1}, \sqrt{\frac{1}{2}\varepsilon_2}, \sqrt{H_1}, \sqrt{H_2}\right)
\tag{15}
$$

In the Equation (15), the relevant parameters are defined as follows:

$$
\begin{cases}
H_1 = -\frac{b_1}{3a_1} + 2\sqrt{-\varsigma_1}\cos\left[\arccos\left(\frac{\chi_1}{(-\varsigma_1)^{3/2}}\right)/3\right] \\[2mm]
H_2 = -\frac{b_2}{3a_2} + 2\sqrt{-\varsigma_2}\cos\left[\arccos\left(\frac{\chi_2}{(-\varsigma_2)^{3/2}}\right)/3\right]
\end{cases}
\tag{16}
$$

$$
\begin{cases}
a_1 = 439 \\
b_1 = -3264\varepsilon_1 - 512\varepsilon_2 \\
c_1 = -37888\varepsilon_1^2 + 3072\varepsilon_1\varepsilon_2 \\
d_1 = 32768\varepsilon_1^2\varepsilon_2 \\
\chi_1 = -\frac{b_1^3}{27a_1^3} - \frac{d_1}{2a_1} + \frac{b_1 c_1}{6a_1^2} \\
\varsigma_1 = \frac{c_1}{3a_1} - \frac{b_1^2}{9a_1^2}
\end{cases}
\quad and \quad
\begin{cases}
a_2 = 5 \\
b_2 = -8\varepsilon_1 - 15\varepsilon_2 \\
c_2 = -80\varepsilon_1^2 + 24\varepsilon_1\varepsilon_2 \\
d_2 = 128\varepsilon_1^2\varepsilon_2 \\
\chi_2 = -\frac{b_2^3}{27a_2^3} - \frac{d_2}{2a_2} + \frac{b_2 c_2}{6a_2^2} \\
\varsigma_2 = \frac{c_2}{3a_2} - \frac{b_2^2}{9a_2^2}
\end{cases}
\tag{17}
$$

3.4. Stability Proofs

Equation (18) is the new Lyapunov function with the following expression:

$$
V = \frac{\xi^T P \xi}{\lambda^4}
\tag{18}
$$

where the parameters ξ and P are defined in Equations (19) and (20), respectively.

$$
\xi = \left[\left|\bar{i}_s\right|^{1/2}sign(\bar{i}_s) \quad \bar{i}_s \quad \overline{E}_s\right]^T
\tag{19}
$$

$$
P = \frac{1}{2}
\begin{bmatrix}
\frac{33}{16}\lambda^2 & \frac{1}{16}\lambda^2 & -\frac{1}{4}\lambda \\
\frac{1}{16}\lambda^2 & \frac{33}{16}\lambda^2 & -\frac{1}{4}\lambda \\
-\frac{1}{4}\lambda & -\frac{1}{4}\lambda & 2
\end{bmatrix}
\tag{20}
$$

Through Equation (18), Equation (21) can be further derived:

$$
\begin{cases}
\dot{V} = \dot{V}_1 + \dot{V}_2 \\
\dot{V}_1 = \xi^T \dot{Q} \xi \\
\dot{V}_2 = 2\xi^T Q \dot{\xi} \\
Q = P/\lambda
\end{cases}
\tag{21}
$$

From Equation (21), the proof of Lyapunov stability can be analyzed for both $\dot{V}_1$ and $\dot{V}_2$.

3.4.1. Analyze $\dot{V}_1$

Write the $\dot{V}_1$ expression:

$$
\dot{V}_1 = \frac{1}{2}\xi^T \left\{ \dot{\lambda}\underbrace{\begin{bmatrix}
-\frac{33}{8}\lambda^{-3} & -\frac{1}{8}\lambda^{-3} & \frac{3}{4}\lambda^{-4} \\
-\frac{1}{8}\lambda^{-3} & -\frac{33}{8}\lambda^{-3} & \frac{3}{4}\lambda^{-4} \\
\frac{3}{4}\lambda^{-4} & \frac{3}{4}\lambda^{-4} & -8\lambda^{-5}
\end{bmatrix}}_{N} \right\} \xi
\tag{22}
$$

$$
= \frac{1}{2}\dot{\lambda}\xi^T N \xi
$$

Based on Equation (13), Equation (23) can be derived, so Equation (22) can be rewritten as Equation (24):

$$
\dot{\lambda} = \dot{h}sign[h] \leq \dot{h} \leq \frac{\sigma}{e^{-|\bar{i}_s|}}
\tag{23}
$$

$$\dot{V}_1 = \frac{1}{2}\dot{\lambda}\xi^T N\xi \leq \frac{1}{2}h\xi^T N\xi \leq \frac{\eta}{2e^{-|\bar{i}_s|}}\xi^T N\xi \tag{24}$$

In Equation (24), because λ is always greater than 0, the N-matrix is positive definite, thus Equation (25) can be obtained:

$$\dot{V}_1 \leq \frac{\sigma}{2e^{-|\bar{i}_s|}}\xi^T N\xi < 0 \tag{25}$$

3.4.2. Analyze $\dot{V}_2$

Write the $\dot{V}_2$ expression:

$$\begin{cases} \dot{V}_1 = 2\dot{\xi}^T Q\xi = -|\bar{i}_s|^{-1/2}\xi^T y_1\xi - \xi^T y_2\xi + y_3^T\xi \\ y_1 = \frac{1}{8}\lambda^{-3}\begin{bmatrix} 17\lambda^2/16 & 0 & -\lambda/4 \\ 0 & 37\lambda^2/16 & -3\lambda/4 \\ -\lambda/4 & -3\lambda/4 & 1 \end{bmatrix} \\ y_2 = \frac{1}{4}\lambda^{-3}\begin{bmatrix} 5\lambda^2/8 & 0 & 0 \\ 0 & 17\lambda^2/16 & -\lambda/4 \\ 0 & -\lambda/4 & 1 \end{bmatrix} \\ y_3^T = \begin{bmatrix} -g/4\lambda^3 & -g/4\lambda^3 & 2g/\lambda^4 \end{bmatrix} \end{cases} \tag{26}$$

Equation (27) is the expression for the new variable η^T:

$$\eta^T = \begin{bmatrix} |\bar{i}_s|^{\frac{1}{2}} & |\bar{i}_s| & |\bar{E}_s| \end{bmatrix} \tag{27}$$

From the variable η^T and relation $|g| \leq \varepsilon_1 + \varepsilon_2|\bar{i}_s|$, Equation (28) can be derived:

$$\begin{cases} -|\bar{i}_s|^{-1/2}\xi^T y_1\xi - \xi^T y_2\xi \leq -|\bar{i}_s|^{\frac{1}{2}}\eta^T y_1\eta - \eta^T y_2\eta \\ y_3^T\xi \leq |\bar{i}_s|^{-1/2}\eta^T \tau_1\eta - \eta^T \tau_2\eta \end{cases} \tag{28}$$

The parameters τ_1 and τ_2 in Equation (28) are defined as follows:

$$\tau_1 = \begin{bmatrix} \varepsilon_1/4\lambda^3 & 0 & \varepsilon_1/\lambda^4 \\ 0 & \varepsilon_2/4\lambda^3 & 0 \\ \varepsilon_1/\lambda^4 & 0 & 0 \end{bmatrix} \tag{29}$$

$$\tau_2 = \begin{bmatrix} \varepsilon_1/4\lambda^3 & 0 & 0 \\ 0 & \varepsilon_2/4\lambda^3 & \varepsilon_2/\lambda^4 \\ 0 & \varepsilon_2/\lambda^4 & 0 \end{bmatrix} \tag{30}$$

From Equations (26) and (28), the scaling equation for $\dot{V}_2$ is given as:

$$\begin{aligned} \dot{V}_2 &= -|\bar{i}_s|^{-1/2}\xi^T y_1\xi - \xi^T y_2\xi + y_3^T\xi \\ &\leq -|\bar{i}_s|^{-1/2}\eta^T(y_1 - \tau_1)\eta - \eta^T(y_2 - \tau_2)\eta \end{aligned} \tag{31}$$

Thus, it can be concluded that $\dot{V}_2 < 0$ is constant when $y_1 - \tau_1$ and $y_2 - \tau_2$ are positive definite matrices. That is, $\dot{V}_2 < 0$ is constant as long as the time-varying gain function λ satisfies Equation (15).

3.4.3. Consider the Following Factual Circumstances

$$\begin{cases} \phi_{\min}\{Q\}\|\xi\|_2^2 \leq V = \xi^T Q\xi \leq \phi_{\max}\{Q\}\|\xi\|_2^2 \\ \phi_{\min}\{y_1 - \tau_1\}\|\eta\|_2^2 \leq \eta^T\{y_1 - \tau_1\}\eta \leq \phi_{\max}\{y_1 - \tau_1\}\|\eta\|_2^2 \end{cases} \tag{32}$$

where max{·} and min{·} denote the maximum and minimum eigenvalues of the matrix, respectively; and $\|\cdot\|_2^2$ denotes the Euclidean norm.

The range of Euclidean paradigms for ξ is given by Equation (32):

$$|\bar{i}_s|^{1/2} \le \|\zeta\|_2 \le \frac{V^{1/2}}{\phi_{\min}^{1/2}\{Q\}} \tag{33}$$

Therefore, Equation (31) can be rewritten as Equation (34):

$$\begin{aligned}
\dot{V}_2 &\le -|\bar{i}_s|^{-1/2}\eta^T\{y_1 - \tau_1\}\eta - \eta^T\{y_2 - \tau_2\}\eta \\
&\le -|\bar{i}_s|^{-1/2}\eta^T\{y_1 - \tau_1\}\eta \\
&\le -|\bar{i}_s|^{-1/2}\phi_{\min}\{y_1 - \tau_1\}\|\eta\|_2^2 \\
&= -|\bar{i}_s|^{-1/2}\phi_{\min}\{y_1 - \tau_1\}\|\zeta\|_2^2 \\
&\le -\frac{\phi_{\min}^{1/2}\{Q\}\phi_{\min}\{y_1-\tau_1\}}{V^{1/2}\phi_{\max}\{Q\}}V \\
&= -\frac{\phi_{\min}^{1/2}\{Q\}\phi_{\min}\{y_1-\tau_1\}}{\phi_{\max}\{Q\}}V^{1/2}
\end{aligned} \tag{34}$$

Equation (35) can be derived from Equation (34):

$$\dot{V} = \dot{V}_1 + \dot{V}_2 < \dot{V}_2 \le -\frac{\phi_{\min}^{1/2}\{Q\}\phi_{\min}\{y_1 - \tau_1\}}{\phi_{\max}\{Q\}}V^{1/2} \tag{35}$$

It is worth noting that the system can converge only when the time-varying gain function λ satisfies the condition Equation (15). This also means that the conditions of Equation (13) need to be satisfied at all times when the value λ is increased to the point where Equation (15) holds, and finally the system is proved to converge based on Equation (35).

Therefore, it can be concluded that $\bar{i}_s$ and $\overline{E}_s$ can converge to zero in finite time and the proof ends. Figure 2 shows the schematic block diagram of AGFSTA–SMO proposed in this paper.

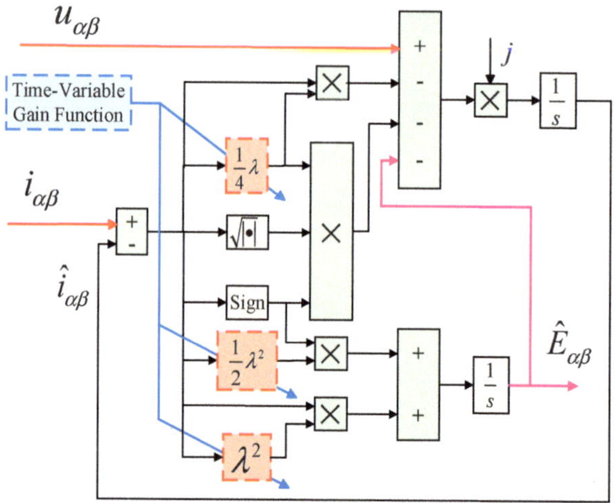

Figure 2. Block diagram of AGFSTA-SMO.

4. Rotor Information Extraction Program

4.1. Synchronized Reference System Filter

Due to the existence of a large number of high harmonics in the extended back EMF waveform observed by the traditional STA–SMO, the observation accuracy will be greatly affected when the motor is running at high speed. In order to improve the observation accuracy and reduce the amplitude of jitter, this paper adopts the SRFF to filter out the harmonic components in the extended back EMF waveform.

The SRFF designed in this paper is based on the characteristics of space vector control of permanent magnet synchronous motor, which firstly uses Park transform to change the back EMF into direct current, then low-pass filtering, and finally uses inverse Park transform to AC quantity. Due to the characteristics of space vector control, this method is not only able to filter out the high-frequency components in the back EMF waveform, but also able to ensure the phase stability of the system [29]. The block diagram of SRFF is shown in Figure 3.

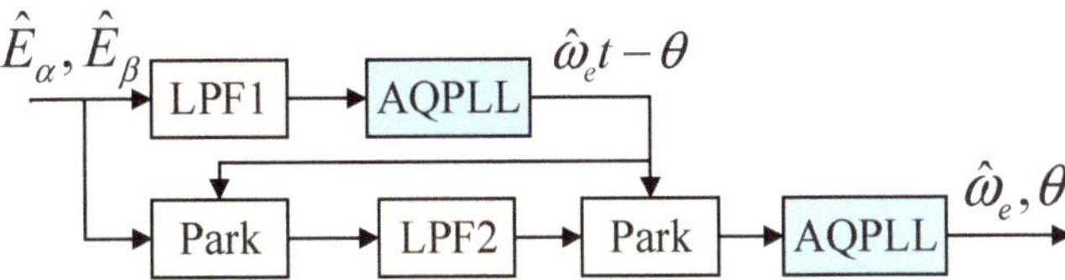

Figure 3. Block diagram of SRFF structure.

The back EMFs $\hat{E}_\alpha$ and $\hat{E}_\beta$ pass through the low-pass filter LPF1 into an adaptive quadrature phase-locked loop [30]. Let the phase lag angle due to LPF1 be $\Delta\theta$. Thus, the extended back EMF can be expressed as a superposition of the fundamental and the higher harmonics, and Equation (36) is the expression for the extended back EMF:

$$\begin{cases} \hat{E}_\alpha(t) = -K_0 \sin(\hat{\omega}_e t) + \sum_{x=2}^{\infty} K_x \sin(x\hat{\omega}_e t + \Delta\theta) \\ \hat{E}_\beta(t) = K_0 \cos(\hat{\omega}_e t) + \sum_{x=2}^{\infty} K_x \sin(x\hat{\omega}_e t + \Delta\theta) \end{cases} \tag{36}$$

where $\hat{\omega}_e t$ is the fundamental phase angle, K_0 is the fundamental amplitude, $x\hat{\omega}_e t$ is the harmonic phase, and K_x is the harmonic amplitude.

The back EMF base wave is Park transformed to give Equation (37):

$$\begin{bmatrix} \hat{E}_d \\ \hat{E}_q \end{bmatrix} = \begin{bmatrix} \cos(\hat{\omega}_e t - \Delta\theta) & \sin(\hat{\omega}_e t - \Delta\theta) \\ -\sin(\hat{\omega}_e t - \Delta\theta) & \cos(\hat{\omega}_e t - \Delta\theta) \end{bmatrix} \begin{bmatrix} -K_0 \sin(\hat{\omega}_e t) \\ K_0 \cos(\hat{\omega}_e t) \end{bmatrix} = \begin{bmatrix} -K_0 \sin(\Delta\theta) \\ K_0 \cos(\Delta\theta) \end{bmatrix} \tag{37}$$

Since K_0 is only related to the electrical angular frequency and the rotor flux linkage, and the error angle $\Delta\theta$ is only related to the electrical angular frequency and the filter LPF1 cutoff frequency, both K_0 and $\Delta\theta$ remain constant after the rotational speed is stabilized, i.e., the base wave of the back EMF is changed to a straight flow. After passing through the low-pass filter LPF2, there is no phase lag in the fundamental wave component of this reaction potential.

Equation (38) can be obtained after Park's inverse transformation:

$$\begin{bmatrix} \hat{E}_\alpha \\ \hat{E}_\beta \end{bmatrix} = \begin{bmatrix} \cos(\hat{\omega}_e t - \theta) & \sin(\hat{\omega}_e t - \Delta\theta) \\ -\sin(\hat{\omega}_e t - \theta) & \cos(\hat{\omega}_e t - \Delta\theta) \end{bmatrix}^{-1} \begin{bmatrix} -K_0 \sin(\Delta\theta) \\ K_0 \cos(\Delta\theta) \end{bmatrix} = \begin{bmatrix} -K_0 \sin(\hat{\omega}_e t) \\ K_0 \cos(\hat{\omega}_e t) \end{bmatrix} \tag{38}$$

From the above derivation, it can be concluded that when the filtering is performed in the d-q axis, the filter does not affect the amplitude phase of the fundamental wave, since the fundamental wave is transformed into a straight flow and also accounts for the unobserved angular lag. The high-frequency harmonics are still AC quantities after Park's transform, so they can be filtered out by the low-pass filter, which in turn reduces system jitter.

4.2. Adaptive Quadrature Phase-Locked Loop

Since the conventional phase-locked loop is affected by its own characteristics during the switching of the motor speed, it ensures the stability of the system while failing to maintain a low phase delay [31]. Therefore, in this paper, the estimation strategy of adaptive quadrature phase-locked loop is used in the rotor information extraction part. Figure 4 shows the schematic diagram of the AQPLL.

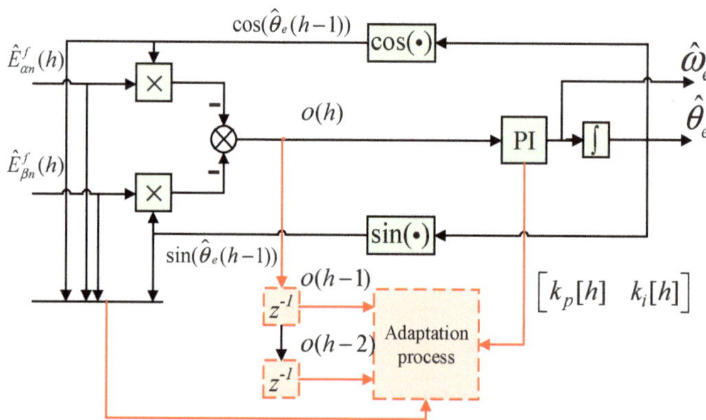

Figure 4. Schematic diagram of adaptive quadrature phase-locked loop.

First, the back EMF of the AGFSTA–SMO output is normalized:

$$\hat{E}_n = \hat{E}_s / \sqrt{E_\alpha^2 + E_\beta^2} \tag{39}$$

where $\hat{E}_n = \begin{bmatrix} \hat{E}_{n\alpha} & \hat{E}_{n\beta} \end{bmatrix}^T$.

After processing through Equation (39), the rotational speed variable will no longer affect the calculation of the phase-locked loop bandwidth. Equation (40) is the transfer function of the adaptive quadrature phase-locked loop:

$$G_{PLL}(s) = \frac{k_p s + k_i}{s^2 + k_p s + k_i} \tag{40}$$

The AQPLL adaptive strategy designed based on stochastic gradient descent method is designed to minimize the squared error of the controller input by adjusting the adaptive gains k_p and k_i. Therefore, the poles of Equation (40) can be replaced by Equation (41):

$$k_p[h] = 2\zeta v[h], k_i[h] = v^2[h] \tag{41}$$

where ζ is the damping factor, h is the discrete time step parameter, and v is the intrinsic frequency, i.e., the AQPLL tuning parameter.

Since ζ is a constant and the phase-locked loop bandwidth is related to v, the adaptive strategy for the AQPLL parameters can be expressed as Equation (42):

$$v[h] = v[h-1] - \iota\frac{1}{2}\left(\frac{\partial o^2[h]}{\partial v[h-1]}\right) \tag{42}$$

where ι is the step parameter that determines the speed of adaptive tuning and o is the input error of the PI controller.

By solving Equation (42), the expression for updating the tuning parameters can be obtained as:

$$v[h] = v[h-1] - \iota w_1[h]w_2[h] = v[h-1] - \Delta v[h] \tag{43}$$

where $\Delta v[h]$ is the step increment, and w_1 and w_2 are partial solutions with the following expressions:

$$\begin{cases} w_1[h] = \hat{E}_\alpha^f[h]\sin(\hat{\theta}[h-1]) - \hat{E}_\beta^f[h]\cos(\hat{\theta}[h-1]) \\ w_2[h] = 2\zeta o[h-1] + T_s v[h-1](o[h-1] - o[h-2]) \end{cases} \tag{44}$$

where $\hat{E}_\alpha^f$ and $\hat{E}_\beta^f$ are the estimated values of the inverse electromotive force after filtering.

5. Simulation and Experimental Verification

In order to verify the advantages of the AGFSTA–SMO algorithm, this paper establishes two different position sensorless control models for permanent magnet synchronous motors in Simulink: a linear gain super-twisting sliding mode observer and a parameter adaptive finite time super-twisting sliding mode observer. The motor rotor information extraction results are analyzed in terms of rotational speed tracking, rotational speed error, rotor position tracking, rotor position error, and sudden load increase/decrease to further validate the advantages of the AGFSTA–SMO algorithm over the LSTA–SMO and conventional STA–SMO algorithms.

5.1. Simulation Analysis

The parameters of the mounted permanent magnet synchronous motor used for the simulation and experimentation are provided by the motor manufacturer, as shown in Table 1. In the simulation process, in order to show the difference between the two observation algorithms more intuitively, the parameters of the current loop and speed loop controllers in the two models are set to be the same, the simulation time of the system is fixed to be 3 s, and the fixed step size is set to be 1 μs.

Table 1. Motor parameters and gain parameters.

Items	Values	Parameters	Values
Stator winding resistance Rs	0.56 Ω	k_1	100
Stator winding inductance L_S	0.62 mH	k_2	400
Flux linkage ψ_f	0.0125 Wb	k_3	100
Rotational inertia J	1.5 kg·cm^2	k_4	50
Pole pairs p	4	λ	0.001
Rated power	250 W	Rated current	7.5 A
Rated toque	0.796 N·m	Rated speed	3000 rpm

As can be seen from Figure 5, the AGFSTA–SMO algorithm has a higher speed tracking accuracy during motor acceleration when the target speeds are 1000 rpm, 1500 rpm, and 2000 rpm, respectively, and the estimated speeds are able to track the target speeds faster and maintain a relatively small jitter amplitude after that. As a result, the AGFSTA–SMO algorithm has a higher speed tracking accuracy and faster convergence despite the different target speeds of the motors.

As can be seen in Figure 6, when the target speed of the motor differs from the actual speed, AGFSTA–SMO is able to adjust the sliding mode gain more quickly, and quickly reduce the error between the estimated speed and the actual speed. When the motor reaches the target speed, the AGFSTA–SMO algorithm limits the sliding mode gain to a certain range based on the motor's speed, which reduces speed overshoot. Therefore, when the motor reaches the target speed, the AGFSTA–SMO algorithm has less fluctuation in the speed error and the absolute value of the error is smaller, which makes it easier to control the motor.

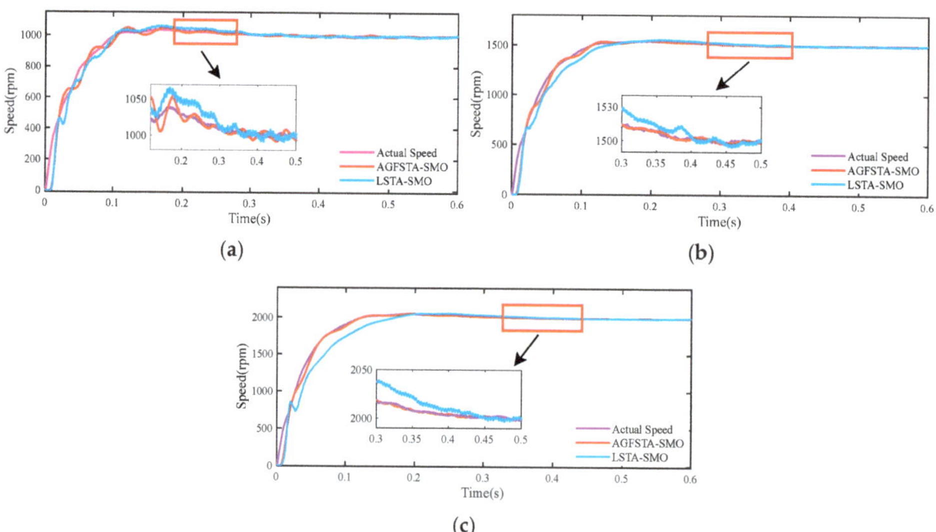

Figure 5. Comparison of estimated speed values with actual speed values for AGFSTA–SMO, LSTA–SMO algorithms. (**a**) 1000 rpm. (**b**) 1500 rpm. (**c**) 2000 rpm.

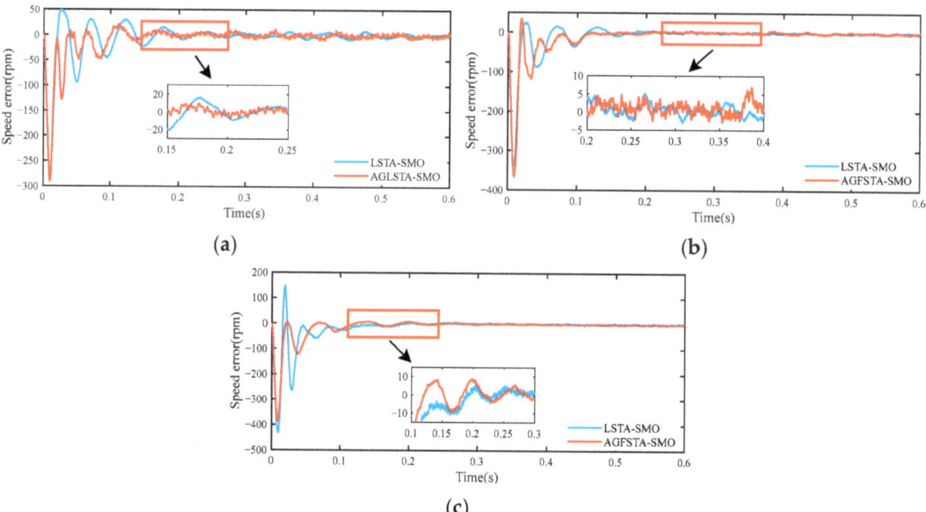

Figure 6. AGFSTA–SMO, LSTA–SMO algorithms estimated RPM value and actual RPM value error. (**a**) 1000 rpm. (**b**) 1500 rpm. (**c**) 2000 rpm.

From the theory of sliding mode control, it can be seen that the system's ability to resist disturbances mainly depends on the sliding mode gain, i.e., when the system encounters disturbances, it can be quickly restored to a stable state by adjusting the sliding mode gain. Figure 7 shows the speed tracking of the motor during sudden load application. As can be seen from Figure 7, when a load perturbation of $T_L = 0.2$ N·m is applied abruptly at 1 s, the estimated speed obtained by the AGFSTA–SMO algorithm is able to track the actual speed more quickly and maintain a smaller error during the recovery to the target speed. Therefore, the tracking performance of the AGFSTA–SMO is superior to the existing LSTA–SMO when the motor is in a loaded condition.

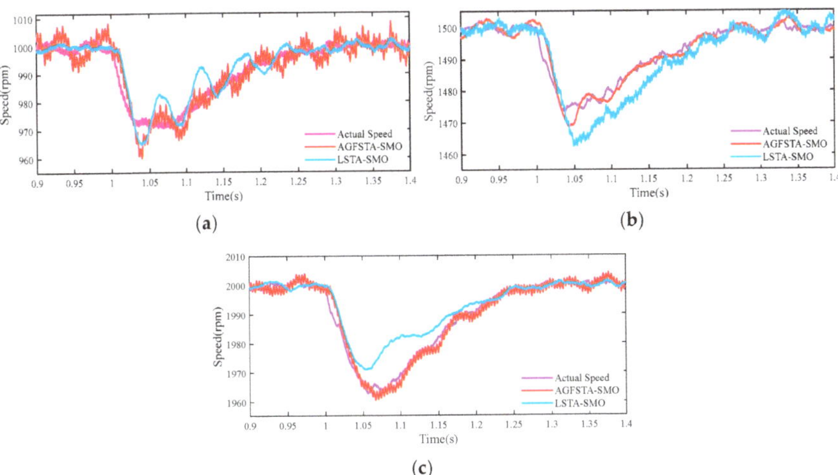

Figure 7. Comparison of estimated and actual speed values of AGFSTA–SMO, LSTA–SMO algorithms with load. (**a**)1000 rpm. (**b**)1500 rpm. (**c**)2000 rpm.

Figure 8 is able to further validate the speed tracking ability of both algorithms when a sudden load is applied. From the results demonstrated in Figure 8, it can be seen that AGFSTA–SMO is able to adjust the sliding mode gain more quickly when there is a sudden change in the operating conditions of the motor, so that the estimated rotational speed quickly tracks the actual rotational speed. It can be concluded that the AGFSTA–SMO observer proposed in this paper has a significant advantage over LSTA–SMO under multiple operating conditions of the motor.

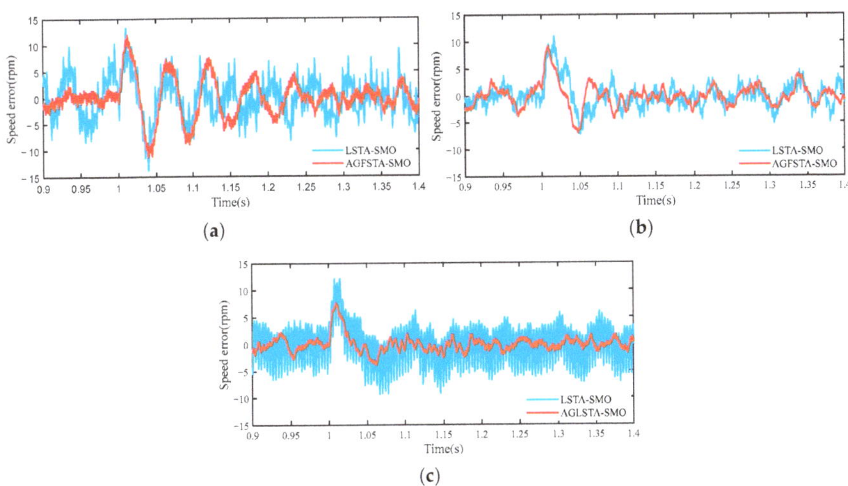

Figure 8. Error between estimated and actual RPM values of AGFSTA–SMO, LSTA–SMO algorithms with load. (**a**) 1000 rpm. (**b**) 1500 rpm. (**c**) 2000 rpm.

As can be seen from Figure 9, the position sensorless control system reaches a steady state at 0.1 s. The rotor position estimated by the algorithm always fluctuates up and down around the true value due to the accelerated frequency change of the position waveform as a result of the motor rotating too fast. It can thus be shown that the AGFSTA–SMO algorithm is always able to accurately estimate the rotor position and stabilize the error within 0.1 rad when the target speed of the motor is different.

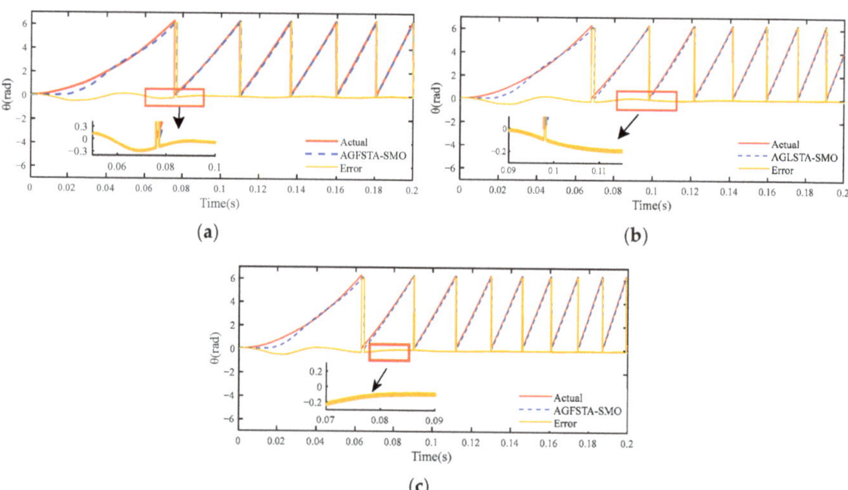

Figure 9. Comparison of estimated rotor position with actual rotor position for AGFSTA–SMO, LSTA–SMO algorithms at no load. (**a**) 1000 rpm. (**b**) 1500 rpm. (**c**) 2000 rpm.

As can be seen from Figure 10, when the load is applied abruptly at 1 s, the actual rotor position changes abruptly, but it is estimated that the rotor position will still maintain the previous motion trend for a short period of time. As a result, the position error increases slightly for a short period of time, then immediately retraces its steps and eventually reaches a steady state. When the motor is adjusted to a stable state, the rotor position error obtained by the AGFSTA–SMO algorithm remains within 0.1 rad, which verifies the feasibility of the algorithm under different working conditions.

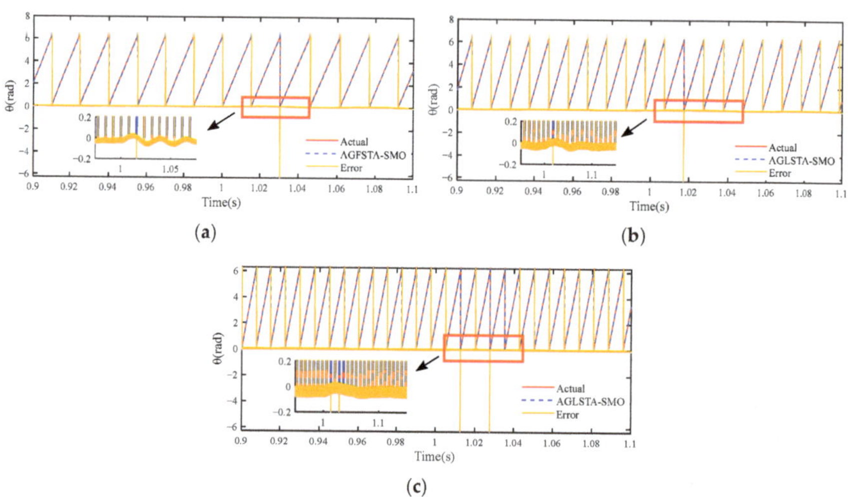

Figure 10. Comparison of estimated rotor position with actual rotor position for AGFSTA–SMO, LSTA-SMO algorithms with load. (**a**) 1000 rpm. (**b**) 1500 rpm. (**c**) 2000 rpm.

From the comparison of the simulation results of the two algorithms, it can be seen that the AGFSTA–SMO algorithm not only has better tracking performance, but also has a higher estimation accuracy. In addition, since the sliding mode gain of AGFSTA–SMO is related to the motor speed, the algorithm is able to respond quickly to the sudden change of the motor condition, avoiding the further expansion of the error in time and ensuring the stability of the control system.

From Figure 11, it can be seen that when the sliding mode gain λ is 0.03, the speed tracking effect is the best, and the tracking ability of the speed is strongest under the loaded working condition. The value of 0.03 for the algorithmic gain is just an example to give the reader an idea of the output results for each magnitude of gain.

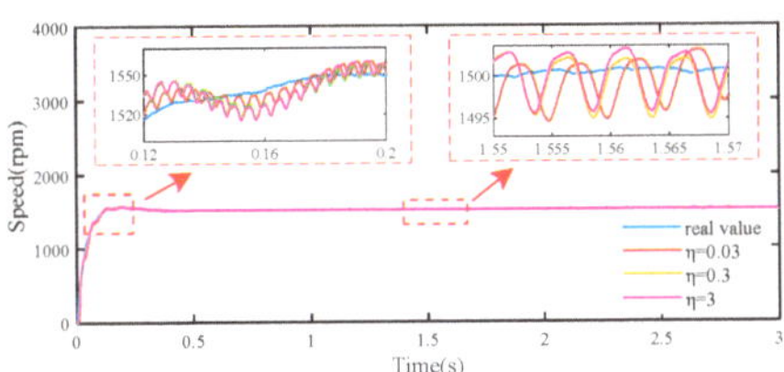

Figure 11. Speed tracking with different gain parameters.

5.2. Experimental Verification Analysis

In order to further test the performance of AGFSTA–SMO, an experimental platform for permanent magnet synchronous motor control system was constructed as shown in Figure 12. In this study, the position sensorless vector control strategy was realized by using an STM32F407 microcontroller.

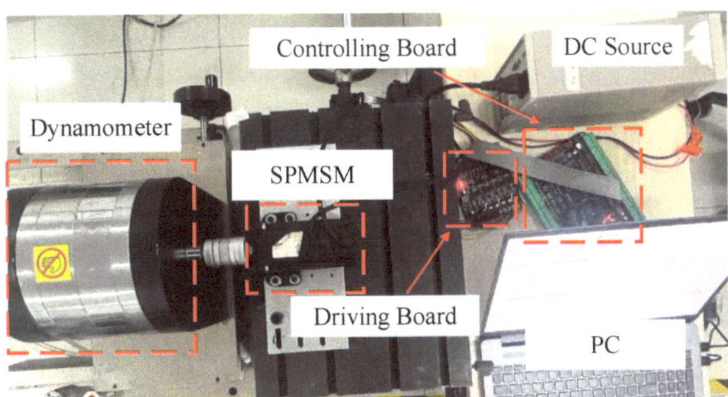

Figure 12. Experimental platform for sensor-less vector control system.

5.2.1. No-Load Experiment

Figure 13 represents the waveform of rotor position when the rotor rises from the stationary state to the target speed. The target speed is set to 1000 rpm, 1500 rpm, and 2000 rpm, and the sliding mode gain λ is set to 0.001.

Figure 13 shows the dynamic performance test results for target speeds of 1000 rpm, 1500 rpm, and 2000 rpm, respectively. Due to the delay in data transmission from the upper computer, the rotor position waveform in Figure 13 only represents the rotor position waveform at a time of approximately 0.1 s. The Hall sensor is used to read the actual rotor position information during the test, which is used to verify the correctness of the estimated rotor position, and the AGFSTA–SMO algorithm is used to estimate the rotor position using AQPLL. As can be seen in Figure 13, the estimates obtained by the AGFSTA–SMO algorithm still float around the true value, except for the errors caused by the mechanical characteristics of the Hall sensors. The relatively small amplitude of the back EMF in the low-speed range with respect to the gain of the AGFSTA–SMO leads to poor accuracy in estimating the rotor position during the low-speed operation of the motor. Therefore, when the motor speed is increased to 2000 rpm, the oscillations in the estimated rotor position are significantly reduced and the estimation accuracy is significantly improved.

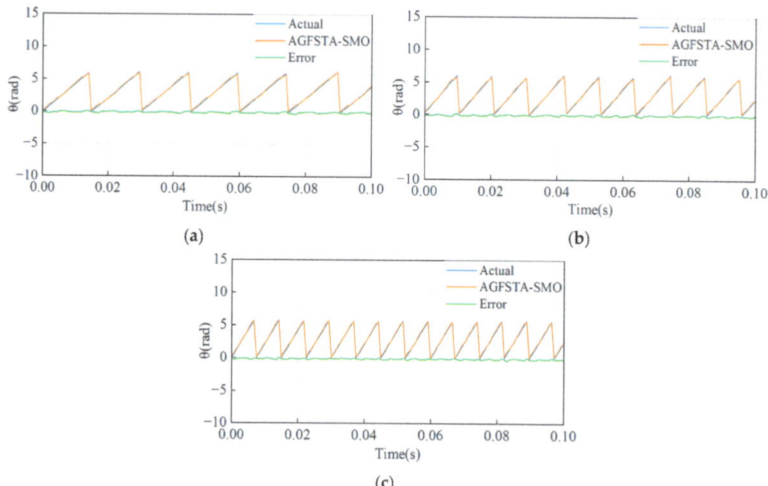

(**a**)

(**b**)

(**c**)

Figure 13. Comparison of estimated and actual values of rotor position at different target speeds. (**a**) 1000 rpm. (**b**) 1500 rpm. (**c**) 2000 rpm.

5.2.2. Comparison of Speed Tracking Performance of LSTA–SMO Algorithm and AGFSTA–SMO Algorithm under Sudden Load Change Conditions

As can be seen in Figure 14, the sudden change in motor load results in increasing the jitter amplitude of the observations due to the constant sliding mode gain of the LSTA–SMO algorithm, which further leads to oscillations in the outputs of the LSTA–SMO, resulting in a decrease in the accuracy of the estimation of the rotor position and speed, and is particularly prominent in the low-speed domain. When the rotational speed is increased to 1500 rpm and 2000 rpm, the LSTA–SMO gain is increased, thus resulting in a faster convergence of the observations and a reduction in the jitter amplitude. As can be seen in Figure 15, when the AGFSTA–SMO algorithm is used, the sliding mode observer gain can be adjusted adaptively according to the motor speed, and the original chattering of the sliding mode can be suppressed. Therefore, the AGFSTA–SMO algorithm has a better performance in terms of jitter amplitude, convergence speed, and jitter frequency when the motor operating conditions change abruptly.

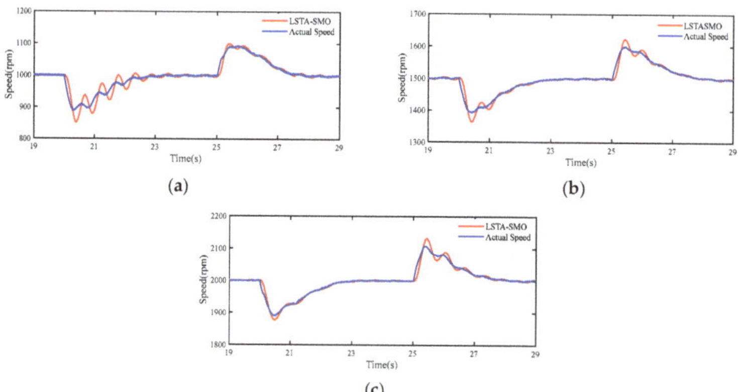

(**a**)

(**b**)

(**c**)

Figure 14. Estimated speed values of LSTA–SMO algorithm vs. actual speed values for different target speeds and sudden load changes. (**a**) 1000 rpm. (**b**) 1500 rpm. (**c**) 2000 rpm.

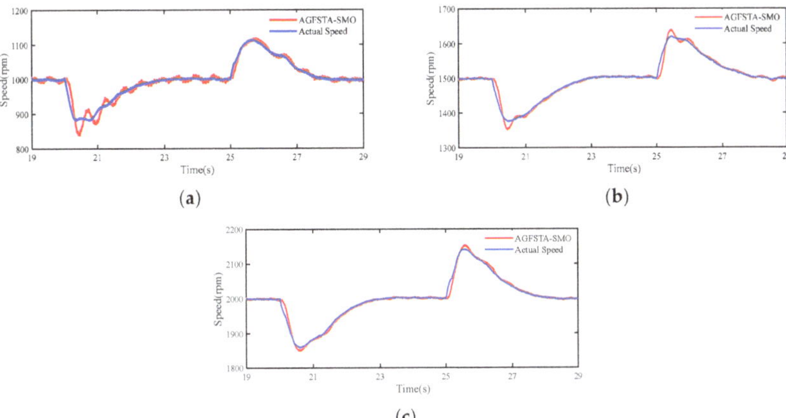

Figure 15. AGFSTA–SMO algorithm estimated speed values vs. actual speed values for different target speeds and sudden load changes. (**a**) 1000 rpm. (**b**) 1500 rpm. (**c**) 2000 rpm.

5.2.3. Performance of AGFSTA–SMO Method for Rotor Position Estimation during Sudden Load Changes

From Figure 16, it can be seen that when the motor is suddenly loaded at 25 s, the error between the estimated rotor position and the actual rotor position increases by 0.05 **rad**, and the recovery time is about 0.6 s. Therefore, the effect of the perturbation of the load on the estimation accuracy can be almost ignored.

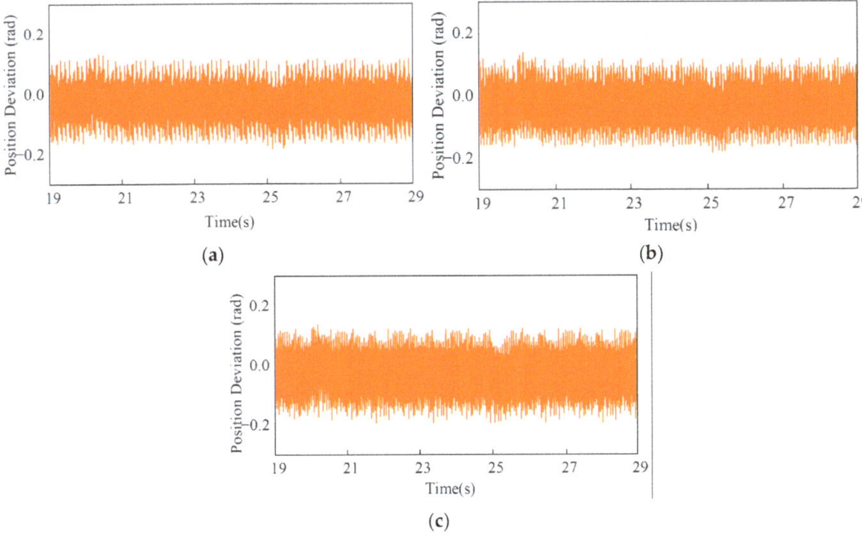

Figure 16. Error between the estimated rotor position and the actual rotor position of the AGFSTA–SMO algorithm for different target speeds and sudden load changes. (**a**) 1000 rpm. (**b**) 1500 rpm. (**c**) 2000 rpm.

6. Conclusions

In this paper, a simple and efficient sensorless control strategy is proposed for the permanent magnet synchronous motor drive control system. Compared with the traditional super-twisting sliding mode observer and the improved super-twisting sliding mode observer, the proposed AGFSTA–SMO is able to improve the sliding mode convergence rate by using the gain adaptive parameter, which solves the problem of the difficulty in parameter adjustment due to the excessive number of gains in the traditional model. It also simplifies the parameter adjustment process by designing the rotational speed-dependent adaptive function to replace the multiple gain parameters in the super-twisting sliding mode observer In addition, the observation accuracy is improved. Meanwhile, this paper reduces the negative impact of harmonic components by using synchronized reference system filters and adaptive quadrature phase-locked loops, which further improves the convergence speed of the observer and reduces the jitter amplitude of the observation results. Finally, the Lyapunov theory was used to demonstrate the steady state nature of the sensorless control system for permanent magnet synchronous motors. Numerous experimental results validated the advantages of the scheme proposed in this paper. In future research, work will be focused on improving the situation of the large error and difficult startup of the downward sliding mode observer in the low-speed domain, and extending the method proposed in this paper to the low-speed domain.

Author Contributions: Conceptualization, J.R. and Y.Z.; methodology, M.L.; software, M.L.; validation, M.L., H.Y. and L.W.; formal analysis, M.L.; investigation, L.W.; resources, J.R.; data curation, H.Y.; writing—original draft preparation, M.L.; writing—review and editing, J.R. and Y.Z.; visualization, H.Y.; supervision, Y.Z.; project administration, J.R.; funding acquisition, J.R. All authors have read and agreed to the published version of the manuscript.

Funding: This work is funded by the National Natural Science Foundation of China. (Approval number: 52177052).

Data Availability Statement: Data are contained within the article.

Conflicts of Interest: The authors declare no conflicts of interest.

References

1. Wang, Y.; Xu, Y.; Zou, J. Sliding-mode sensorless control of PMSM with inverter nonlinearity compensation. *IEEE Trans. Power Electron.* **2019**, *34*, 10206–10220. [CrossRef]
2. Gong, C.; Hu, Y.; Gao, J.; Wang, Y.; Yan, L. An improved delay-suppressed sliding-mode observer for sensorless vector-controlled PMSM. *IEEE Trans. Ind. Electron.* **2019**, *67*, 5913–5923. [CrossRef]
3. Wang, Y.; Wang, X.; Xie, W.; Dou, M. Full-speed range encoderless control for salient-pole PMSM with a novel full-order SMO. *Energies* **2018**, *11*, 2423. [CrossRef]
4. Apte, A.; Joshi, V.A.; Mehta, H.; Walambe, R. Disturbance-observer-based sensorless control of PMSM using integral state feedback controller. *IEEE Trans. Power Electron.* **2019**, *35*, 6082–6090. [CrossRef]
5. Volpato Filho, C.J.; Xiao, D.; Vieira, R.P.; Emadi, A. Observers for high-speed sensorless pmsm drives: Design methods, tuning challenges and future trends. *IEEE Access* **2021**, *9*, 56397–56415. [CrossRef]
6. Yin, Z.; Zhang, Y.; Cao, X.; Yuan, D.; Liu, J. Estimated position error suppression using novel PLL for IPMSM sensorless drives based on full-order SMO. *IEEE Trans. Power Electron.* **2021**, *37*, 4463–4474. [CrossRef]
7. Chen, Z.; Zhang, X.; Zhang, H.; Liu, C.; Luo, G. Adaptive sliding mode observer-based sensorless control for SPMSM employing a dual-PLL. *IEEE Trans. Transp. Electrif.* **2021**, *8*, 1267–1277. [CrossRef]
8. Ni, Y.; Shao, D. Research of improved mras based sensorless control of permanent magnet synchronous motor considering parameter sensitivity. In Proceedings of the 2021 IEEE 4th Advanced Information Management, Communicates, Electronic and Automation Control Conference (IMCEC), Chongqing, China, 18–20 June 2021; Volume 4, pp. 633–638.
9. Zerdali, E.; Barut, M. The comparisons of optimized extended Kalman filters for speed-sensorless control of induction motors. *IEEE Trans. Ind. Electron.* **2017**, *64*, 4340–4351. [CrossRef]
10. Nurettin, A.; İnanç, N. Sensorless vector control for induction motor drive at very low and zero speeds based on an adaptive-gain super-twisting sliding mode observer. *IEEE J. Emerg. Sel. Top. Power Electron.* **2023**, *11*, 4332–4339. [CrossRef]
11. Benevieri, A.; Formentini, A.; Marchesoni, M.; Passalacqua, M.; Vaccaro, L. Sensorless control with switching frequency square wave voltage injection for SPMSM with low rotor magnetic anisotropy. *IEEE Trans. Power Electron.* **2023**, *38*, 10060–10072. [CrossRef]
12. Xu, W.; Qu, S.; Zhao, L.; Zhang, H. An improved adaptive sliding mode observer for middle-and high-speed rotor tracking. *IEEE Trans. Power Electron.* **2020**, *36*, 1043–1053. [CrossRef]
13. Zuo, Y.; Lai, C.; Iyer, K.L.V. A review of sliding mode observer based sensorless control methods for PMSM drive. *IEEE Trans. Power Electron.* **2023**, *38*, 11352–11367. [CrossRef]
14. Ren, N.; Fan, L.; Zhang, Z. Sensorless PMSM control with sliding mode observer based on sigmoid function. *J. Electr. Eng. Technol.* **2021**, *16*, 933–939. [CrossRef]

15. Jin, D.; Liu, L.; Lin, Q.; Liang, D. Sensorless Control Strategy of PMSM with Disturbance Rejection Based on Adaptive Sliding Mode Control Law. *IEEE Trans. Transp. Electrif.* **2023**, *10*, 5424–5438. [CrossRef]
16. Wang, C.; Gou, L.; Dong, S.; Zhou, M.; You, X. Sensorless control of IPMSM based on super-twisting sliding mode observer with CVGI considering flying start. *IEEE Trans. Transp. Electrif.* **2021**, *8*, 2106–2117. [CrossRef]
17. Ge, Y.; Yang, L.; Ma, X. A harmonic compensation method for SPMSM sensorless control based on the orthogonal master-slave adaptive notch filter. *IEEE Trans. Power Electron.* **2021**, *36*, 11701–11711. [CrossRef]
18. Chen, L.; Jin, Z.; Shao, K.; Wang, H.; Wang, G.; Iu, H.H.; Fernando, T. Sensorless Fixed-time Sliding Mode Control of PMSM Based on Barrier Function Adaptive Super-Twisting Observer. *IEEE Trans. Power Electron.* **2023**, *39*, 3037–3051. [CrossRef]
19. Yang, C.; Song, B.; Xie, Y.; Zheng, S.; Tang, X. Adaptive Identification of Nonlinear Friction and Load Torque for PMSM Drives via a Parallel-Observer-Based Network with Model Compensation. *IEEE Trans. Power Electron.* **2023**, *38*, 5875–5897. [CrossRef]
20. Jiang, T.; Ni, R.; Gu, S.; Wang, G. A study on position estimation error in sensorless control of PMSM based on back EMF observation method. In Proceedings of the 2021 24th International Conference on Electrical Machines and Systems (ICEMS), Gyeongju, Republic of Korea, 31 October–3 November 2021; pp. 1999–2003.
21. Chen, Y.; Li, M.; Gao, Y.W.; Chen, Z.Y. A sliding mode speed and position observer for a surface-mounted PMSM. *ISA Trans.* **2019**, *87*, 17–27. [CrossRef]
22. Wang, B.; Shao, Y.; Yu, Y.; Dong, Q.; Yun, Z.; Xu, D. High-order terminal sliding-mode observer for chattering suppression and finite-time convergence in sensorless SPMSM drives. *IEEE Trans. Power Electron.* **2021**, *36*, 11910–11920. [CrossRef]
23. Yang, Z.; Ding, Q.; Sun, X.; Lu, C.; Zhu, H. Speed sensorless control of a bearingless induction motor based on sliding mode observer and phase-locked loop. *ISA Trans.* **2022**, *123*, 346–356. [CrossRef] [PubMed]
24. Xu, W.; Qu, S.; Zhang, H. A composite sliding-mode observer with PLL structure for a sensorless SPMSM system. *Int. J. Electr. Power Energy Syst.* **2022**, *143*, 108510. [CrossRef]
25. Naderian, M.; Markadeh, G.A.; Karimi-Ghartemani, M.; Mojiri, M. Improved sensorless control strategy for IPMSM using an ePLL approach with high-frequency injection. *IEEE Trans. Ind. Electron.* **2023**, *71*, 2231–2241. [CrossRef]
26. Hou, Q.; Ding, S.; Dou, W.; Jiang, L. Estimated Position Error Suppression Using Quasi-Super-Twisting Observer and Enhanced Quadrature Phase-Locked Loop for PMSM Sensorless Drives. *IEEE Trans. Circuits Syst. II Express Briefs* **2024**, *71*, 3865–3869. [CrossRef]
27. Zhu, W.; Li, S.; Du, H.; Yu, X. Nonsmooth observer-based sensorless speed control for permanent magnet synchronous motor. *IEEE Trans. Ind. Electron.* **2022**, *69*, 13514–13523. [CrossRef]
28. Moreno, J.A.; Osorio, M. A Lyapunov approach to second-order sliding mode controllers and observers. In Proceedings of the 2008 47th IEEE conference on decision and control, Cancun, Mexico, 9–11 December 2008; pp. 2856–2861.
29. Lin, S.; Zhang, W. An adaptive sliding-mode observer with a tangent function-based PLL structure for position sensorless PMSM drives. *Int. J. Electr. Power Energy Syst.* **2017**, *88*, 63–74. [CrossRef]
30. Novak, Z.; Novak, M. Adaptive PLL-based sensorless control for improved dynamics of high-speed PMSM. *IEEE Trans. Power Electron.* **2022**, *37*, 10154–10165. [CrossRef]
31. Liu, G.; Zhang, H.; Song, X. Position-estimation deviation-suppression technology of PMSM combining phase self-compensation SMO and feed-forward PLL. *IEEE J. Emerg. Sel. Top. Power Electron.* **2020**, *9*, 335–344. [CrossRef]

actuators

MDPI

Article

Development of a 3 kW Wind Energy Conversion System Emulator Using a Grid-Connected Doubly-Fed Induction Generator

Boussad Boukais [1]**, Koussaila Mesbah** [1]**, Adel Rahoui** [1,2]**, Abdelhakim Saim** [3,*]**, Azeddine Houari** [3]
and Mohamed Fouad Benkhoris [3]

[1] L2CSP Laboratory, Mouloud Mammeri University of Tizi-Ouzou, BP 17 RP, Tizi-Ouzou 15000, Algeria;
boussad.boukais@ummto.dz (B.B.); koussaila.mesbah@ummto.dz (K.M.); a.rahoui@enstp.edu.dz (A.R.)

[2] Preparatory Classes Department, Ecole Nationale Supérieure des Travaux Publics Francis Jeanson,
Rue Sidi Garidi BP 32 Vieux Kouba, Algiers 16051, Algeria

[3] Institut de Recherche en Énergie Électrique de Nantes Atlantique (IREENA), Nantes Université, UR 4642,
F-44600 Saint-Nazaire, France; azeddine.houari@univ-nantes.fr (A.H.);
mohamed-fouad.benkhoris@univ-nantes.fr (M.F.B.)

* Correspondence: abdelhakim.saim@univ-nantes.fr

Abstract: This paper presents the design and performance evaluation of an experimental platform that emulates the static and dynamic behavior of a 3 kW Wind Energy Conversion System (WECS). The platform includes a wind turbine emulator (WTE) using a separately excited DC motor (SEDCM) as the prime mover, coupled with a grid-connected doubly-fed induction generator (DFIG). This setup enables comprehensive laboratory studies of a WECS without the need for large-scale field installations. A novel inertia compensation strategy is implemented to ensure the SEDCM accurately replicates the power and torque characteristics of a real wind turbine across various wind profiles. The DFIG was chosen for its high efficiency at variable wind speeds and its reduced power converter requirements compared to other generators. The control strategy for the DFIG is detailed, highlighting its performance and seamless integration within the system. Unlike most studies focusing on generators connected to simple loads, this research considers a grid-connected system, which introduces additional challenges and requirements. This study thoroughly investigates the grid-connected converter, addressing specific demands for grid connection and ensuring compliance with grid standards. Experimental results validate the effectiveness of the emulator, demonstrating its potential as a key tool for optimizing wind turbine control strategies and real-time algorithm validation, and enhancing the performance and reliability of renewable energy systems.

Keywords: wind energy conversion system (WECS); wind turbine emulator (WTE); grid-connected power systems; doubly-fed induction generator (DFIG); direct power control (DPC)

Citation: Boukais, B.; Mesbah, K.;
Rahoui, A.; Saim, A.; Houari, A.;
Benkhoris, M.F. Development of a
3 kW Wind Energy Conversion
System Emulator Using a
Grid-Connected Doubly-Fed
Induction Generator. *Actuators* **2024**,
13, 487. https://doi.org/10.3390/
act13120487

Academic Editor: Dong Jiang

Received: 21 October 2024
Revised: 15 November 2024
Accepted: 26 November 2024
Published: 29 November 2024

1. Introduction

Renewable energy is critical to the global energy transition, with wind power emerging as one of the most promising sources for electricity generation. Among green energy alternatives, wind power stands out due to its zero emissions and widespread availability [1,2]. Recent advancements in wind energy conversion systems (WECSs) have driven a rapid expansion of this clean and cost-effective energy source worldwide over the past few decades [3–5].

Maximizing WECS efficiency requires extensive research into turbine design, generator integration, and grid connection. However, real-time testing of actual wind turbines is often infeasible in research environments due to their large physical footprint, high costs, and reliance on intermittent wind conditions. Wind turbine emulators address these challenges by enabling both simulation and hardware-based emulation of WECSs. While simulation is a fundamental tool for WECS analysis and design [6], hardware emulation provides a more realistic testing environment, allowing for the evaluation of control strategies, power converter topologies, and the effects of wind variability on grid integration. Consequently, research laboratories are increasingly focused on developing comprehensive emulators that replicate the aerodynamic, mechanical, and electrical behaviors of real wind turbines [7].

WECS emulators can be constructed using only power electronic converters, known as electronic emulators [8,9]. However, to better replicate real WECS behavior, most emulators consist of two mechanically coupled machines: one motor simulating wind torque, commonly referred to as a wind turbine emulator (WTE), and the other acting as an electrical generator. The choice of motor and generator is a critical design consideration, and numerous studies have been conducted on this topic. For the prime mover, some studies use permanent magnet synchronous motors (PMSMs) [10,11] or switched reluctance motors (SRMs) [12], which offer high energy efficiency and precise torque control. However, induction motors (IMs) and DC motors are the most commonly used options.

IMs are widely used due to their robustness, moderate cost, and ability to operate in diverse environments with minimal maintenance. They can handle overloads and provide a wide range of speeds and torques, making them suitable for wind turbine emulation under variable wind conditions. IM-based WTEs are discussed in [13–21]. For instance, [13] uses a 7.5 kW IM to replicate the static characteristics of a wind turbine, while [14] employs a 1.5 kW squirrel-cage motor to emulate the dynamic behavior of a wind turbine. Other works focus on maximum power point tracking (MPPT) algorithms [15], tower shadow effects [16], and wind shear [17]. Authors in [18,19] present systems that emulate the behavior of small-scale wind turbines. Study [20] explores power and voltage fluctuations as well as flicker emissions, while [21] examines the impact of short-circuit faults on the electrical and mechanical components of a fixed-speed wind turbine.

On the other hand, DC motors are widely appreciated for their simplicity of control and ability to provide consistent torque over a broad speed range. Their straightforward modeling and implementation make them a popular choice for test platforms in research and rapid prototyping applications. Additionally, they are cost-effective, requiring fewer power electronics components than IMs. Separately excited DC motors (SEDCMs) and permanent magnet DC motors have been widely used in the literature [22–33]. They are often preferred for WTEs because of the direct relationship between armature current and torque. Two control strategies—speed control and torque control—were proposed by [23]. In [24], a refined torque control strategy was introduced to account for harmonic torques caused by wind gradients and tower shadow effects, while [25] employed a three-phase thyristor rectifier instead of a chopper for DC motor control. Advanced techniques for optimizing the PI controller parameters were presented in [26–28], and ref. [29] replaced the PI controller with a fuzzy logic controller. An open-loop control method that eliminates the need for a processor or complex control systems was proposed in [30]. In [31], a WTE using a DC motor controlled by an FPGA-based chopper was developed, and a Hardware-in-the-Loop (HIL) implementation of a DC motor-based WTE was presented in [32]. A 300W DC motor was used to emulate the dynamic characteristics of a 5 MW wind turbine in [33].

Studies on WECS emulators have employed various generators, including permanent magnet synchronous generators (PMSGs), induction generators (IGs), and doubly-fed induction generators (DFIGs). PMSGs are favored for their high efficiency and superior reliability, owing to the absence of mechanical components such as the brush-collector system and gearboxes in high-pole machines. These generators do not require separate excitation, making them suitable for small-scale designs. Emulators using PMSGs are described in [19,26,29,33–37]. However, the high cost of permanent magnets is a significant concern, especially in large wind turbines. Additionally, a full-scale power converter is required to connect the generator to the grid. IGs, on the other hand, offer advantages such as lower unit cost and high robustness. However, they require an external excitation circuit and reactive power from the grid. Furthermore, they need a synchronization relay and a full-scale power converter for grid connection. Studies using IGs are detailed in [14–17,20,21,38].

By contrast, DFIGs are the most commonly used generators in WECSs, primarily because they only require a converter rated at approximately 30% of the nominal power, while allowing a speed variation of 60% around the synchronous speed [1]. This leads to reduced costs at higher overall efficiency. DFIGs are also valued for their ability to control reactive power, making them more suitable for high-power applications. Despite this, few studies on WTEs have employed DFIGs [36].

The design of a WTE must overcome the challenge of replicating the slow dynamic behavior of a real wind turbine on a test platform using low-inertia machines. This challenge is further complicated by the fact that the inertia of the motor-generator system (WTE) is significantly lower than that of an actual wind turbine. To address this, two primary approaches are proposed in the literature. The first is rigid inertia compensation using a mechanical flywheel [37,39]. The second, more flexible approach, dynamically adjusts the motor torque based on the difference between the inertia of the WTE and that of the wind turbine. This method, which employs a control algorithm to adapt the drive torque [36,40–42], offers several advantages, including no need for hardware modifications, ease of implementation, and precise tuning. This flexible approach is one of the key improvements discussed in this paper.

Research on grid-connected wind turbine emulators using DFIGs are crucial due to the growing demand for efficient integration of intermittent energy sources. A grid-connected emulator not only replicates the real operating conditions of a wind turbine but also facilitates the testing of various control strategies. These experiments are vital for assessing grid stability in response to sudden and often unpredictable fluctuations in wind power, while minimizing the risk of instability or imbalance. Additionally, they offer deep insights into DFIG performance during grid disturbances, such as voltage dips or load variations, to enhance system resilience. There is limited research on grid-connected emulators [35,43]. For instance, [35] employs a PMSMG generator, while [43] explores DFIG grid connection with unrealistic wind profiles and a unidirectional converter. In this work, however, a back-to-back bidirectional converter and a realistic wind profile model are used for grid connection experiments, representing a significant improvement.

This work provides a comprehensive overview of the development of a complete WECS emulator test bench. The system uses an SEDCM as the prime mover, controlled via a DC-DC buck converter to simulate the aerodynamic behavior of a wind turbine, and a 3 kW grid-connected DFIG as the generator. The DFIG allows for independent regulation of active and reactive power exchanged with the grid, while a synchronous reference frame phase-locked loop (SRF-PLL) ensures precise grid synchronization. The system implementation was carried out using dSPACE DS1104 hardware system, operated through ControlDesk software (version 3.7.1).

Key features of the developed test bench include the following:

- The capability to simulate and test the system under various wind profiles.
- Real-time validation of advanced control strategies for managing active and reactive power.
- Online implementation and evaluation of different MPPT algorithms.
- Testing of different grid synchronization techniques.
- A highly valuable tool for the WECS industry, research laboratories, and as an educational platform for teaching wind turbine operation, control, and dynamic behavior.

2. Overall Scheme of the WECS

The studied WECS, shown in Figure 1, consists of a wind turbine coupled with a DFIG connected to the grid. The DFIG, is a widely used generator in wind energy systems due to its ability to operate efficiently over a range of wind speeds. The DFIG is interfaced with the grid via two converters: the Rotor-Side Converter (RSC) and the Grid-Side Converter (GSC). The RSC controls the rotor voltages, ensuring optimal power extraction from the wind, while the GSC regulates the DC bus voltage and ensures that power is injected into the grid with a unity power factor. The control strategy employed in the DFIG is based on direct power control (DPC), which decouples the active and reactive power components.

In the synchronous reference frame (*dq*-frame), the active power is controlled through the *d*-axis by adjusting the rotor speed to follow the maximum power point tracking (MPPT) algorithm. The *q*-axis controls the reactive power, ensuring that the system operates at a unity power factor by minimizing reactive power exchange with the grid. The system also incorporates phase-locked loops (PLLs) to ensure precise synchronization with the grid by continuously tracking the phase angle of the grid voltage. This guarantees smooth integration and stable operation under varying wind conditions.

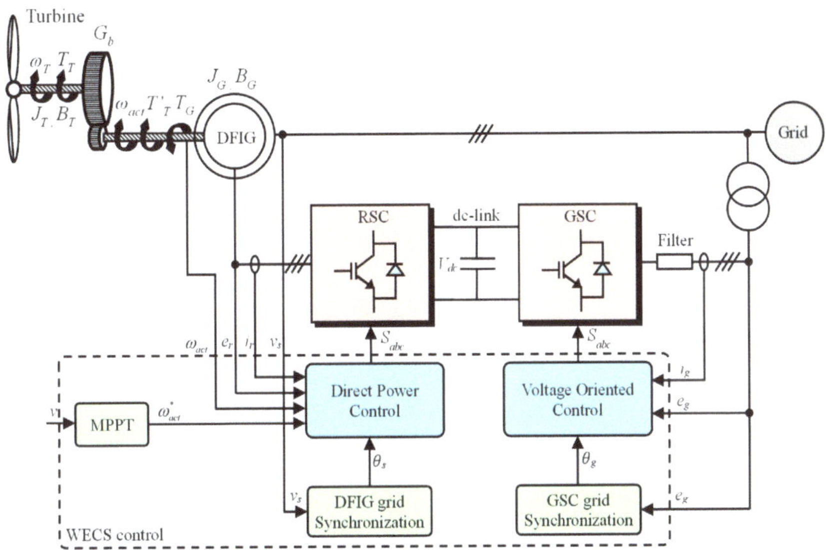

Figure 1. DFIG control strategy diagram for variable-speed wind turbines.

Overall, the WECS is designed to maximize energy capture from the wind, ensure efficient power transfer to the grid, and maintain stable operation, even during dynamic wind conditions.

2.1. Extracted Wind Power

Energy production is achieved by extracting the wind kinetic energy as it passes through the wind turbine. The mechanical power P_T extracted from the wind and available on the turbine shaft is then expressed as follows:

$$P_T = \frac{1}{2}\rho S v^3 C_p(\lambda) \tag{1}$$

where ρ is the air density, v is the wind speed at the blades, S is the swept area by the blades of length R such that $S = \pi R^2$, and $C_p(\lambda)$ is the power coefficient which depends only on λ for small power wind turbines with a fixed pitch angle.

λ represents the tip-speed ratio, defined as the ratio of the wind turbine's peripheral speed $v_T = \omega_T R$ to the wind speed v, given by

$$\lambda = \frac{\omega_T R}{v}. \tag{2}$$

The λ value that yields a maximum power coefficient C_{pmax} is λ_{opt}, which corresponds to the Betz limit, formulated by the German physicist Albert Betz in 1919 [44]. According to this limit, a wind turbine can extract a maximum of 59.3% of the available wind power. Although wind turbine manufacturers try to approach this limit, the current C_{pmax} is around 40%, and the Betz limit is never reached in practice. Each wind turbine is characterized by a $C_p = f(\lambda)$ curve that describes its efficiency. This characteristic is either provided by the manufacturer or determined experimentally. In this study, the $C_p(\lambda)$ curve shown in Figure 2 corresponds to experimental measurements of an actual wind turbine obtained from [45]. Due to its non-linear nature, Matlab's 'Basic Fitting' tool is used to interpolate this characteristic. The obtained expression is given by

$$C_p(\lambda) = \begin{aligned} &-9.727 \times 10^{-7}\lambda^9 + 0.00004554\lambda^8 - 0.00088816\lambda^7 + 0.009302\lambda^6 - 0.056085\lambda^5 \\ &+ 0.19293\lambda^4 - 0.35191\lambda^3 + 0.30148\lambda^2 - 0.070868\lambda + 0.0039702. \end{aligned} \tag{3}$$

The $C_p = f(\lambda)$ curve has an optimal point ($\lambda_{opt} = 6.71$, $C_{pmax} = 0.41$), which corresponds to the maximum efficiency of the turbine. At this point, the maximum mechanical power can be extracted from the available wind power. Table 1 presents the studied wind turbine parameters.

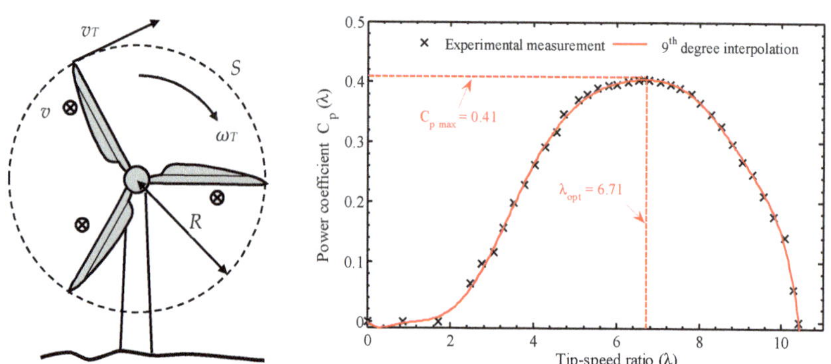

Figure 2. Power coefficient $C_p(\lambda)$ curve of real and simulated wind turbine.

Table 1. Wind turbine parameters [45].

Parameters	Values
Rated Power P_n (kW)	3
Rotor radius R (m)	4.5
Gearbox ratio G_b	24
Moment of inertia J_T (Kgm^2)	7
Performance Coefficient C_{pmax}	0.41
Tip-speed ratio λ_{opt}	6.71
Wind speed max v_{max} (m/s)	5.71
Wind speed min v_{min} (m/s)	3

Knowing the rotational speed ω_T of the turbine, the mechanical torque T_T available on the turbine's slow shaft can be expressed as follows [15]:

$$T_T = \frac{P_T}{\omega_T} = \frac{1}{2}\rho S \frac{v^3}{\omega_T} C_p(\lambda). \tag{4}$$

By substituting ω_T from (2) into (4), it is shown that torque T_T depends on λ.

$$T_T = \frac{P_T}{\omega_T} = \frac{1}{2}\rho \pi R^3 \frac{v^2}{\lambda} C_p(\lambda). \tag{5}$$

2.2. Wind Turbine Mechanical Model

The physical system consists of a wind turbine on a slow shaft, coupled to the generator's fast shaft via a gearbox with gain G_b, which adjusts the primary shaft's speed to match the generator's speed. Indeed, the low rotational speed of the wind turbine does not allow for efficient electrical power generation with conventional generator. To derive the mathematical model linking the turbine's rotational speed ω_T and torque T_T to the generator's speed ω_{act} and torque T'_T, let J_T and B_T represent the turbine's moment of inertia and friction coefficient, and $J_G = 0.037$ kg·m^2 and $B_G = 0.012$ Nm·s/rad represent the generator's inertia and friction coefficient. Assuming negligible mechanical losses in the gearbox, the following is obtained:

$$T_T = G_b T'_T, \qquad \omega_T = \omega_{act}/G_b. \tag{6}$$

Based on these assumptions, a mechanical model consisting of two masses is obtained. The total equivalent inertia J_{TG} and the total friction coefficient B_{TG} of the wind turbine system, viewed from the generator rotor, are calculated as follows:

$$\begin{cases} J_{TG} = J_G + J_T/G_b^2 \\ B_{TG} = B_G + B_T/G_b^2 \end{cases}. \tag{7}$$

The equation of motion that determines the mechanical speed ω_{act} from the driving torque T'_T at the generator rotor and the electromagnetic torque T_G developed by the generator is given by [15,45]

$$J_{TG}\frac{d}{dt}\omega_{act} + B_{TG}\omega_{act} = T'_T - T_G. \tag{8}$$

The corresponding transfer function $G_{TG}(s)$ is as follows:

$$G_{TG}(s) = \frac{\omega_{act}(s)}{T'_T(s) - T_G(s)} = \frac{1}{J_{TG}s + B_{TG}}. \tag{9}$$

In this model, the friction coefficient B_T is neglected due to the very low ratio of $B_T/G_b{}^2$. This assumption simplifies the system dynamics.

2.3. MPPT-Based Wind Turbine Control

This article focuses on MPPT operation. The wind turbine starts operating at a minimum wind speed v_{min} to extract maximum power until the wind reaches the maximum speed v_{max}, corresponding to the generator's nominal mechanical power P_{Gn}. To maximize extracted power, the turbine must operate at the point corresponding to C_{pmax} and λ_{opt} as shown in Figure 3a. The optimal torque is obtained by substituting C_p with C_{pmax} and λ with λ_{opt} in (5), resulting in

$$T_{T-opt} = \frac{P_{Tmax}}{\omega_T} = \frac{1}{2}\rho\pi R^3 \frac{v^2}{\lambda_{opt}}C_{pmax}. \tag{10}$$

For wind speeds between $v_{min} = 3$ m/s and $v_{max} = 5.71$ m/s, corresponding to 109.9 and 204.2 rad/s ($\pm 30\%\ \omega_{Gn}$), respectively, the extracted power can be maximized by applying the corresponding optimal torque. For wind speeds exceeding $v_{max} = 5.71$ m/s, these values exceed the nominal ratings of the turbine and generator, making MPPT infeasible and necessitating limiting devices. Figure 3b illustrates that to extract maximum power, the mechanical torque must be optimal, which does not always coincide with the turbine's maximum torque across different wind speeds.

Various MPPT strategies exist with and without rotational speed control. This work adopts an MPPT approach that adjusts the generator's electromagnetic torque to set the rotational speed to a reference, enabling maximum power extraction. The method's scheme is shown in Figure 4. Mechanical power is maximized when the coefficient C_p reaches its maximum value at the optimal tip-speed λ_{opt}. Thus, the reference rotational speed of the generator is calculated as follows [15]:

$$\omega_{act}^* = \frac{\lambda_{opt}v}{R}G_b. \tag{11}$$

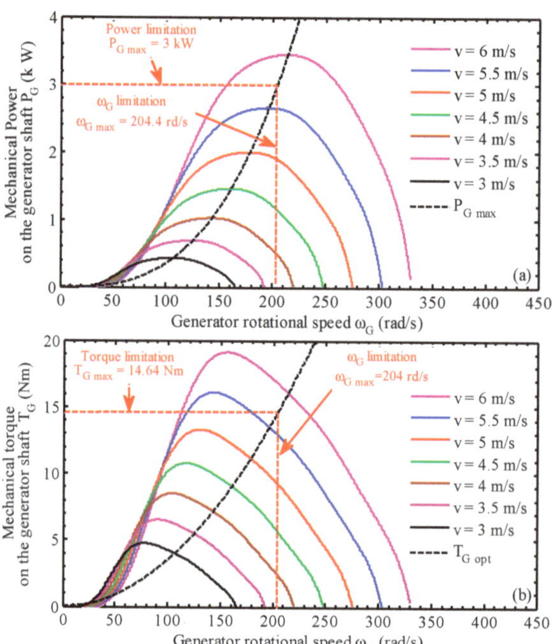

Figure 3. Wind turbine characteristics for different wind speed: (**a**) mechanical power on DFIG shaft and maximal power; (**b**) mechanical torque and optimal torque.

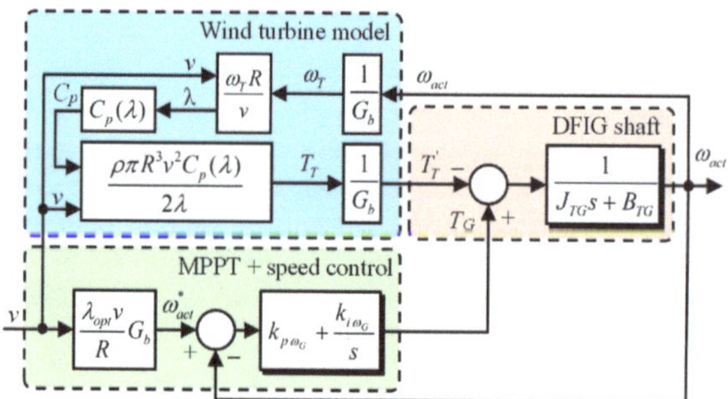

Figure 4. MPPT control with rotational speed feedback.

To regulate the rotational speed to the setpoint determined by the MPPT, a proportional-integral (PI) controller is employed, characterized by proportional gain $k_{p\omega G}$ and integral gain $k_{i\omega G}$ [46]. The mechanical torque T'_T is considered a compensable disturbance since its value can be estimated using (5) and (6). The design of the controller is based on the mechanical model (9) and the corresponding closed-loop transfer function is given by

$$G_{\omega G}(s) = \frac{\omega_{act}(s)}{\omega^*_{act}(s)} = \frac{(k_{p\omega G}/k_{i\omega G})s + 1}{(J_{TG}/k_{i\omega G})s^2 + ((k_{p\omega G} + B_{TG})/k_{i\omega G})s + 1}. \tag{12}$$

By identifying with a 2nd-order system having a damping ratio $\zeta = 1$ and a cutoff frequency $\omega_n = 4.8/\mathrm{tr}$ [46], where t_r is the desired response time, the following results are obtained:

$$\begin{cases} k_{i\omega G} = \omega_n^2 J_{TG} \\ k_{p\omega G} = 2\zeta \omega_n J_{TG} - B_{TG} \end{cases}. \tag{13}$$

For a wind energy system, the response time t_r should ideally be at least ten times faster than the period of the wind speed variations to ensure that the controller can adjust swiftly to dynamic changes. In the wind profile studied, variations occur in a period of approximately 1 s. Thus, selecting t_r = 100 ms allows the controller to respond quickly and accurately to these fluctuations. Accordingly, the values of $k_{i\omega G}$ = 113.24 and $k_{p\omega G}$ = 4.70 are derived to optimize the system's performance. This setup effectively balances the need for rapid adaptation with stable control.

3. SEDCM-Based WTE

The WTE allows for detailed laboratory studies of WECSs without the need for large-scale field installations. The objective is to replicate the mechanical characteristics of a real wind turbine as illustrated in Figure 5. The proposed WTE uses a torque-controlled SEDCM to mimic the mechanical torque of the wind turbine. However, since the mechanical parameters of the wind turbine and the SEDCM are different, the torque produced by the SEDCM differs from that of the actual turbine. To address this, a new compensation method is presented in this section to ensure accurate emulation.

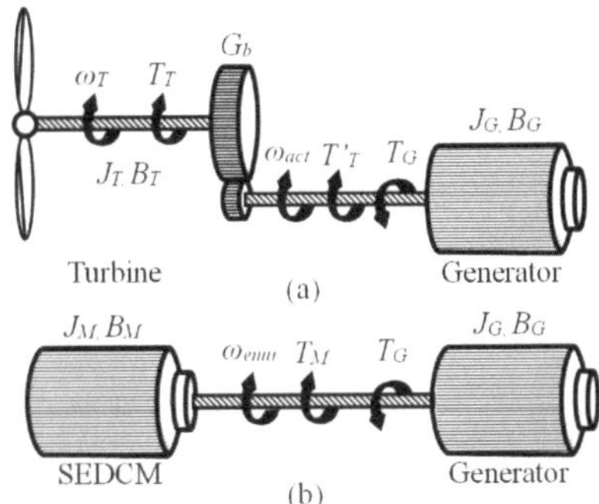

Figure 5. Wind turbine mechanical model: (**a**) wind energy conversion system and (**b**) WTE.

3.1. Electromagnetic Torque Controller Design for the SEDCM

The electromagnetic torque $T_M = K_M I_a$ of the SEDCM is regulated by adjusting the armature current I_a, with K_M being the torque constant. To accurately emulate the wind turbine torque, the SEDCM torque setpoint T^*_M must be continuously calculated. The influence of the electromotive force E is compensated for in the control system, resulting in the 1st-order open-loop transfer function $G_{M-OL}(s)$ as follows:

$$G_{M-OL}(s) = \frac{T_M}{U_a} = \frac{G_a}{\tau_a s + 1}. \tag{14}$$

where $G_a = K_M / R_a$ is the static gain, $\tau_a = L_a / R_a$ is the armature time constant, and L_a and R_a are the armature inductance and resistance of the SEDCM, respectively.

By using a PI controller with proportional parameter k_{pM} and integral parameter k_{iM}, and compensating for the pole L_a / R_a with the zero k_{pM}/k_{iM} of the controller [46], the closed-loop transfer function $G_{M-CL}(s)$ remains 1st order and is expressed as follows:

$$G_{M-CL} = \frac{T_M(s)}{T^*_M(s)} = \frac{1}{(R_a / k_{iM} K_M)s + 1} = \frac{1}{\tau_{cl} s + 1}. \tag{15}$$

where $\tau_{cl} = R_a / (k_{iM} K_M)$ is the closed-loop time constant.

The gain values of the controller are finally obtained as

$$\begin{cases} k_{iM} = R_a / (\tau_{cl} K_M) \\ k_{pM} = L_a / (\tau_{cl} K_M) \end{cases}. \tag{16}$$

The PI controller computes the reference voltage U^*_a to be applied to the SEDCM's armature to achieve the desired reference torque. The control loop requires a faster response time than the wind variations and the mechanical time constants of the actual system. To match the open-loop system dynamics, the time constant τ_{cl} is set equal to τ_a, giving $\tau_{cl} = 6.5$ ms, corresponding to a response time of $t_{rcl} = 19.5$ ms. This value provides the necessary dynamic response for the torque to reach its setpoint efficiently. Furthermore, t_{rcl} is sufficiently smaller than the speed control loop response time ($t_r = 100$ ms), ensuring adequate separation between the control loops. The resulting PI controller parameters are $k_{iM} = 75.18$ and $k_{pM} = 0.488$.

Since the DC motor's mechanical parameters differ from those of the wind turbine, T_M will also differ. Therefore, the reference torque T^*_M must be adjusted so that T_M equals T'_T. This is achieved by introducing a compensation term to T^*_M, which is detailed in the following section.

3.2. Electromagnetic Torque Compensation Loop for the SEDCM

The mechanical model of the SEDCM directly coupled to the generator [45] is given by

$$(J_M + J_G)\frac{d\omega_{emu}}{dt} + (B_M + B_G)\omega_{emu} = T_M - T_G. \tag{17}$$

where ω_{emu} is the rotational speed of the SEDCM, identical to the DFIG's speed, and J_M and B_M represent the SEDCM's inertia and friction coefficient, respectively.

The model described in (17) differs from the actual system outlined in (8), necessitating compensation for friction and inertia to achieve accurate emulation. This compensation involves adjusting the SEDCM torque to replicate the effects of the turbine's inertia and friction coefficient. By comparing the mechanical models in (8) and (17), it can be deduced that both systems will be equivalent if, under the same resisting torque T_G developed by the DFIG, the acceleration of both systems is identical. This leads to

$$T_M - J_{MG}\frac{d\omega_{emu}}{dt} - B_{MG}\omega_{emu} = T'_T - J_{TG}\frac{d}{dt}\omega_{act} - B_{TG}\omega_{act}. \tag{18}$$

where $J_{MG} = J_M + J_G$ and $B_{MG} = B_M + B_G$ represent the total inertia and friction coefficient of the SEDCM coupled to the DFIG, respectively.

From this equality, the compensation torque T_{com}, representing the difference between the turbine torque T'_T and the SEDCM torque T_M, is expressed as

$$T_{com} = T'_T - T_M = (J_{TG} - J_{MG})\frac{d\omega_{emu}}{dt} + (B_{TG} - B_{MG})\omega_{emu}. \tag{19}$$

The resulting 1st-order transfer function is given by

$$\frac{\omega_{emu}}{T_{com}} = \frac{\omega_{emu}}{T'_T - T_M} = \frac{1}{(J_{TG} - J_{MG})s + (B_{TG} - B_{MG})}. \tag{20}$$

To ensure accurate emulation of the real system, T_{com} must be rapidly established, with a response time shorter than other mechanical time constants. Various techniques have been proposed in the literature to address this issue [15,36,45]. In [45], the authors directly exploit (19) for compensation torque estimation. However, the inclusion of a derivative in this equation presents practical implementation challenges. To avoid this issue, [15] replaced the derivative with a high-pass filter (HPF) followed by a low-pass filter (LPF). This solution, while effective, requires precise parameter tuning and complicates the implementation. Moreover, these approaches do not account for turbine and generator friction, resulting in static errors. Works in [36,42] introduced methods that consider friction effects. Authors in [42] proposed a fourth-order filter to mitigate the derivative issue. However, this technique introduces a delay that must be compensated for. To estimate T_{com}, [36] proposed two methodologies: 1-DOF and 2-DOF. These methods estimate the rotational speed based on the difference between turbine and generator torques. The resulting error between the measured and estimated speeds is then utilized to determine the WTE torque reference. Although the 1-DOF technique employs a PI controller, it remains sensitive to mechanical parameter inaccuracies. In contrast, the 2-DOF technique incorporates a pre-filter for speed estimation and a lead-lag controller. Despite its stability, this approach introduces an additional speed estimation loop, increasing complexity. In this study, a method exploiting a straightforward PI controller is proposed to estimate T_{com} directly from the error between the measured speed and the estimated speed by the MPPT. This eliminates the need for a separate speed and fast shaft torque estimation. The conceptual framework of this method is depicted in Figure 6.

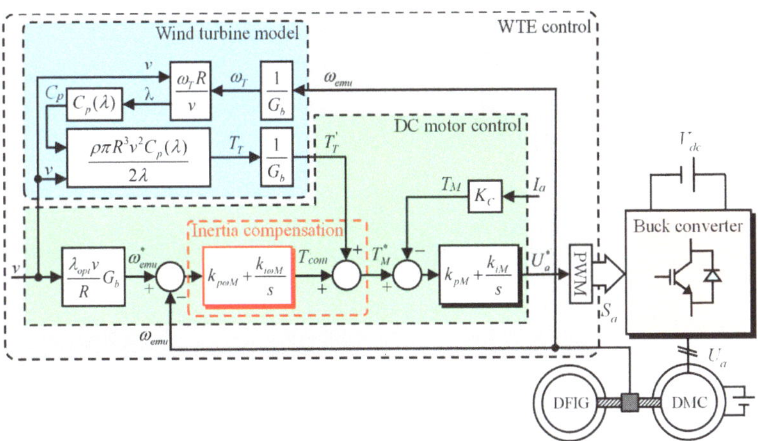

Figure 6. WTE control scheme.

The compensation torque T_{com} is derived using a PI controller with proportional $k_{p\omega M}$ and integral $k_{i\omega M}$ gains calculated from the transfer function in (20) as

$$\frac{\omega_{emu}(s)}{\omega_{emu}^*(s)} = \frac{(k_{p\omega M}/k_{i\omega M})s + 1}{((J_{TG} - J_{MG})/k_{i\omega M})s^2 + ((B_{TG} - B_{MG}) + k_{p\omega M}/k_{i\omega M})s + 1}. \tag{21}$$

The system is identified as a 2nd-order system with a damping ratio $\zeta = 1$ and a natural frequency $\omega_n = 4.8/t_r$ [46], where t_r is the desired response time, leading to the following:

$$\begin{cases} k_{i\omega M} = (J_{TG} - J_{MG})\omega_n^2 \\ k_{p\omega M} = 2\zeta(J_{TG} - J_{MG})\omega_n - (B_{TG} - B_{MG}) \end{cases}. \tag{22}$$

To achieve precise compensation, the response time of the function must be equal to or faster than that of the speed control loop, while remaining slower than the response times of the DFIG power and SEDCM torque control loops. Additionally, for robust stability, the zero must be placed sufficiently far from the origin and decoupled from the poles of function (21). A response time of $t_r = 100$ ms is therefore sufficient to ensure both stability and optimal dynamic performance.

For a response time of $t_r = 100$ ms, the controller gains are $k_{i\omega M} = 100.56$ and $k_{p\omega M} = 4.17$. Finally, the estimated compensation torque T_{com} is added to the mechanical torque T'_T to obtain the reference torque T^*_M of the SEDCM, as illustrated in Figure 6.

3.3. Dynamic Behavior of SEDCM-Based WTE

To demonstrate the effectiveness of the proposed WTE, a simulation comparison in the Matlab/Simulink environment is performed for various quantities of the actual and emulated wind turbine using the parameters from Tables 1 and 2. This test will verify if the emulator's performance is equivalent to that of a wind turbine. The system is controlled by the MPPT strategy with rotational speed feedback control.

Table 2. SEDCM parameters.

Parameters	Values
Rated power P_{Mn} (kW)	3
Rated speed N_{Mn} (rpm)	1500
Rated armature voltage U_a (V)	220
Rated field voltage U_{ext} (V)	120
Rated armature current I_a (A)	16
Maximum field current $I_{ext\text{-}max}$ (A)	1.7
Moment of inertia J_M (Kg.m^2)	0.0558
Viscous friction coefficient B_M (Nm·s/rad)	0.018
Armature resistance R_a (Ω)	1
Armature inductance L_a (mH)	6.447
Electromagnetic torque constant K_M (Nm/A)	1.33

Figure 7 shows the simulation results obtained using a random wind profile with an average speed of 4 m/s, as illustrated in Figure 7a. Figure 7b–f show that the results from the WTE precisely match those of the actual system. The error between the two systems is minimal, validating the emulator's effective operation. Figure 7e shows that the difference between the electromagnetic torque developed by the SEDCM T_M and that produced by the turbine T'_T equals the compensation torque T_{com} calculated by the proposed method. Consequently, as Figure 7d,f show, the tip-speed ratio λ and the power coefficient C_p are perfectly controlled at their optimal values. This indicates that the SEDCM dynamics accurately replicate the behaviour of the wind turbine, with the mechanical torque on the DFIG shaft being precisely compensated for, as depicted in Figure 7c and detailed in Figure 7e.

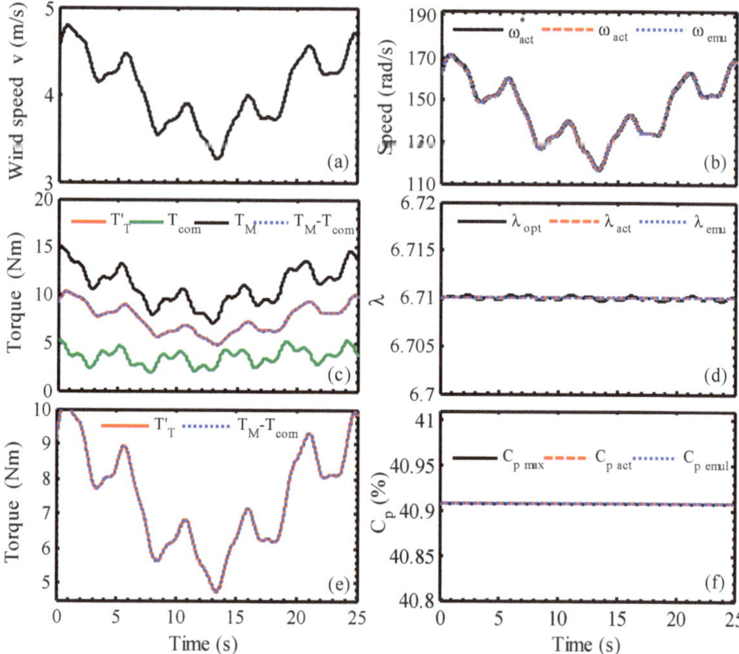

Figure 7. Simulation results of actual and emulated wind turbine under MPPT control with rotational speed feedback: (**a**) wind profile, (**b**) rotational speed, (**c**) torque, (**d**) tip-speed ratio λ, (**e**) zoom of emulated torque $T_M - T_{com}$ and turbine torque T'_T, and (**f**) power coefficient C_p.

4. DFIG Direct Power Control

To simplify the control of the DFIG, the three-phase system is transformed into a two-phase *dq* reference frame using Park's transformation. This allows the direct and quadrature components to be separated, enabling independent control of active and reactive power. The DFIG equations in the *dq* frame, aligned with the stator flux rotating at ω_s, and for rotor electrical angular speed ω_r, are expressed as

$$
\begin{aligned}
V_{sdq} &= R_s I_{sdq} + \frac{d\phi_{sdq}}{dt} + j\omega_s \phi_{sdq} \\
V_{rdq} &= R_r I_{rdq} + \frac{d\phi_{rdq}}{dt} + j\omega_r \phi_{rdq}
\end{aligned}
\tag{23}
$$

where $V_{sdq} = [V_{sd}\ V_{sq}]^T$ and $V_{rdq} = [V_{rd}\ V_{rq}]^T$ are the stator and rotor voltages in the *dq* frame, and $I_{sdq} = [I_{sd}\ I_{sq}]^T$ and $I_{rdq} = [I_{rd}\ I_{rq}]^T$ are the corresponding currents. R_s and R_r are the stator and rotor resistances, respectively.

The stator and rotor fluxes ϕ_{sdq} and ϕ_{rdq} are related to the currents by

$$
\begin{aligned}
\phi_{sdq} &= L_s I_{sdq} + L_m I_{rdq} \\
\phi_{rdq} &= L_r I_{rdq} + L_m I_{sdq}
\end{aligned}
\tag{24}
$$

where L_s, L_r, and L_m are the stator, rotor, and mutual inductances, respectively.

The electromagnetic torque T_{em}, as a function of the stator flux and rotor currents, is given by

$$
T_{em} = \frac{p L_m}{L_s} \left(\phi_{sdq} \times I_{sdq} \right)
\tag{25}
$$

where p is the number of pole pairs, and the symbol $\times$ denotes the cross-product of the vectors ϕ_{sdq} and I_{sdq}.

The stator active power P_s and reactive power Q_s are given by

$$
\begin{aligned}
P_s &= V_{sd} I_{sd} + V_{sq} I_{sq} \\
Q_s &= V_{sq} I_{sd} - V_{sd} I_{sq}.
\end{aligned}
\tag{26}
$$

4.1. DFIG Inner Control Loop

To control the power generated by the DFIG coupled with a wind turbine, independent control of active and reactive power is used. This is achieved by deriving equations that link the rotor voltage generated by the RSC to the stator's active and reactive power output. Assuming a balanced sinusoidal voltage supply and neglecting the stator resistance R_s, the chosen *dq* reference frame is oriented such that the stator voltage space vector aligns along the *d*-axis ($V_{sd} = -\omega_s \phi_{qs}$ and $V_{sq} = 0$), while the stator flux space vector aligns along the *q*-axis ($\phi_{sq} = \phi_s$ and $\phi_{sd} = 0$). Under these conditions, the relationships between stator and rotor currents and fluxes can be derived from (24) as follows:

$$
\begin{aligned}
I_{sdq} &= -\frac{L_m}{L_s} I_{rdq} + \frac{1}{L_s} \phi_{sdq} \\
\phi_{rdq} &= \sigma L_r I_{rdq} + \frac{L_m}{L_s} \phi_{sdq}
\end{aligned}
\tag{27}
$$

where $\sigma = 1 - (L_m^2/(L_s L_r))$ is the leakage coefficient.

Substituting (27) into (23), the rotor voltage model in terms of the rotor currents is derived:

$$
V_{rdq} = \sigma L_r \frac{dI_{rdq}}{dt} + (R_r + j\omega_r \sigma L_r) I_{rdq} - \frac{\omega_r L_m}{L_s} \phi_{sdq}.
\tag{28}
$$

By replacing the stator currents from (27) into (26), the following can be written:

$$
\begin{aligned}
P_s &= -\frac{L_m}{L_s} V_{sd} I_{rd} \\
Q_s &= \frac{L_m}{L_s} V_{sd} I_{rq} - \frac{V_{sd}}{L_s} \phi_{sq}.
\end{aligned}
\tag{29}
$$

Similarly, the expression for the electromagnetic torque T_G is expanded as follows:

$$T_G = -\frac{pL_m}{L_s}\frac{P_s}{\omega_s}.$$ (30)

Substituting the rotor currents from (29) into (28) yields the rotor voltage model as a function of stator power:

$$\begin{bmatrix} V_{rd} \\ V_{rq} \end{bmatrix} = \frac{\sigma L_s L_r}{L_m V_{sd}}\frac{d}{dt}\begin{bmatrix} -P_s \\ Q_s \end{bmatrix} + \left(\frac{R_r L_s}{L_m V_{sd}} + j\frac{\omega_r \sigma L_r L_s}{L_m V_{sd}}\right)\begin{bmatrix} -P_s \\ Q_s \end{bmatrix} - \frac{\phi_{sq}}{L_m}\begin{bmatrix} \omega_r L_r \\ -R_r \end{bmatrix}.$$ (31)

The DFIG model given by (31) shows that independent control of the stator active and reactive powers can be obtained. The coupling terms $\omega_r \sigma L_s L_r Q_s/(L_m V_{sd})$ and $\omega_r \sigma L_s L_r P_s/(L_m V_{sd})$, as well as the perturbation terms $\omega_r \sigma L_s L_r/(L_m V_{sd})$ and $R_r \phi_{sq}/L_m$, are compensated to independently control both active and reactive powers using two PI controllers. Therefore, the same open-loop transfer function is obtained for both active and reactive powers.

$$G_{OL-DFIG} = \frac{k_{im}}{s}\left(\frac{k_{pm}}{k_{im}}s + 1\right)\frac{A}{1 + \sigma \tau_r s}$$ (32)

where $A = L_m V_{ds}/(L_s R_r)$ and $\tau_r = L_r/R_r$.

For tuning the PI controller parameters k_{pm} and k_{im}, the pole-zero compensation method is used, where the controller zero is set equal to the system pole, i.e., $k_{pm}/k_{im} = \sigma \tau_r$. This method is efficient for a first-order transfer function. The controller gains are calculated based on the DFIG parameters and the closed-loop time constant τ_{cl} as follows:

$$G_{CL-DFIG} = 1/(\tau_{cl}s + 1).$$ (33)

where the time constant $\tau_{cl} = 1/(Ak_{im})$.

The closed-loop transfer function is a first-order system with a unity gain, resulting in zero steady-state error. Thus, the controller gains are

$$k_{im} = \frac{1}{A\tau_{cl}}, \quad k_{pm} = k_{im}\tau_{cl}.$$ (34)

To maintain the same closed-loop dynamics as the open-loop system, τ_{cl} is set equal to the open-loop time constant $\tau_{ol} = \sigma \tau_r$. Thus, $\tau_{cl} = 6.7$ ms and the obtained regulator parameters are $k_{im} = 0.787$ and $k_{pm} = 0.0039$.

To achieve a unit power factor operation of the DFIG, the stator reactive power is set to zero ($Q_s^* = 0$). This is accomplished using a PI controller that computes the required rotor voltage V^*_{rq}.

4.2. DFIG Speed Control Loop

The outer control loop regulates the DFIG's rotational speed using the MPPT algorithm presented in Section 2.3 (Figure 4). The transfer function and controller gains are given by (12) and (13). The stator active power reference P^*_s is calculated using (30), with the reference electromagnetic torque T^*_G provided by the MPPT speed controller.

The overall DFIG control strategy is illustrated in Figure 8. The parameters of the DFIG are provided in Table 3, with the rotor parameters referenced to the stator side.

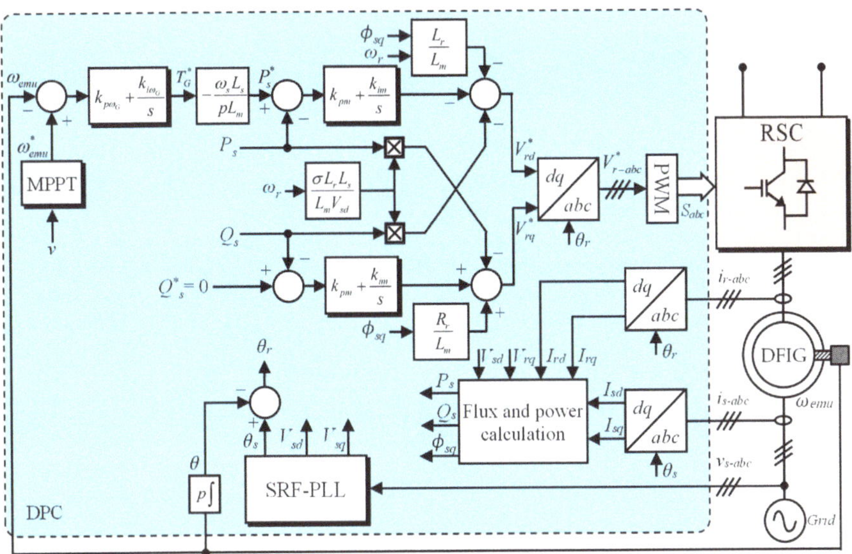

Figure 8. Overall DPC scheme for the DFIG.

Table 3. DFIG parameters.

Parameters	Values
Rated power P_{Gn} (kW)	3
Rated speed ω_{Gn} (rd/s)	148.17
Rated stator frequency f_{sn} (Hz)	50
Rated stator voltage U_{sn} (V)	400
Rated rotor voltage U_{rn} (V)	180
Rated stator current I_{sn} (A)	6.3
Rated rotor current I_{rn} (A)	10.8
Moment of inertia J_G (kg·m^2)	0.037
Viscous friction coefficient B_G (Nm·s/rad)	0.012
Stator resistance R_s (Ω)	1.5
Rotor resistance R_r (Ω)	0.6
Stator inductance L_s (mH)	220
Rotor inductance L_r (mH)	59
Mutual inductance L_m (mH)	110

4.3. DFIG Grid Synchronization

To maintain DFIG control performance during grid frequency ω_s variations, a synchronous reference frame PLL (SRF-PLL) is used, as shown in Figure 9. The linearized model of the SRF-PLL with normalized input voltages can be derived as in [47].

$$G_{CL-PLL} = \frac{\theta_{s-est}}{\theta_s} = \frac{k_{i-pll} + k_{p-pll}s}{s^2 + k_{p-pll}s + k_{i-pll}}. \tag{35}$$

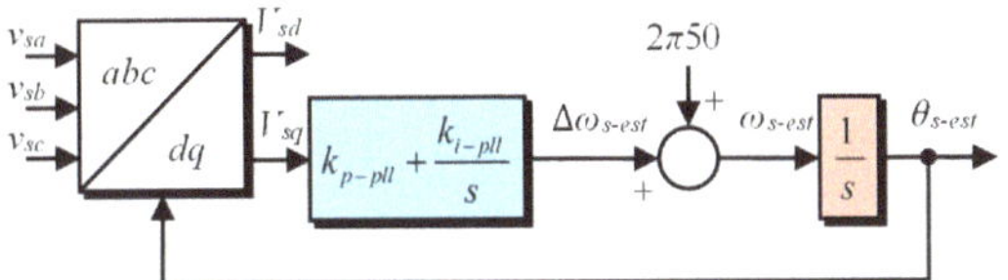

Figure 9. Block diagram of SRF-PLL.

By identifying this model to a second-order system, the PI controller parameters k_{p-pll} and k_{i-pll} are obtain as follows:

$$\begin{cases} k_{i-pll} = \omega^2_{c-pll} \\ k_{p-pll} = 2\zeta\omega_{c-pll}. \end{cases} \tag{36}$$

As the PLL is nonlinear, designing parameters analytically is difficult. While a linearized small-signal model, as in [47], can yield better results, parameters can also be fine-tuned through simulation [48]. In this case, the damping factor is set to $\zeta = 1$ to reduce overshoot during transients. The cutoff frequency is set to $\omega_{c-pll} = 2\pi\cdot 15$ rad/s, balancing dynamic response and noise/harmonic immunity on the estimated angle θ_{s-est}. This results in $k_{p-pll} = 188.5$ and $k_{i-pll} = 8882.6$. Figure 9 shows the structure of the three-phase SRF-PLL.

5. GSC Control

This section presents the Grid-Side Converter (GSC), used to connect the DFIG rotor to the grid. The converter is connected to the grid through an inductance L with an internal resistance R. The converter model in the dq reference frame expressed by (37) and (38) is based on the electrical circuits shown in Figure 10.

$$L\frac{d}{dt}I_{gdq} = (-R - j\omega_g L)I_{gdq} + E_{gdq} - V_{dq} \tag{37}$$

$$C\frac{dV_{dc}}{dt} = \frac{3}{4}\left(S_d I_{gd} + S_q I_{gq}\right) - I_L \tag{38}$$

where $E_{gdq} = [E_{gd}\ E_{gq}]^T$ are the grid voltages and $I_{gdq} = [I_{gd}\ I_{gq}]^T$ are the grid currents in the dq-axes. $V_{gdq} = [V_{gd}\ V_{gq}]^T$ are the converter input voltages, and S_d and S_q represent the IGBT switching states. V_{dc} and C refer to the DC bus voltage and capacitance, while I_L is the DC circuit intermediate branch current.

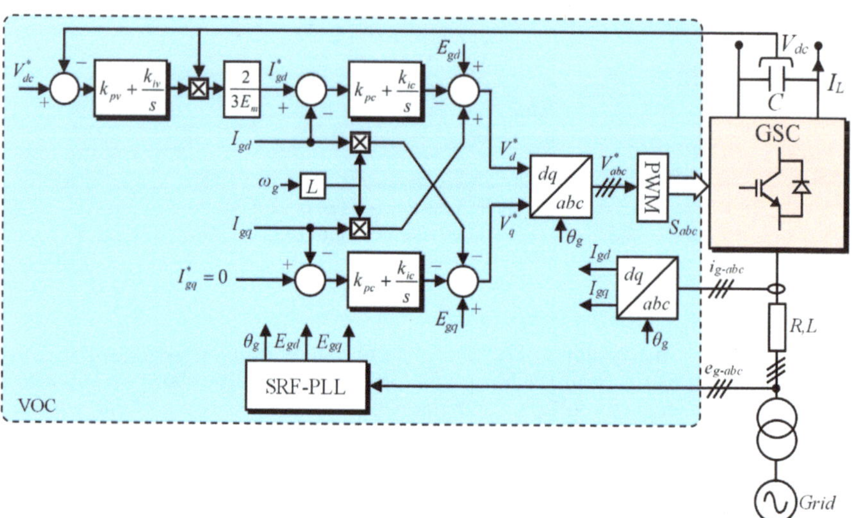

Figure 10. Overall VOC scheme for the GSC.

The active and reactive power, P_g and Q_g, in the dq frame are

$$\begin{cases} P_g = \frac{3}{2}\left(E_{gd}I_{gd} + E_{gq}I_{gq}\right) \\ Q_g = \frac{3}{2}\left(E_{gq}I_{gd} - E_{gd}I_{gq}\right). \end{cases} \tag{39}$$

The Voltage Oriented Control (VOC) strategy is selected for the GSC based on the criteria of simplicity, robustness, and signal quality. This method, developed by analogy with the vector control of electrical machines, is widely used in industrial applications. Its main objectives are to keep the DC bus voltage constant, regardless of power flow direction, and to maintain a unit power factor at the grid connection point [49,50]. To meet these objectives, the dq-axis currents are controlled by an internal loop, while the DC bus voltage V_{dc} is controlled by an external loop. Additionally, a PLL is used to estimate the grid voltage angle θ_g for Park's transformation. The next section provides details on the VOC strategy blocks.

5.1. Current Control Loop

The dq-frame converter model allows independent control of each axis using two separate PI controllers. Compensating for the coupling terms $\omega_g L I_{gq}$ and $\omega_g L I_{gd}$, as well as disturbances E_{gd} and E_{gq} on the dq axes, results in similar open-loop transfer functions G_{OL-c}, including the PI controller for both axes.

$$G_{OL-c} = \frac{k_{pc}s + k_{ic}}{s} \cdot \frac{1}{Ls + R} \tag{40}$$

where k_{pc} and k_{ic} are the proportional and integral gains of the current controller.

To cancel the open-loop pole (L/R), the controller zero is set as $k_{pc}/k_{ic} = L/R$. The closed-loop transfer function G_{CL-c} simplifies to

$$G_{CL-c} = \frac{I_{gd}}{I_{gd}^*} = \frac{G_{iOL}}{1 + G_{iOL}} = \frac{1}{(R/k_{ic})s + 1}. \tag{41}$$

The pole compensation method is used to calculate k_{pc} and k_{ic}, which is straightforward to apply to a first-order system. The system's cutoff frequency ω_{cc} is defined as $\omega_{cc} = k_{ic}/R$, and the response time ensuring that the currents (I_{gd}, I_{gq}) reach 95% of their reference values is set to $t_{rc} = 3/\omega_{cc}$. The parameters k_{pc} and k_{ic} are then calculated as follows:

$$\begin{cases} k_{ic} = 3R/t_{rc} \\ k_{pc} = 3L/t_{rc} \end{cases}. \tag{42}$$

To ensure that the current quickly converges to its reference value, the closed-loop dynamics are accelerated by a factor of 10, resulting in $t_{rc} = 3L/10R = 3\,\text{ms}$. Consequently, the PI controller parameters are set to $k_{ic} = 1000$ and $k_{pc} = 10$.

5.2. DC Voltage Control Loop

The external loop keeps the DC bus voltage V_{dc} constant despite the current I_L variations. This is done by adjusting the reference current amplitude to control the active power flow between the grid and the DC bus. For high accuracy, a PI regulator is often used to minimize the error between V^*_{dc} and V_{dc}. Since the current (internal) loop is faster than the voltage (external) loop, the current loop's transfer function does not affect voltage loop stability and can be considered as a unit gain in steady state. Assuming that the disturbance caused by I_L is fully compensated for by the PI controller, the DC voltage control scheme can be simplified.

From the simplified diagram in Figure 11, the open-loop transfer function is

$$G_{OL-v}(s) = \frac{k_{pv}s + k_{iv}}{s} \frac{1}{Cs} \tag{43}$$

where k_{pv} and k_{iv} are the proportional and integral gains of the DC voltage controller.

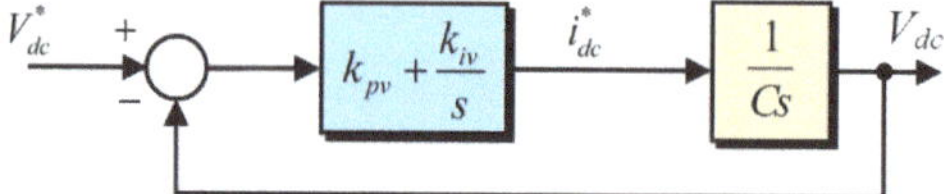

Figure 11. Simplified DC bus voltage regulation.

The closed-loop transfer function of the system is given by

$$G_{CL-v} = \frac{V_{dc}}{V_{dc}^*} = \frac{(k_{iv}/C)\left((k_{pv}/k_{iv})s + 1\right)}{s^2 + (k_{pv}/C)s + (k_{iv}/C)} = \frac{k(\tau s + 1)}{s^2 + 2\zeta\omega_{cv}s + \omega_{cv}^2}. \tag{44}$$

The PI controller parameters are determined using pole placement to balance voltage loop dynamics and minimize current overshoot. For the closed-loop transfer function in (44), with poles placed for a damping factor $\zeta = 0.7$ and cutoff frequency $\omega_{cv} = 3/t_{rv}$, the controller gains k_{iv} and k_{pv} are given by

$$\begin{cases} k_{iv} = C\omega_{cv}^2 \\ k_{pv} = 2\zeta C\omega_{cv} \end{cases} \tag{45}$$

To ensure effective decoupling between the dynamics of the inner and outer control loops, the response time of the DC voltage control loop is set to be ten times that of the current loop, yielding $t_{rv} = 30$ ms. Using the GSC parameters provided in Table 4, the resulting PI controller parameters are $k_{iv} = 33$ and $k_{pv} = 0.46$.

Table 4. GSC parameters.

Parameters	Values
Phase grid voltage (Vrms)	55
Grid voltages frequency f (Hz)	50
DC-link capacitor C (mF)	3.3
Nominal ac filter resistance R (Ω)	1
Nominal ac filter inductance L (mH)	10

6. Experimental Results

To validate the grid-connected WECS with the WTE shown in Figure 12, a test bench was set up, as detailed in Figure 13. Its main components are as follows:

- A 3 kW DFIG, with parameters listed in Table 3, is controlled via a PWM converter connected to its rotor winding (RSC). The DPC algorithm is implemented through the RSC using a dSPACE DS1104 board.
- A 3 kW SEDCM, with parameters listed in Table 2, is used to emulate a wind turbine of the same power, as detailed in Table 1. The motor's torque is controlled by regulating the armature current through a DC-DC buck converter. The emulator's control is implemented using the same dSPACE DS1104 board.
- The GSC is a PWM rectifier connected to the grid through an inductive filter and an isolation transformer. This GSC is controlled to provide a 220 V DC voltage at a unity power factor to the DC bus, which manages the rotor voltages of the DFIG. Additionally, it ensures the connection of the DFIG's rotor side to the grid. The GSC is controlled using the VOC method, implemented through a second dSPACE DS1104 board. The GSC parameters are listed in Table 4.

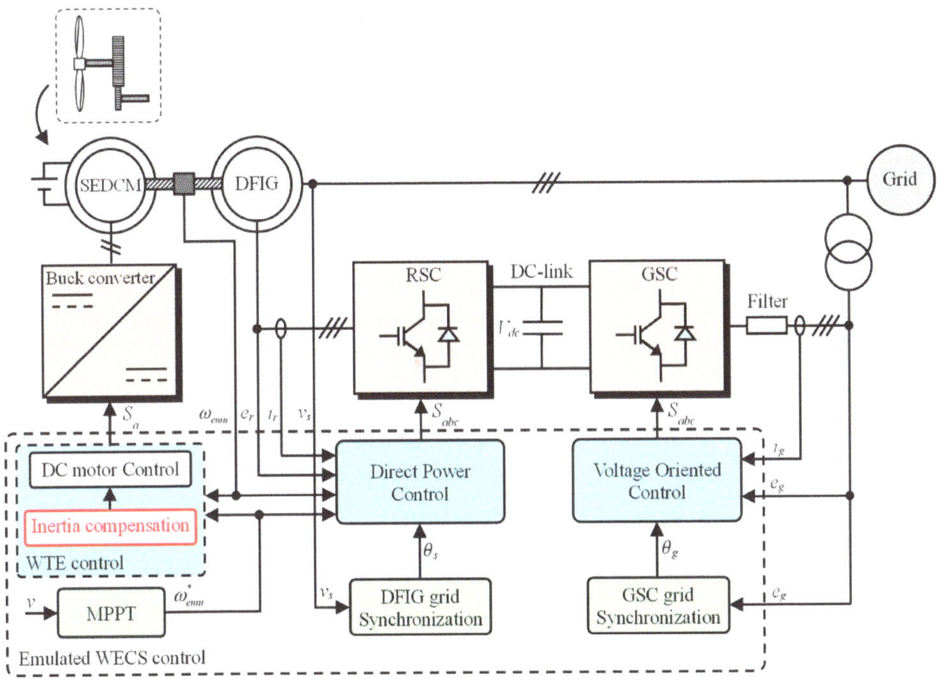

Figure 12. Overall control scheme of the proposed WTE for a DFIG-based WECS.

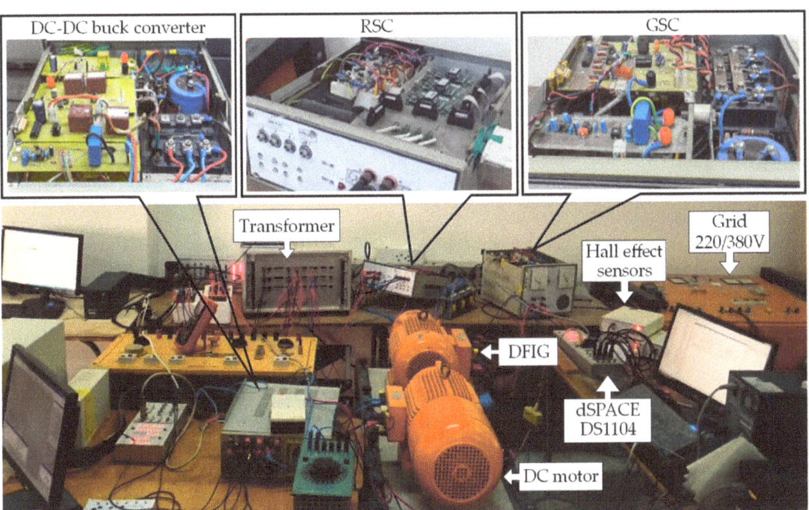

Figure 13. Overview of the test bench.

Current and voltage measurements are performed using Hall effect sensors from LEM (LA 50P and LV 25P). Finally, speed measurements are taken via a 5000-point incremental encoder installed on the DFIG. The control algorithms of the DC-DC buck converter, the RSC, and the GSC are implemented with a switching frequency f_w of 7.5 kHz and a sampling period T_s of 100 μs.

6.1. WTE Results

Figure 14 presents the experimental results of the WTE, obtained for the same wind profile used in the simulation, which is illustrated in Figure 14a. The experimental results show a very strong similarity to those of the simulation in Figure 7. In Figure 14b, the overlap between the reference speed signal and the measured speed is evident. The results for the torque delivered by the turbine model T'_T, the torque of the SEDCM T_M, and the compensation torque T_{com}, estimated using the proposed technique, are shown in Figure 14c. For comparison, the difference $T_M - T_{com}$ is also added to the figure, demonstrating a perfect overlap with T'_T, confirming the high accuracy of the T_{com} estimation method.

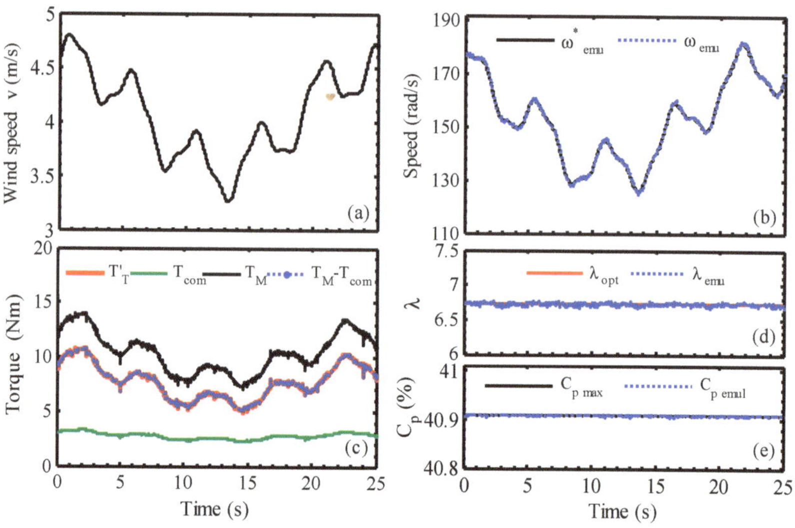

Figure 14. Experimental results for the wind turbine emulator: (**a**) wind speed profile, (**b**) rotational speed, (**c**) mechanical torque provided by the SEDCM, reference torque, and compensation torque, (**d**) tip-speed ratio λ, and (**e**) theoretical and practical power coefficient C_p.

Similarly, the effectiveness of the compensation and control is validated by the results in Figure 14d,e, where the instantaneous values of λ_{opt} and C_{pmax} are overlaid on their theoretical values. The similarity between the experimental and simulation results confirms the effectiveness of the proposed torque compensation and WTE control.

6.2. DFIG Results

Figures 15 and 16 illustrate the experimental results obtained after the DFIG was started and synchronized with the grid, under the previously defined conditions. The results are presented within the framework of rotational speed control.

Figure 15 shows the results for the stator quantities. The voltages applied to the DFIG stator are depicted in Figure 15a. It can be observed that V_{sd} is 380 V, while V_{sq} is zero, which aligns with the expectations of the stator's PLL, where the stator voltage is oriented along the direct axis. Figure 15b demonstrates good synchronization between the stator voltage and the phase current "a" with the current amplitude being proportional to the active power injected into the grid. Figure 15c presents the active and reactive powers at the stator, overlaid with their setpoints, where the chosen reference for reactive power is zero. This confirms that the reactive power control loop is functioning correctly. Likewise, the stator active power P_s precisely follows its reference, which is always negative, indicating the DFIG operates in generator mode. Finally, Figure 15d illustrates the stator currents, showing that I_{sq} is kept at zero since it is proportional to the reactive power Q_s, while I_{sd} is negative, reflecting the active power P_s.

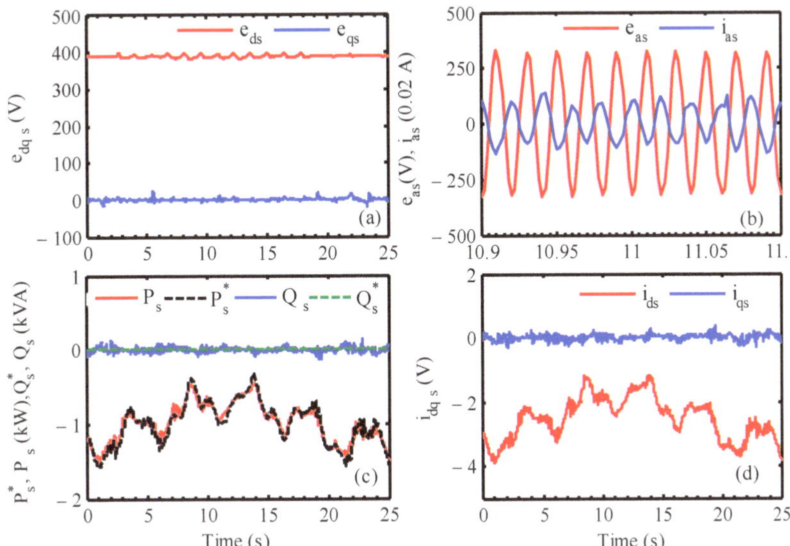

Figure 15. Experimental results for the DFIG stator quantities: (**a**) *dq*-axes voltage components, (**b**) phase *a* current and voltage, (**c**) active and reactive power and their references, and (**d**) *dq*-axes current components.

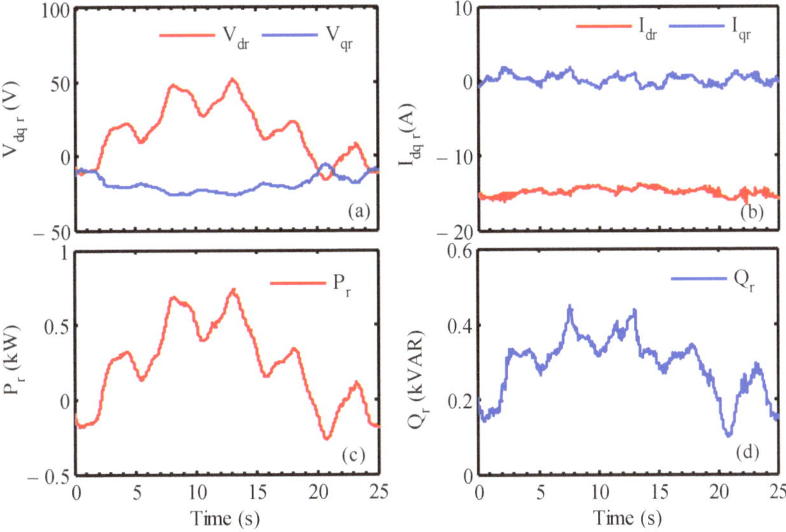

Figure 16. Experimental results for the DFIG rotor quantities: (**a**) *dq*-axes voltage components, (**b**) dq-axes current components, (**c**) active power, and (**d**) reactive power.

Figure 16 presents the experimental results for the rotor quantities. Figure 16a shows the control voltages V_{dr} and V_{qr}, where V_{dr} follows the shape of P_s, being negative in super synchronous mode and positive in sub synchronous mode, while V_{qr} takes very low values, ensuring decoupling and compensation. Figure 15b illustrates the rotor currents I_{rd} and I_{rq}. It is observed that I_{rq} remains very small regardless of the DFIG speed, as it is only necessary for decoupling linked to active power P_s. In contrast, I_{rd} varies depending on the power P_s delivered to the grid. Figure 15c,d show the rotor active and reactive power, respectively. The active power P_r oscillates around an average value of approximately

250 W, representing the power dissipated due to Joule heating in the machine's resistances. The variable part of P_r corresponds to the electrical power produced by the rotor in super synchronous mode or absorbed by the rotor in super synchronous mode. The reactive power Q_r remains positive in both modes as it is used to magnetize the DFIG.

6.3. GSC Results

Figure 17 shows the experimental results obtained for the GSC. In Figure 16a, the DC bus voltage V_{dc} oscillates around 220 V with a ripple rate of 5 to 10%. Figure 17b presents the waveform of the three-line currents i_g from the grid, with variable amplitudes depending on the current I_{dc}, which regulates the V_{dc} voltage. Figure 17c shows that the I_{gdq} currents closely follow their setpoints imposed by the internal control loops. In this figure, the quadrature current I_{gq} is zero to ensure unity power factor operation, as this current represents the reactive power Q_g. Similarly, the direct current I_{gd} reflects the active power P_g injected into the grid. The waveforms for active power P_g and reactive power Q_g are shown in Figure 17f. Figure 17e presents the direct and quadrature grid voltage components E_{gdq}, while Figure 17d shows the phase b alternating voltage component E_{gb} and the line current (multiplied by five) of the same synchronized phase in opposition.

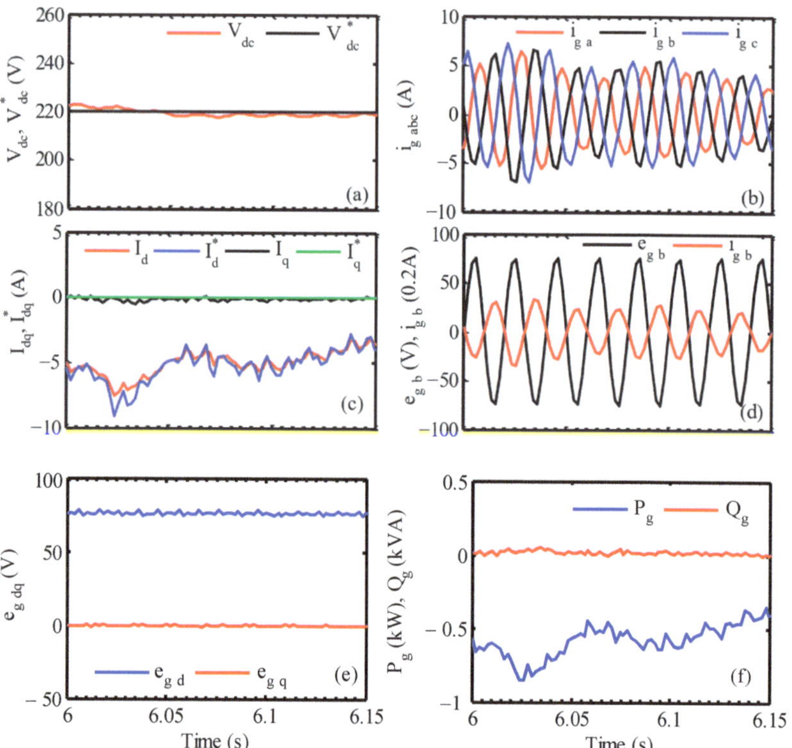

Figure 17. Experimental results for the GSC: (**a**) DC bus voltage and its reference, (**b**) three-phase currents, (**c**) *dq*-axes current components with their references, (**d**) voltage and current of phase *b*, (**e**) *dq*-axes voltage components, and (**f**) active and reactive power.

The experimental results for the GSC demonstrate that the DC bus voltage is well regulated around 220 V, with efficient injection of active power into the grid and reactive power maintained at zero. These results indicate that the GSC's performance meets expectations and design specifications, ensuring optimal integration with the electrical grid. Additionally, synchronization of the grid-side voltages is successfully achieved.

7. Conclusions

This paper presented the development of a comprehensive 3 kW WECS emulator with a grid-connected DFIG, enabling detailed real-time analysis of wind energy systems. The emulator successfully replicates the mechanical behavior of wind turbines using an SEDCM while allowing for independent control of active and reactive power exchanged with the grid via the DFIG. The experimental results demonstrate that the system operates effectively under various wind profiles, accurately validating the implemented MPPT strategy and grid synchronization method. This emulator offers significant potential for future research in WECS control strategies and grid integration, providing a valuable tool for the development of sustainable wind energy solutions.

Author Contributions: Conceptualization, B.B., A.R., K.M. and A.S.; methodology, B.B., A.R., K.M. and A.S.; software, B.B., A.R. and K.M.; investigation, B.B., A.R., A.H. and K.M.; resources, A.S.; writing, B.B., A.R. and K.M.; writing—review, B.B., A.S., A.H. and M.F.B.; supervision, M.F.B. All authors have read and agreed to the published version of the manuscript.

Funding: This research received no external funding.

Data Availability Statement: The data are available on request from the corresponding author.

Conflicts of Interest: The authors declare no conflicts of interest.

References

1. Blaabjerg, F.; Teodorescu, R.; Liserre, M.; Timbus, A.V. Overview of Control and Grid Synchronization for Distributed Power Generation Systems. *IEEE Trans. Ind. Electron.* **2006**, *53*, 1398–1409. [CrossRef]
2. Chhipa, A.A.; Vyas, S.; Kumar, V.; Joshi, R.R. Role of Power Electronics and Optimization Techniques in Renewable Energy Systems. In *Intelligent Algorithms for Analysis and Control of Dynamical Systems*; Kumar, R., Singh, V.P., Mathur, A., Eds.; Springer: Singapore, 2021; pp. 167–175, ISBN 978-981-15-8045-1.
3. Maheshwari, Z.; Kengne, K.; Bhat, O. A comprehensive review on wind turbine emulators. *Renew. Sustain. Energy Rev.* **2023**, *180*, 113297. [CrossRef]
4. Rajendran, S.; Diaz, M.; Devi, V.S.K.; Jena, D.; Travieso, J.C.; Rodriguez, J. Wind Turbine Emulators—A Review. *Processes* **2023**, *11*, 747. [CrossRef]
5. Fadaeinedjad, R.; Moallem, M.; Moschopoulos, G. Simulation of a Wind Turbine with Doubly Fed Induction Generator by FAST and Simulink. *IEEE Trans. Energy Convers.* **2008**, *23*, 690–700. [CrossRef]
6. Gan, C.; Todd, R.; Apsley, J.M. Drive System Dynamics Compensator for a Mechanical System Emulator. *IEEE Trans. Ind. Electron.* **2015**, *62*, 70–78. [CrossRef]
7. Son, J.-Y.; Ma, K. Wind Energy Systems. *Proc. IEEE* **2017**, *105*, 2116–2131. [CrossRef]
8. Wollz, D.H.; da Silva, S.A.O.; Sampaio, L.P. Real-time monitoring of an electronic wind turbine emulator based on the dynamic PMSG model using a graphical interface. *Renew. Energy* **2020**, *155*, 296–308. [CrossRef]
9. Previero, G.R.; da Silva, S.A.O.; Sampaio, L.P.; Wollz, D.H. Electronic Emulator of a Wind Energy System with Full-Scale Power Converter and Permanent Magnet Synchronous Generator. *J. Control Autom. Electr. Syst.* **2024**, *35*, 574–587. [CrossRef]
10. Yan, J.; Feng, Y.; Dong, J. Study on dynamic characteristic of wind turbine emulator based on PMSM. *Renew. Energy* **2016**, *97*, 731–736. [CrossRef]
11. Chen, P.-Y.; Hu, K.-W.; Lin, Y.-G.; Liaw, C.-M. Development of a Prime Mover Emulator Using a Permanent-Magnet Synchronous Motor Drive. *IEEE Trans. Power Electron.* **2018**, *33*, 6114–6125. [CrossRef]
12. Schechner, K.; Hackl, C.M. Scaling of the Drive Train Dynamics of Large-Scale Wind Turbine Systems for Real-Time Emulation in Small-Scale Laboratory Setups. *IEEE Trans. Ind. Electron.* **2019**, *66*, 6779–6788. [CrossRef]
13. Kojabadi, H.; Chang, L.; Boutot, T. Development of a Novel Wind Turbine Simulator for Wind Energy Conversion Systems Using an Inverter-Controlled Induction Motor. *IEEE Trans. Energy Convers.* **2004**, *19*, 547–552. [CrossRef]
14. Sajadi, A.; Rosłaniec, Ł.; Kłos, M.; Biczel, P.; Loparo, K.A. An emulator for fixed pitch wind turbine studies. *Renew. Energy* **2016**, *87*, 391–402. [CrossRef]
15. Castelló, J.; Espí, J.M.; García-Gil, R. Development details and performance assessment of a Wind Turbine Emulator. *Renew. Energy* **2016**, *86*, 848–857. [CrossRef]
16. Gan, L.K.; Shek, J.K.H.; Mueller, M.A. Modeling and Characterization of Downwind Tower Shadow Effects Using a Wind Turbine Emulator. *IEEE Trans. Ind. Electron.* **2017**, *64*, 7087–7097. [CrossRef]
17. Mohammadi, E.; Fadaeinedjad, R.; Naji, H.R.; Moschopoulos, G. Investigation of Horizontal and Vertical Wind Shear Effects Using a Wind Turbine Emulator. *IEEE Trans. Sustain. Energy* **2019**, *10*, 1206–1216. [CrossRef]
18. Mohammadi, E.; Fadaeinedjad, R.; Naji, H.R. Platform for design, simulation, and experimental evaluation of small wind turbines. *IET Renew. Power Gener.* **2019**, *13*, 1576–1586. [CrossRef]

19. Martínez-Márquez, C.I.; Twizere-Bakunda, J.D.; Lundback-Mompó, D.; Orts-Grau, S.; Gimeno-Sales, F.J.; Seguí-Chilet, S. Small Wind Turbine Emulator Based on Lambda-Cp Curves Obtained under Real Operating Conditions. *Energies* **2019**, *12*, 2456. [CrossRef]

20. Mohammadi, E.; Fadaeinedjad, R.; Moschopoulos, G. Investigation of power quality and structural loads for two-bladed wind turbines with rigid and teetered rotors using a wind turbine emulator. *IET Electr. Power Appl.* **2020**, *14*, 2707–2716. [CrossRef]

21. Mohammadi, E.; Fadaeinedjad, R.; Moschopoulos, G. An electromechanical emulation-based study on the behaviour of wind energy conversion systems during short circuit faults. *Energy Convers. Manag.* **2020**, *205*, 112401. [CrossRef]

22. Battaiotto, P.; Mantz, R.; Puleston, P. A Wind Turbine Emulator Based on a Dual DSP Processor System. *IFAC Proc. Vol.* **1995**, *28*, 249–254. [CrossRef]

23. Diop, A.D.; Nichita, C.; Belhache, J.J.; Dakyo, B.; Ceanga, E. Error Evaluation for Models of Real Time Wind Turbine Simulators. *Wind. Eng.* **2000**, *24*, 203–221. [CrossRef]

24. Lopes, L.A.; Lhuilier, J.; Khokhar, M.F.; Mukherjee, A. A Wind Turbine Emulator That Represents the Dynamics of the Wind Turbine Rotor and Drive Train. In Proceedings of the 2005 IEEE 36th Power Electronics Specialists Conference, Dresden, Germany, 16 June 2005; pp. 2092–2097.

25. Chinchilla, M.; Arnaltes, S.; Rodriguez-Amenedo, J.L. Laboratory Set-up for Wind Turbine Emulation. In Proceedings of the 2004 IEEE International Conference on Industrial Technology, IEEE ICIT'04, Hammamet, Tunisia, 8 December 2004; pp. 553–557.

26. Premkumar, K.; Vishnupriya, M.; Sudhakar Babu, T.; Manikandan, B.V.; Thamizhselvan, T.; Nazar Ali, A.; Rabiul Islam, M.; Kouzani, A.Z.; Parvez Mahmud, M.A. Black Widow Optimization-Based Optimal PI-Controlled Wind Turbine Emulator. *Sustainability* **2020**, *12*, 10357. [CrossRef]

27. Hazzab, A.; Gouabi, H.; Habbab, M.; Rezkallah, M.; Ibrahim, H.; Chandra, A. Wind turbine emulator control improvement using nonlinear PI controller for wind energy conversion system: Design and real-time implementation. *Int. J. Adapt. Control Signal Process.* **2023**, *37*, 1151–1165. [CrossRef]

28. Bhayo, M.A.; Mirsaeidi, S.; Koondhar, M.A.; Chandio, S.; Tunio, M.A.; Allasi, H.L.; Aziz, M.J.A.; Idris, N.R.N. An Experimental Hybrid Control Approach for Wind Turbine Emulator. *IEEE Access.* **2023**, *11*, 58064–58077. [CrossRef]

29. Benzaouia, S.; Mokhtari, M.; Zouggar, S.; Rabhi, A.; Elhafyani, M.L.; Ouchbel, T. Design and implementation details of a low cost sensorless emulator for variable speed wind turbines. *Sustain. Energy Grids Netw.* **2021**, *26*, 100431. [CrossRef]

30. Martinez, F.; Herrero, L.C.; de Pablo, S. Open loop wind turbine emulator. *Renew. Energy* **2014**, *63*, 212–221. [CrossRef]

31. Moussa, I.; Bouallegue, A.; Khedher, A. New wind turbine emulator based on DC machine: Hardware implementation using FPGA board for an open-loop operation. *IET Circuits Devices Syst.* **2019**, *13*, 896–902. [CrossRef]

32. Munteanu, I.; Bratcu, A.I.; Bacha, S.; Roye, D.; Guiraud, J. Hardware-in-the-Loop-based Simulator for a Class of Variable-speed Wind Energy Conversion Systems: Design and Performance Assessment. *IEEE Trans. Energy Convers.* **2010**, *25*, 564–576. [CrossRef]

33. Farag, W.A. Accurate wind energy harvest assessment using a comprehensive ultra-large-wind-turbine emulation. *J. Eng. Res.* **2023**, *11*, 123–133. [CrossRef]

34. Hussain, J.; Mishra, M.K. An Efficient Wind Speed Computation Method Using Sliding Mode Observers in Wind Energy Conversion System Control Applications. *IEEE Trans. Ind. Appl.* **2020**, *56*, 730–739. [CrossRef]

35. Krim, Y.; Abbes, D.; Krim, S.; Mimouni, M.F. A second-order sliding-mode control for a real time emulator of a wind power system synchronized with electrical network. *Int. Trans. Electr. Energy Syst.* **2019**, *29*, e12051. [CrossRef]

36. Nair, R.; Narayanan, G. Emulation of Wind Turbine System Using Vector Controlled Induction Motor Drive. *IEEE Trans. Ind. Appl.* **2020**, *56*, 4124–4133. [CrossRef]

37. Arıfujjaman, M.; Iqbal, M.T.; Quaicoe, J.E. Emulation of a Small Wind Turbine System with a Separately-Excited Dc Machine. *IU-J. Electr. Electron. Eng.* **2008**, *8*, 569–579.

38. Behera, P.K.; Mendi, B.; Sarangi, S.K.; Pattnaik, M. Robust wind turbine emulator design using sliding mode controller. *Renew. Energy Focus* **2021**, *36*, 79–88. [CrossRef]

39. Park, J.-Y.; Lee, J.-K.; Oh, K.-Y.; Lee, J.-S.; Kim, B.-J. Design of Simulator for 3MW Wind Turbine and Its Condition Monitoring System. In Proceedings of the International MultiConference of Engineers and Computer Scientists, IMECS, Hong Kong, China, 17–19 March 2010; pp. 930–933.

40. Jing, B.; Qun, L.; Qiang, L.; Zaiyu, C.; Yuchong, H.; Minghui, Y. Inertia compensation for wind turbine emulator based on anticipation deviation suppression. *Electr. Power Syst. Res.* **2023**, *224*, 109751. [CrossRef]

41. Weijie, L.; Minghui, Y.; Rui, Z.; Minghe, J.; Yun, Z. Investigating Instability of the Wind Turbine Simulator with the Conventional Inertia Emulation Scheme. In Proceedings of the 2015 IEEE Energy Conversion Congress and Exposition (ECCE), Montreal, QC, Canada, 20–24 September 2015; pp. 983–989.

42. Yin, M.; Li, W.; Chung, C.Y.; Chen, Z.; Zou, Y. Inertia compensation scheme of WTS considering time delay for emulating large-inertia turbines. *IET Renew. Power Gener.* **2017**, *11*, 529–538. [CrossRef]

43. Abdeddaim, S.; Betka, A.; Drid, S.; Becherif, M. Implementation of MRAC controller of a DFIG based variable speed grid connected wind turbine. *Energy Convers. Manag.* **2014**, *79*, 281–288. [CrossRef]

44. Coelho, P. The Betz limit and the corresponding thermodynamic limit. *Wind. Eng.* **2023**, *47*, 491–496. [CrossRef]

45. Monfared, M.; Kojabadi, H.M.; Rastegar, H. Static and dynamic wind turbine simulator using a converter controlled dc motor. *Renew. Energy* **2008**, *33*, 906–913. [CrossRef]

46. Aström, K.J.; Hägglund, T. *PID Controllers: Theory, Design, and Tuning*; ISA—The Instrumentation, Systems and Automation Society: Research Triangle Park, NC, USA, 1995; ISBN 978-1-55617-516-9.
47. Golestan, S.; Guerrero, J.M.; Vasquez, J.C. Three-Phase PLLs: A Review of Recent Advances. *IEEE Trans. Power Electron.* **2017**, *32*, 1894–1907. [CrossRef]
48. Rahoui, A.; Boukais, B.; Mesbah, K.; Otmane-Cherif, T. Neural Networks Based Frequency-Locked Loop for Grid Synchronization Under Unbalanced and Distorted Conditions. In Proceedings of the 2020 International Conference on Electrical Engineering (ICEE), Istanbul, Turkey, 25–27 September 2020; pp. 1–5.
49. Mesbah, K.; Rahoui, A.; Boukais, B. Neural Network Quadrature Signal Generator-Based Virtual Flux Estimation for Predictive Control of PWM Converters. *IEEE Trans. Ind. Electron.* **2024**, *71*, 4785–4794. [CrossRef]
50. Rahoui, A.; Mesbah, K.; Boukais, B.; Otmane-Cherif, T. Frequency-Adaptive Neural Network Based Virtual Flux Estimation for Sensorless Control of PWM Converters Under Unbalanced Conditions. In Proceedings of the 2020 International Conference on Electrical Engineering (ICEE), Istanbul, Turkey, 25–27 September 2020; pp. 1–6.

MDPI AG

Grosspeteranlage 5

4052 Basel

Switzerland

Tel.: +41 61 683 77 34

Actuators Editorial Office

E-mail: actuators@mdpi.com

www.mdpi.com/journal/actuators